2015中国建筑院校境外交流优秀作业集

全国高等学校建筑学专业教育评估委员会
中国建筑学会建筑教育评估分会

中国建筑工业出版社

中国建筑学会建筑教育评估分会

中国建筑学会建筑教育评估分会在住房和城乡建设部人事司的直接领导和推动下，经中国科学技术协会、民政部批准，于2012年10月在北京成立。

中国建筑学会建筑教育评估分会的主要任务是与全国高等学校建筑学专业教育评估委员会、全国高等学校建筑学学科专业教学指导委员会协调工作，共同搭建一个活动平台，开展国际及港澳台地区学术交流与合作，参与建筑学专业教育国际互认《堪培拉建筑教育协议》有关活动，维护我国公民参加国外建筑师注册时享有与本国建筑专业教育背景同等地位的权利；跟踪建筑专业教育和学科的国际、国内发展趋势，动态分析建筑专业教育现状，努力促进建筑专业教育学术繁荣和技术进步，为中国建筑院校学术研讨和交流提供服务。

建筑教育评估分会成立以来，召开了三次理事会，健全了组织机构和工作制度，承担了堪培拉建筑教育协议中国秘书处工作，印制了《堪培拉建筑学教育协议》文件汇编，成功举办了两次中国建筑院校学生国际交流作业展暨优秀国际交流作业确认活动并出版《2013中国建筑院校学生境外交流优秀作业集》，积极组织策划了中英建筑学生工作坊和海峡两岸建筑院校学术交流等教学交流活动。

建筑教育评估分会秘书处设在中国建筑学会国际部，不定期出刊中国建筑学会建筑教育评估分会简讯。

附：中国建筑学会建筑教育评估分会第一届理事会名单

与世界全面接触的中国建筑教育

——《2015 中国建筑院校境外交流优秀作业集》序言

2015 年 3 月 21 ~ 22 日，由中国建筑学会建筑教育评估分会主办，天津大学建筑学院承办的中国建筑学会建筑教育评估分会第一届第四次全体理事会会议在天津召开。中国建筑学会建筑教育评估分会朱文一理事长、王建国副理事长、张百平秘书长，顾勇新副秘书长、王晓京副秘书长以及全体理事和教师代表参加了本次会议，会议同时还邀请了住房和城乡建设部人事司副巡视员赵琦、专业人才与培训处调研员高延伟等嘉宾。会议同期，在天津大学建筑学院二层展厅举办了“2015 年中国建筑院校境外交流作业展暨优秀境外交流作业奖”的评选活动。按照惯例，本次评选的优秀作业由中国建筑工业出版社结集出版，以期对一年来的国内建筑院校的境外交流情况进行总结，并让国内更多的建筑院校和广大读者参考和学习。

本次活动的评选委员包括中国建筑学会副理事长兼秘书长周畅，分会理事长、清华大学建筑学院教授朱文一，分会副理事长、东南大学建筑学院院长王建国，天津大学建筑学院院长张颀，《堪培拉协议》主席、前国际建协（UIA）副主席、韩国著名建筑师赵诚重（Sungjung Chough），天津市建筑设计院院长、全国工程勘察设计大师刘景樑，天津华汇工程建筑设计有限公司总建筑师、全国工程勘察设计大师周恺，中国建筑东北设计研究院有限公司总建筑师王洪礼，中国建筑设计研究院执行总建筑师汪恒，中国建筑科学研究院建筑设计院总建筑师薛明，同济大学建筑与城市规划学院副院长黄一如，大连理工大学建筑与艺术学院院长范悦，沈阳建筑大学建筑学院院长张伶伶，南京大学建筑与城市规划学院院长丁沃沃，湖南大学建筑学院副院长柳肃，以及建筑教育评估分会常务理事孔宇航、刘临安、孙澄、孙一民、李保峰、单军、赵万民、赵继龙、罗卿平等人。

参选作品来自国内建筑院校 2013-2014 年度（2013 年 9 月 ~ 2014 年 8 月）的境外交流设计，分本科生组、研究生组和国际竞赛组进行。在收到的 37 所建筑院校的 227 份作业中，有境外交流课程作业 186 份（包括本科生作业 124 份、研究生作业 62 份），还有学生国际竞赛作品 41 份。评选活动仅对境外交流课程作业进行确认和评审，国际竞赛作品只确认不评审。经过评选前的技术审查，共有 223 份作业进入到评审环节。通过两轮评委投票，以简单多数选出 186 份作品（本科生作业 101 份、研究生作业 49 份、国际竞赛 36 份）获得“中国建筑院校境外交流优秀作业”，占本次全部参赛作业的 83.4%；另外选出 49 份作业（包括本科生作业 30 份、研究生作业 19 份）获得“中国建筑院校境外交流优秀作业特别奖”，占本次全部参赛作业的 22%。这也是该活动首次设立这一奖项。综观本年度“中国建筑院校境外交流优秀作业”的评选情况，大体可以总结出以下特点：

1. 国际交流日趋常态化，合作覆盖面继续扩大

本次评选活动的参与院校较往年持续增加。平均每所院校提交了 6.13 份作业，相较去年提高了 8%。

交流的广度方面，合作国际教学活动的境外学校达到 90 余所，分布于欧洲、美洲、大洋洲、亚洲和非洲等地区。其中，美洲地区 24 所，包括美国哈佛大学、耶鲁大学、康奈尔大学、普林斯顿大学、加利福尼亚大学伯克利分校、加利福尼亚大学洛杉矶分校、密歇根大学、加拿大 UBC 等名校；欧洲地区 38 所，覆盖英、法、德、奥、瑞、荷、意、西等 13 个国家，包括英国 AA 学院、伦敦大学、德国达姆施塔特工业大学、德国柏林工业大学、奥地利维也纳理工大学、荷兰代尔夫特理工大学、法国拉维莱特建筑学院、凡尔赛大学、斯特拉斯堡大学、意大利米兰理工大学、威尼斯建筑大学、丹麦皇家美术学院等；大洋洲地区 2 所，包括澳大利亚新南威尔士大学、新西兰 UNITEC 理工大学等；亚洲地区有 27 所，

其中来自日、韩、新等国的13所大学，包括日本东京大学、东京工业大学、早稻田大学、韩国全南大学、新加坡国立大学等；来自港澳台地区有13所，包括香港中文大学、台湾东海大学、中原大学、逢甲大学、成功大学、淡江大学等；非洲地区有1所，为南非开普敦大学。

近年来，很多国内院校一方面将国际联合设计工作坊嵌入到常规设计课程中，通过调整教学框架，使国际交流成为常态化教学的一部分；另一方面，在保持传统的国际交流合作的同时，仍在不断地拓展交流渠道，增强交流的覆盖度和多样性。目前国内建筑院校与美国、欧洲、亚洲和港台等地知名大学的教学交流已进入了体系化的稳定阶段。

2. 教学合作向专业纵深发展，合作领域更趋多元

自2008年《堪培拉协议》（Canberra Accord）签署以来，国内院校和国际知名大学的建筑学院建立起了广泛的交流渠道。建筑院校的对外交流完成了从“请进来”到“走出去”的蜕变，交流方式和合作领域更加多元。中国建筑教育和教学质量也逐渐得到国际同行的关注和认可。

一个明显的变化是，此前在合作项目的选择上大多使用国内的场地，但此次出现了许多在境外合作院校所在城市选地的情况，这说明国际交流的基础已经发生了微妙的变化，不仅是满足国外师生对“中国问题”的好奇，而是中外师生以一种更加平衡的视角讨论建筑学本体的问题。例如，清华大学的交流项目就是选址在新南威尔士州的北邦迪（North Bondi），讨论如何结合当地悠久的居住传统创造高品质的社区和景观。天津大学的交流项目选址在美国加利福尼亚州的洛杉矶，聚焦美国西部的汽车文化，探讨汽车博物馆与城市、社区的互动网络。这些都说明了国际交流的纽带更加稳定和成熟，并向着专业纵深的方向发展。

从本次交流作品不难看出，交流院校对城市及其亚文化给予了更多的关注。例如，哈尔滨工业大学的项目选址在台湾中坜市，探讨街区的演变过程以及如何自下而上地生成高密度社区的问题；中央美术学院的项目选址在瑞士小镇布鲁嫩（Brunen），讨论风景旅游区旅馆的建筑结构、空间与景观的一体化组织。华侨大学的项目结合当地屋顶自发搭建的现象，提出将这一行为转化为拓展城市和社区空间的触媒。

3. 交流机制日渐完善，全面接触正在孕育

国际交流形式的日益丰富使更多的学生有机会在本科期间完成工作室互访、短期留学、联合毕业设计等项目。除此之外，日益普遍的暑期学校或暑期工作坊显然也是借鉴了国际通行的经验。在短短的不到十年时间，中国建筑院校对外交流的交流机制已日渐完善和成熟。

与此同时，互聘教授、长期和短期相结合的工作坊，也使教师之间的交流达到了前所未有的高度，对国内建筑教学体系的变化产生了深远的影响。国内本科生在欧美高水平建筑院校攻读硕士、博士学位的人数也在持续增加。随着这些留学回国的年轻学者进入国内高校任教，国际交流的底层架构更加坚实和成熟，并且有望形成良性的循环。随着研究与教学的进一步融合，联合工作室的设计主题更趋多元化，几乎涵盖了当前所有的热点问题，如城市更新、历史建筑保护、绿色建筑、数字化建筑等，问题涉及城市设计、景观设计、建筑设计、装置艺术等多个学科领域，充分展现出建筑学在国际化和学科交叉过程中迸发出的活力。

综上所述，中国建筑院校境外交流优秀作业展不仅记录了中国建筑院校的国际化进程，也见证了中国建筑教育在变革与转型时期的各种探索。衷心希望展览能越办越好，给我们带来更多的启示和惊喜。

天津大学建筑学院院长　张　颀
天津大学建筑学院副院长　许　蓁
2016.1.15

目　录

与世界全面接触的中国建筑教育

1. 共享天空
Sharing in the Sky 10
2. 新陈代谢——从眷村到新城的渐变与更新计划
Metabolism 12
3. 东艺，西衍——澳门内港码头片区更新活化计划
Terminal Area Update Activation Plan 14
4. 传统街区的生长
The Growth of Traditional Blocks 16
5. 城市再生
Urban Regeneration 18
6. 运河岸边的城市更新——拱墅区多媒体中心设计
Renovation along Grand Canal 20
7. 律动河流——郑州滨水文化休闲区城市设计
River Beat 22
8. 南京地铁马群站城市设计
Nanjing Subway Station Urban Design 24
9. 公园 · 城市
Living in the Park 26
10. 澳门望德堂文化创意区近代平民排屋“屋顶触媒”
Roof Catalyst 28
11. 拱之森林——佛罗伦萨市场设计
Arch Forest 30
12. 回舍——北院门小客栈设计
Old House 32
13. 鱼儿园
The Aqua Garten 34
14. 济南花鸟鱼虫水族市场设计
Space to Activate 36
15. TransferJET 创业中心——基于空间流动模式的办公空间设计
TransferJET Entrepreneurial Base—Knowledge
Space Design based on the Flow of Information 38
16. 东京共享住宅
Share with Light 40

17. 生长的乌托邦
A Growing Utopia 42
18. 城市裂缝
Crevice in the City 44
19. 缝合——黟县际村村落改造与建筑设计
Suture—Village Reconstruction 46
20. 街 & 巷
Street & Lane 48
21. 万花景——分型理论下的汽车博物馆设计
Kaleidoscope—Automobile Museum Design 50
22. 柔性改造——漕运博物馆和丰备义仓地区城市设计
Centle Change—Urban Redevelopment of Fengbei Refief Area, Suzhou 52
23. 北邦迪公寓设计
North Bondi Apartment Design 54
24. 街巷
Streets Alleys 56
25. 空角实心
Corner Dancing 58
26. 生态—形态的重组——漕运博物馆和丰备义仓地区城市设计
Eco-Morphology Reordering 60
27. 木“扇”亭——当代“传统木榫卯”设计
Wooden “Fan” Pavilion........ 62
28. 结构与空间探究
Focus on Structure-Hotel 64
29. 街区进化论——基于中坜市中原校区周边街区的演变研究
Block Evolution Theory 66
30. 双重否定
Double Negative 68
31. 保护与再生——以桂峰村清代茶楼为例
The Protection and Regeneration of Traditional Architecture — Teahouse in Guangxi Village in Qing Dynasty as an Example 70
32. 分类学——边界重构
Taxonomy 72
33. 模糊场域的界定
Claim the Territory by Chaos 74
34. 3.11 东日本震灾复兴计划——岩手县大槌町社区信息交流中心设计
3.11 East Japan Earthquake Recovery Plan—Design of Iwate Tsuchi Community Information Exchange Center 76
35. 工业遗产 & 社区复兴
Industrial Heritage & Community Renewal 78

36. 多因素控制下的城市更新设计
Urban Renewal Design Under Multi-Factor Control 80
37. 山水舞台——岩手县大槌町社区信息交流中心设计
Landscape Stage—Design of Iwate Tsuchi Community Information
Exchange Center 82
38. 荔枝湾南——广东文化艺术保育和创新基地设计
South of Litchi Gulf — Design of Cultural and Art Conservation and
Innovation Base in Guangdong Province 84
39. 墙之礼乐——西安博物院城市设计及重点建筑设计
The Wall of the Ritual — Xi'an City Museum Design 86
40. 矛盾的共生
The Contradiction of Symbiosis 88
41. 好奇的格子
Cabinet of Curiosity 90
42. 行与市
Traffic and City 92
43. 变废为宝之城
Post-Waste Cities 94
44. 生态链接，产业重塑——国际化绿色生态城市规划设计
Rebuild the Chain—Green-Eco-Sustainable Town Design toward Glocalization ... 96
45. 城墙博古架
Modern Frame—The Design of a Connection for the Gap of Nanjing Wall 98
46. 共享居室——2050 年的曼哈顿
Shared Living Rooms—Manhattan 2050 100
47. 生态乡村，活力再造——国际化绿色生态城市规划设计
Green-Eco-Sustainable Town Design toward Glocalization 102
48. 步行者天堂——城市步行系统节点城市设计
Dalian Road — Siping Road and Dalian Road Intersection 104
49. “迹”
Unfolding Harbor Sequence 106
50. 织·补
Weaving of the Old Community 108
51. 儿童博物馆
The Children's Museum 110
52. 循环农场——城市立体农场设计
Farm Symbiosis the Urban Farm Design 112
53. 出乎意料的城市——巴别城
Unexpected City: Babel City 114
54. 节点激活——黟县际村村落改造与建筑设计
Nodes & Activation: The Village Reformation and Architectural Design in Jicun...... 116

55. 德班城市光纤
City Fiber of Durban 118
56. 出乎意料的城市——城市天空之剧
Unexpected City: City Dramas with Sky 120
57. 相信城市
Believe-in City 122
58. 畅通城市
Fluent City 124
59. 上与下
On and Under 126
60. 新陈代谢与重生
Combine Regeneration and Metabolism 128
61. 重树标志
Re-Symbol 130
62. 移动城市，移动生活
Movable City Mobile Life........ 132
63. 寺曾相识——西安博物院城市设计及重点建筑设计
Urban Design & Architecture Design of Small Goose Pagoda Area 134
64. 渗透城市
Osmosis City 136
65. 缝合——釜山车站连接体设计
Sewing—Design of Connecting Body of Busan Station 138
66. 水天一线——中东铁路百年江桥的新生
Between the Sky and Rive: Rebirth of the Centurial Railway Bridge 140
67. 澳门路环岛老船厂更新设计
Macau Coloane Old Shipyard Update 142
68. 被消失的故居
The Disappeared House 144
69. 缝合
Sew up 146
70. 浮动网络
Floating Web 148
71. 新的天际线
A New Sky-line 150
72. 城市边缘的复兴
The Regeneration of Edge Louisville 152
73. 平凡事件 vs 复合站点
Ordinary Events VS Complex Sites 154
74. 缝合城市——南非德班沃里克枢纽站地区改造城市设计
Suture the City: Urban Design for the Ethekweni 156

75. 辙上建造
Architecture on Wheels 158
76. 激活 · 腾空 · 漂浮各地
Activating-Blanking-Floating Anywhere 160
77. 延——小客栈设计
Inheriting the Small Inn Design 162
78. 小站那些事
Those Things in the Bus Stop 164
79. 出乎意料的城市——元胞螺旋城
Unexpected City: Spiral City with Cellular Automata 166
80. 虫洞
Wormhole 168
81. 织城为家——探索城市空白空间的再利用
Knit City Space into House: Explore the Reuse of Blank Zones in Cities........ 170
82. 被涉及
Get Involved 172
83. 救赎——后工业废弃地的垂直农场转型
Salvation: The Transform from Derserted Land of the Post Industrial Vertical Farm........ 174
84. 植城静息
City Coexistent with Breath 176
85. 城市尊严
Dignity of Human 178

2015 年中国建筑院校境外交流优秀作业名单 180
中国建筑学会建筑教育评估分会第一届第四次理事会代表名单 193
后 记........ 195

SHARING IN THE SKY

By introducing spaces in various weights,the house created a dialoge between free columns and human behaviours.At the same time,the project added a new blood to the noisy and crowded urban landscape.

HISTORY

IN THE MEIJI PERIOD

It developed as the center of a pleasure quarter

Great Kanto Earthquake have moved to Kagurazaka.

THE PRESENT TIME

The residence,the building,and the office are intermingled.

CONCEPT

SITE ANALYSIS

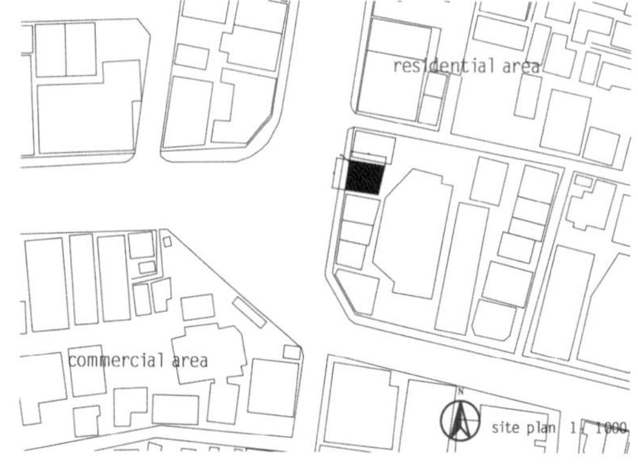

Sunlight in the site

City interface

the future height

the main light surface

Transport in the site

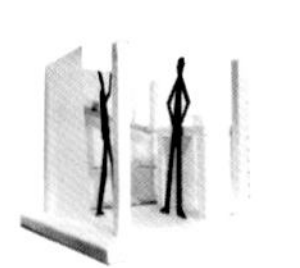

Possible Activities

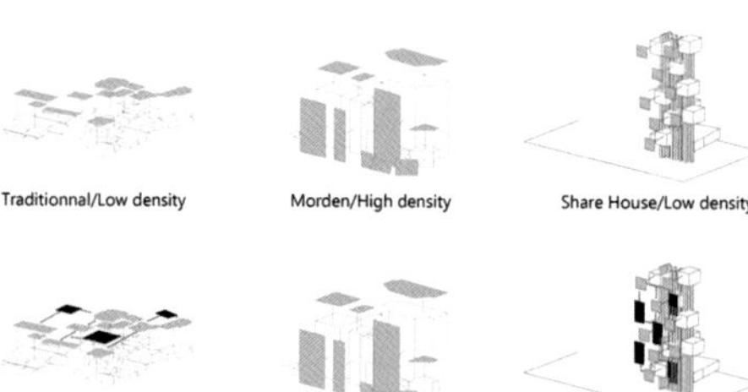

Traditionnal/Low density

Morden/High density

Share House/Low density

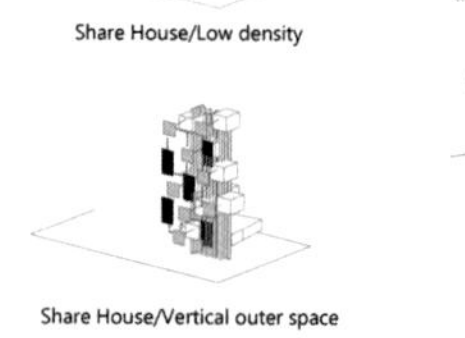

Traditionnal/ outer space

Morden/ no suitable space

Share House/Vertical outer space

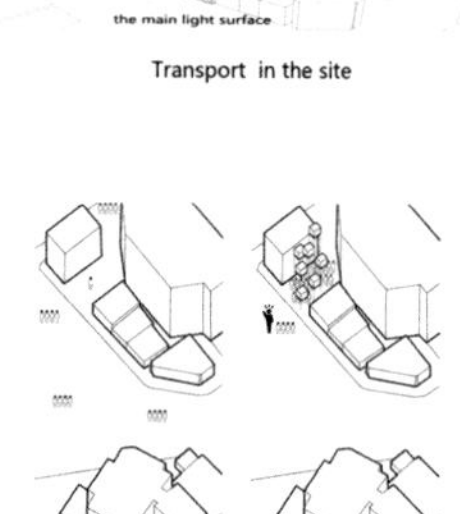

作品名称：共享天空

Sharing in the Sky

院校名：天津大学
设计人：Martin Rasmus，唐奇靓，李桃
指导教师：张昕楠，王绚
课程名称：住宅工作营
作业完成日期：2014 年 12 月
对外交流对象：日本女子大学，淡江大学

SHARING IN THE SKY

In the numerous factors that influence residence,this project selects sharing space——a factor which is extremely important to distinguish share house from normal apartment.

By introducing spaces in various weights,the house created a dialoge between free columns and human behaviours.At the same time,the project added a new blood to the noisy and crowded urban landscape.

In the urban jungle,Tokyo,the increasingly shrinking living space and the raising of pubilc trust make sharing become a new way of life.The share house focus on the intersection of residential life.As a carrier of life,the space promotes the sharing behaviors

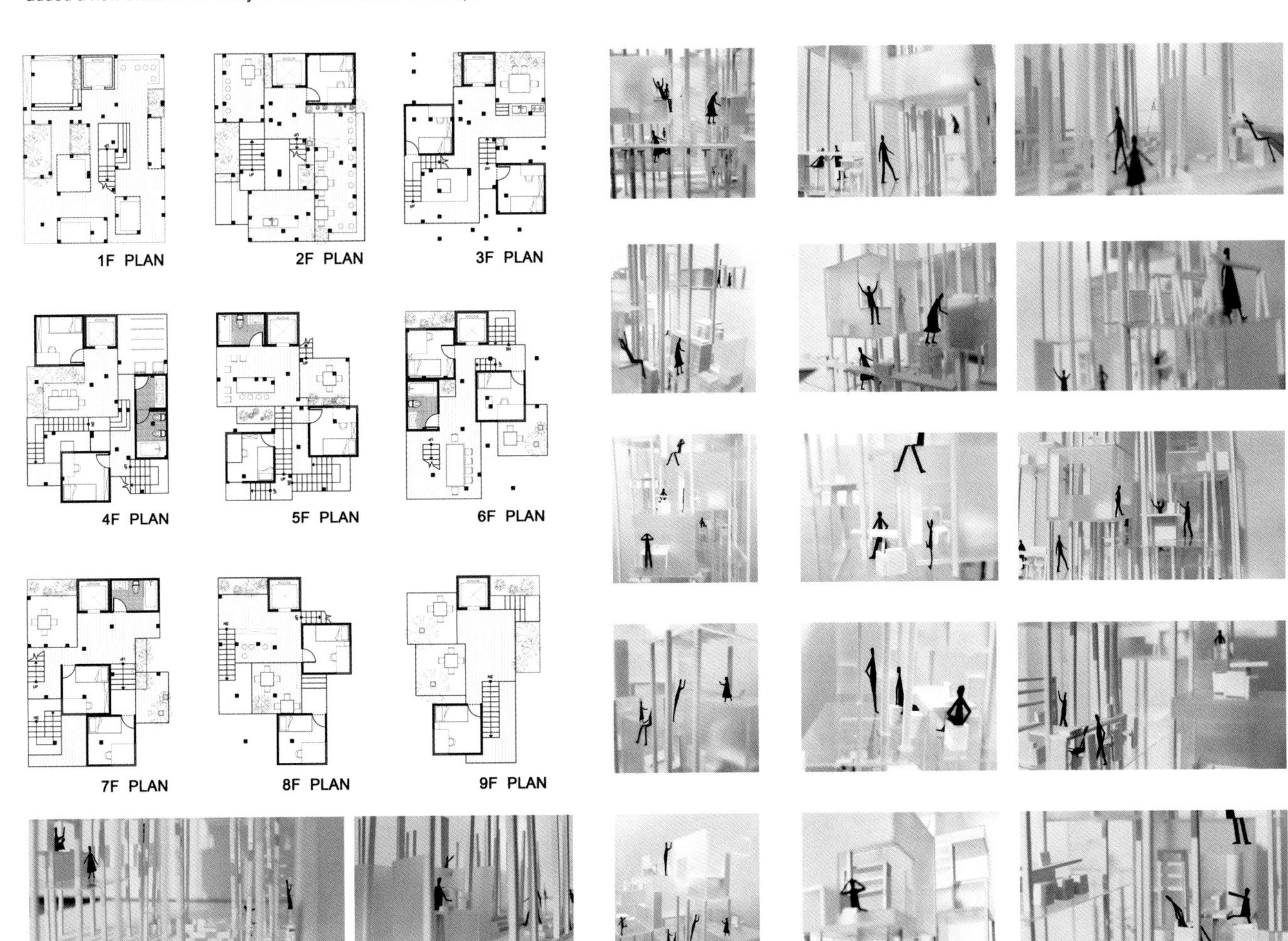

University: Tianjin University
Designer: Martin Rasmus, Tang Qiliang, Li Tao
Tutors: Zhang Xinnan, Wang Xuan
Course Name: Share House
Finished Time: Dec. 2014
Exchange Institute: Japan Women's University, Tamkang University

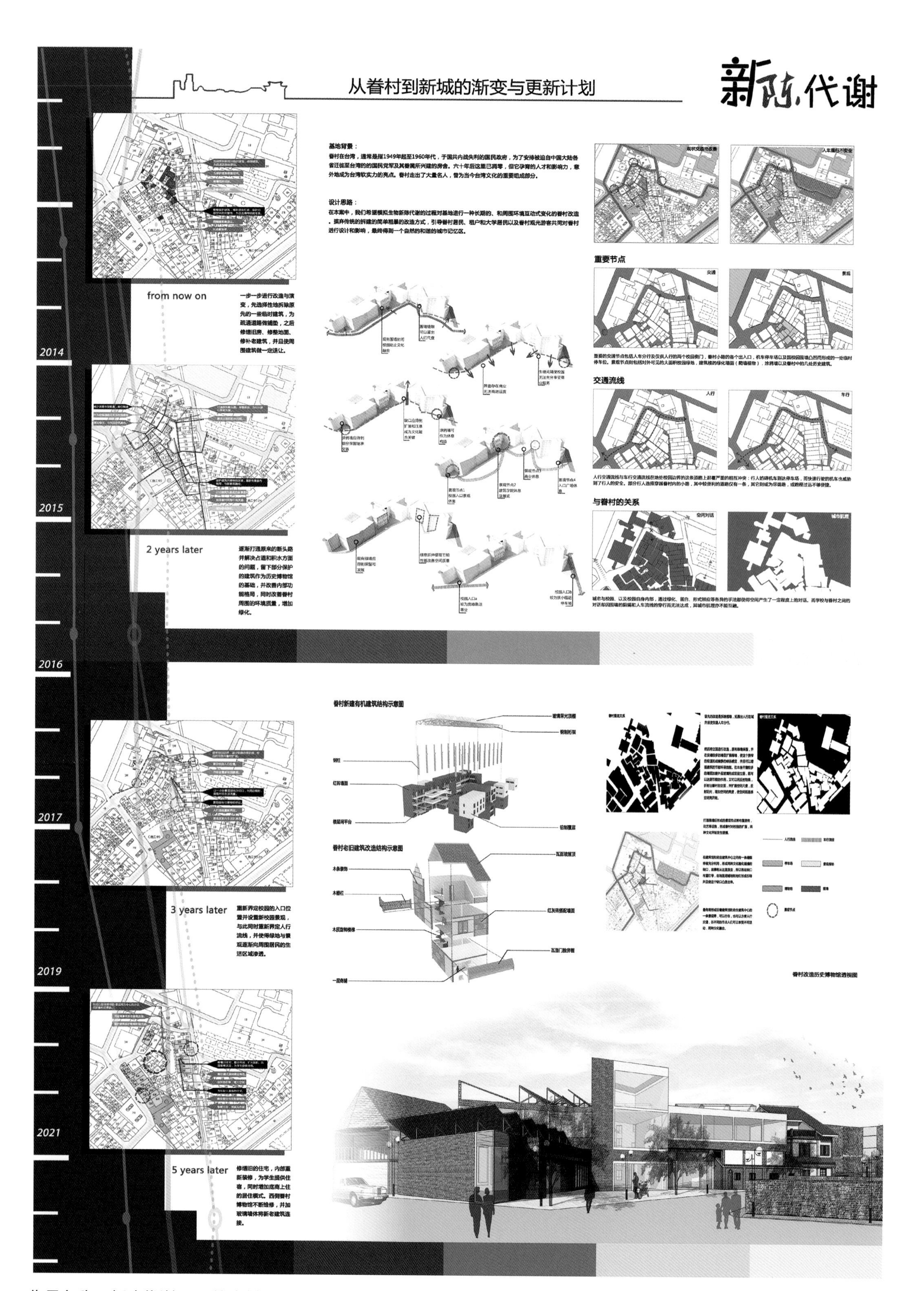

作品名称：新陈代谢——从眷村到新城的渐变与更新计划
Metabolism

院校名：哈尔滨工业大学
设计人：陈永祥，骆盛韬，尤昊博，王渡，逄佳音
指导教师：薛名辉，董宇，张姗姗，邵郁，喻肇青，张华荪
课程名称：开放式研究型建筑设计
作业完成日期：2014 年 03 月
对外交流对象：中原大学

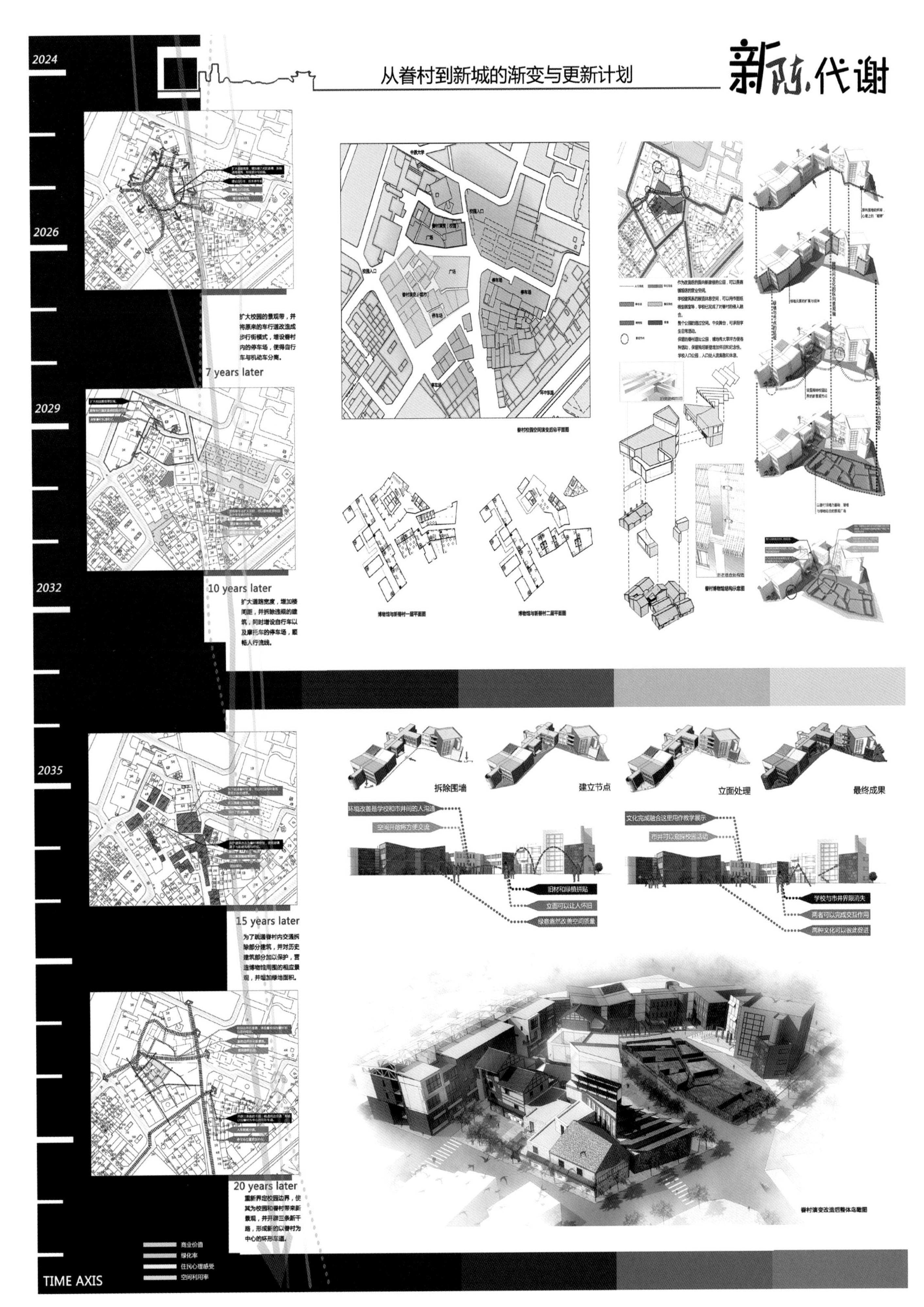

University: Harbin Institute of Technology
Designer: Chen Yongxiang, Luo Shengtao, You Haobo, Wang Du, Pang Jiayin
Tutor: Xue Minghui, Dong Yu, Zhang Shanshan, Shao Yu, Yu Zhaoqing, Zhang Huasun
Course Name: Internationsal Joint Design Studio
Finished Time: Mar. 2014
Exchange Institute: Chung Yuan Christian University

東藝，西衍

澳門內港碼頭片區更新改造企劃

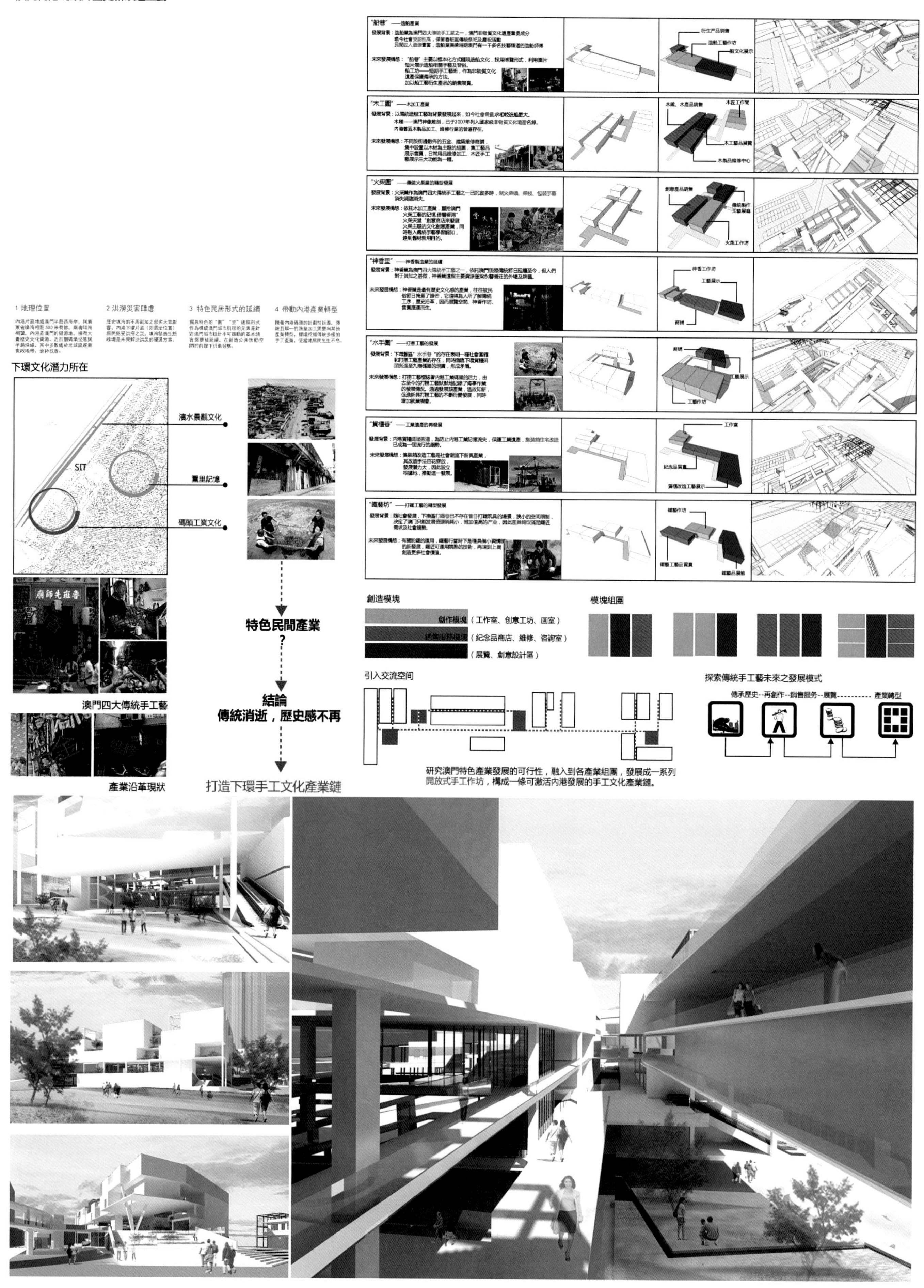

作品名称：东艺，西衍——澳门内港码头片区更新活化计划

Terminal Area Update Activation Plan

院校名：华侨大学

设计人：程坦

指导教师：费迎庆，郑剑艺

课程名称：澳门内港滨水区码头更新与城市设计

作业完成日期：2014年06月

对外交流对象：澳门土地工务运输局，澳门文化局

東藝，西衍

設計說明

設計地段為下環區原5號碼頭區域。
待貨櫃碼頭全部搬遷后，以貨櫃碼頭工業遺址為前提背景，尊重歷史，以内港居民為主要服務對象，深入挖掘澳門特色傳統手工文化產業，并結合下環區實際情況，通過園里的方式創造出不同的產業組團，各產業組團加以聯繫，形成一條特色手工文化產業鏈，打造一系列"開放工廠"，創造就業機會，促進内港片區的產業轉型，從而起到活化内港的作用。最終實現東部居民區技藝向西部衍進發展，即為作品名稱《東藝西衍》。

設計中需要解決的問題以下幾個：
· 為下環區居民創造公共空間
· 填海築堤，解決下環水患嚴重的現狀
· 延续澳门特色城市肌理
· 加強东部民居与西部海岸线的联系

方案分期建设步骤

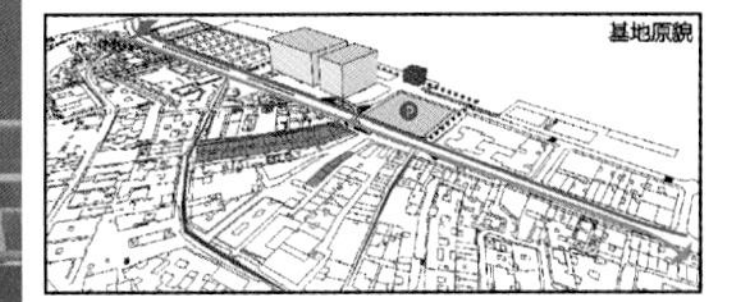

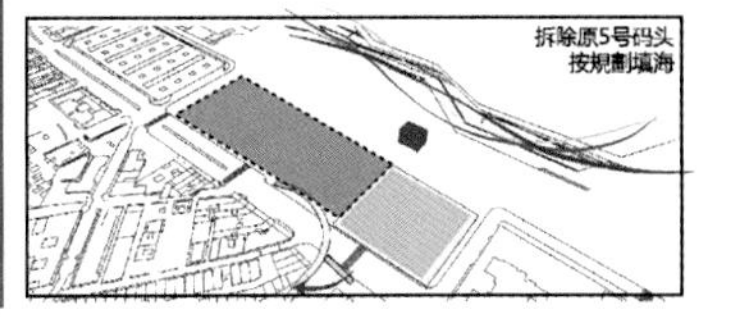

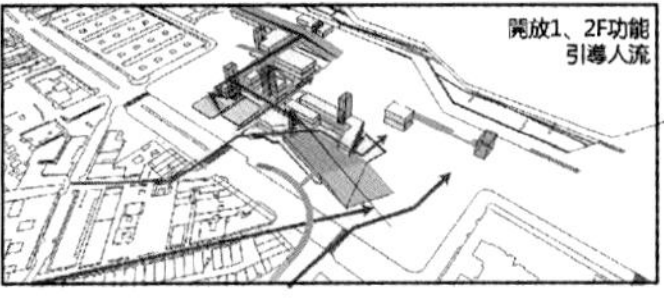

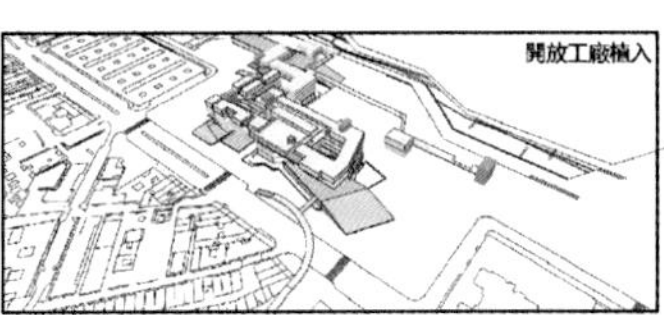

现状图解

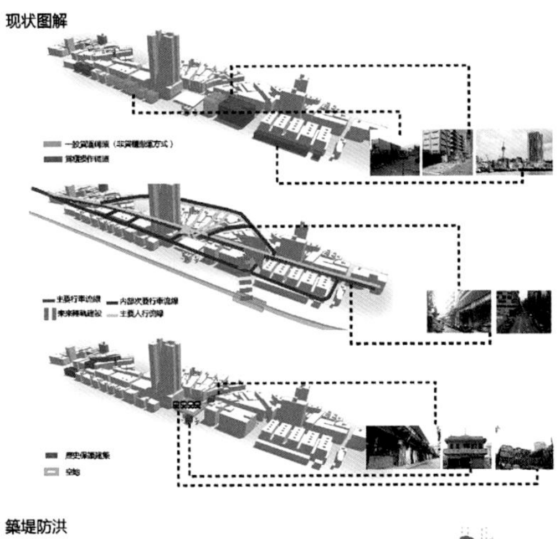

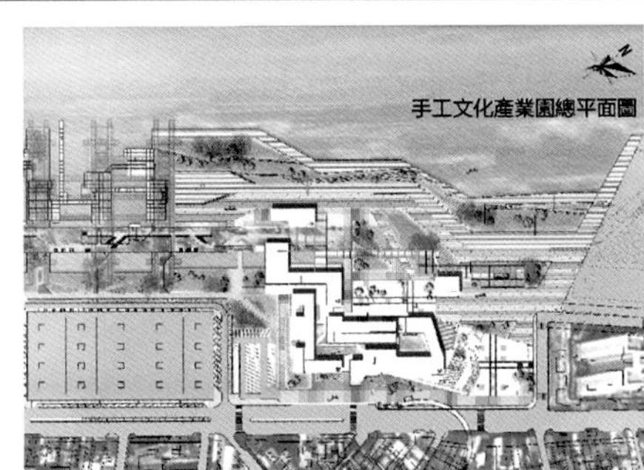

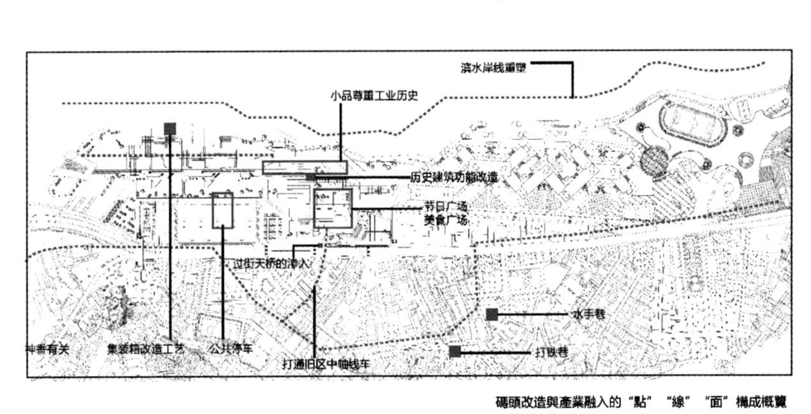

碼頭改造與產業融入的"點" "線" "面"構成概覽

築堤防洪

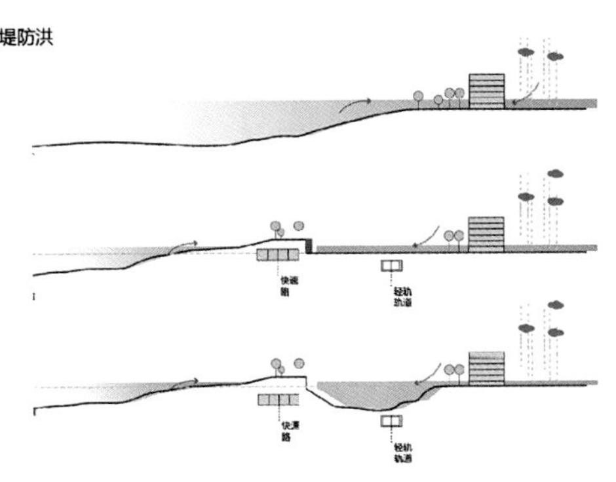

西南立面 1：500

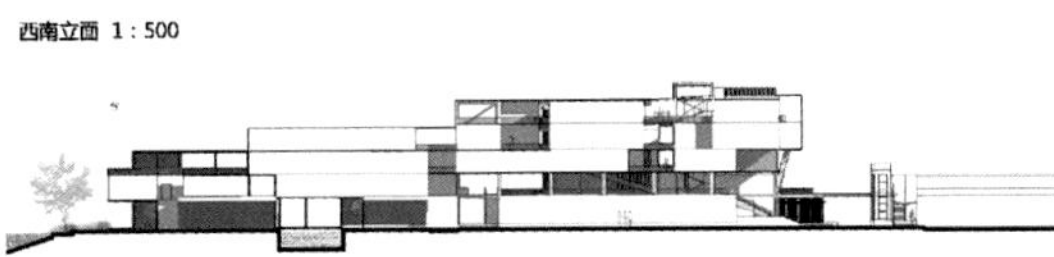

東南立面 1：500

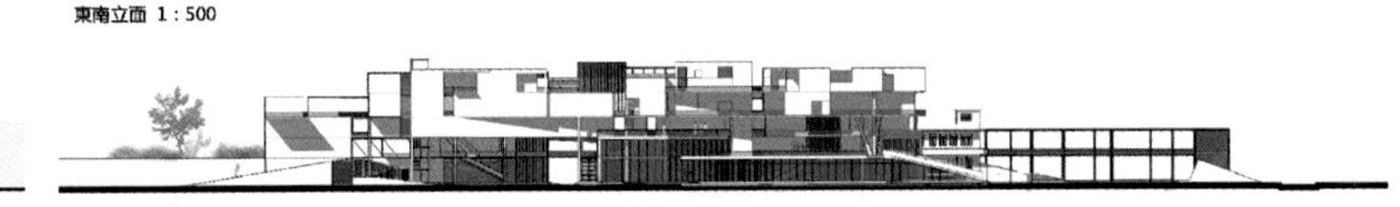

University: Huaqiao University
Designer: Cheng Tan
Tutor: Fei Yingqing, Zheng Jianyi
Course Name: Renovation and urban design for wharf of Macao harbor
Finished Time: Jun. 2014
Exchange Institute: The Land, Public Works and Transport Bureau of the Macao，Cultural Bureau of the Macau

煽动空间

——传统街区的生长

本课题为中西联合毕业设计，合作方为西班牙六所顶级建筑院校之一的巴塞罗那 THE SCHOOL OF ARCHITECTURE LA SALLE, RAMON LLULL UNIVERSITY。

课题以南京南捕厅历史文化街区为研究对象，探讨在旧城有机更新过程中，如何在原有的社会、经济及建成环境肌理中置入新的混合功能，以激发城市公共生活及公共空间的活力。在实地调研的基础上，与西方同学合作提出可行性策划与概念性规划成果，进而展开相应的规划与建筑设计研究。

关注现代建筑与传统文化的关系，理解传统街区物质形态与生活形态，以及传统文化中的精神追求，病放在当下的环境中思考；了解设计所在社区的需求。通过社区公共建筑的建设，丰富社区公共生活，增加社区凝聚力；理解传统街区的形成、生长和演变过程。研究建筑与城市肌理之间的互动关系；分析传统街区的各种建筑类型，以及适应现代功能的各种可能性。研究复杂的行为流线，符合的空间功能、特殊的结构形式与城市建筑类型之间互动的设计方法。

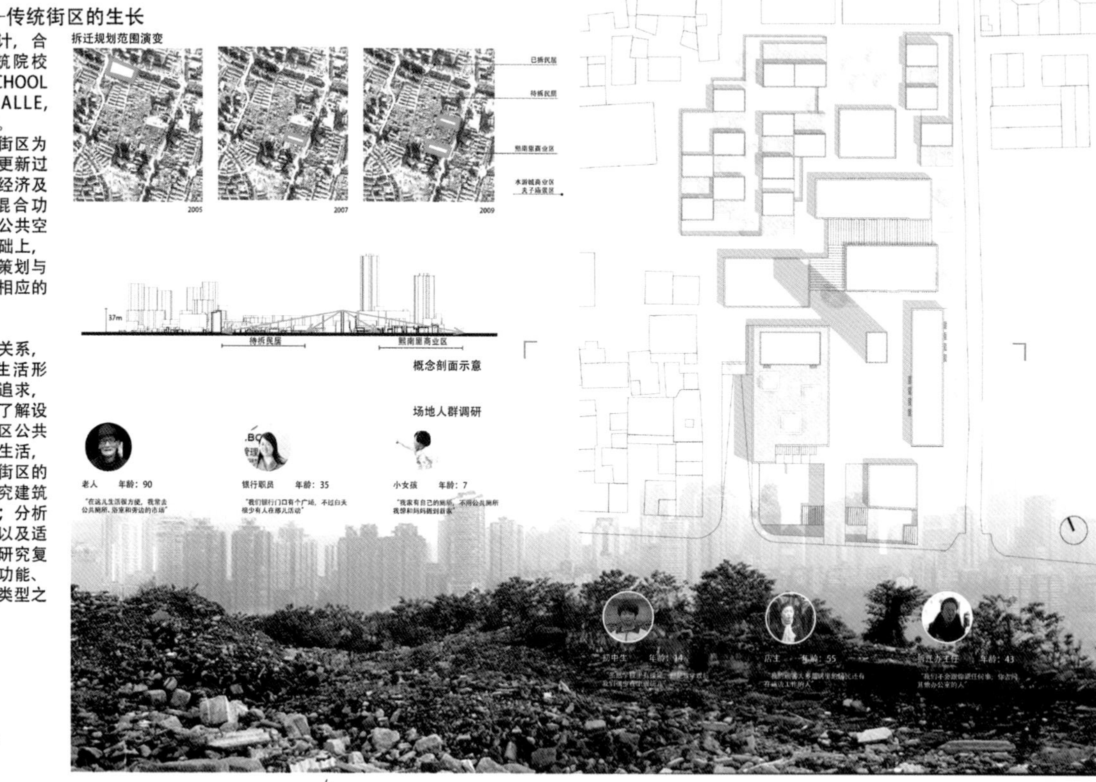

研究方法

Research Methodology

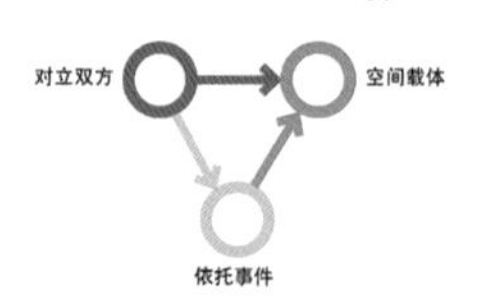

案例研究 Otaru Unga

小樽运河是日本北海道小樽市的一处地标。昔日运河里层拥挤着无数装卸货物的船板。后来由于战争令工业倒退，丧失了运河的功用，因此有一半长度的运河被填平作马路使用。如今这些仓库建筑都改成了玻璃工艺品商店、茶馆、餐厅和大型商铺，成为了一处旅游胜地。

对立双方

政府决定将公有水面填埋，市民发现杂物取出后，景观良好。1973年12月，24位小樽市民组成了"小樽运河保存协会"，自此开启了往后十年引起全国舆论热烈关注的小樽运河保护活动。

空间载体

为了请更多的市民来到运河周围，参与为共同学习运河的重要性而持续举办环境学习活动，如从1978年开始的小樽港都庆典，在广场和历史性建筑物中举办相声或演讲会等活动。

依托事件

"小樽再生论坛"是1985年小樽运河问题结束后，为结合市民力量，推动小樽社区营造而成立的团体。他们向市长和道知事提出陈情书、建议书之外，还举办演讲、系列座谈会、小樽街屋散步等活动。

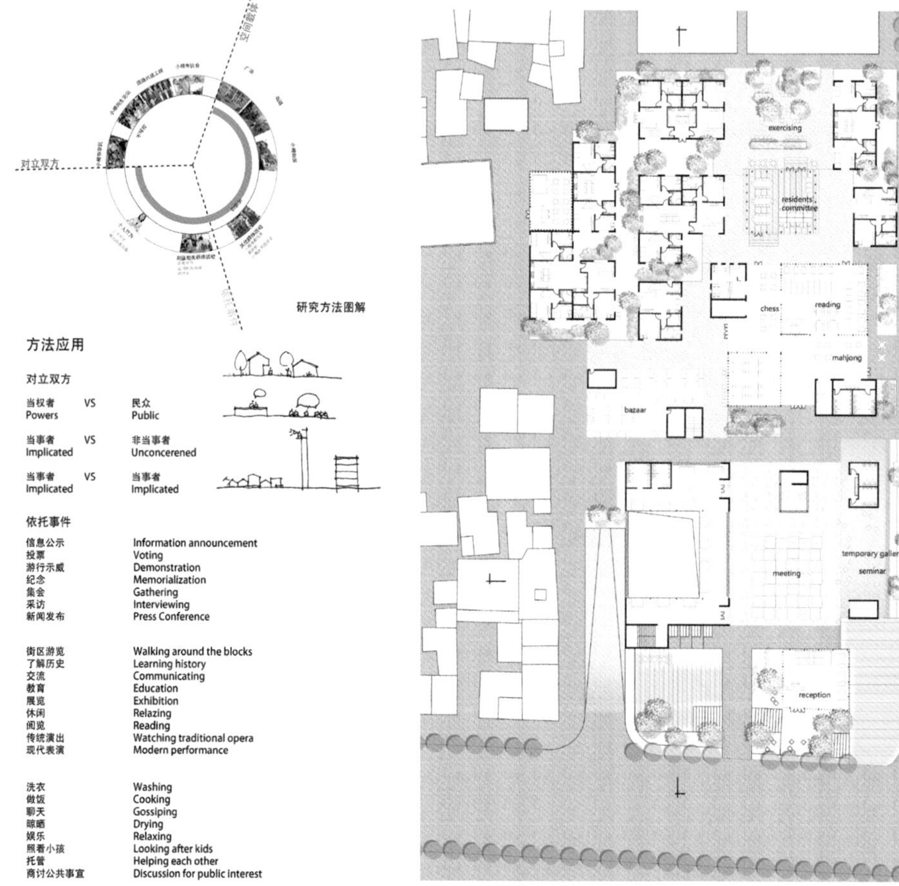

对立双方
研究方法图解
方法应用
对立双方
当权者 Powers VS 民众 Public
当事者 Implicated VS 非当事者 Unconcerened
当事者 Implicated VS 当事者 Implicated
依托事件
信息公示 Information announcement
投票 Voting
游行示威 Demonstration
纪念 Memorialization
集会 Gathering
采访 Interviewing
新闻发布 Press Conference
街区游览 Walking around the blocks
了解历史 Learning history
交流 Communicating
教育 Education
展览 Exhibition
休闲 Relazing
阅览 Reading
传统演出 Watching traditional opera
现代表演 Modern performance
洗衣 Washing
做饭 Cooking
聊天 Gossiping
晾晒 Drying
娱乐 Relaxing
照看小孩 Looking after kids
托管 Helping each other
商讨公共事宜 Discussion for public interest
exercising
residents committee
chess
reading
mahjong
bazaar
temporary gallery
seminar
meeting
reception

作品名称：传统街区的生长

The Growth of Traditional Blocks

院校名：东南大学
设计人：崔百合
指导教师：龚恺，鲍莉，刘捷，Josep Ferrando，Jaime Font
课程名称：东南大学—La salle 建筑学院联合设计课程
作业完成日期：2014 年 06 月
对外交流对象：巴塞罗那 La salle 建筑学院

煽动空间 ——传统街区的生长

南捕厅历史文化街区属于敏感的拆迁地带，场地周边面临着不可忽略的历史建筑保护、民居被强拆、拔地而起的商业街区等诸多矛盾。本设计在满足新社区生活需求和城市生活需求之余，试图在空间上体现，甚至激化这种社会矛盾，引发人们对于这类问题的思考。

场地西、北两面紧邻待拆民居区域；东面隔街是一、二两期新中式熙南里商业街区，南部面临城市干道。故由南至北依次设置为‘城市文化区’——‘社区公共活动区’——‘社区生活区’。

为减弱传统曲艺中心的较大建筑体量对西面民居的影响，设置了地下剧场，同时解放了屋面层提供更多室外空间为前来的游客和当地居民使用；地面层为观望塔围合了集会广场，以刺激相应的公共活动产生。社区居住主要面向场地的主要群体——无工作的中老年人群，为其设置了室外空间丰富的小户型，为其子女的临时居住也预留了可能性。公共服务类建筑周边的种植小广场为中老年人社交、商讨拆迁等公共事宜提供了友好的空间。

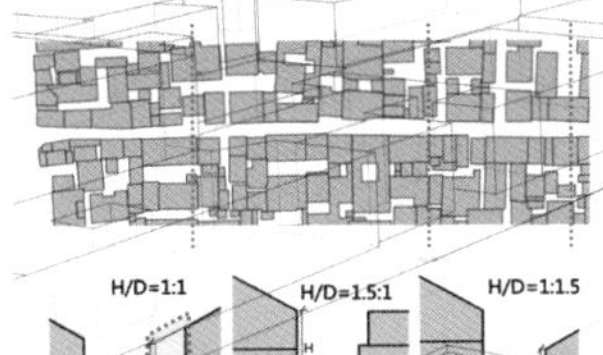

历史文化街区室外空间形态

狭窄而不规则的街巷空间促成了尺度、形态丰富的公共空间以及曲折迂回的入户空间，既形成了居民小群体之间活动的场所，也保证了大多数非沿街住户的私密性。外来人群在场地中行走，虽然没有明确的围墙阻隔，却也很难干扰到住户的居住生活。

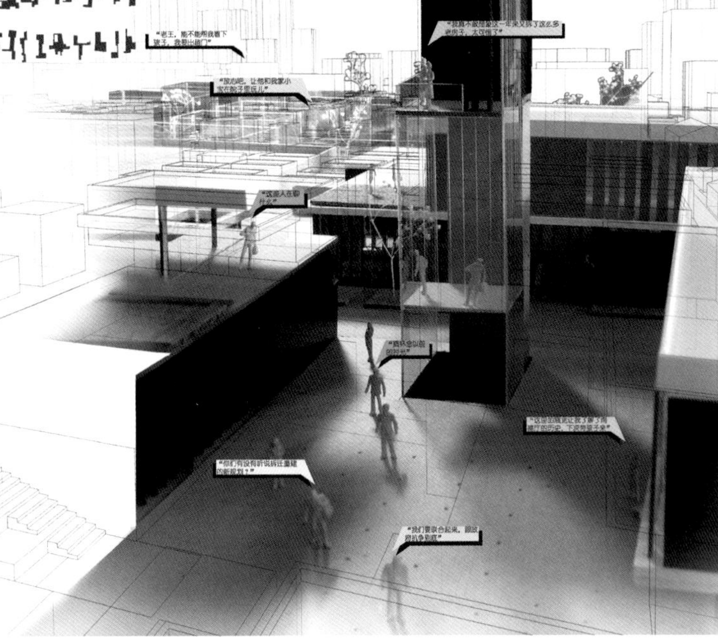

住宅分析

传统住宅虽然有丰富的外部空间和私密的入户方式，但由于住宅被分割，每户的实际居住面积极小。设计通过对居民实际居住需求的调研，设计了适合独居老人和两代人同住的可组合式户型。建筑肌理与周边住宅相似，并以外部空间相互联系。

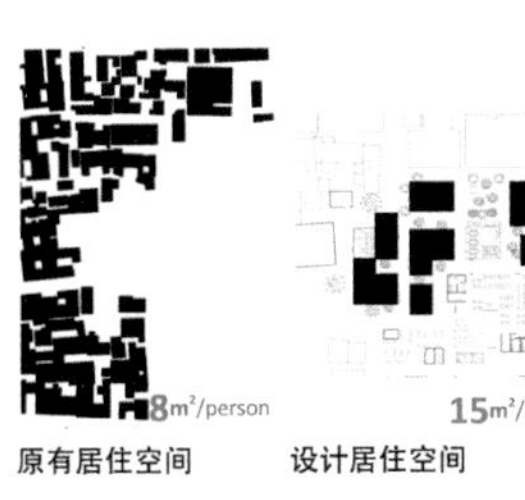

原有居住空间　设计居住空间

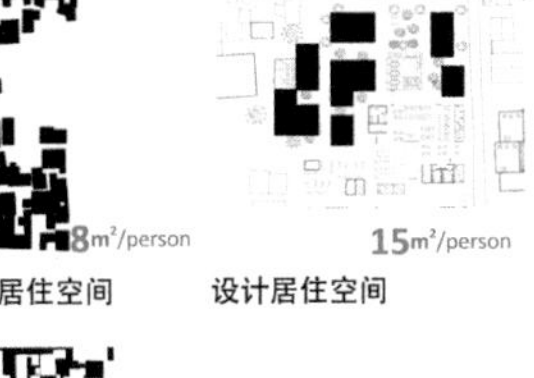

原有交往空间　设计交往空间

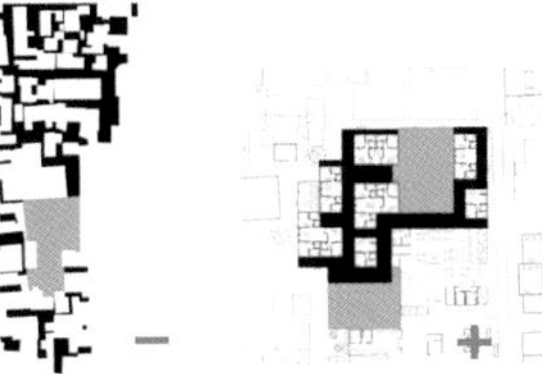

原有入户方式　设计入户方式

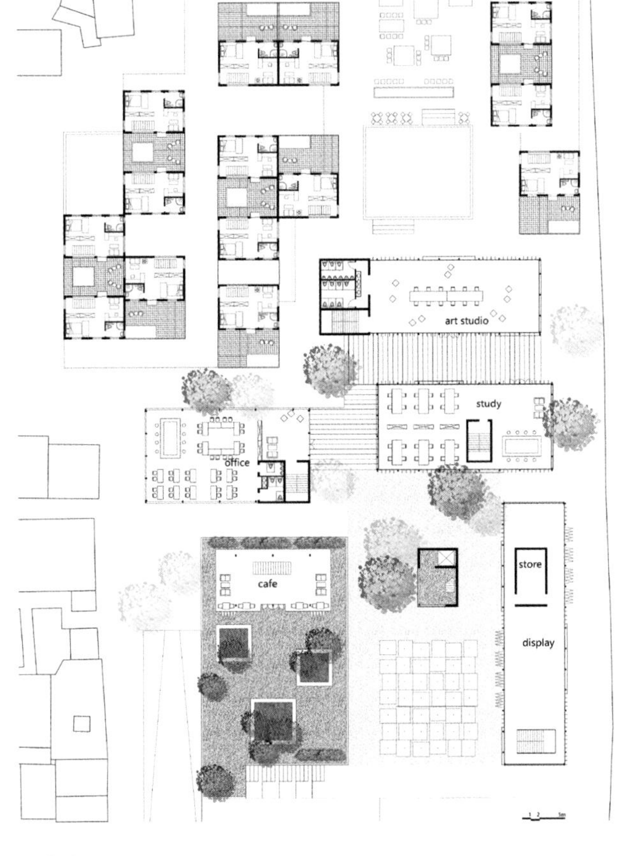

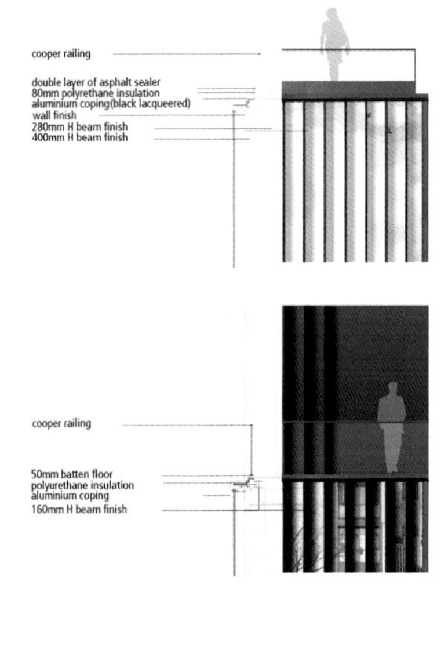

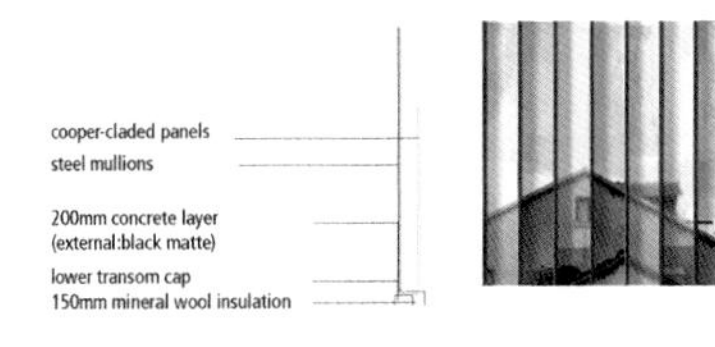

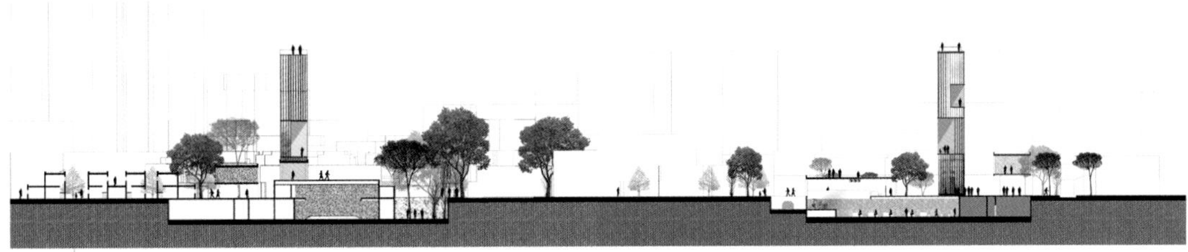

University: Southeast University
Designer: Cui Baihe
Tutor: Gong Kai, Bao Li, Liu Jie, Josep Ferrando, Jaime Font
Course Name: Southeast University-The School of Architecture La Salla Joint Design Studio
Finished Time: Jun. 2014
Exchange Institute: The School of Architecture La Salle, Ramon Llull University

URBAN REGENERATION

The Rebuilding Of The Old Block In Jinan

Current Problem & Logic Of Design

LOCATION

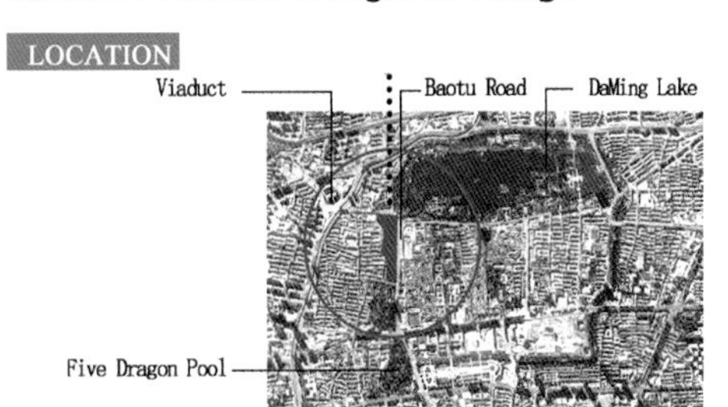

Jinan is located in the Midwest of shandong province, in the north side of mount tai, north across the Yellow River, and surrounded by mountains and water. Jinan have a large number of springs, referred to as "water city", At the same time, the city with more than 700 natural inspiration,constitute the jinan "every family have springs and willows" unique spring landscape.

AREA LINK NODE ANALYSIS

The surrounding landscape and roads net in site

The surrounding function distribution

MAIN PROBLEMS & MAIN MEASURES

Surrounding Environment

The residential areas in the side of east and west have no connectionwith square area .The residentialareas are quite closed

Clear the streets in the sides east and west. Enhance the connection with each other

Haven't taken full advantage of the stream of people in the scenic spots on the north and south areas in the site.

Design a north-southern main road ; In order to introduce tourists ; Add more vigour to side; Strengthen the communication of both areas.

SITE ANALYSIS

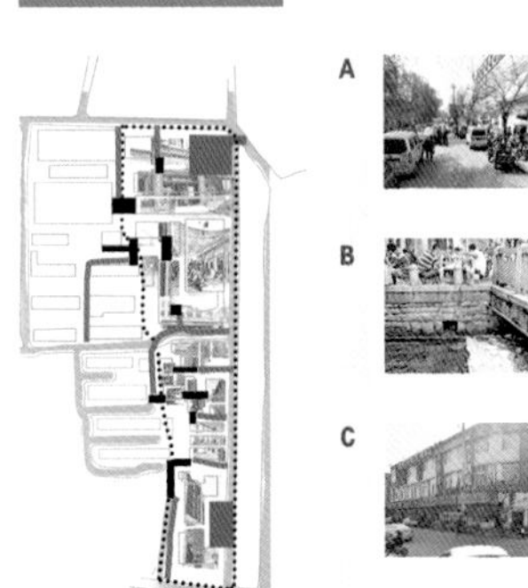

The current situation of site:

A: The jam in the interior roads.
B: The terrible view around river.
C: The lack of public opening space.
D: Tconfused interior functions.
The noise environment.

STRATEGIES

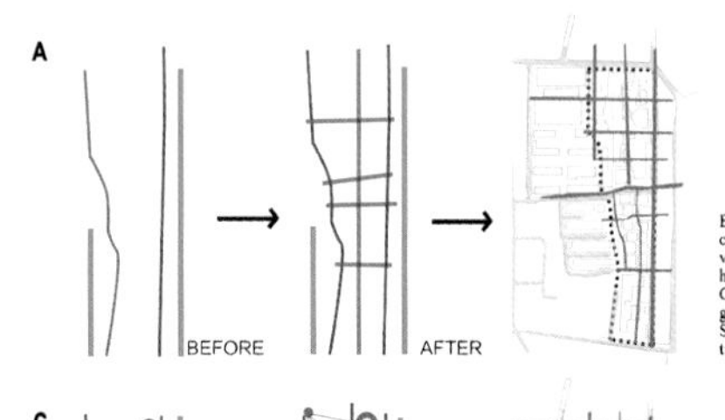

Enhance the connection vertically and horizontally; Clear the original roads; Strength the trafficability.

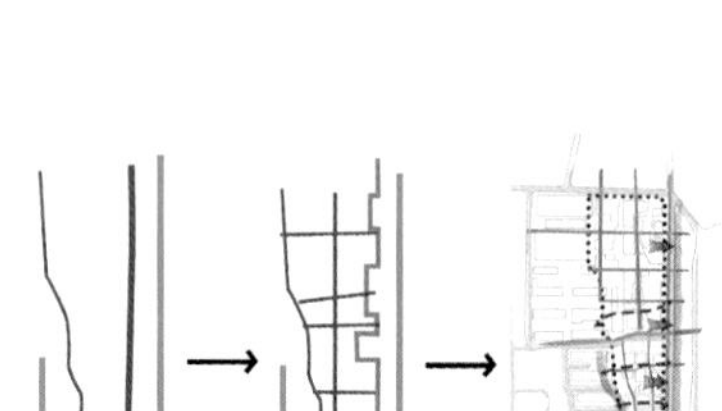

Innovate the riverway space; Add water loving platform; Penetrate the waterview into the interior of site.

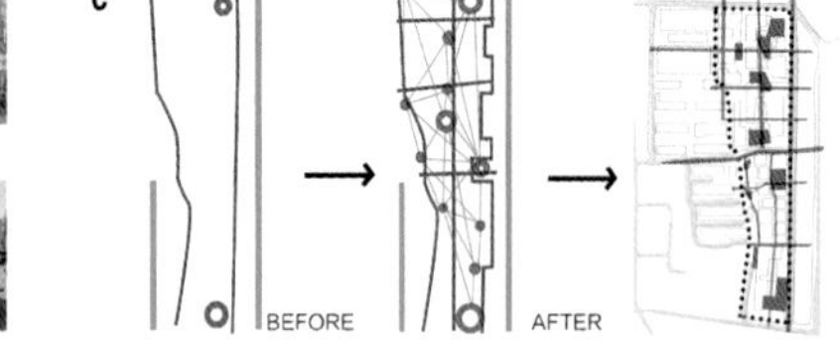

Increase vigour and lively areas;Introduce the north-western steam of people into site; Activate the community.

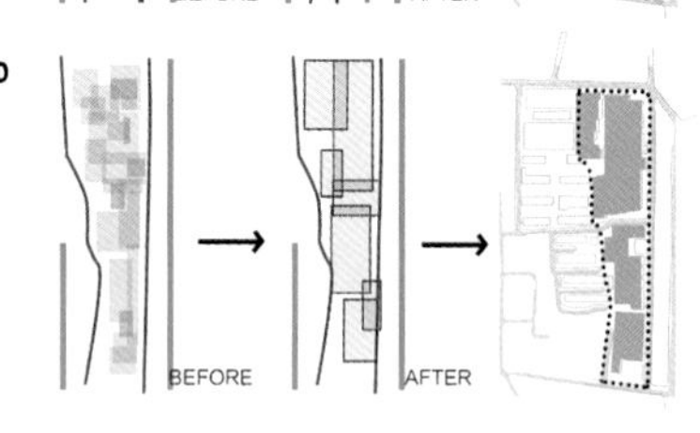

Fix the original function; Shape more suitable surrounding environment and new functional establishment.

ELEMENT & POPULATION STRUCTURE

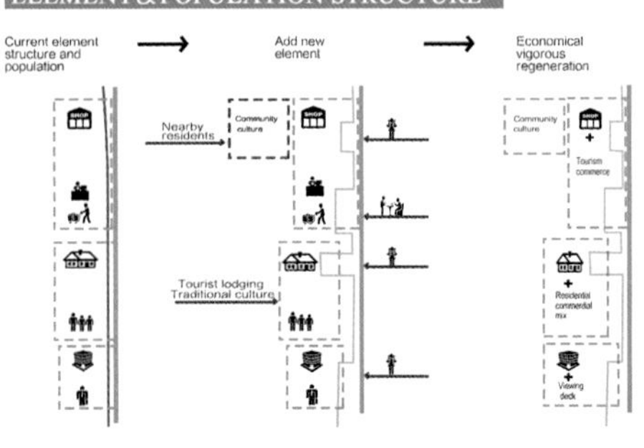

BUILDING CLASSIFICAION

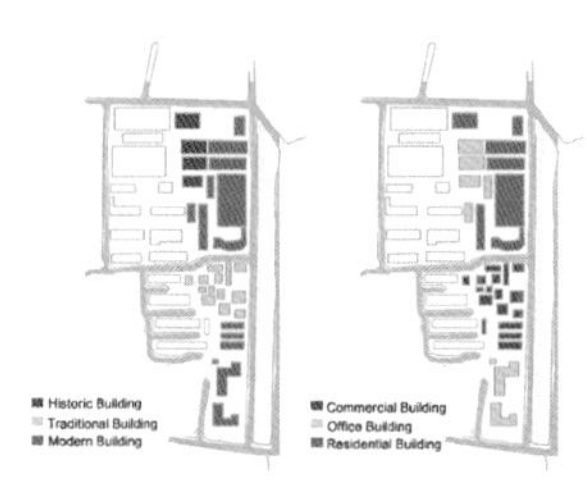

CONSTRUTION TIME

ORIGINAL FUNCTION

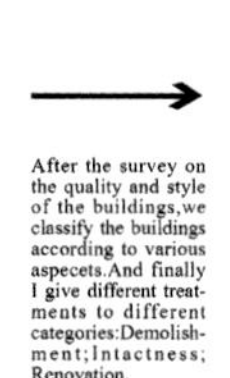

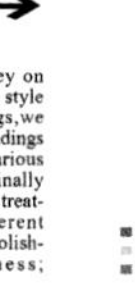

After the survey on the quality and style of the buildings,we classify the buildings according to various aspects.And finally I give different treatments to different categories:Demolishment;Intactness; Renovation.

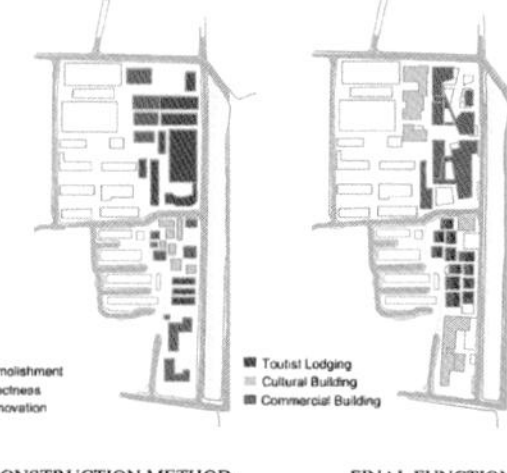

RECONSTRUCTION METHOD

FINAL FUNCTION

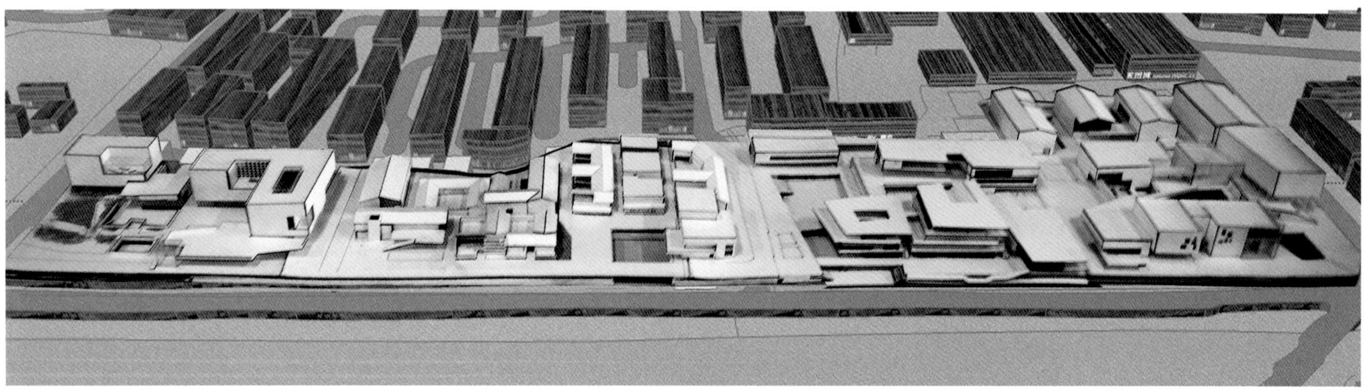

作品名称：城市再生
Urban Regeneration

院校名：山东建筑大学
设计人：崔雅婧，李珏，王梦真，梁正蕾
指导教师：陈林，周忠凯，Tony Van Raat
课程名称：山东建筑大学—新西兰 UNITEC 理工学院联合设计课程教学 2
作业完成日期：2014 年 06 月
对外交流对象：新西兰 UNITEC 理工学院

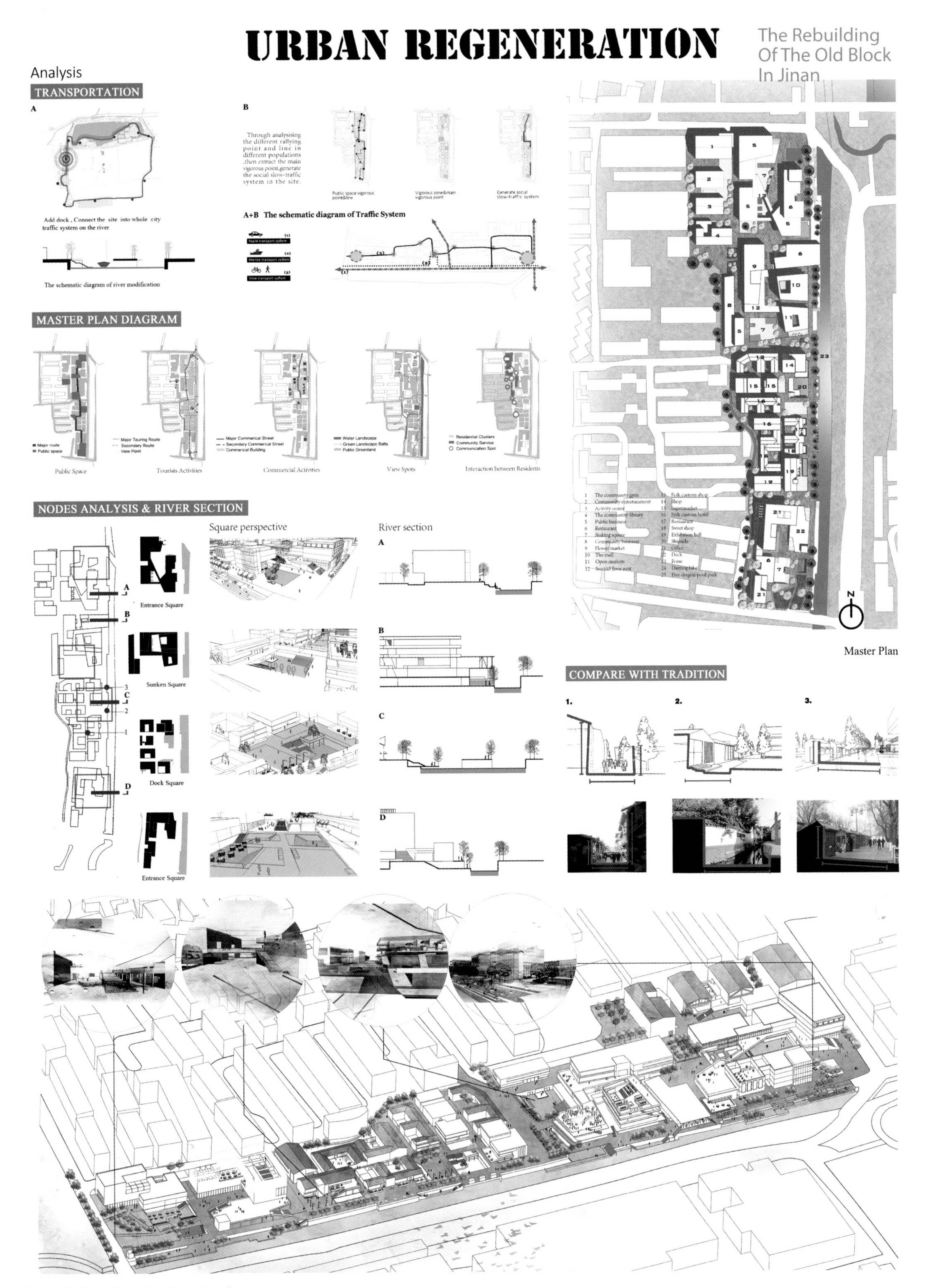

University: Shandong Jianzhu University

Designer: Cui Yajing, Li Jue, Wang Mengzhen, Liang Zhenglei

Tutor: Chen Lin, Zhou Zhongkai, Tony van Raat

Course Name: Shandong Jianzhu University - UNITEC Institute of Technology Joint Design Studio 2

Finished Time: Jun. 2014

Exchange Institute: Unitec Institute of Technology

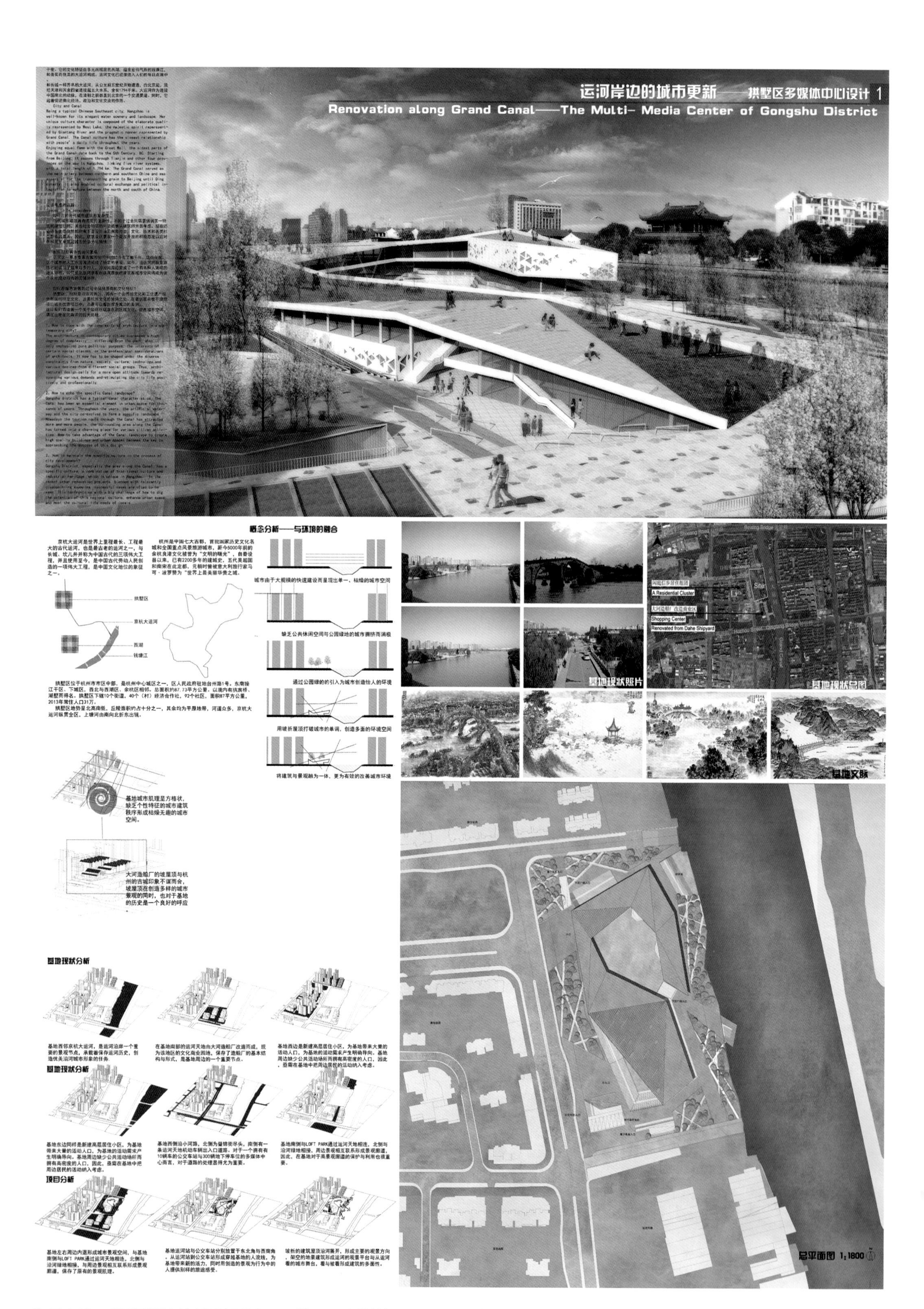

作品名称：运河岸边的城市更新——拱墅区多媒体中心设计

Renovation along Grand Canal

院校名：浙江大学

设计人：邓奥博

指导教师：陈林，王晖

课程名称：联合毕业设计

作业完成日期：2014 年 06 月

对外交流对象：西班牙马德里圣帕保罗大学

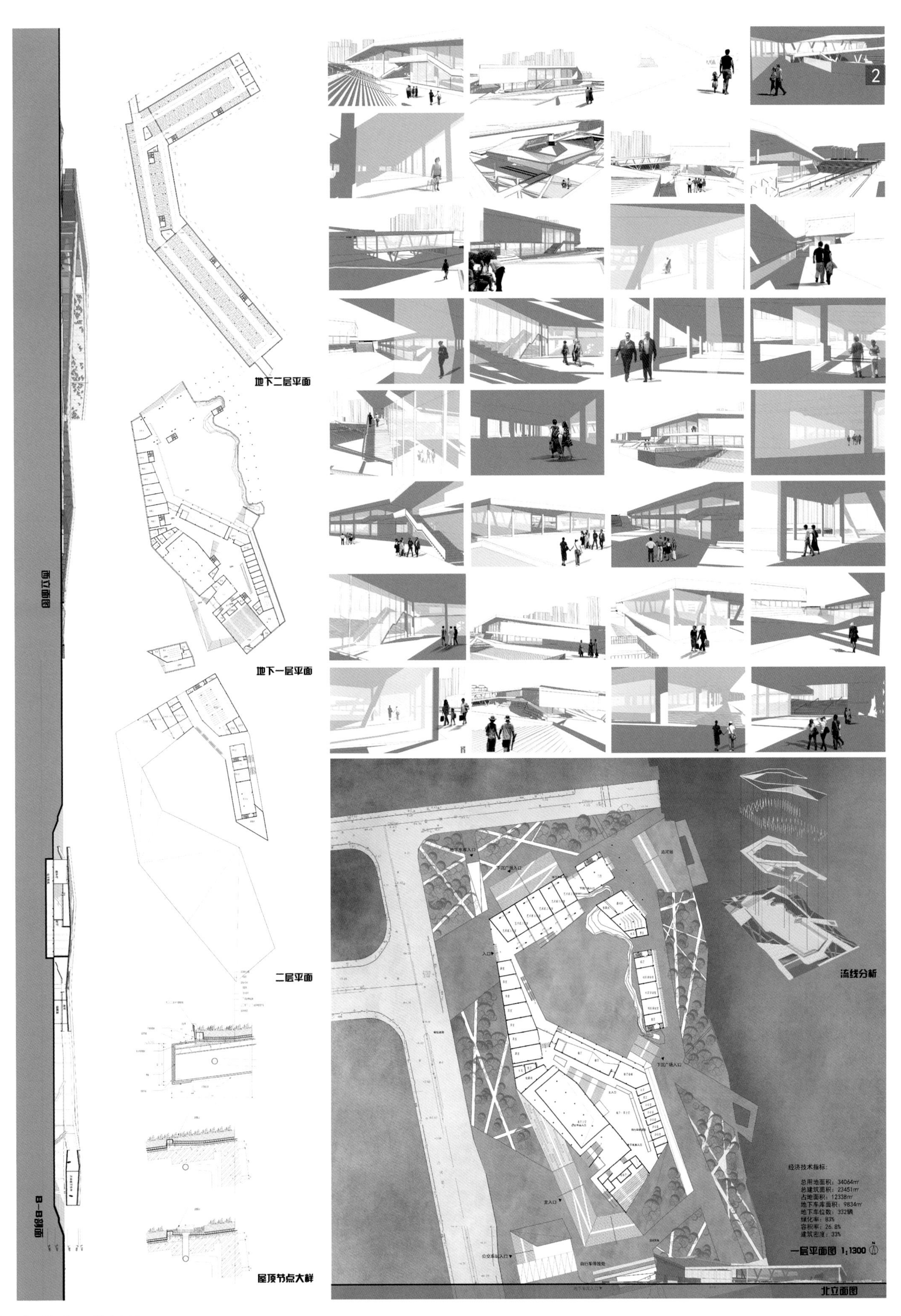

University: Zhejiang University
Designer: Deng Aobo
Tutor: Chen Lin,Wang Hui
Course Name: Zhejiang University-San Pablo CEU Joint Design Studio
Finished Time: Jun. 2014
Exchange Institute: CEU San Pablo University

1 RIVER BEAT

CELEBRATING ZHENGZHOU'S PAST, PRESENT, AND FUTURE

Wetland Park
The ecological wetland park will be a treasure for citizens' leisure needs

Community Center
The landscape ribbon cordinate various programs and provide convenience for citizens.

Community Performance Center
The outdoor performance space has economic effects as well as meets people's need.

Waterfall District
The waterfall district not only serves as an education center but also as a place for people to rest.

Cultural Center Lake Dstrict
Around the lake, Pedestrain and plazas provide a gathering space with a landmark.

KEY NODES ANALYSIS

CIRCULATION ANALYSIS

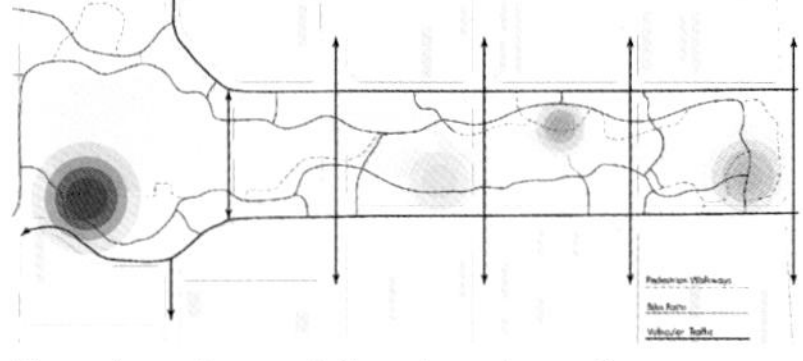

The primary transportation elements on site serve as a regulating grid within which buildings can slide on the north-south axis, creating a variety of relationships between building, street, and water. There is no clear boundary between the water, the landscape, and the buildings.

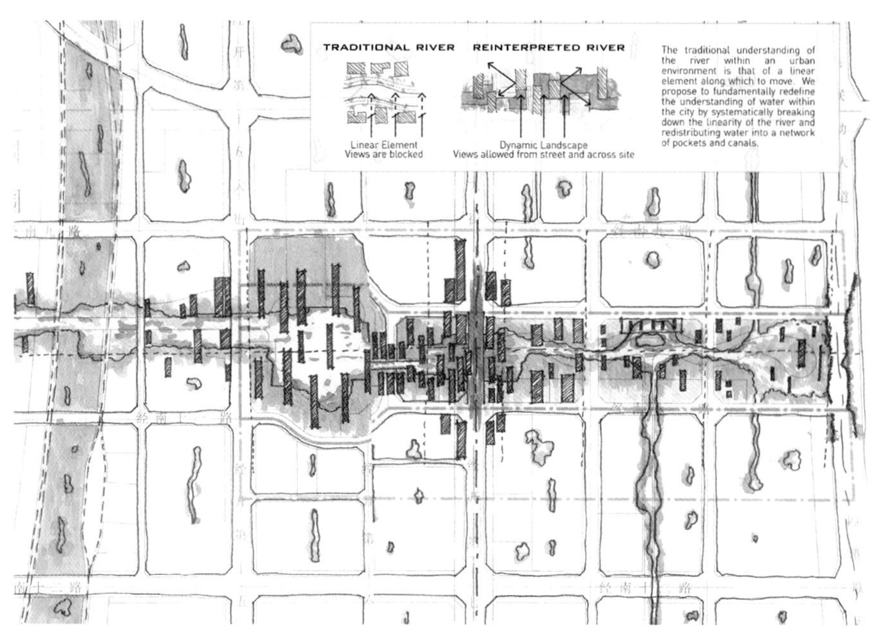

TRADITIONAL RIVER
REINTERPRETED RIVER
Linear Element
Views are blocked
Dynamic Landscape
Views allowed from street and across site
The traditional understanding of the river within an urban environment is that of a linear element along which to move. We propose to fundamentally redefine the understanding of water within the city by systematically breaking down the linearity of the river and redistributing water into a network of pockets and canals.

SITE ANALYSIS

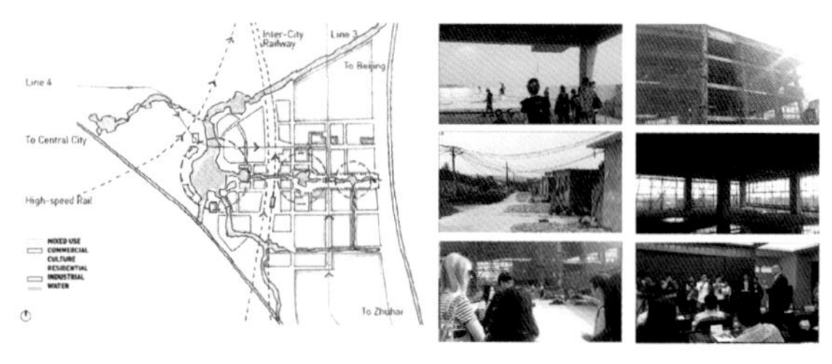

CONCEPTUAL STRATEGY

The ambitions of this proposal are twofold:
1. We seek to link the blocks to the north and south of the site through the design of a water collection and dispersal system, which creates a dynamic, walkable landscape influenced by seasonal conditions and fluctuations.
2. We seek to educate both visitors and residents on sustainability by envisioning the site as an experiential journey through the harvest, consumption, and production of energy.

The site can be understood as four unique districts of use unified by a similar language. These districts work to engage the surrounding environment within the larger context of the Binhe district. Redefined elements are overlaid with landscape and built form to generate a layered environment of water, circulation, and program.

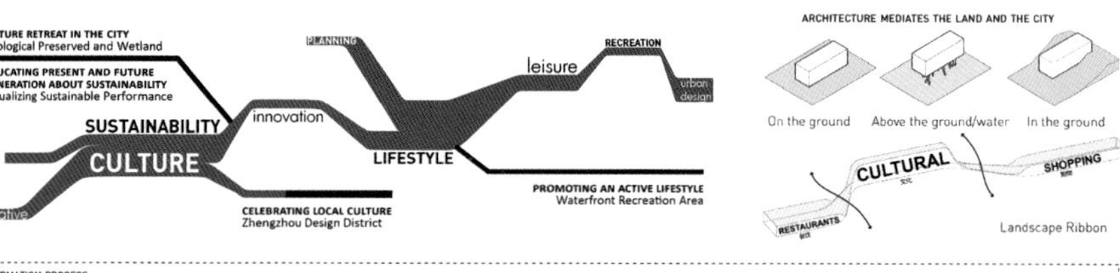

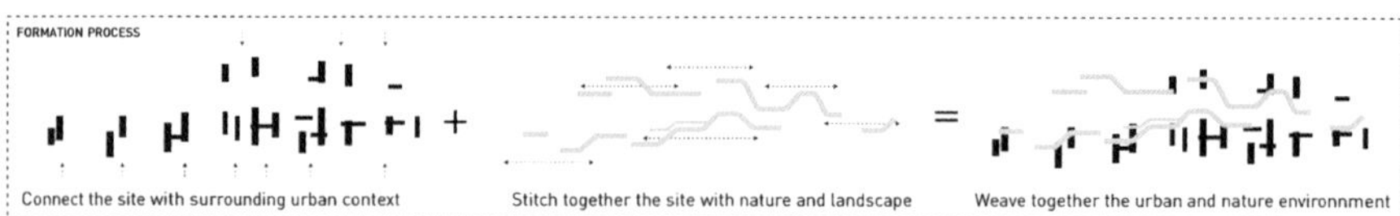

PROGRAM ANALYSIS

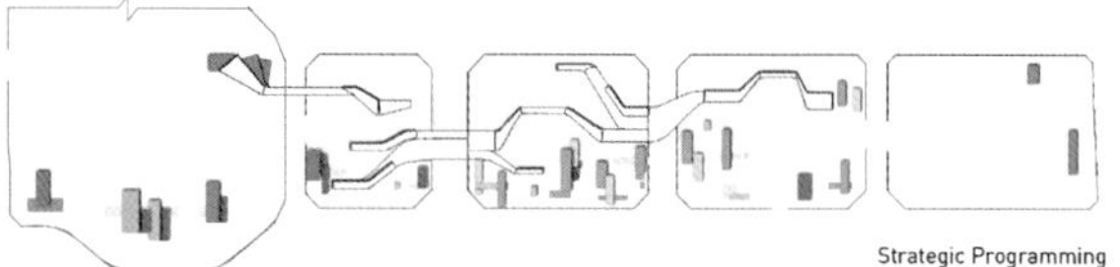

PROPOSED MATERPLAN

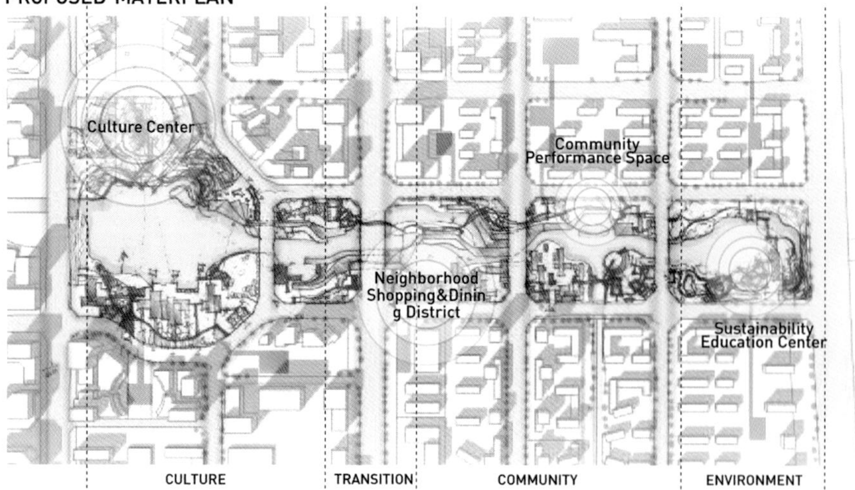

作品名称：律动河流——郑州滨水文化休闲区城市设计
River Beat

院校名：北京建筑大学
设计人：谷筝，刘玲
指导教师：Lars Grabner，丁奇，王佐
课程名称：北京建筑大学—美国密歇根大学—思朴国际设计工作营
作业完成日期：2014 年 05 月
对外交流对象：美国密歇根大学

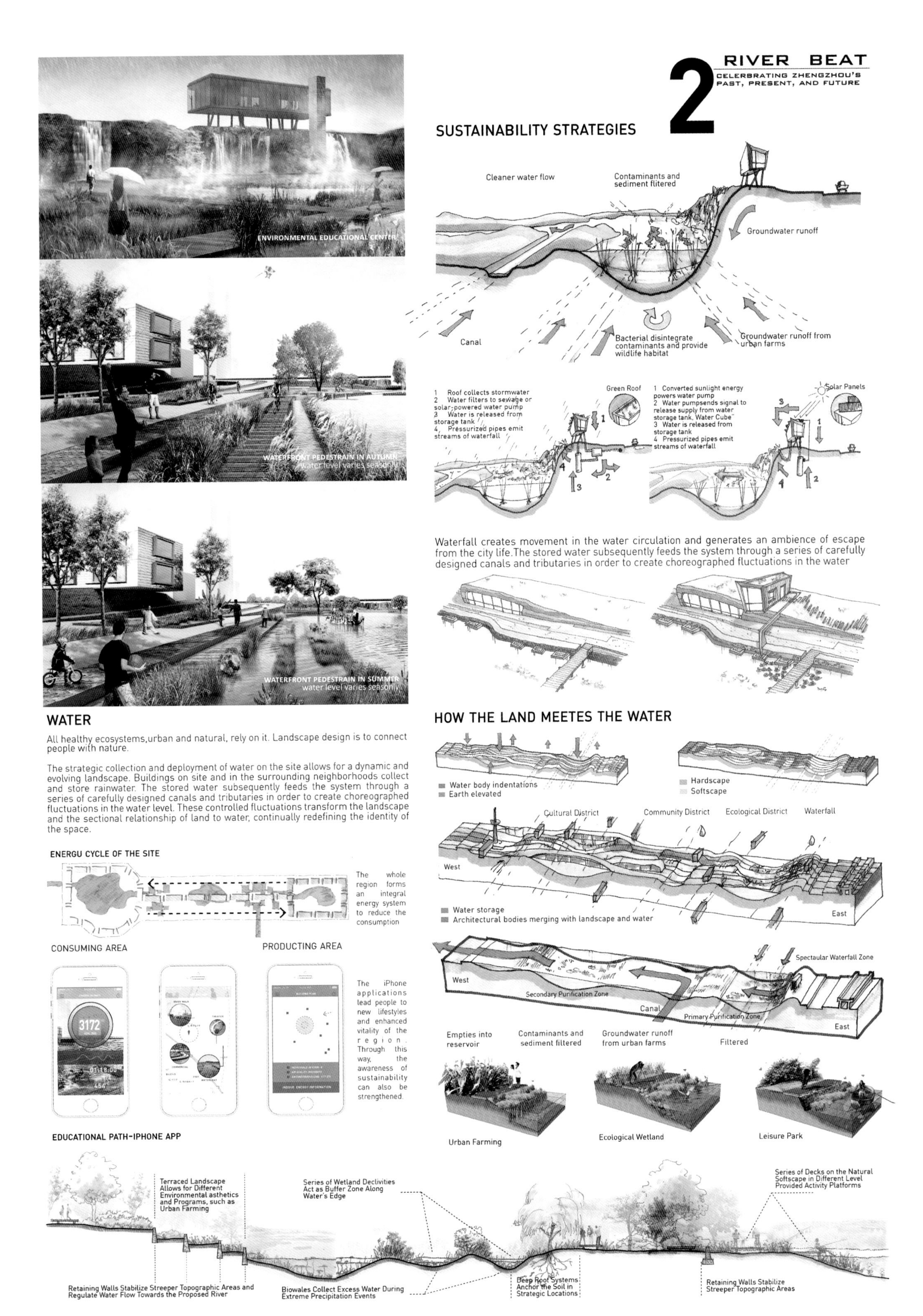

University: Beijing University of Civil Engineering and Architecture
Designer:Gu Zheng, Liu Ling
Tutor: Lars Grabner, Ding Qi, Wang Zuo
Course Name: Workshop project with University of Michigan
Finshed Time: May 2014
Exchange Institute: University of Michigan

都市の成长 南京地铁马群站城市设计

本课题为四年级中日联合课程设计，课题选址于南京马群地铁站南侧地块，拟建的马群换乘中心是南京首个综合交通枢纽，承担地铁、公交、私家车、自行车、出租车等多种交通方式的综合换乘，建成后将成为城东地区进入南京主城的主枢纽。联合教学由日本京都大学教授以及三菱地所设计设计师与东南大学教师联合授课。课题结合日本东京地铁主要站点的实地考察和日本建筑师的现地讲解，让学生对城市设计有更为切身的认识，了解从关怀人的行为出发进行城市设计和建筑设计的思维方法，建立以人为本的，城市设计与建筑设计一体化的设计思想。

本方案希望从人文关怀及具体建设周期等层面为此类大型综合体开发项目提供一种新的思路。从城市设计中的"时间"维度入手，捕捉基地周边城中村居民生活环境变迁与大型综合体建设之间的时序关系，并以此为出发点，提出"分期建设"的概念。本方案通过对建设周期的具体规划和针对性设计，为城中村居民融入城市的过程提供缓冲，同时使综合体建设更具适应性和灵活性。在结构上主要使用规则柱网的框架结构来适应各个时期增改建的需要。

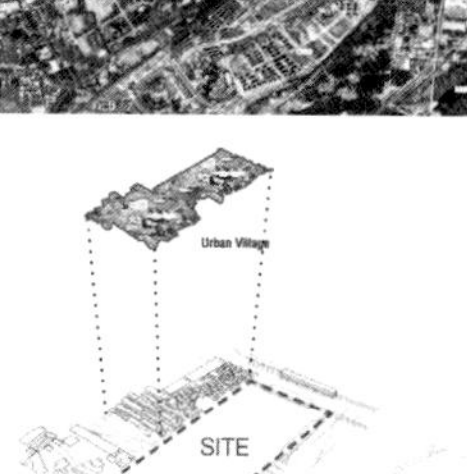

问题及解决方法 Problems&Solution

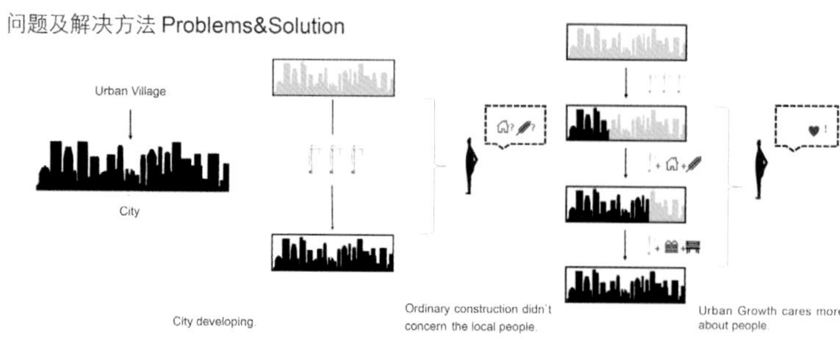

生长 The Growth

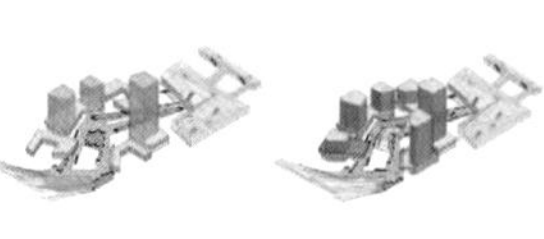

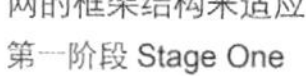

第一阶段 Stage One

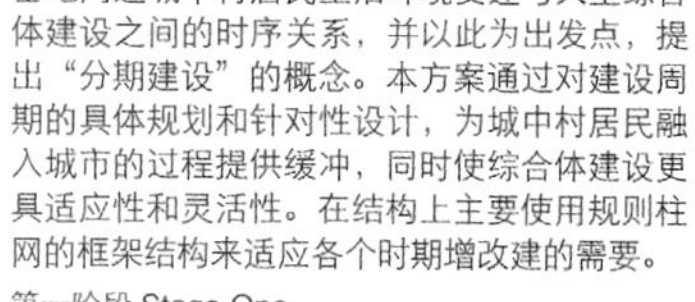

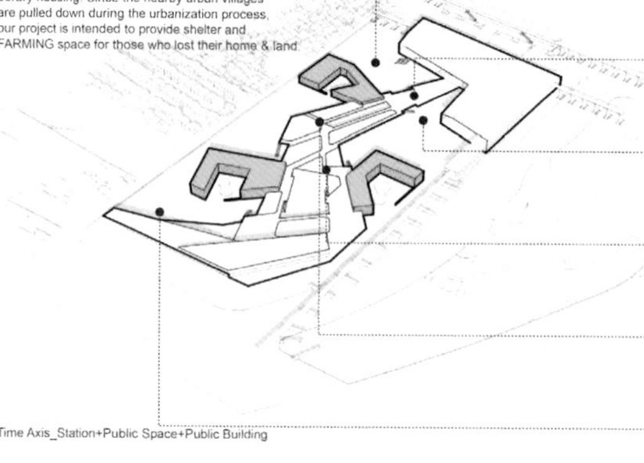

Time Axis is a witness to the changes on the site during the years.
On stage 1, three of the annexes are built as temporary housing. Since the nearby urban villages are pulled down during the urbanization process, our project is intended to provide shelter and FARMING space for those who lost their home & land.
Time Axis_Station+Public Space+Public Building

第二阶段 Stage Two

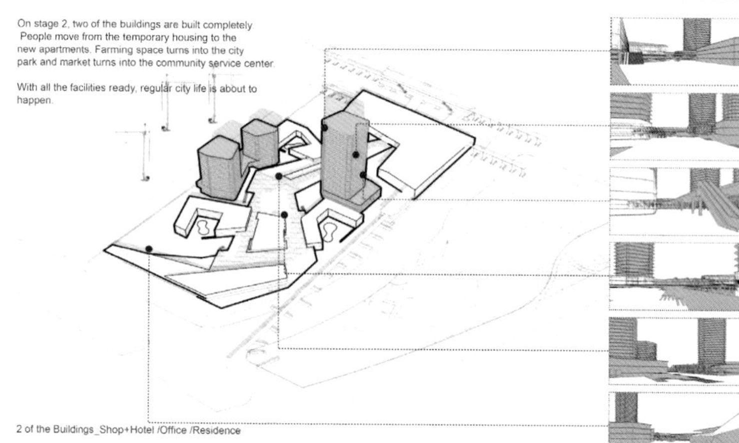

On stage 2, two of the buildings are built completely. People move from the temporary housing to the new apartments. Farming space turns into the city park and market turns into the community service center.
With all the facilities ready, regular city life is about to happen.
2 of the Buildings_Shop+Hotel /Office /Residence

第三阶段 Stage Three

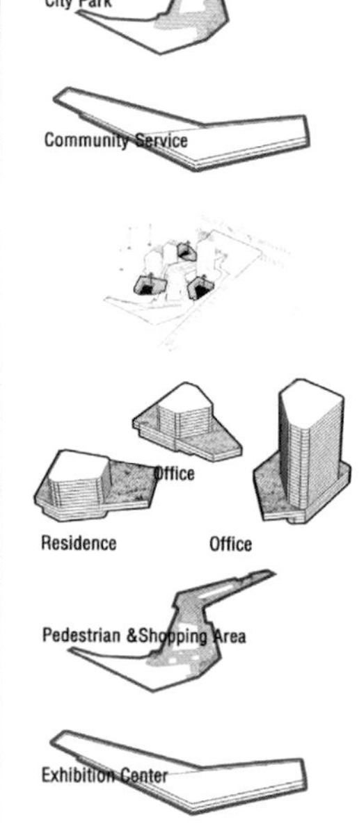

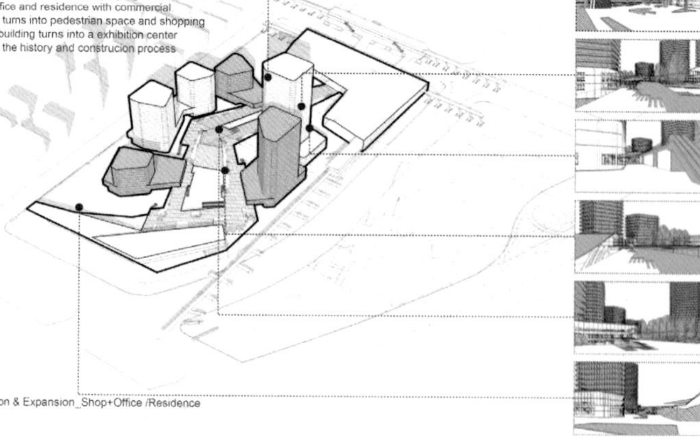

On stage 3, the first three annexes are reconstructed.
Temporary housing turns into the preprogrammed functions_ office and residence with commercial. Public space turns into pedestrian space and shopping area. Public building turns into a exhibition center which shows the history and construction process on the site.
Reconstruction & Expansion_Shop+Office /Residence

作品名称：南京地铁马群站城市设计
Nanjing Subway Station Urban Design

院校名：东南大学
设计人：管睿，商琪然
指导教师：唐芃，沈旸，宗本顺三，惠良隆二
课程名称：东南大学设计课程教学（本科四年级）
作业完成日期：2014 年 08 月
对外交流对象：日本京都大学，三菱地所

URBAN GROWTH
URBAN DESIGN OF NANKING METRO MAQUN STATION

This design project is a China-Japan joint studio. The proposed Maqun transfer station will be the first comprehensive transportation terminal in Nanjing for visitor entering the major city from the east side. The studio is taught by professors from Kyoto university, architects of Mitsubishi Estate and professors from Southeast university. The visiting of metro stations in Tokyo is aimed to enhance the students` understanding of urban design, and make them build the idea "people-oriented" in both urban and architecture design.

Concerning humanistic care and construction period,this design aims to provide an inspiration for such kind of urban complex projects. We started with consideration of the dimension of "time" . We also seized the sequential relationship between the change of living conditions of the local people and the site, leading to the idea of stage construction. Through these ideas we tried to provide buffer for local people to integrate into the city.

University: Southeast University
Designer: Guan Rui, Shang Qiran
Tutors: Tang Peng, Shen Yang, Munemoto Junzo, Era Ryuji
Course Name: Southeast University Design Studio
Finished Time: Aug. 2014
Exchange Institute: Kyoto University, Mitsubishi Estate

CITY ANALYSIS

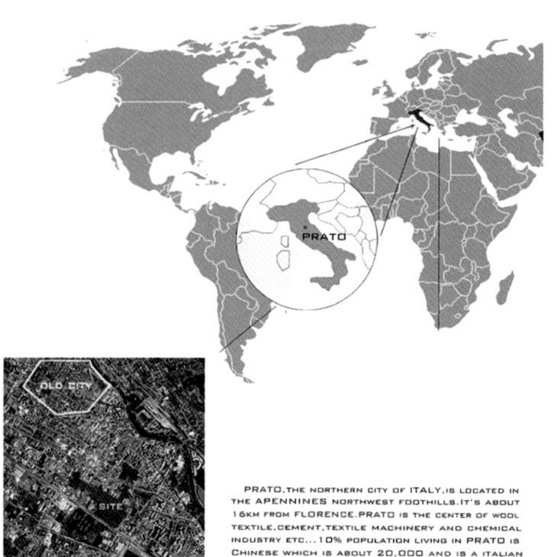

PRATO,THE NORTHERN CITY OF ITALY,IS LOCATED IN THE APENNINES NORTHWEST FOOTHILLS.IT'S ABOUT 16KM FROM FLORENCE.PRATO IS THE CENTER OF WOOL TEXTILE,CEMENT,TEXTILE MACHINERY AND CHEMICAL INDUSTRY ETC...10% POPULATION LIVING IN PRATO IS CHINESE WHICH IS ABOUT 20,000 AND IS A ITALIAN CITYOF MOST CHINESE IMMIGRANTS.

THE PHOTO OF OLD CITY

THE PHOTO OF NEW CITY

LIVING IN THE PARK

公園·城市

PARK·CITY

ANLYSIS OF THE SITE AROUND

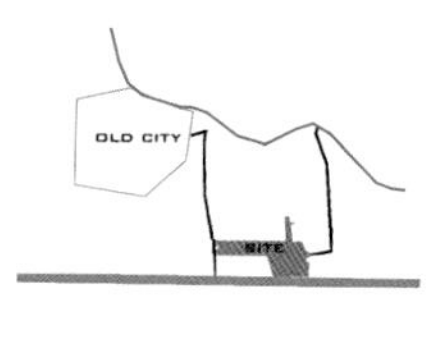

THERE IS CLOSE RELATION BETWEEN THE BASE AND THE OLD CITY AS WELL AS THE RIVER

THE MAIN FUNCTIONAL ZONE CONNECTED BY THE EXPRESSWAY TO THE SOUTH OF THE BASE

THE EXPRESSWAY AND THE SURROUNDING ROAD NETWORK CONVENIENTLY LINK THE DIFFERENT FUNCTIONAL ZONES

THE OPEN SPACE SURROUNDING THE BASE IS SCATTERED AND OBSOLETE,NOT HAVING BEEN MADE GOOD USE OF

CHALLENGE OF THE SITE

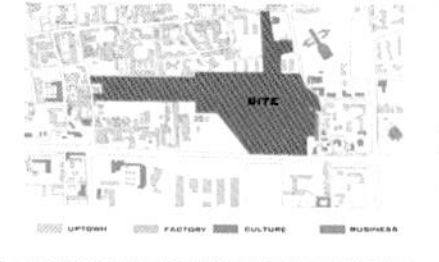

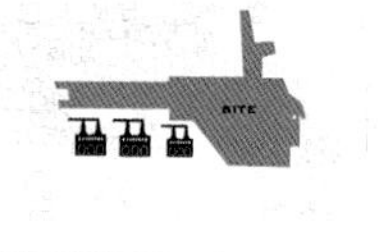

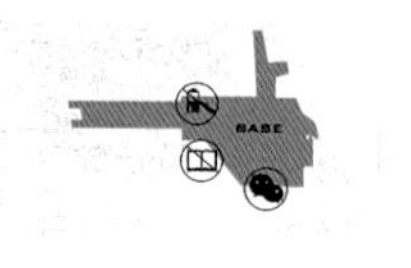

THE EXPRESSWAY OBSTRUCTS THE DEVELOPMENT OF THE CITY,ISOLATING THE SOUTHERN AND NORTHERN SIDES AND THE ROAD

THE FUNCTIONALITY SURROUNDING THE BASE IS MONOTONOUS

THE FACTORY SOUTHWEST OF THE BASE HAS HAD NEGATIVE EFFECTS ON IT

THE BASE FALLS SHORT OF INTERIOR FUNCTIONALITYLACKING CULTURAL, RECREATIONAL,AND COMMUNICATIONAL

MAIN STRATEGIES OF THE SITE

- CREATE AN AXIS TO CONNECT THE NORTH AND SOUTH AREA WHERE A FAST ROAD BREAK THE COMMUNICATION.

+

- CREATE AN AREA SERVICING THE OLD CITY. IMPROVE THE FUNCTION WHICH THE OTHER AREAS DO NOT HAVE FOR THE OLD CITY.BUILD 4 FOUNCTIONAL "THEME PARKS" FOR THE PROPLE LIVING AROUND.

+

- RETAIN THE CULTIVATED LAND THE SITE USED HAD.CREAT A NEW CITY AGRICULTURE FOR THE WHOLE CITY AND ATTRACT MORE.

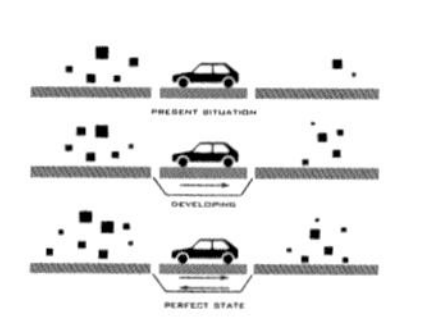

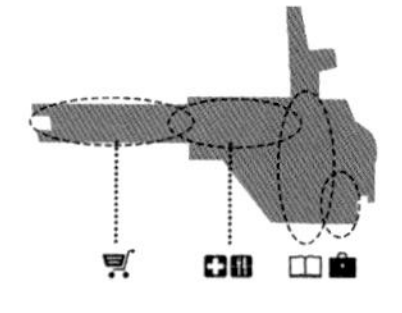

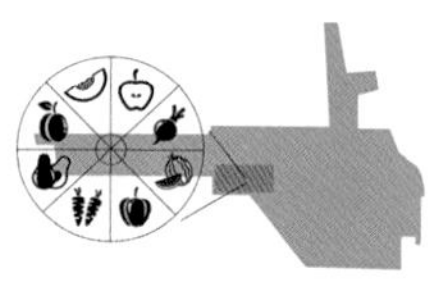

SPECIFIC METHODS OF THE SITE DESIGN

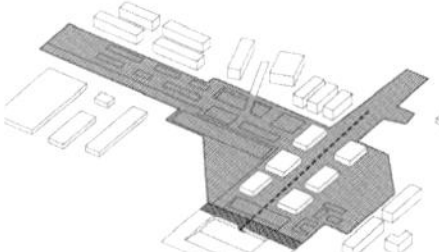

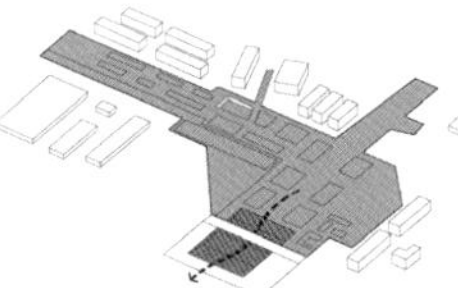

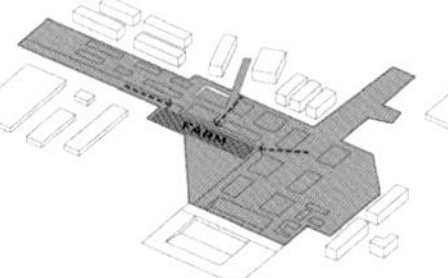

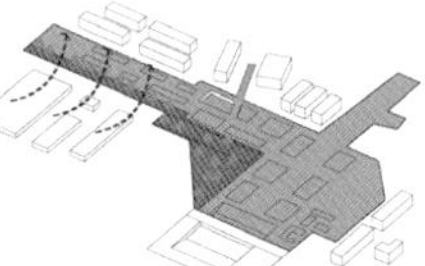

ENCLOSE THE BUILDINGS TO MAKE A VISION VESTUBULE CORRIDOR FOR LINKING THE AREAS AT BOTH SIDES OF THE EXPRESSWAY

LOCATE THE MARKET,WHICH FUNCTIONS AS THE CORE NODE, WHERE THE INTERNAL AND EXTERNAL RELATIONS TAKE PLACE

MAKE A FUNCTIONAL STREAMLINE TO LINK THE RELATED FUSCTIONAL ZONE

LEAD THE TRAFFIC UNDERGROUND,AND ENABLE PEOPLE TO CORSS THE ROAD WITHOUT DESTROYING THE CONTINUITY OF THE EXPRESSWAY

SET A CITY FARM TO ENHANCE THE VIGOR OF THE SURROUNDING MARKET INSIDE THE BASE

SET GREEN SPACE SYSTEM TO ISOLATE NEGATIVE FACTORS AND TO PROVIDE PEOPLE WITH A LEISYRE STREAMLINE

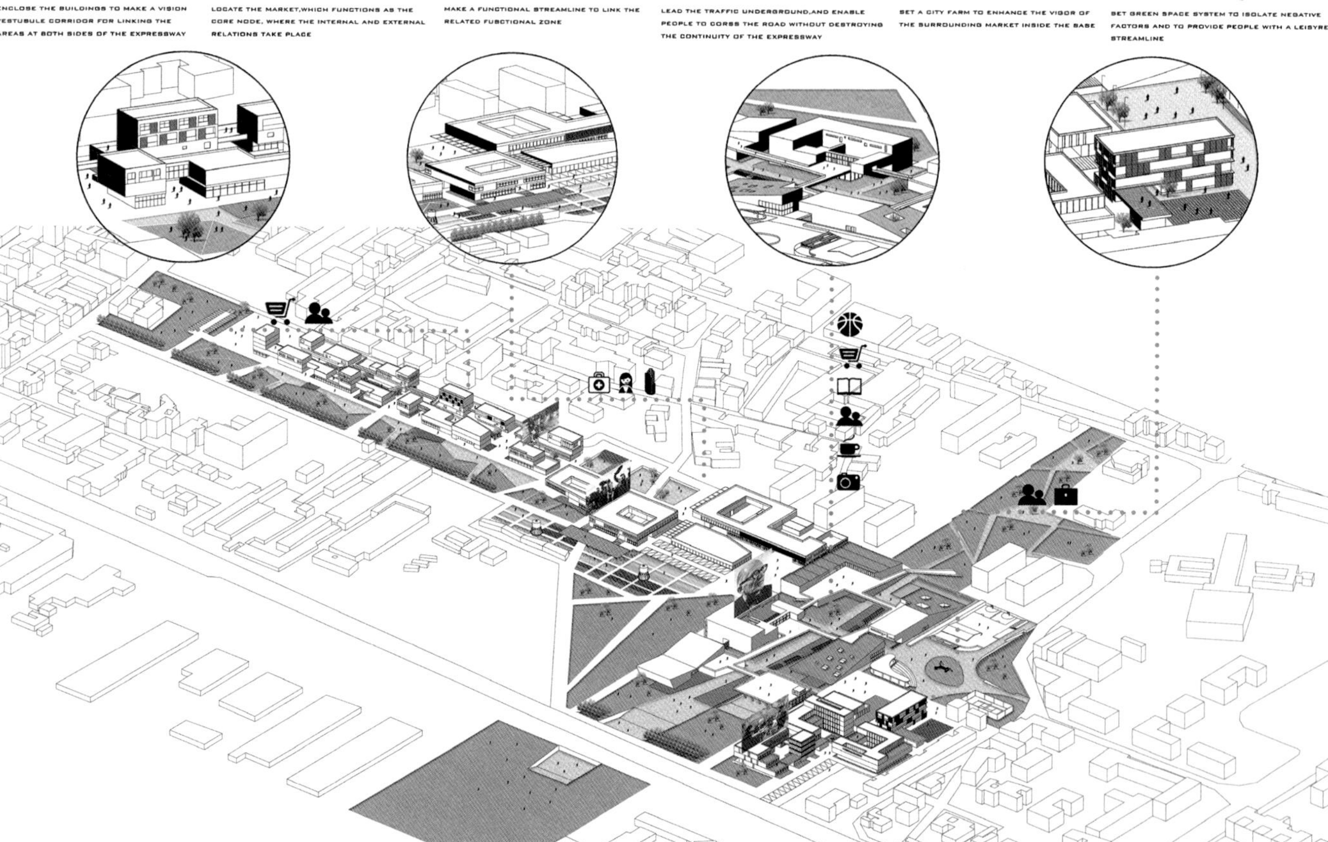

作品名称：公园·城市

Living in the Park

院校名：山东建筑大学

设计人：韩昆衡，骆逸，孙宁晗，李闻达

指导教师：周忠凯，陈林，Tony Van Raat

课程名称：山东建筑大学—新西兰 UNITEC 理工学院联合设计课程教学 2

作业完成日期：2014 年 06 月

对外交流对象：新西兰 UNITEC 理工学院

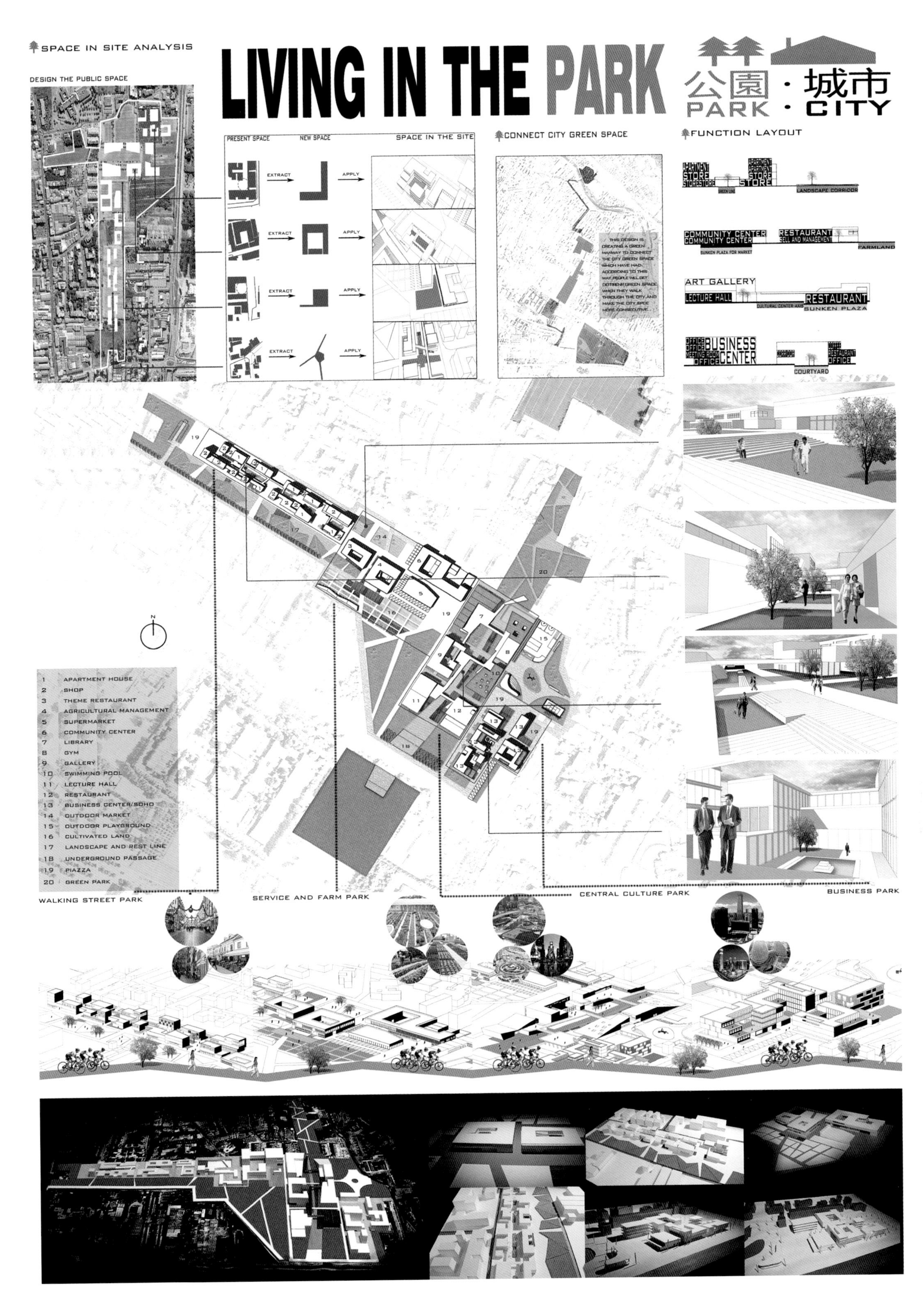

University: Shandong Jianzhu University
Designer: Han Kunheng, Luo Yi, Sun Ninghan, Li Wenda
Tutor: Zhou Zhongkai, Chen Lin, Tony Van Raat
Course Name: Shandong Jianzhu University - UNITEC Institute of Technology Joint Design Studio2
Finished Time: Jun. 2014
Exchange Institute: Unitec Institute of Technology

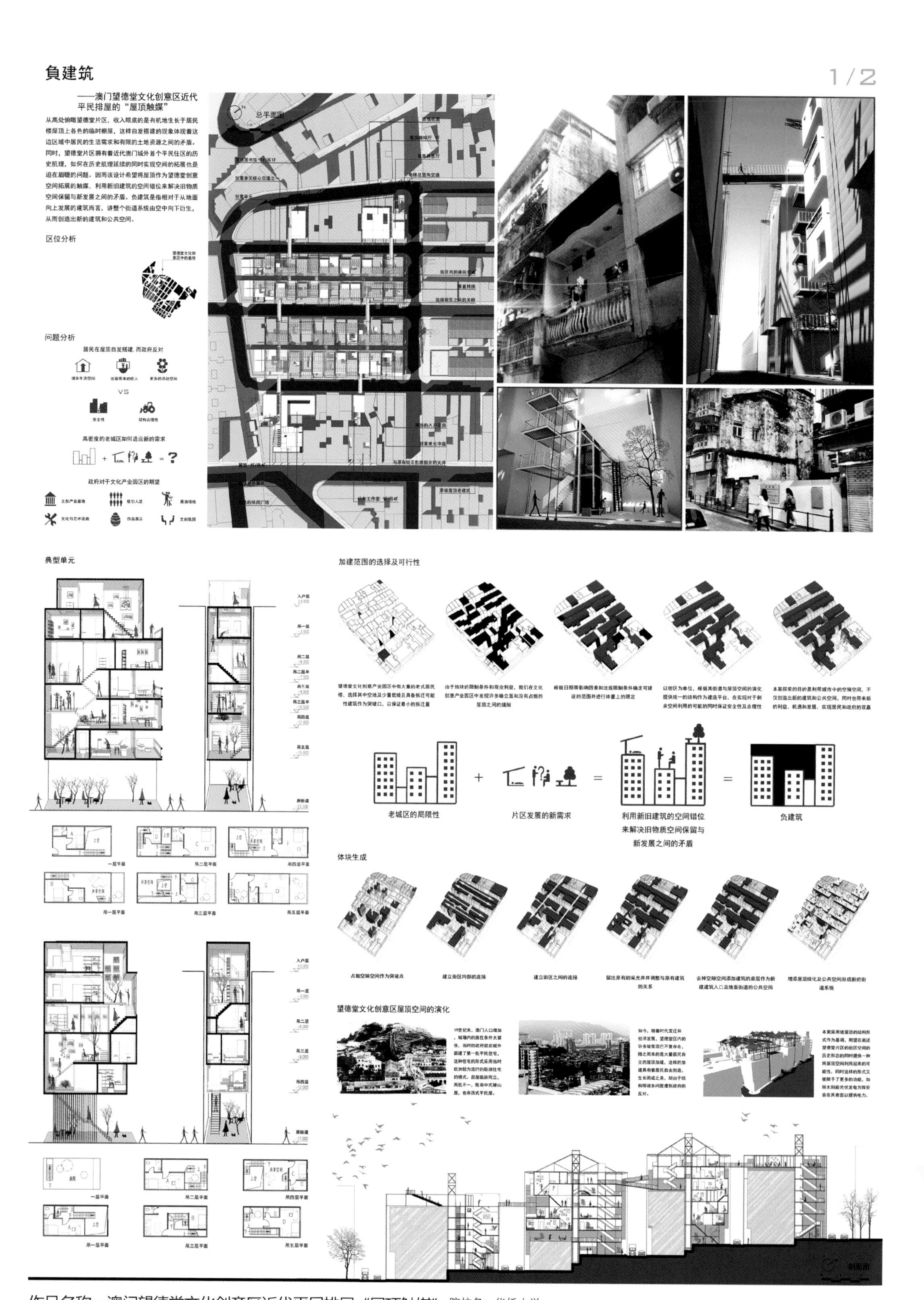

作品名称：澳门望德堂文化创意区近代平民排屋“屋顶触媒”
Roof Catalyst

院校名：华侨大学
设计人：霍颜颜
指导教师：郑剑艺
课程名称：澳门塔石片区城市更新与建筑设计
作业完成日期：2014 年 06 月
对外交流对象：澳门土地工务运输局，澳门文化局

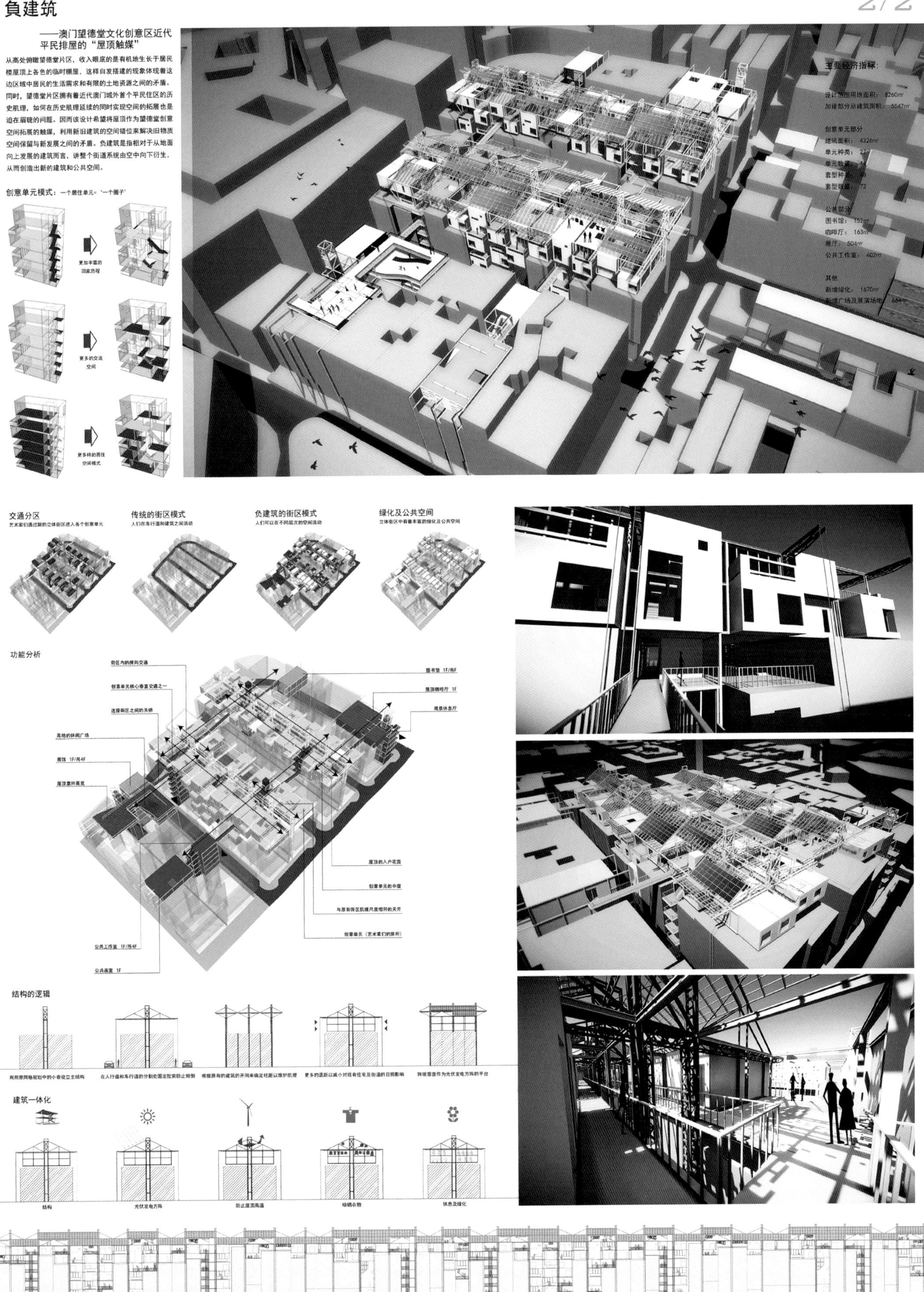

University: Huaqiao University
Designer: Huo Yanyan
Tutor: Zheng Jianyi
Course Name: Urban renewal and architectural design of Macao Tap-seac area
Finished Time: Jun. 2014
Exchange Institute: The Land, Public Works and Transport Bureau of the Macao，Cultural Bureau of the Macau

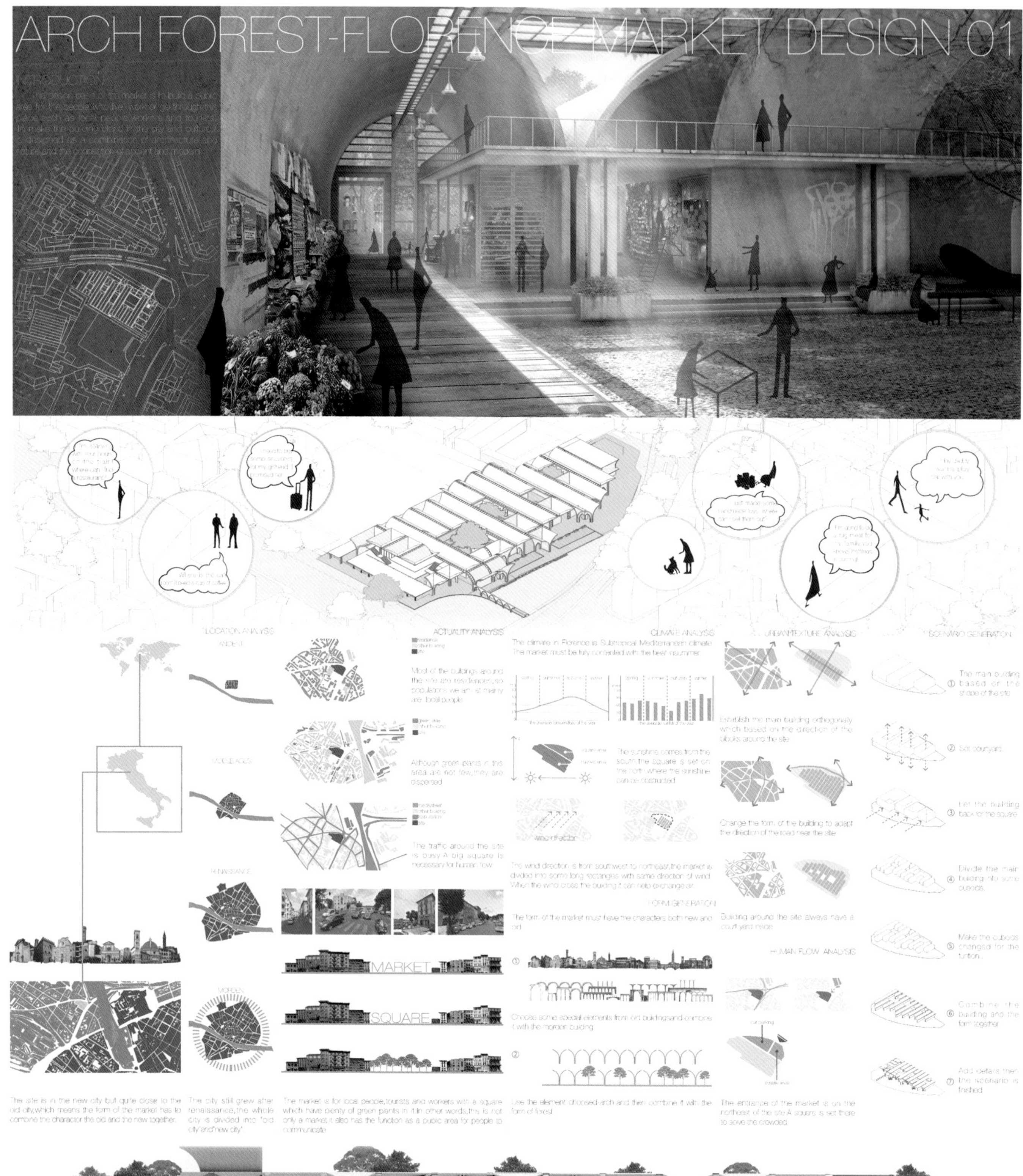

作品名称：拱之森林——佛罗伦萨市场设计
Arch Forest

院校名：山东建筑大学
设计人：贾鹏
指导教师：房文博，慕启鹏，王月涛，Daniella da Silva
课程名称：山东建筑大学—新西兰 UNITEC 理工学院联合设计课程教学 1
作业完成日期：2014 年 06 月
对外交流对象：新西兰 UNITEC 理工学院

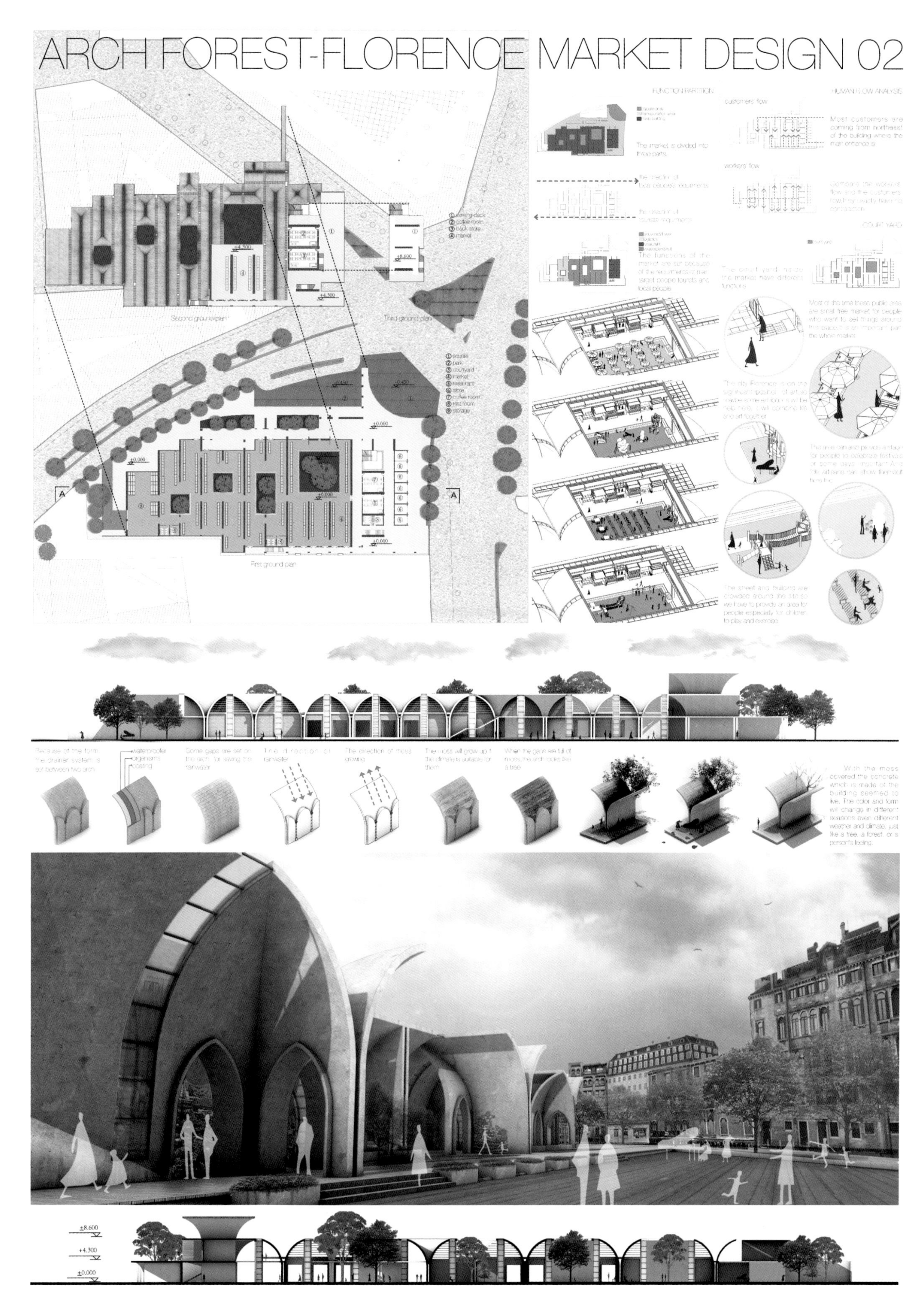

University: Shandong Jianzhu University
Designer: Jia Peng
Tutor: Fang Wenbo, Mu Qipeng, Wang Yuetao, Daniella da Silva
Course Name: Shandong Jianzhu University - New Zealand UNITEC Institute of Technology Joint Design Studio 1
Finished Time: Jun. 2014
Exchange Institute: New Zealand UNITEC Institute of Technology

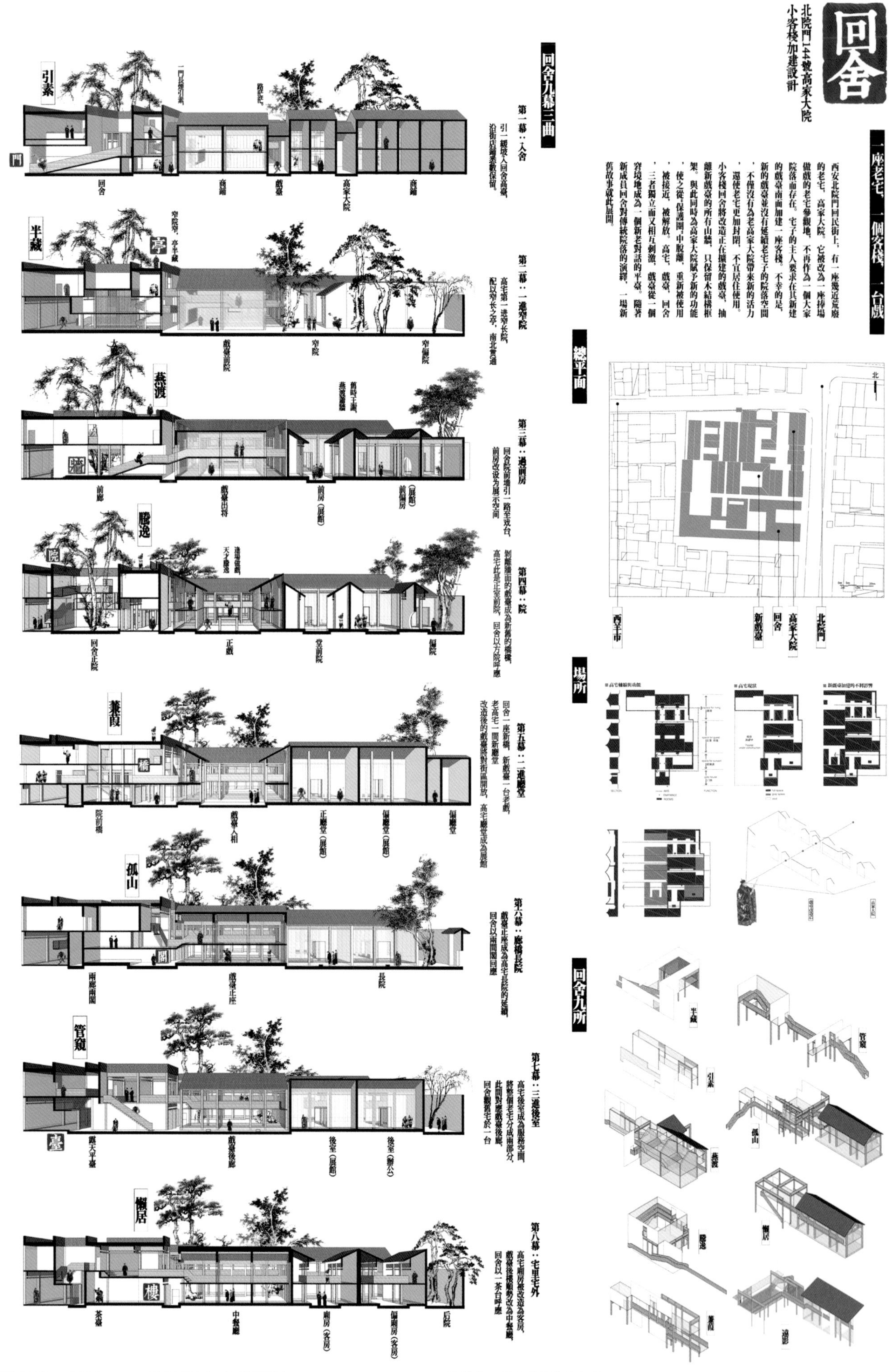

作品名称：回舍——北院门小客栈设计
Old House

院校名：西安建筑科技大学
设计人：蒋一汉
指导教师：刘克成，丸山欣也，吴瑞，王毛真，蒋蔚，俞泉
课程名称：北院门小客栈设计
作业完成日期：2014 年 07 月
对外交流对象：日本早稻田大学

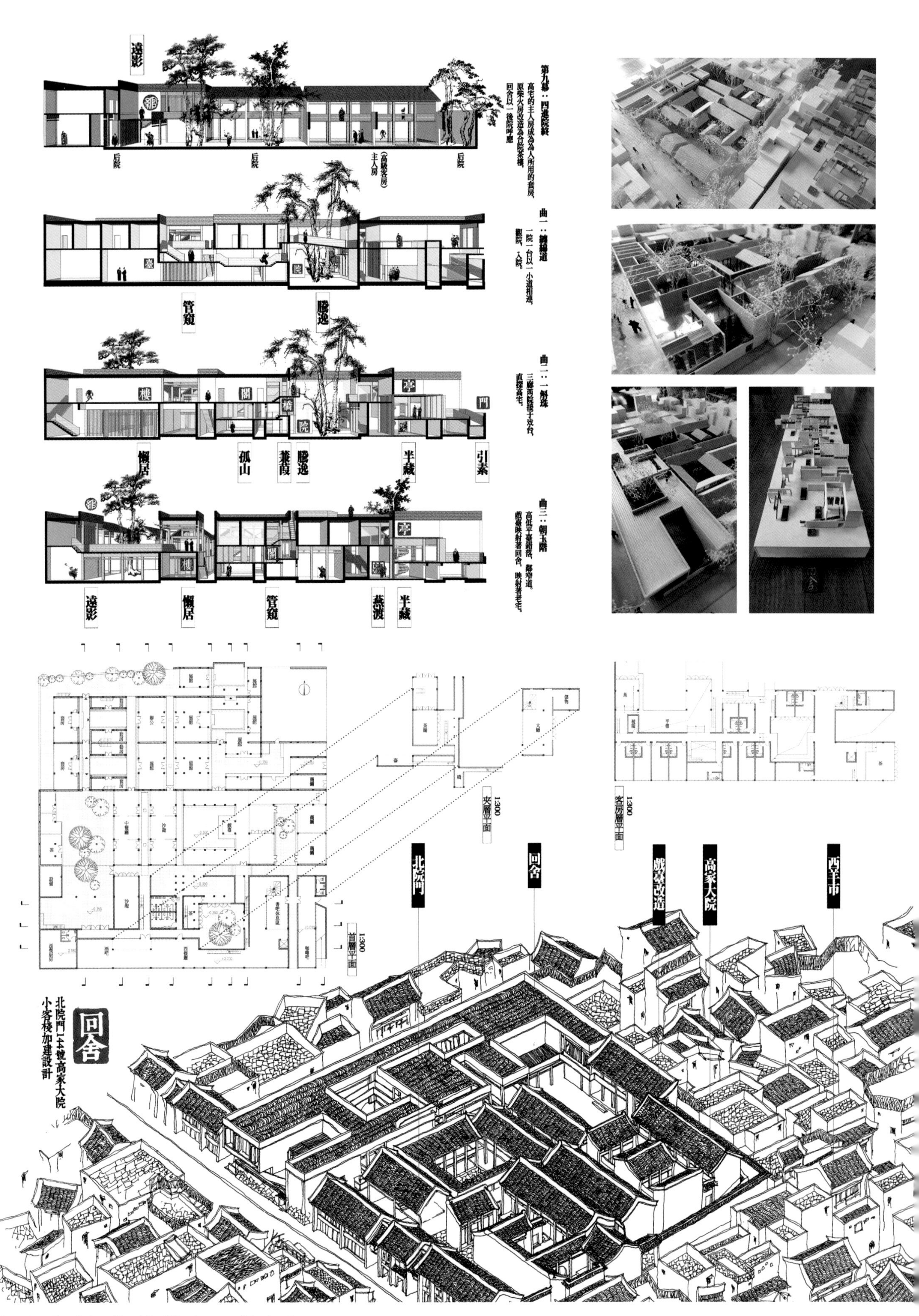

University: Xi'an University of Architecture and Technology
Designer: Jiang Yihan
Tutor: Liu Kecheng, Maruyama Kinya, Wu Rui, Wang Maozhen, Jiang Wei, Yu Quan
Course Name: Xi'an University of Architecture and Technology-Waseda University Joint Design Studio
Finished Time: Jul. 2014
Exchange Institute: Waseda University

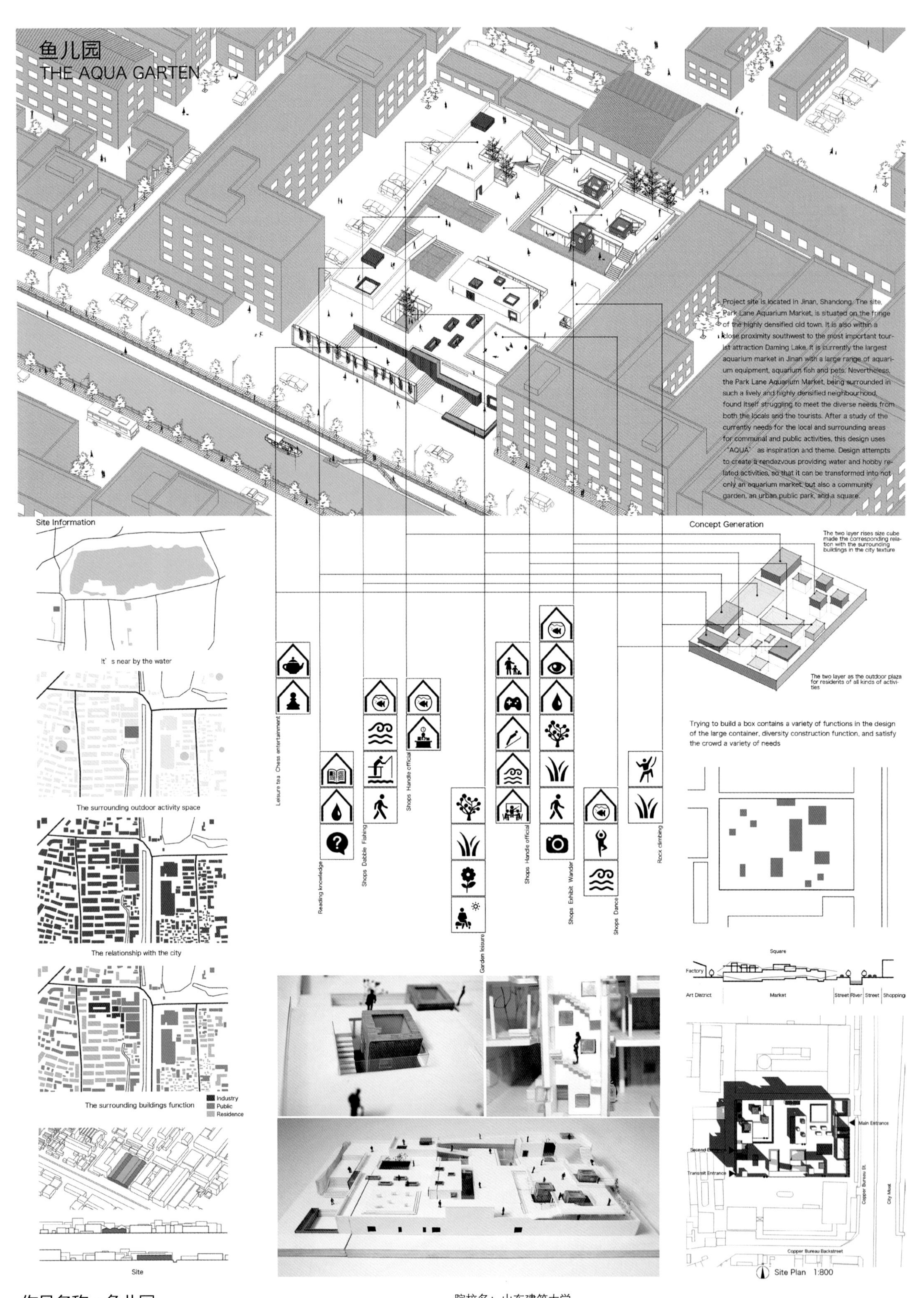

作品名称：鱼儿园
The Aqua Garten

院校名：山东建筑大学
设计人：孔德硕
指导教师：夏云，石涛，Gihan Karunatne
课程名称：山东建筑大学——新西兰 UNITEC 理工学院联合设计课程教学 1
作业完成日期：2014 年 06 月
对外交流对象：新西兰 UNITEC 理工学院

鱼儿园
THE AQUA GARTEN

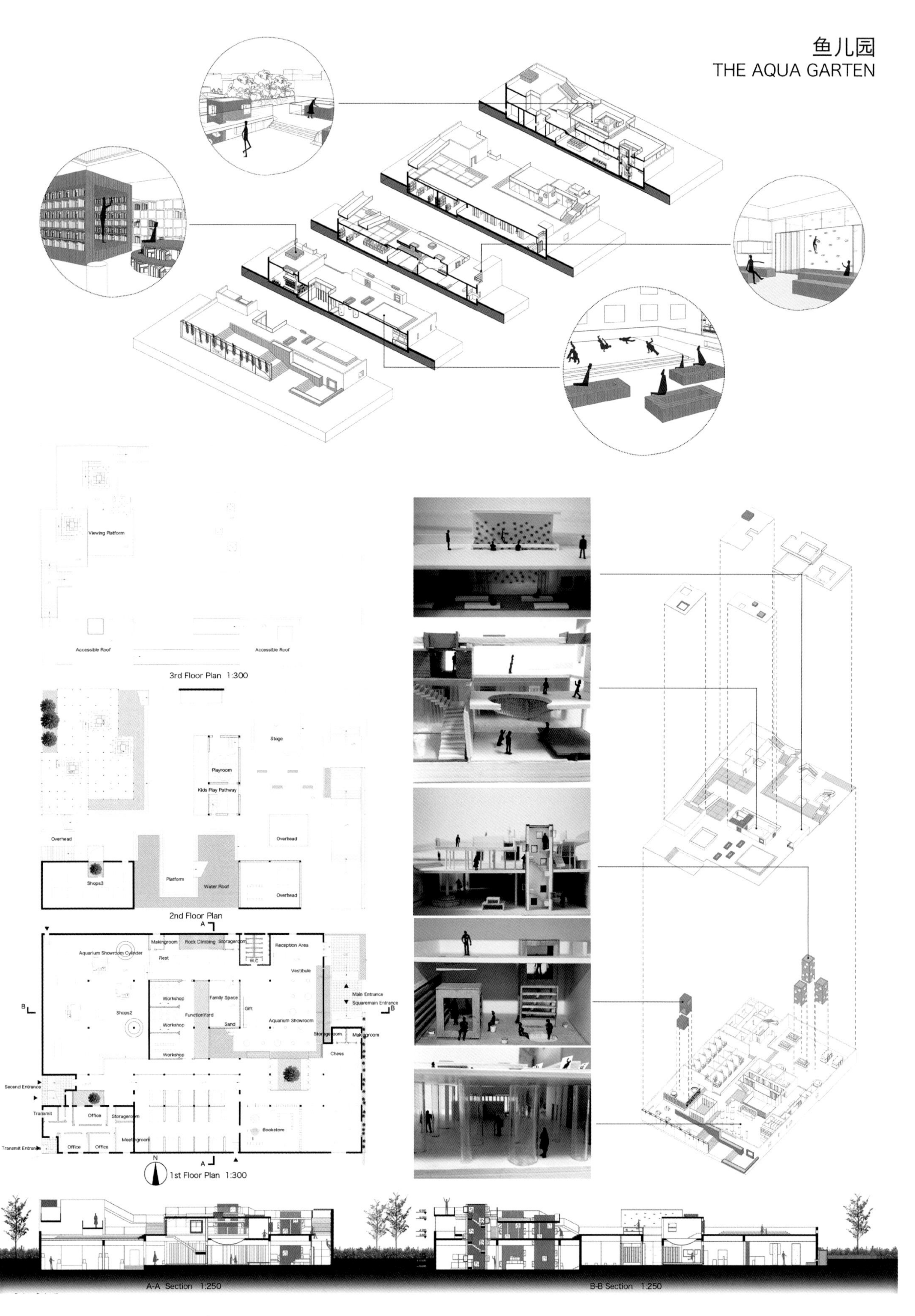

University: Shandong Jianzhu University
Designer: Kong Deshuo
Tutor: Xia Yun, Shi Tao, Gihan Karunatne
Course Name: Shandong Jianzhu University - New Zealand UNITEC Institute of Technology Joint Design Studio1
Finished Time: Jun. 2014
Exchange Institute: New Zealand UNITEC Institute of Technology

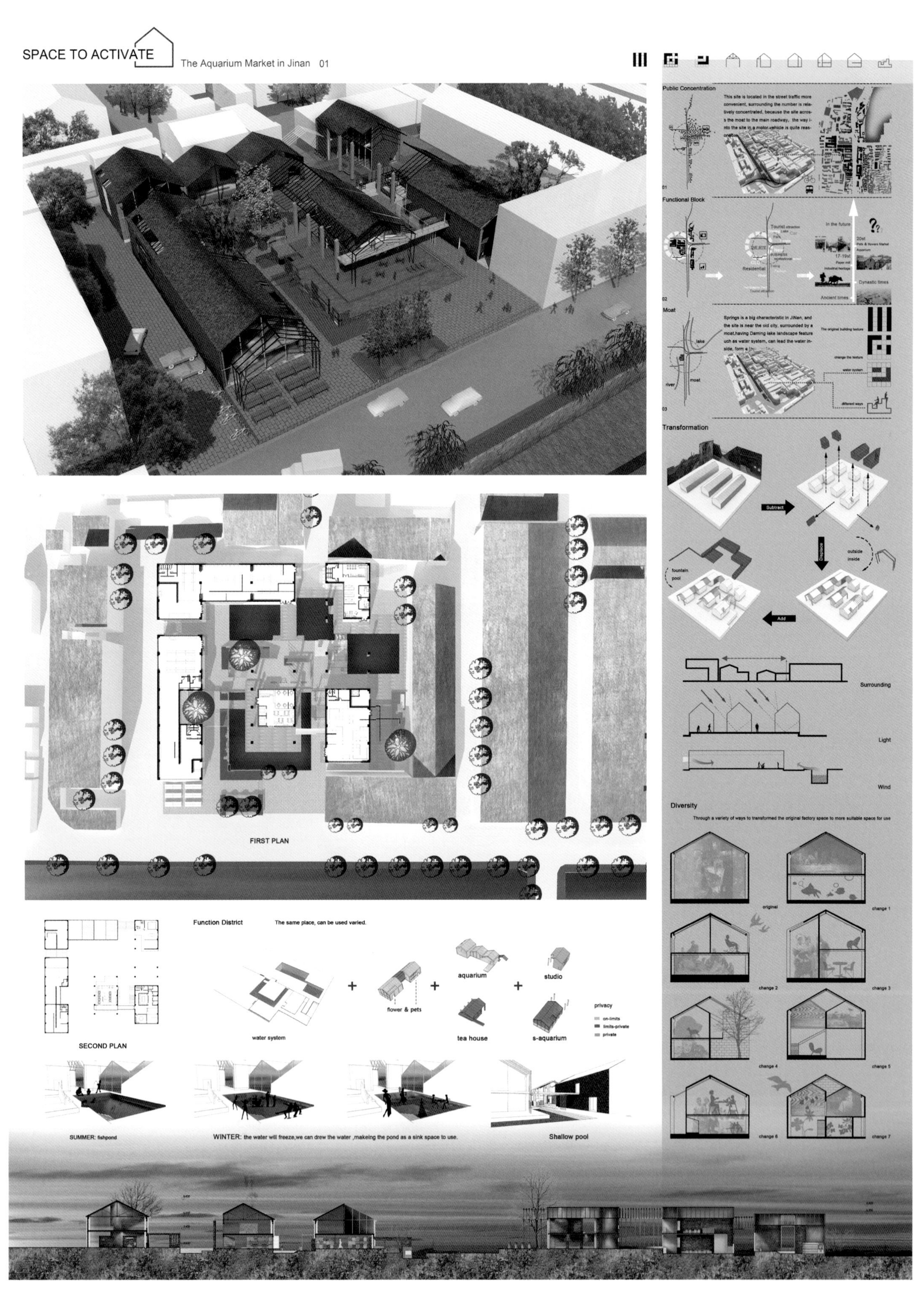

作品名称：济南花鸟鱼虫水族市场设计
Space to Activate

院校名：山东建筑大学
设计人：李亚男
指导教师：金文妍，刘文，Tony Burge
课程名称：山东建筑大学—新西兰 UNITEC 理工学院联合设计课程教学 1
作业完成日期：2014 年 06 月
对外交流对象：新西兰 UNITEC 理工学院

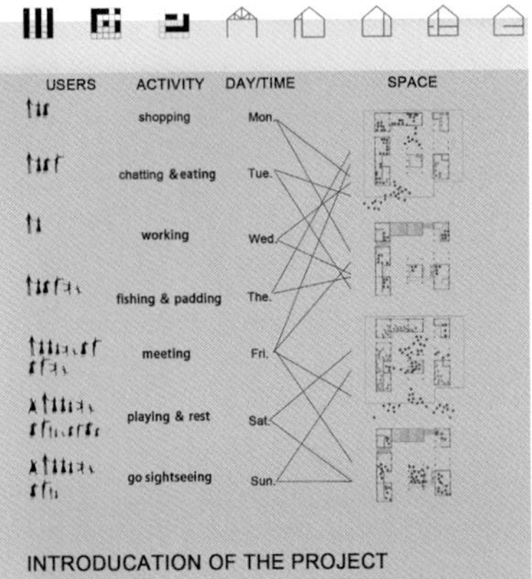

INTRODUCATION OF THE PROJECT

On the base of the investigation and analysis, before the industrial heeritage is the pets market ,flower market and the aquarium.In fact, the market rarely play the meaning of public building,In the protection of industrial heritage andstimulate public space lively, using the originalstructure, etc. Through the space such as material handling-contrast makes the past and present ,environmental protection more prominent,vas well as promoting the jinan local culture idea.

01 In a new way to choose and buy ornamental fish through cell phones and other advanced equipment , for example, you can pay the money by scaning the pets.while enjoy door to door delivery treatment. Outdoor pond extends into the interior, not only improve the indoor environment, also increased the interesting about watching fish.

02 A square is full of fun meeting the needs of the pet and his master playing together.

03 The different material can show different characteristics and effect. Using the original brick and concrete, combined with glass which has extremely strong ontemporary feeling and coated paint wood species, contrast and complement each other.

Streamline

Public

staff & consumer

Second Plan

First Plan

car

bicycle

Because the site belongs to public place faceing on one side of the street, just hope poeple can just by walking into here logistics channels are used on both sides of the street.

Shadow Analysis

TIME:September 1

9:00a.m.-15:00p.m.

9:00-10:00

10:00-11:00

11:00-13:00

13:00-14:00

14:00-15:00

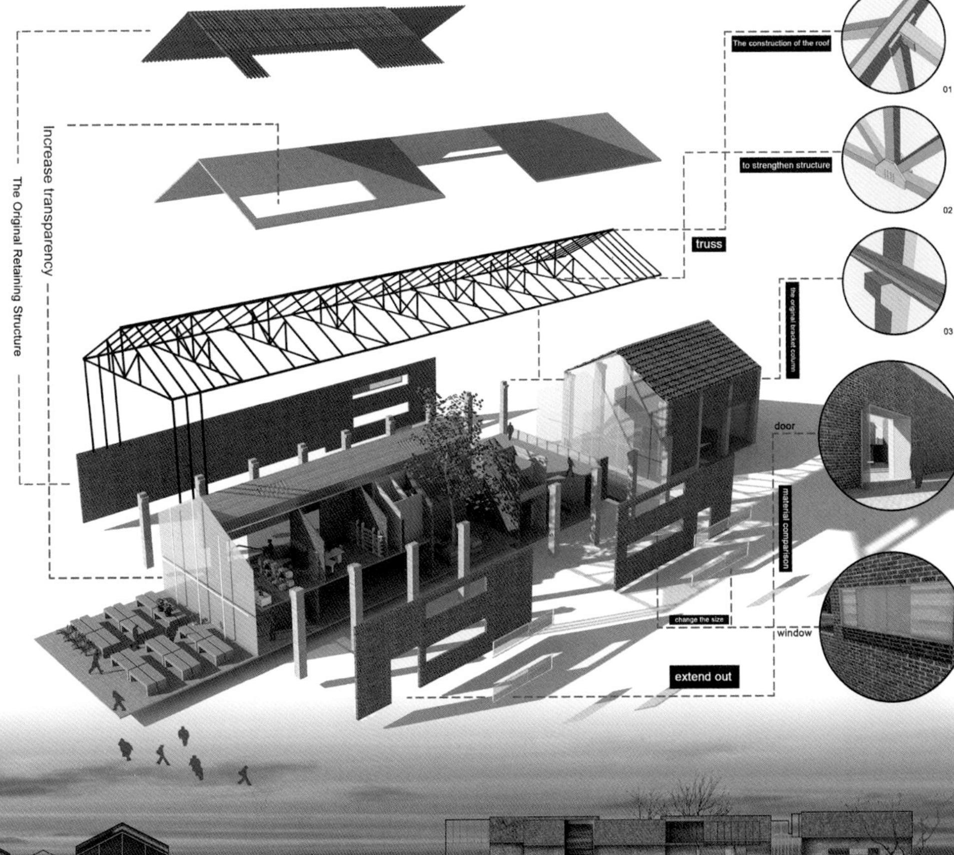

University: Shandong Jianzhu University
Designer: Li Yanan
Tutor: Jin Wenyan, Liuwen, Tony Burge
Course Name: Shandong Jianzhu University-New Zealand UNITEC Institute of Technology Jonit Design Studio 1
Finished Time: Jun. 2014
Exchange Institute: New Zealand UNITEC Institute of Technology

TransferJET ENTREPRENEURIAL BASE

—KNOWLEDGE SPACE DESIGN BASED ON THE FLOW OF INFORMATION

Generation of the Concept

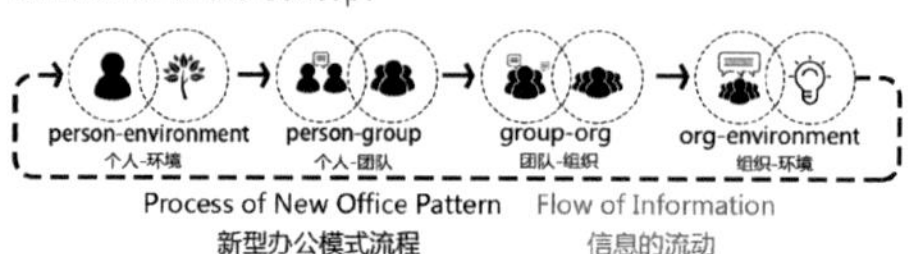

Starting from personal work and basing on enviroment of inner and outer space,we rearrange the process of information flowing to form a new kind of woking pattern.District of high information density is chosen as personal office which increase the contact to inside according to the flow.

Development of the Knowledge Space

We pay attention to the flow of information in order to find a new architecture typlogy which can make the office work become more efficient and rational.Just like the technology of TransferJet in IT which would change the world.

Design Description

Office of the process is essentially the process of information transmission, the design of the new era of mining process information transfer and synchronize work processes and information communicated through the process and the information density around the boundary of the largest personal workspace as a way to penetrate inward.

Plan of Conception

二层平面 Second Floor 1:400

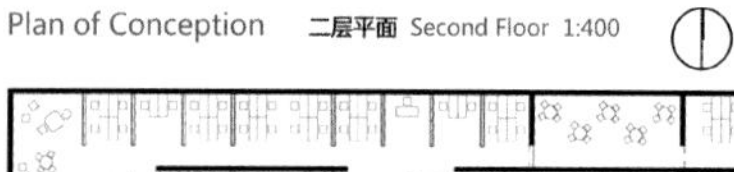

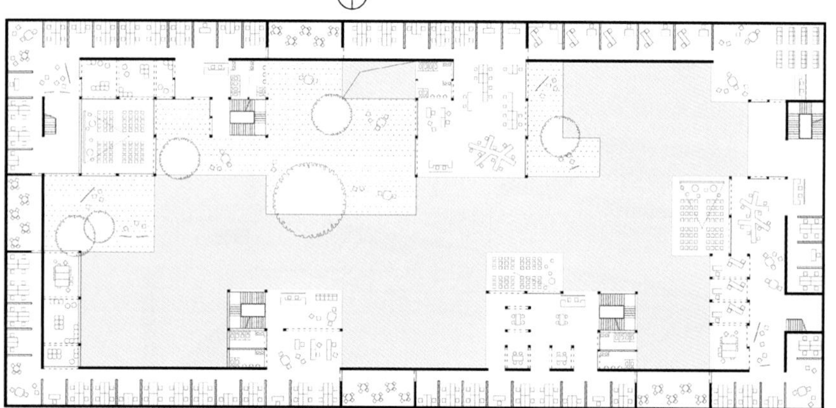

Site Plan

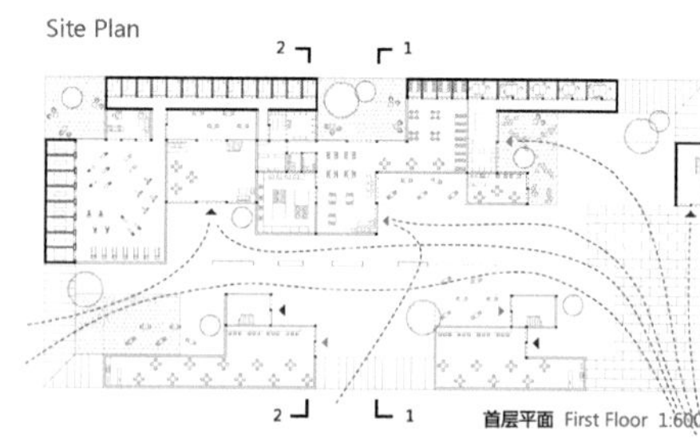

Considering about the environment,where people passing by most frequently is chosen as open boundary.there are much more grey and service space was arranged to fulfill needs of people flow

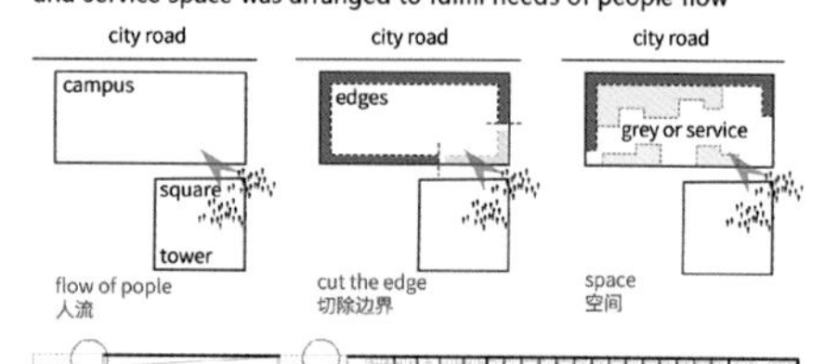

Analysis of Plan

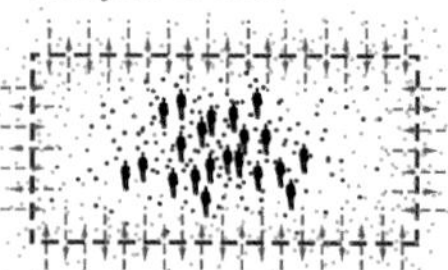

Step 1 Flow of information inside & outside

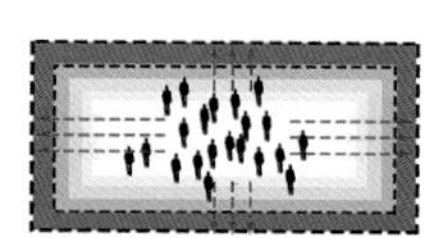

Step 2 Personal office to edges for more information

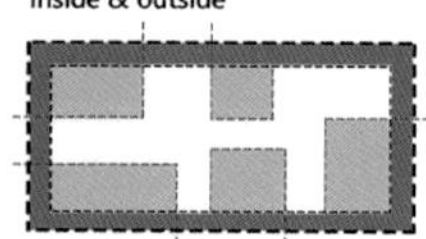

Step 3 Knowledge space extending inward

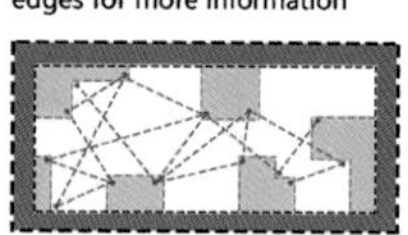

Step 4 Influence between knoeledge space

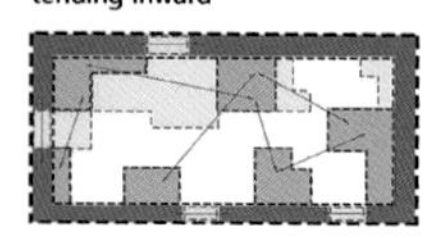

STEP 5 Variety air-space for connection

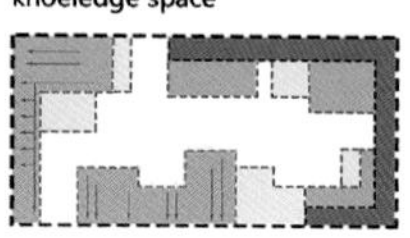

STEP 6 Broke the edges of plans to make special space

Information circulation

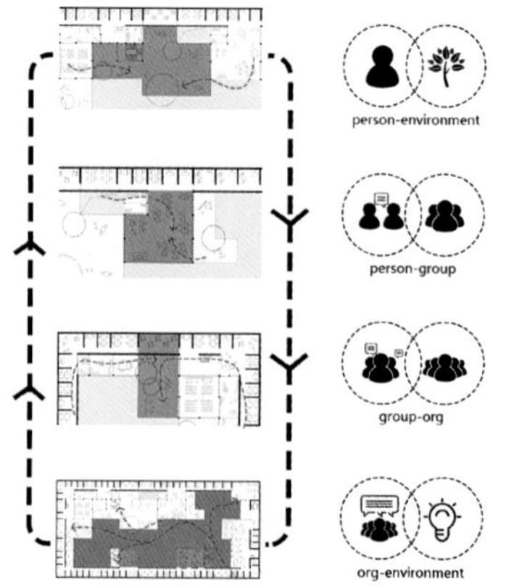

By the role of simulation in office to draft there kinds of working form,and according to what it need to arrange space.

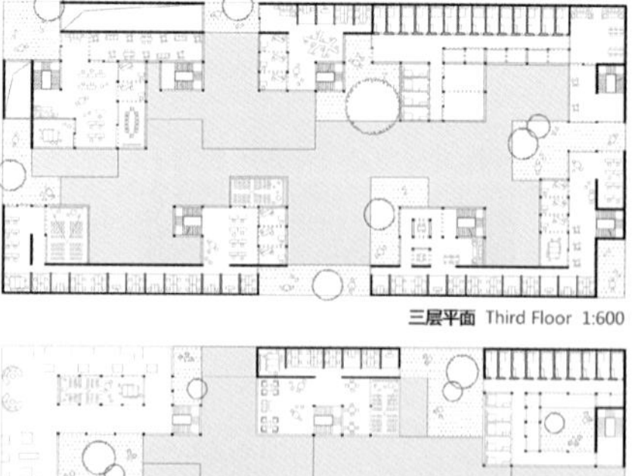

作品名称：TransferJET创业中心——基于空间流动模式的办公空间设计
TransferJET Entrepreneurial Base —Knowledge Space Design based on the Flow of Information

院校名：哈尔滨工业大学
设计人：李彦儒，徐子博，滕东霖
指导教师：陈旸，邢凯，唐康硕，Giorgio Ponzo
课程名称：国际联合设计（本科三年级）
作业完成时间：2014 年 07 月
对外交流对象：荷兰代尔夫特理工大学

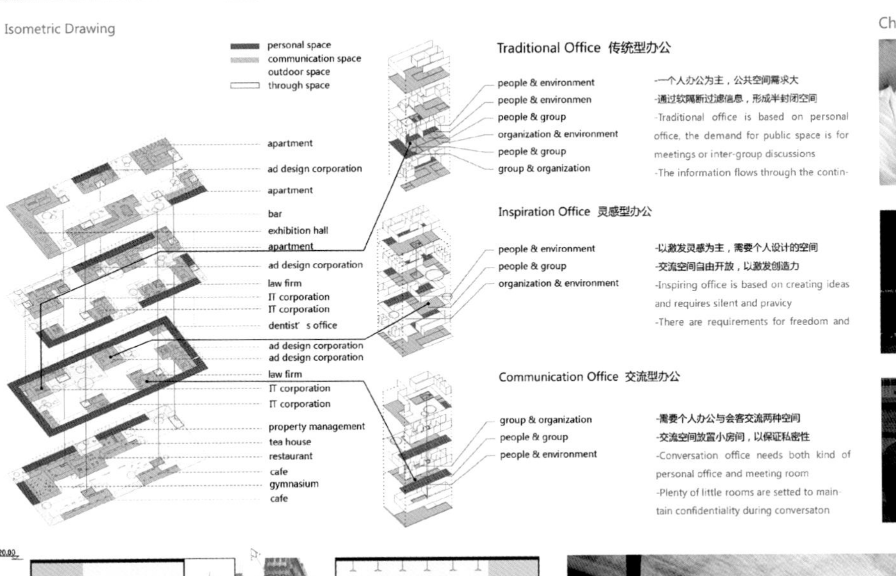

Characters

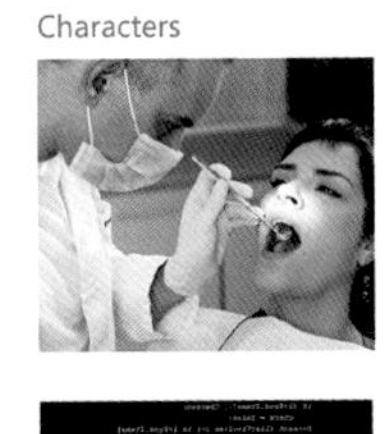

Dentist 牙医

- Loves to has a cup of tea in the afternoon
  喜爱下午茶
- Works in his own clinic in campus
  在园区中自己的诊所工作
- Enjoys reading in the evening
  喜爱在晚间阅读
- Has a family of three people
  三口之家的一员

Young Designer 青年设计师

- Always stays up all night and gets up late
  晚睡早起
- Bachelor who wants to have a girlfriend
  渴望爱情的单身汉
- Shy and difficult to communicate with
  害羞而难以沟通
- Coffee and fast-food addict
  咖啡和快餐的拥趸

Young Female Lawyer 青年女律师

- Talks to lots of people every day
  工作以交流为主
- Goes to bar with lost of alchool
  夜店达人
- Outgoing,popular in campus
  外向而受欢迎
- Fresh graduate
  职场新鲜人

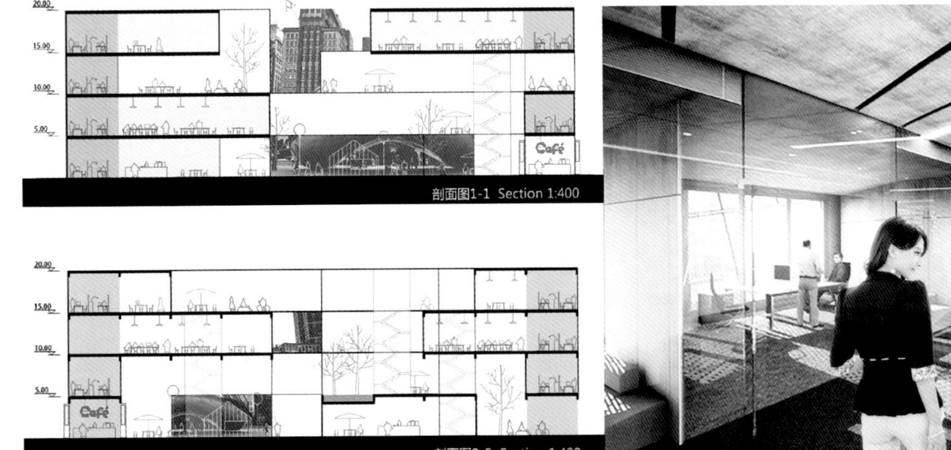

University: Harbin Institute of Technology
Designer: Li Yanru, Xu Zibo, Teng Donglin
Tutor: Chen Yang, Xing Kai, Tang Kangshuo, Giorgio Ponzo
Course Name: International Joint Design Studio
Finished Time: Jul. 2014
Exchange Institute: Delft University of Technology

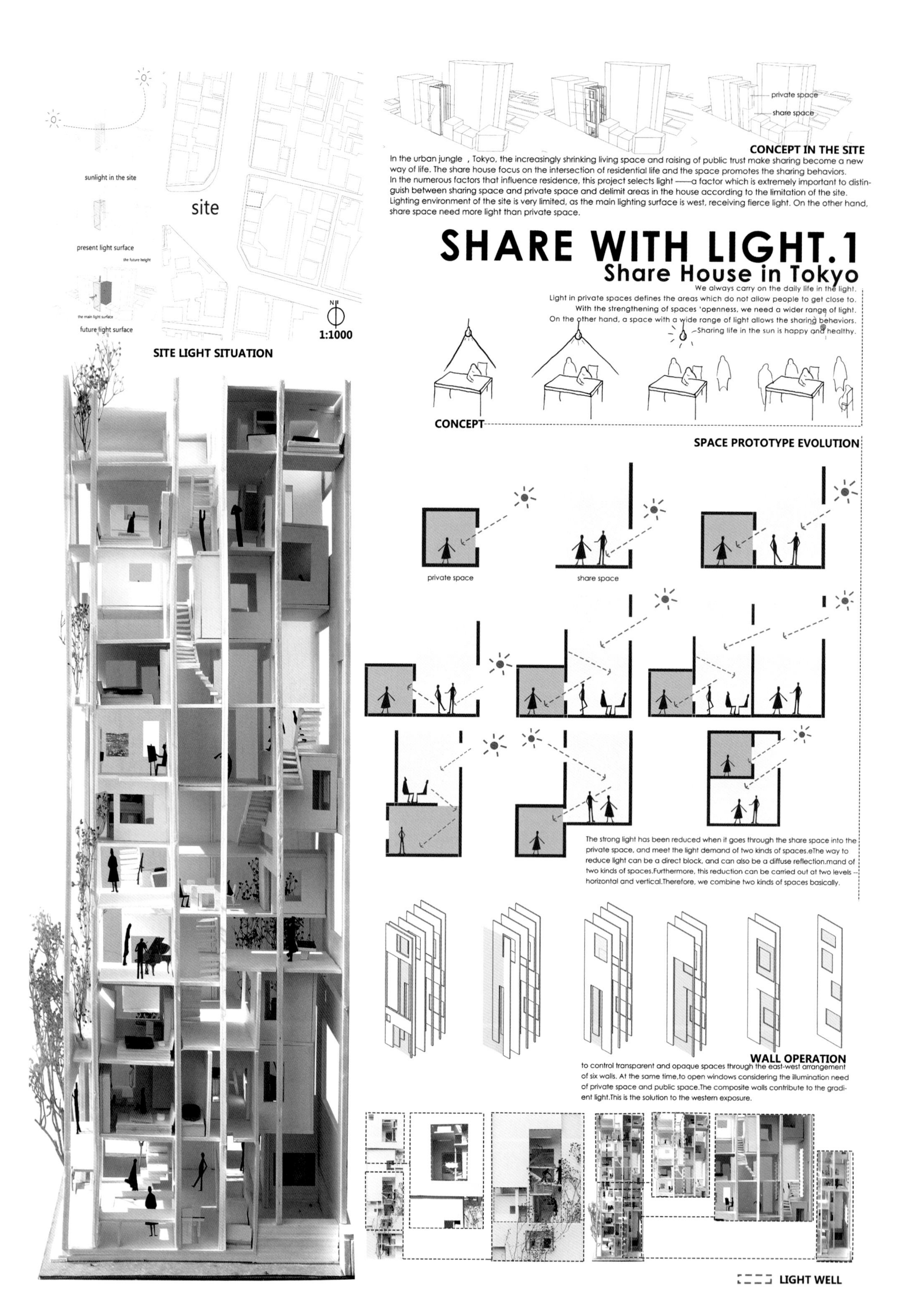

作品名称：东京共享住宅
Share with Light

院校名：天津大学
设计人：林碧虹，于安然
指导教师：王绚，张昕楠
课程名称：住宅工作营
作业完成日期：2014 年 12 月
对外交流对象：日本女子大学，淡江大学

Share House in Tokyo

SHARE WITH LIGHT.2

We always carry on the daily life in the light.
Light in private spaces defines the areas which do not allow people to get close to.
With the strengthening of spaces 'openness, we need a wider range of light.
On the other hand, a space with a wide range of light allows the sharing behaviors.
Sharing life in the sun is happy and healthy.

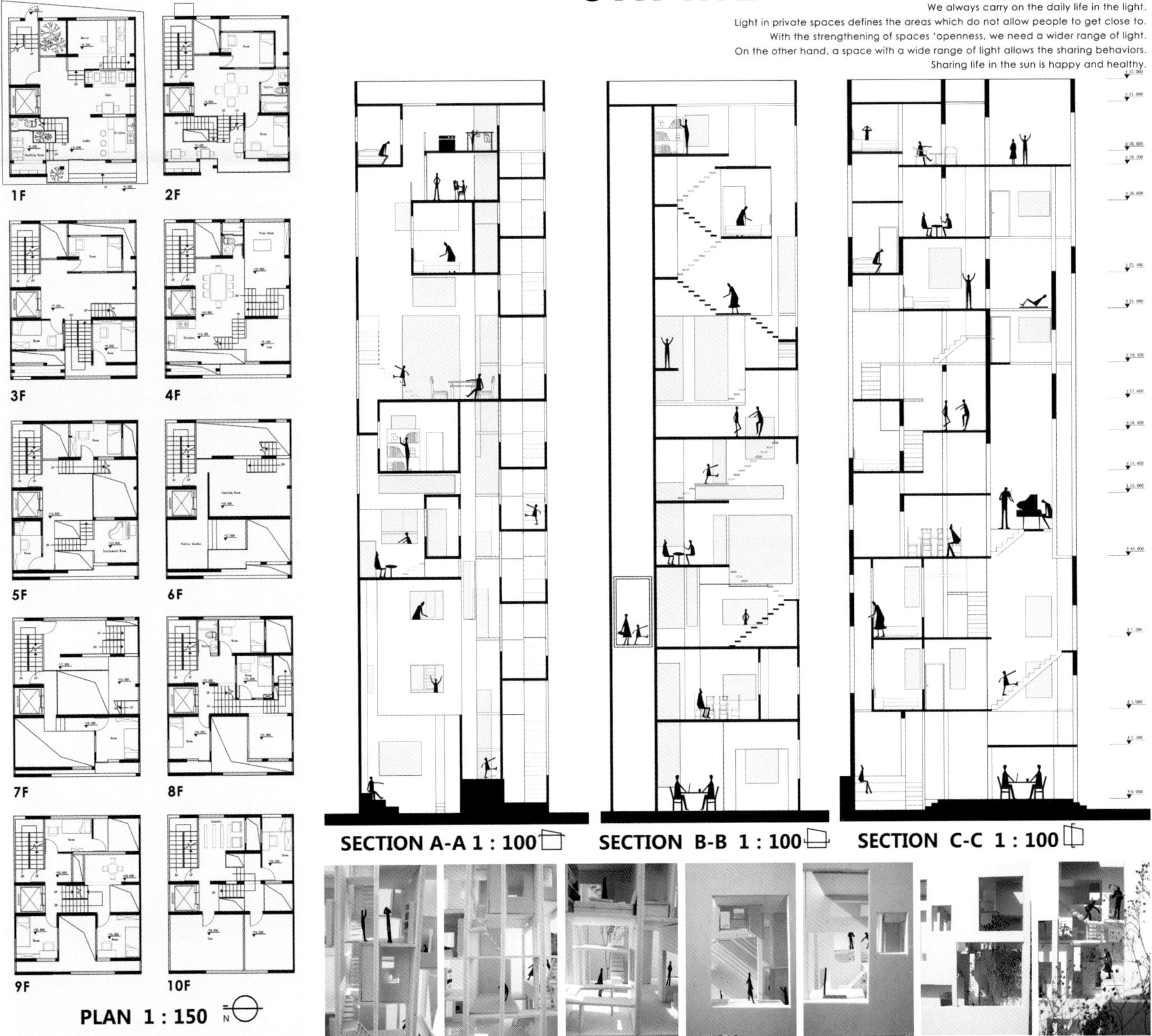

University: Tianjin University
Designer: Lin Bihong, Yu Anran
Tutor: Wang Xuan, Zhang Xinnan
Course Name: Share House
Finished Time: Dec. 2014
Exchange Institute: Japan Women's University, Tamkang University

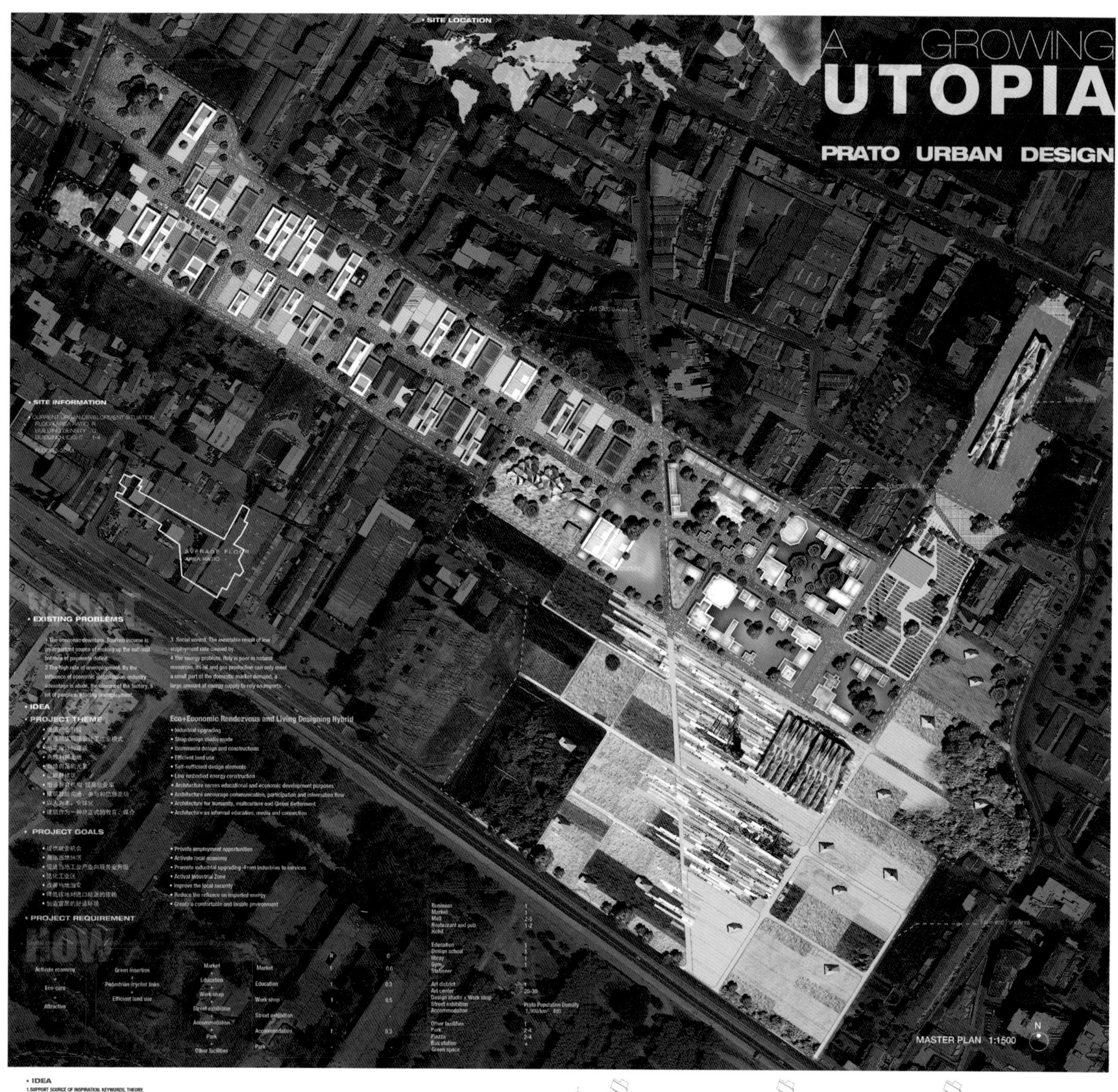

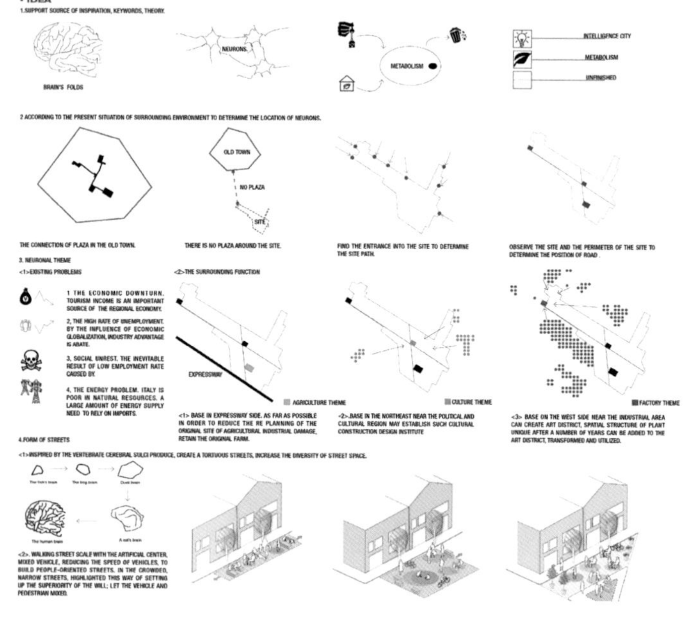

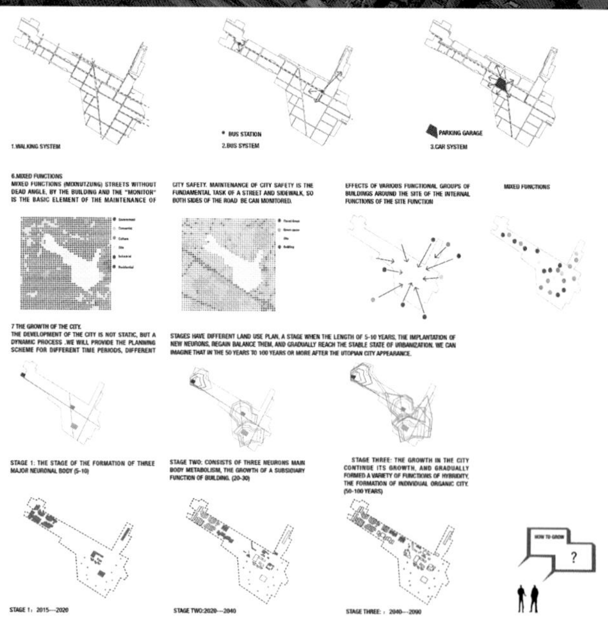

作品名称：生长的乌托邦
A Growing Utopia

院校名：山东建筑大学
设计人：刘丹笛，林晓宇，包依凡，鞠婧
指导教师：慕启鹏，房文博，Francesco Collotti
课程名称：山东建筑大学—新西兰 UNITEC 理工学院联合设计课程教学 2
作业完成日期：2014 年 06 月
对外交流对象：新西兰 UNITEC 理工学院

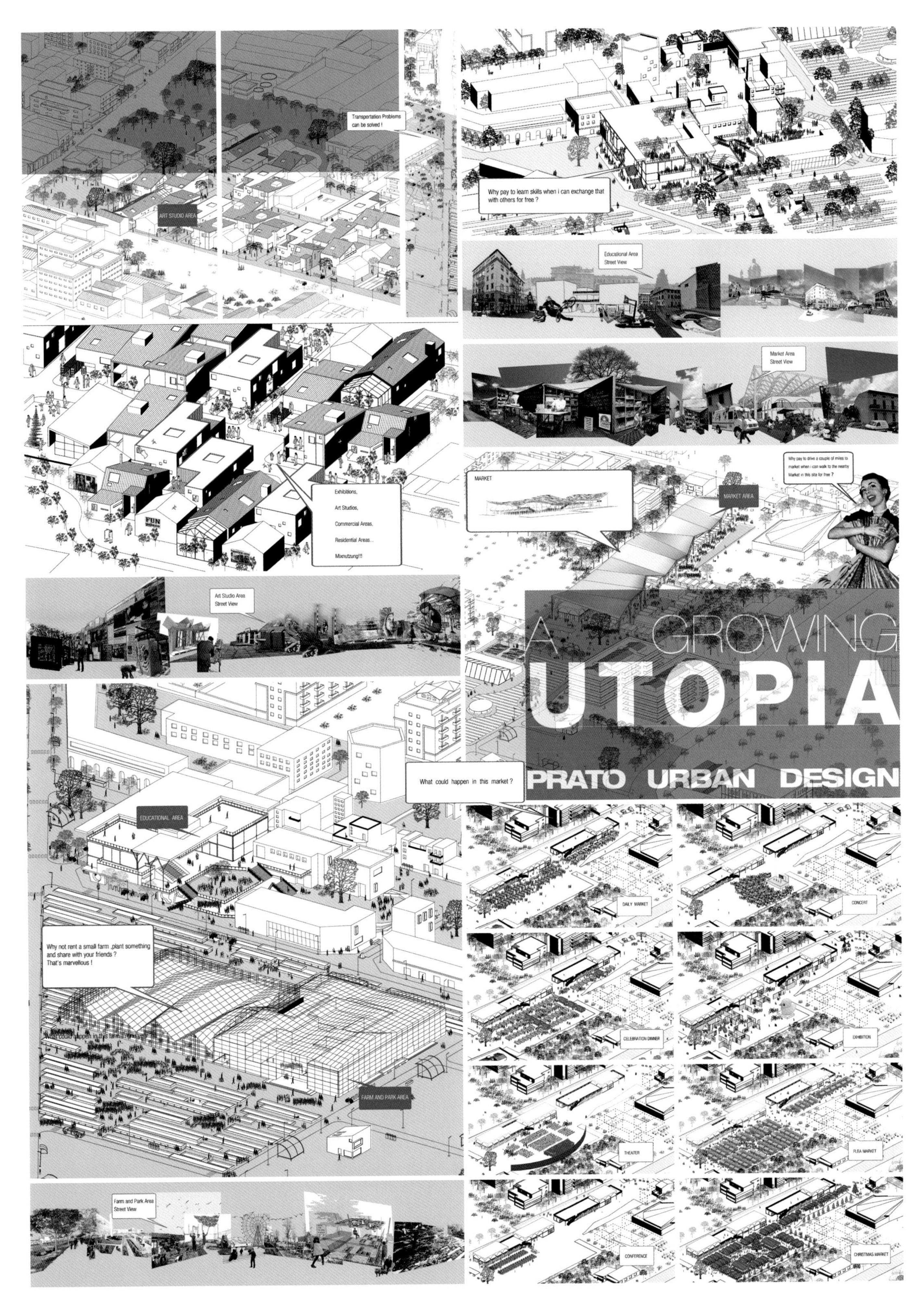

University: Shandong Jianzhu University
Designer: Liu Dandi, Lin Xiaoyu, Bao Yifan, Ju Jing
Tutor: Mu Qipeng, Fang Wenbo, Francesco Collotti
Course Name: Shandong Jianzhu University - New Zealand UNITEC Institute of Technology Joint Design Studio2
Finished Time: Jun. 2014
Exchange Institute: New Zealand UNITEC Institute of Technology

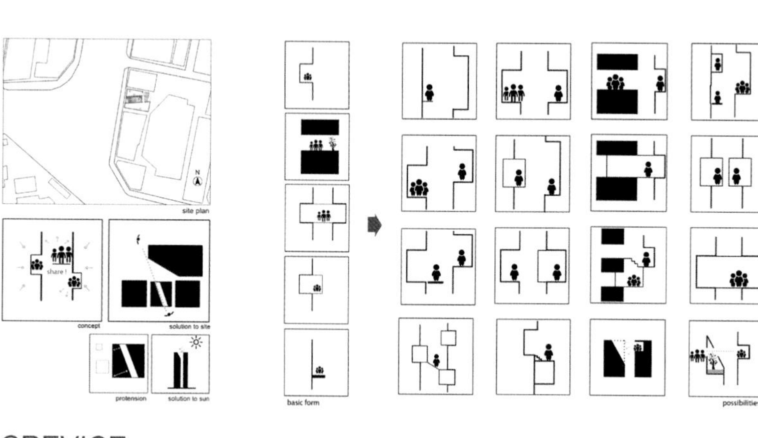

CREVICE in the city

-- A CREVICE THAT CONNECTS.

Strangeness often makes people feel embarrassed in the Share House or Youth Hostel. It is hard for people to make clear what activities suites them and they even do not know what sharing activities is here . In this share house, people communicating and sharing activities,which takes place in and out of the two-meter-wide' gap . The multi-group sharing activities take place on the platform floating in the cracks, and also occur in the semi-open space of both sides of the gap. The gap opens up the line of sight between the front and back street. Therefore, the guests can easily find the most suitable sharing activities for themselves, and then join in it selectively.

We list five kinds of basic models about the possibility of crack's forming . In turn, it can be combined into twenty prototypes. After interviewing 15 residents, we extract the special requirements of Li Wen Shuang , Chen Mo, He Tao, YANG Zhao, and choose the room for their own need in twenty prototypes. When form the model, we respect the requirements of everyone for their needs of the layers and faces.

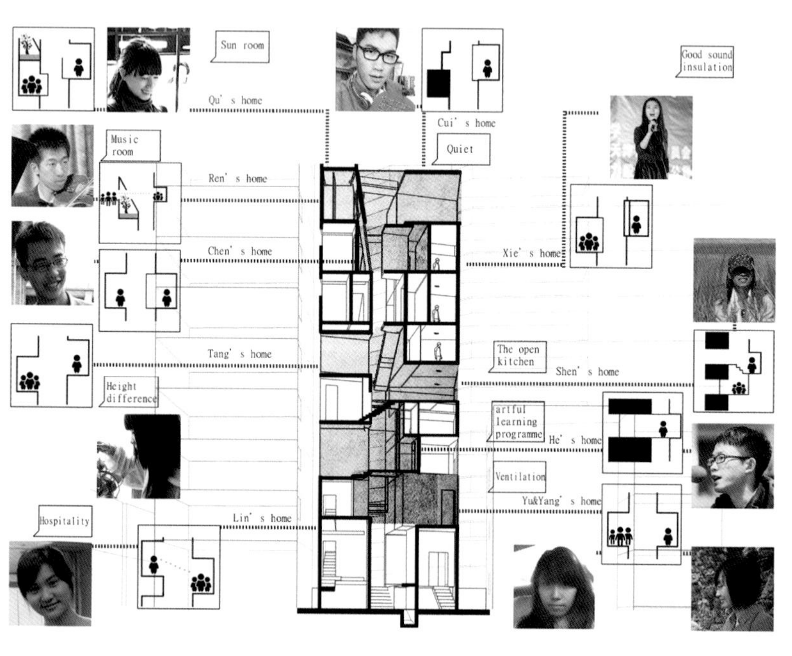

作品名称：城市裂缝

Crevice in the City

院校名：天津大学

设计人：祁山，何涛

指导教师：王绚，张昕楠

课程名称：住宅工作营

作业完成日期：2014年12月

对外交流对象：日本女子大学，淡江大学

CREVICE in the city

The part of building plane has been cut,which opens the line of urban sight and the flow.The whole building passes on the style and features of city.And the crack which is filled with green and citizens become the scene of the city's vertical face.
We list five kinds of basic models about the possibility of crack's forming . In turn, it can be combined into twenty prototypes. After interviewing 15 residents, we extract the special requirements of Li Wen Shuang , Chen Mo, He Tao, YANG Zhao, and choose the room for their own need in twenty prototypes. When form the model, we respect the requirements of everyone for their needs of the layers

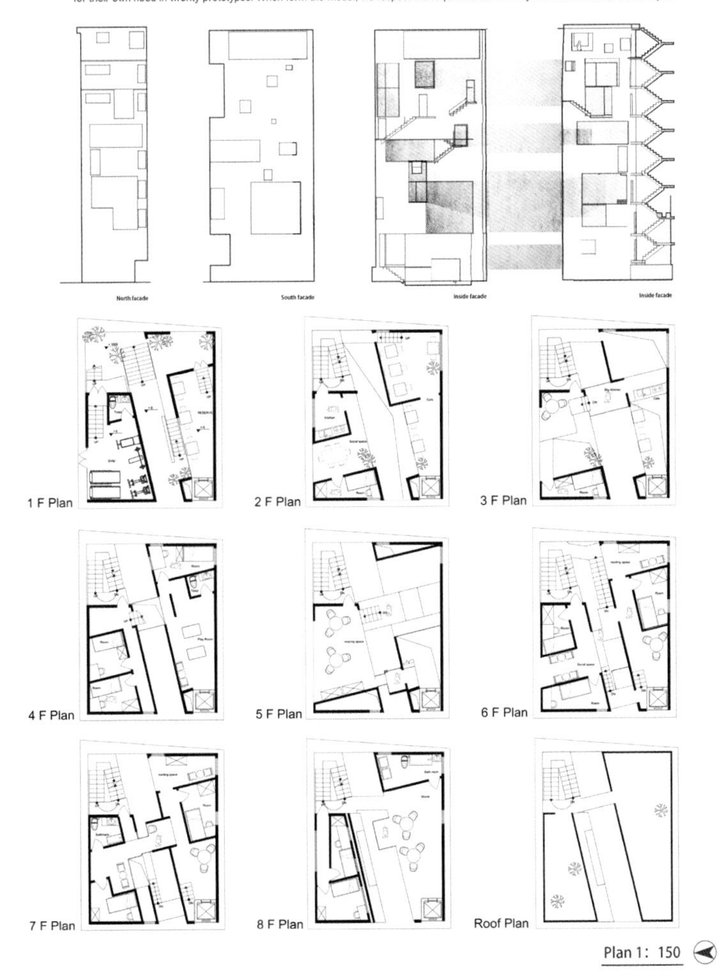

University: Tianjin University
Designer: Qi Shan, He Tao
Tutor: Wang Xuan, Zhang Xinnan
Course Name: Share House
Finished Time: Dec. 2014
Exchange Institute: Japan Women's University, Tamkang University

缝合 ——黟县际村村落改造与建筑设计

一边是作为历史展示的宏村，一边是现代商业拔地而起的水泥森林——水墨宏村，而之间的际村，经历了千百年的兴衰轮回，正面临着艰难的抉择，生存还是衰亡？中国农村的复兴之路是否无法摆脱生态的恶化与景观的颓败？

规划层面，将问题的核心聚焦于际村与水墨宏村之间模棱两可的中间地带，首先将这片区域的地面恢复为农业景观，同时还“缝合”了因为商业开发而断裂的生态水系，使际村再一次回到与自然和谐共生的关系中。而在这片区域的地下设计了现代基础设施的功能中枢，线性的地下停车场，供水供电等管线系统，为际村未来的发展预留着足够的空间。

建筑设计层面，选择了这条规划绿链的中部节点，连接了商业区和地下基础设施以及宏村入口的狭长区域，试图通过建筑和场所的设计从人文层面进一步“缝合”际村和新的商业居住区。

自然生态调研
Natural Ecology Research

新的商业开发项目和快速路建设切断了际村原有的山脉和水系，使得原本与村落和谐共生的生态系统支离破碎。
The estate development project and the construction of highway cut off the orignial mountains and rivers,broke the balance between village and the nature.

水系原状
Original State of The River system

水系现状
Current State of The River system

基础设施调研
Infrastructure Research

际村作为宏村旅游的大后方，其基础设施的建设却处于相当落后的状态。除了停车位的匮乏，村内给水、排水、污水、电力、电信、有线电视管线均是个人根据需要随意架设，无序混乱且存在很大安全隐患。
The infrastructure of Jicun is at a rather backward state.Except the lack of parking aeras,the water supply, drainage, sewage, electricity, telecommunications, cable television lines were just set up according to the needs of individuals without any scientific plan.

际村电力系统现状
Electric System of Jicun

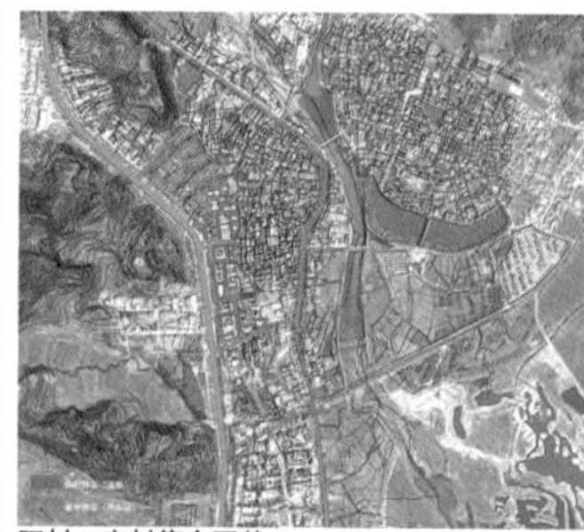

际村、宏村停车现状
Parking Area of Jicun and Hongcun

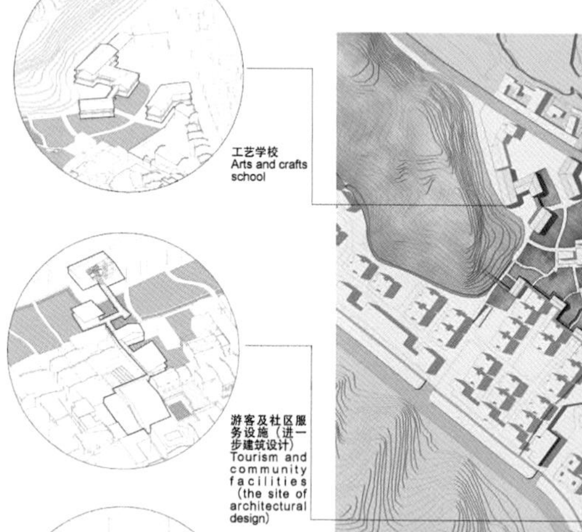

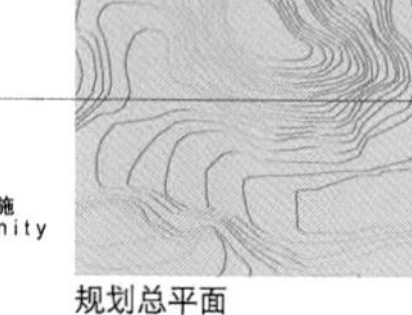

规划总平面
Masterplan for The Village Renovation

地下停车场
Underground Parking System

作品名称：缝合——黟县际村村落改造与建筑设计
Suture — Village Reconstruction

院校名：东南大学
设计人：邵星宇
指导教师：夏兵

作业完成日期：2014 年 07 月
对外交流对象：2014 Vectorworks 学生设计奖

SEWING ——Village Renovation and Architectural Design in Jicun village, Yixian

On the eastern side of Jicun village is Hongcun village — a historical "museum",on the western side is the new estate development project, which way to follow?the old village now should make its decision.

In the village renovation plan,we focus on the ambiguous place between business project and old village, making it a new "pivot" of the whole area.Firstly,change this place back to the agricultural landscape,"sew"the rivers cut off by estate construction.Secondly,we have planned a modern infrastructure under the fields,including an underground car park,piping systems of water and electricity. This will be the fundation of Jicun village's future development.

Then, we choose the middle part of the "pivot" as the site for architectural design.Linking the new commercial facilities ,the underground park and the entrance of Hongcun,the buildings and open places will furtherly "sew" the activities of different people ——the Aboriginal people, the new residents,the tourists…inhabitants' own will.

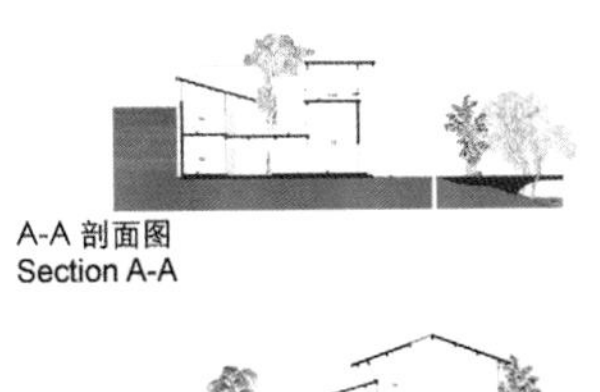

A-A 剖面图
Section A-A

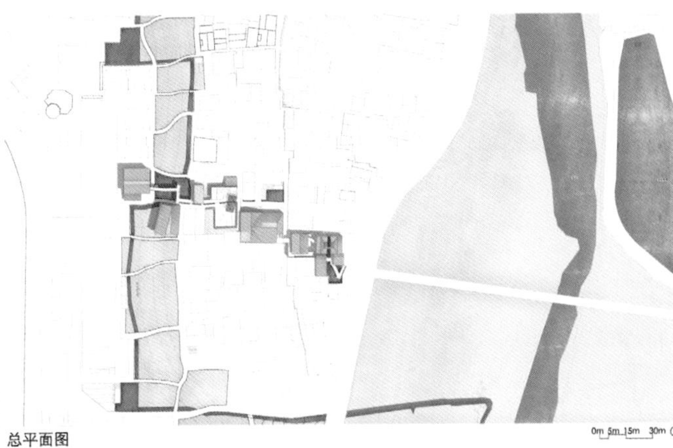

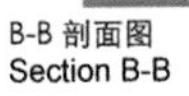

B-B 剖面图
Section B-B

总平面图
Master Plan

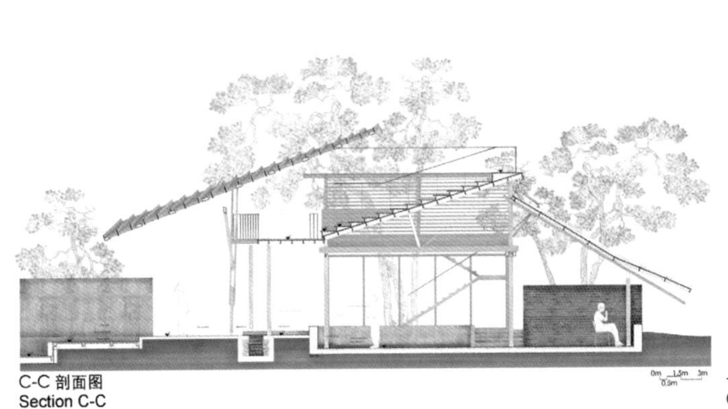

C-C 剖面图
Section C-C

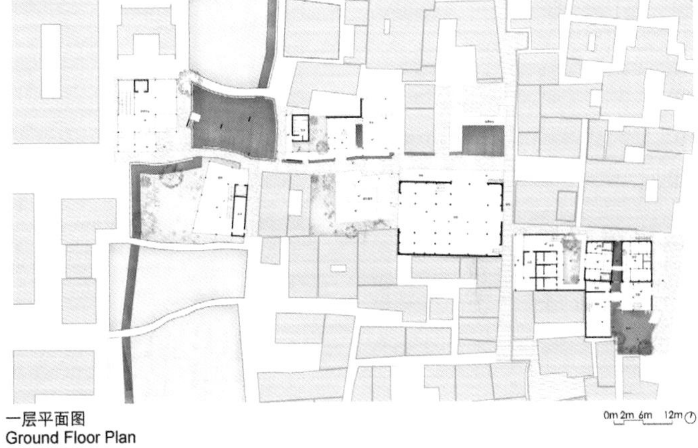

一层平面图
Ground Floor Plan

University: Southeast University
Designer: Shao Xingyu
Tutor: Xia Bing
Finished Time: Jul. 2014
Exchange Institute: 2014 Vectorworks Design Scholarship

STREET & LANE

A 【FIELD WORK REPORT】

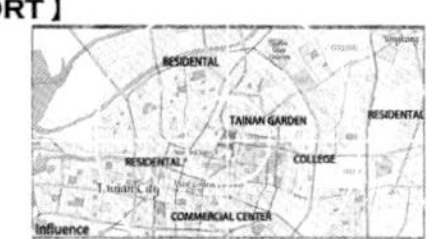

The site is located opposite the TaiNan Garden in Southern Taiwan. The site is close to the university and high school campuses, whose commercial space shows a significant temporal variation, relying on student groups' rise and fall. Meanwhile, compared with the surrounding existing commercial space, we found what makes the site commercial space more active.

The site experienced complex historical changes. It has a living form of longhouse and street space from Ming and Qing Dynasties, which has influenced the current consuming ways of shopping along the street so far. Meanwhile, the Arsenal grocery market during the war period has transformed to commercial soace relying on the school district.

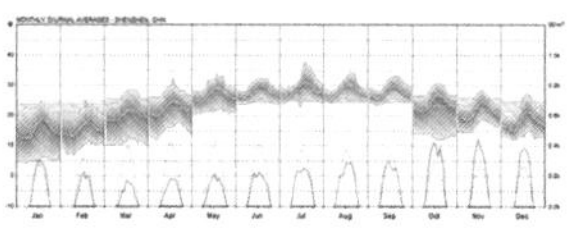

According to the Ecotect Anaylise, the local climate has a feature of high temperature and high humidity. Therefore, natural ventilation is an important method to improve the physical environment.

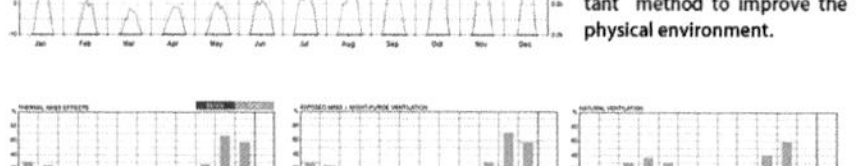

B 【TARGET ANAYLISE】

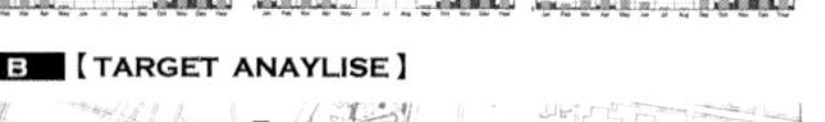
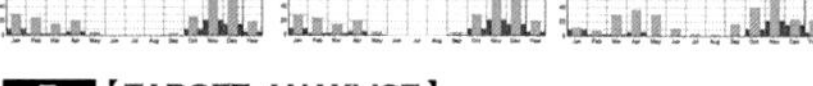

SOHO
Small Office Home Office.
17%

FAMILY
Fathe and mother with a child.
11%

Foreigners
Who work in this city.
9%

Business
Working in surrounding olace.
14%

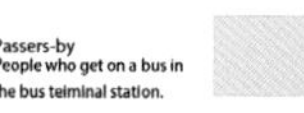

Passers-by
People who get on a bus in the bus teiminal station.
23%

Students
Who studied nearby.
25%

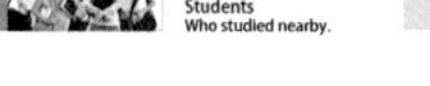

C 【CONCEPT ANAYLISE】

Streets and Lanes play an important role of media soace as it hosts various events. Based on the Inheritance of old street texture, the design aims to activate street life, complicate and diverse the behavior that occurs in commercial space and create integration of multii-consumption patterns by using prints of city.

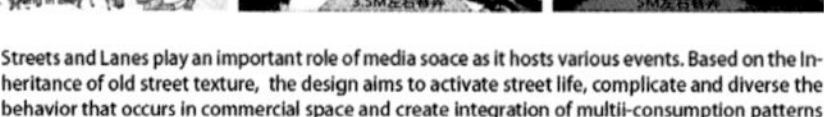
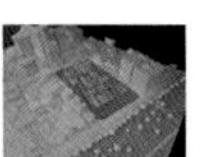

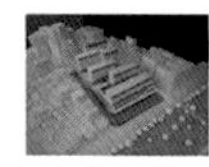

During the designing process, we came up with an idea that the whole building may divide into four levels in order to avoid life between different kind of people.

The first and sescond level are commercial levels, which contains different kind of shops. It is organised according to the surrounding streets' texture. Meanwhile, there are various of functions on the ground floor in order to satisfied different people's need.

The third level is community level. People living in the site may experience various kind of social activities, such as reading, singing, communciating, walking or running and so on. It is designed as a semi-public and semi-private level.

The last level is living level, which is a totally private level. According to different kind of target, building plan divided into soho, family, hostel and hotel four parts.

【EXTERIOR PESPECTIVE】

E 【SITE ANAYLISE】

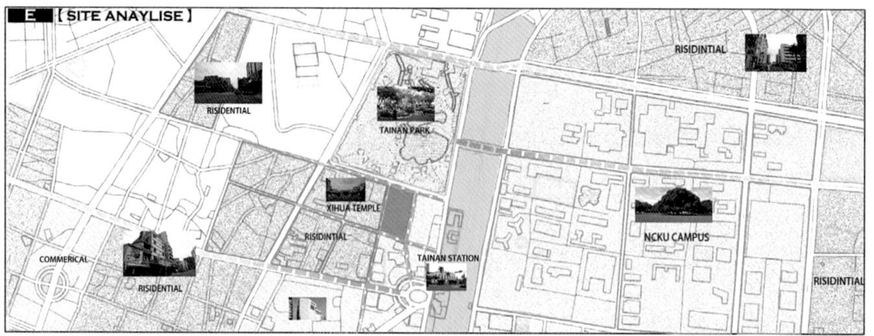

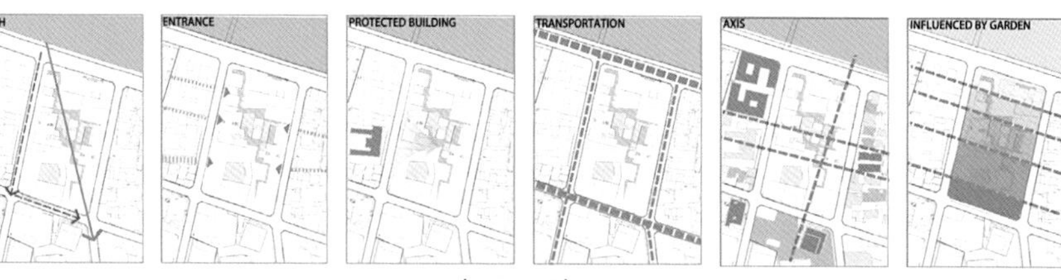

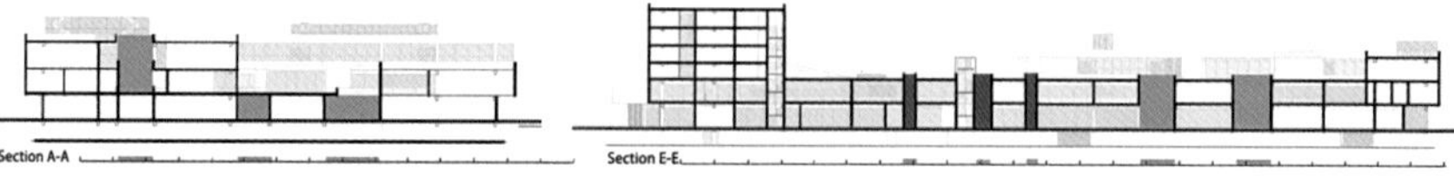

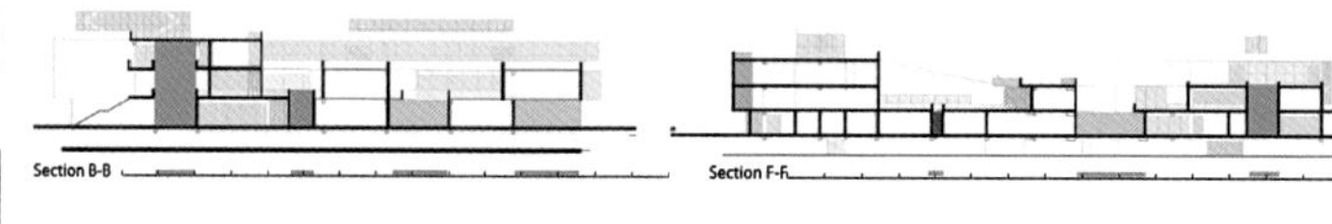

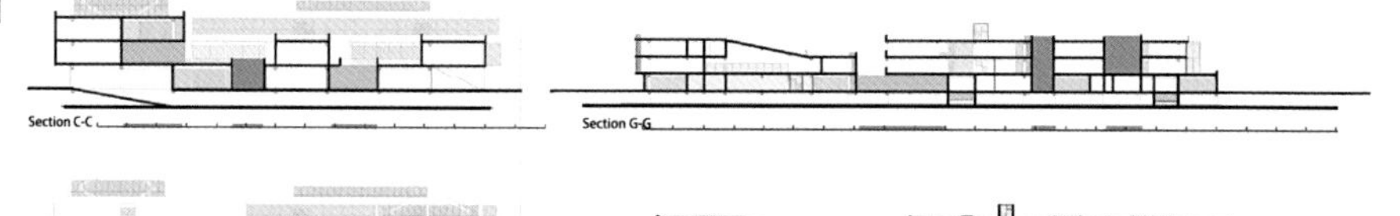

Streets and lanes are the symbol of Tainan, therefore we tried to add those elements into this design. Through different sections it is easily observed that different size of lanes lead to a different experience of space, and different-sized lanes connect each other make the space more interesting.

D 【ELEVATION】

作品名称：街&巷
Street & Lane

院校名：大连理工大学
设计人：王博伦，王雨杨
指导教师：赵建铭，吴亮
课程名称：Hybrid Design（本科三年级）
作业完成日期：2014 年 06 月
对外交流对象：台湾成功大学

HYBRID DESIGN 2

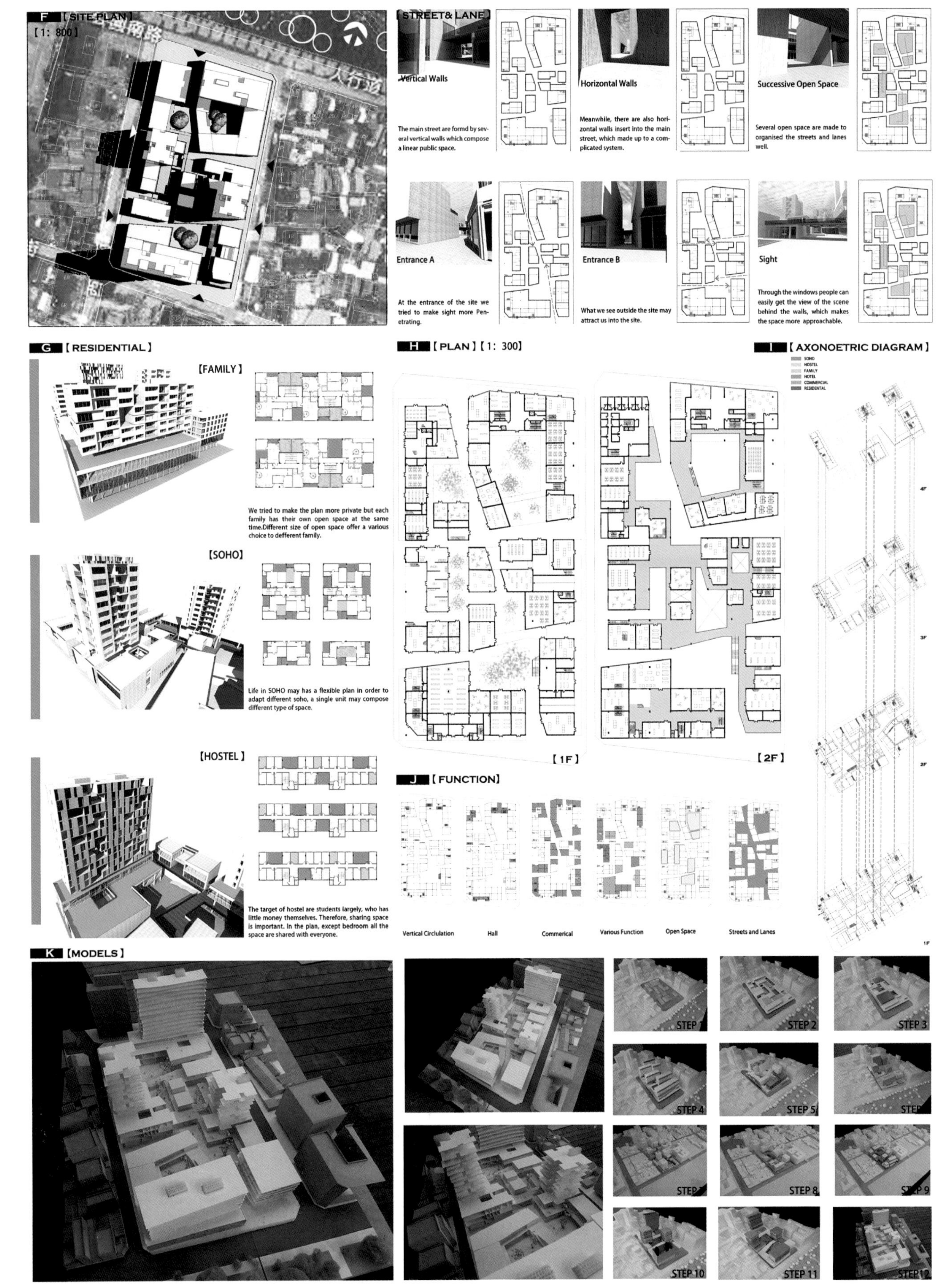

University: Dalian University of Technology
Designer: Wang Bolun, Wang Yuyang
Tutor: Zhao Jianming, Wu Liang
Course Name: Hybrid design
Finished Time: Jun. 2014
Exchange Institute: Cheng Kung University

作品名称：万花景——分型理论下的汽车博物馆设计
Kaleidoscope — Automobile Museum Design

院校名：天津大学
设计人：王雪睿，张知
指导教师：许蓁，盛强
课程名称：天津大学—加利福尼亚大学洛杉矶分校联合设计课程教学
作业完成时间：2014 年 12 月
对外交流对象：加利福尼亚大学洛杉矶分校

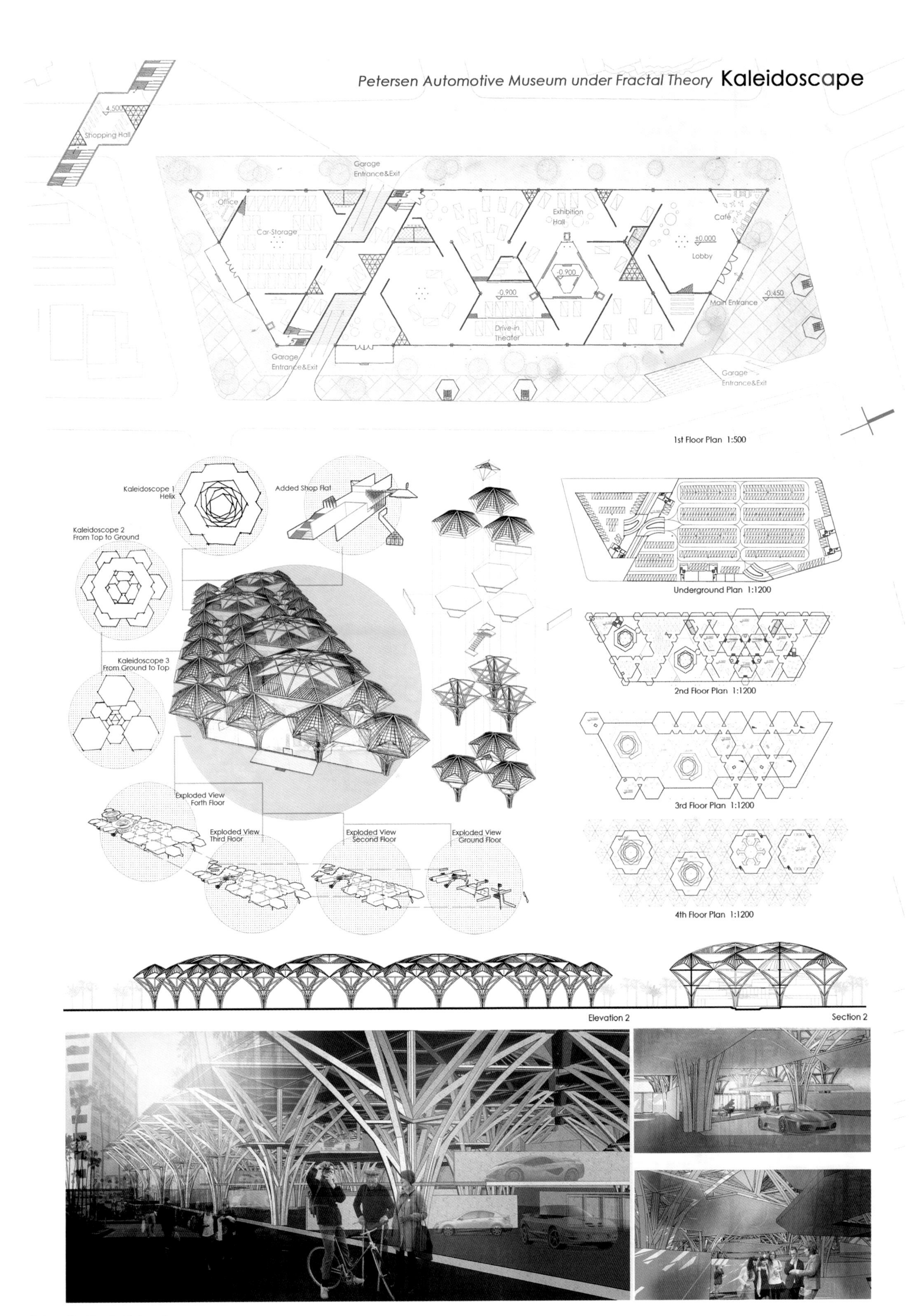

University: Tianjin University
Designer: Wang Xuerui, Zhang Zhi
Tutor: Xu Zhen, Sheng Qiang
Course Name: Tju-Ucla Joint Design Studio
Finished Time: Dec. 2014
Exchange Institute: University of California, Los Angeles

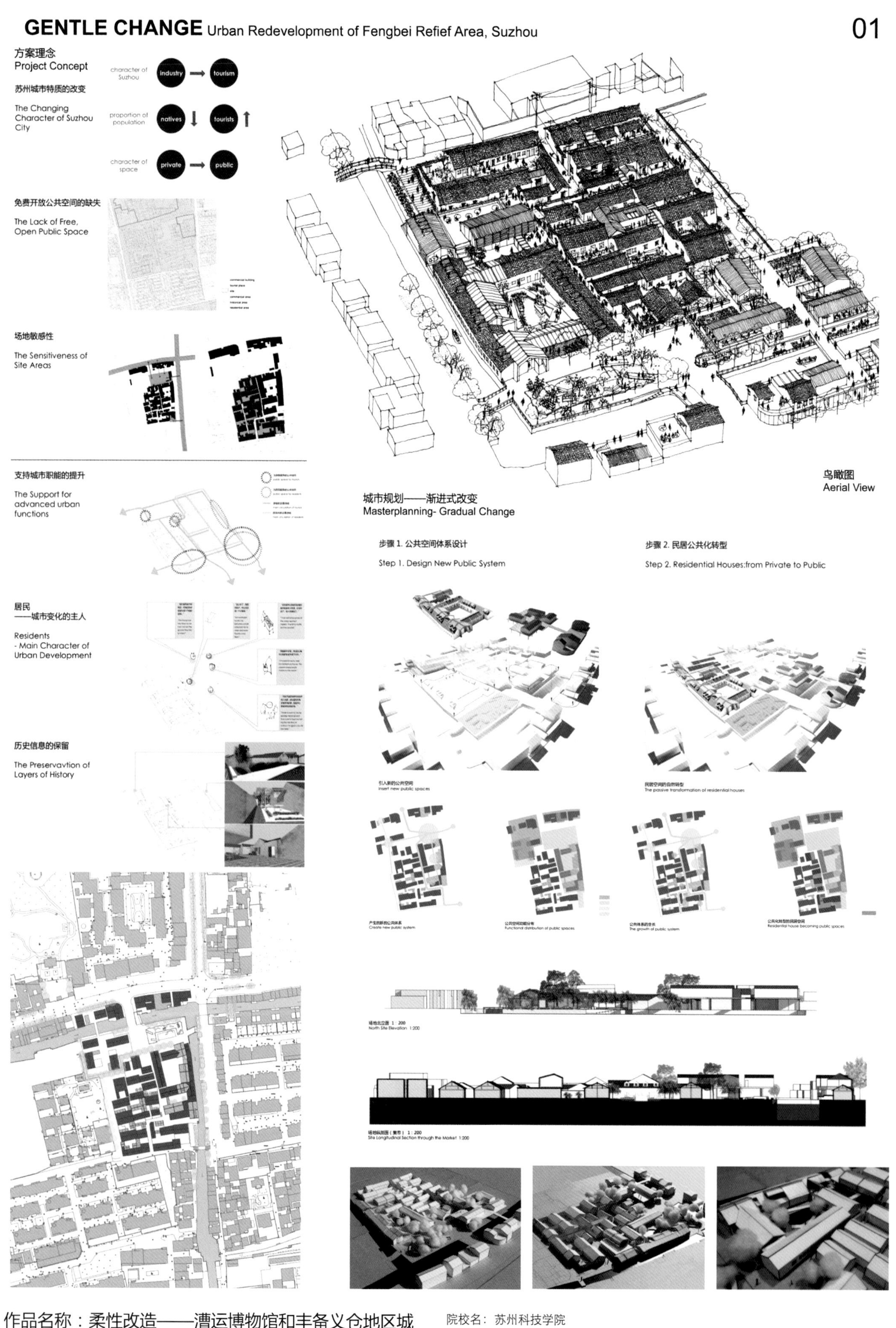

作品名称：柔性改造——漕运博物馆和丰备义仓地区城市设计

Centle Change—Urban Redevelopment of Fengbei Refief Area, Suzhou

院校名：苏州科技学院
设计人：徐佳，David Whiteworth，程欣韵，赵文哲
指导教师：胡莹，夏健，马骏华，许亦农，龚恺，王路，谢鸿权
课程名称：澳大利亚新南威尔士大学—东南大学—苏州科技学院—清华大学联合跨文化设计课
作业完成日期：2014 年 05 月
对外交流对象：澳大利亚新南威尔士大学

GENTLE CHANGE Urban Redevelopment of Fengbei Refief Area, Suzhou

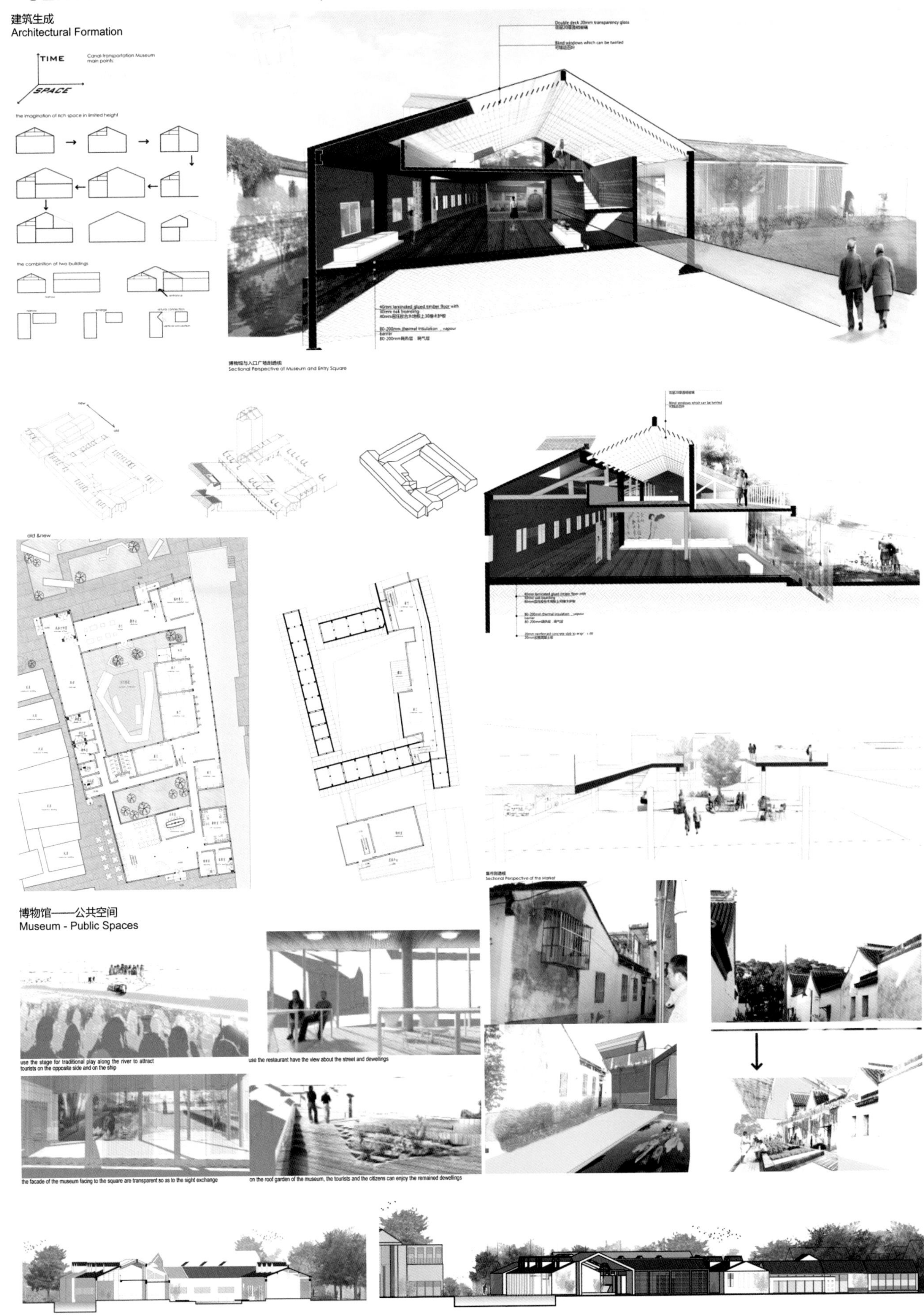

University: Suzhou University of Science and Technology
Designer: Xu Jia, David Whiteworth, Cheng Xinyun, Zhao Wenzhe
Tutor: Hu Ying, Xia Jian, Ma Junhua, Xu Yinong, Gong Kai, Wang Lu, Xie Hongquan
Course Name: UNSW-SEU-SUST-Tsinghua Joint Cross-Cultural Design Studio
Finished Time: May 2014
Exchange Institute: University of New South Wales

NORTH BONDI APARTMENT

Context / Site Generated Design for a Living Environment

The project site in is in North Bondi, a locality of Waverley Council that has a long history of apartment living dating from the 1930s. As part of a neighbourhood a short distance from the beach with shops and apartments traditionally scattered amongst houses, the rationale for a mixed use development here is clear.
Projects will generallybe generated from mixed or hybrid briefs incorporating more than one function or building type. The studio will explore contextual design, including urban patterns, as well as building design and detailed consideration of the technical resolution of an aspect of the building to a high level of resolution.

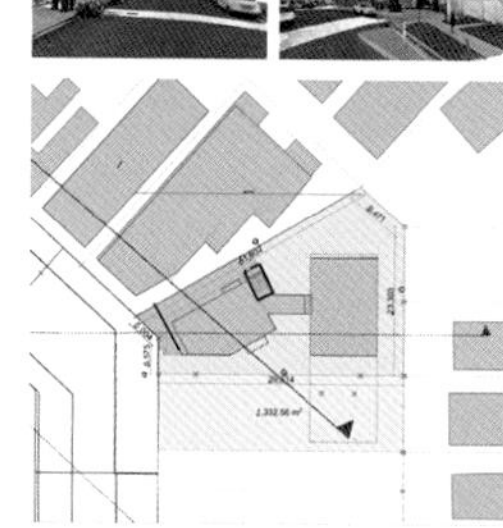

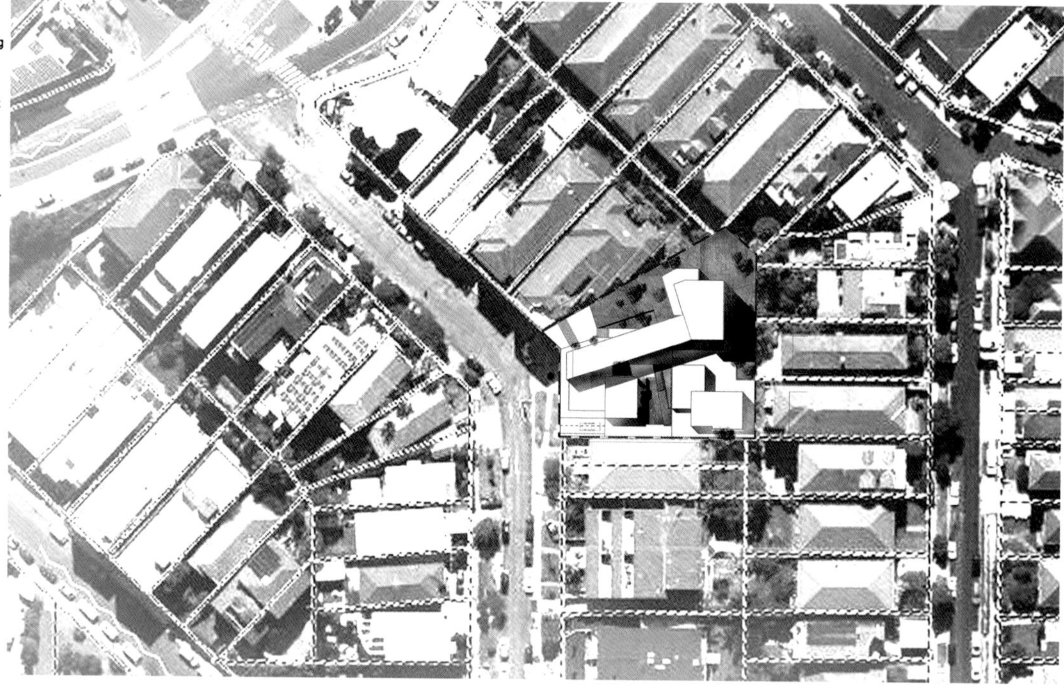

SITE STUDIES

The existing building at 108 was probably constructed in the late sixties and is typical of the period and representative of the area height and density.The current height is estimated at about 13 meters and density /FSR of about 1.23 :1.Parking is provided by internal road access to open areas below and around the north block.The geography of the site provides a number of challenges with respect to slope and as a triangular site at the apex of grid shift in street alignment.

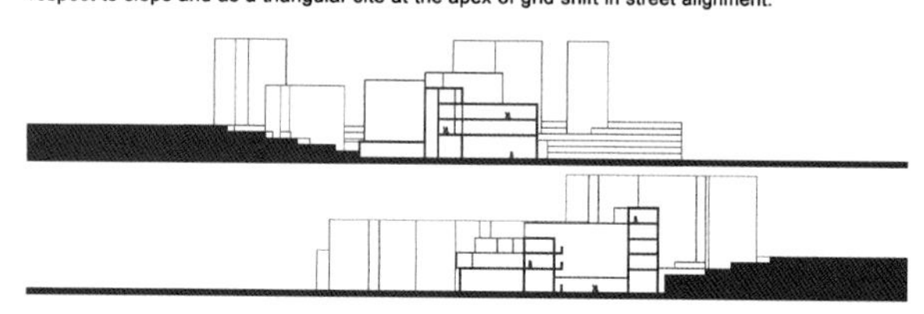

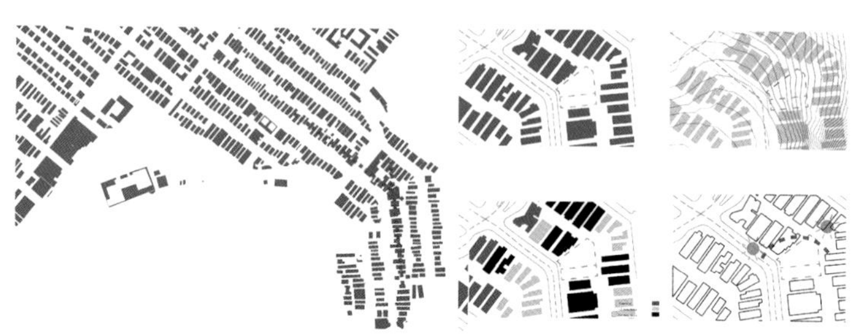

PERCEDENT STUDY NAPIER STREET HOUSING

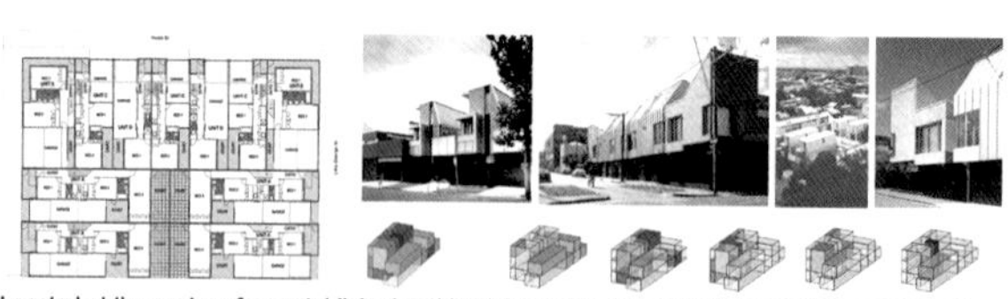

Located at the centre of an established residential precinct, this project responds to a variety of neighbouring building types ranging from historic terraces and row houses to converted industrial buildings and post-war housing towers. It takes the linear disposition of the traditional terrace housing – formal rooms at the front, narrow services rooms into the depth of the block, indented side light court and rear pocket garden – and reconfigures it vertically.

Bondi Beach,1 kilometre long, receives many visitors throughout the year

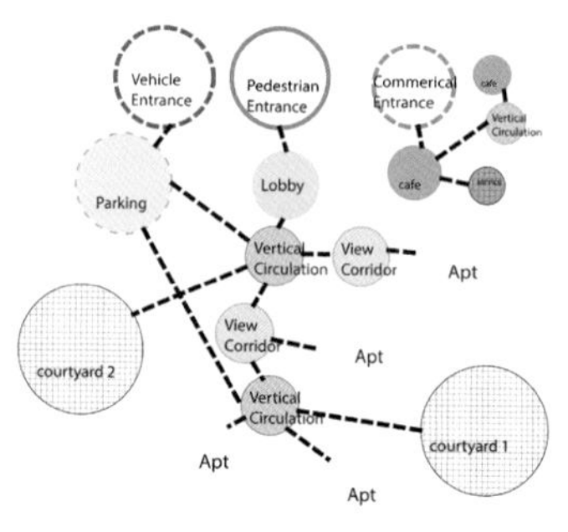

located in a typical Australia residential area

great view, bad sunlight and steep slope

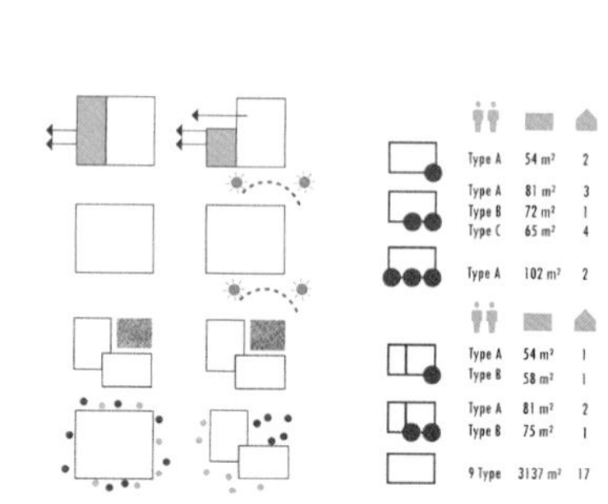

COMPOSITION STUDIES

protopye generated by sunlight, view and privacy

apartment types differed in volulme and rooms

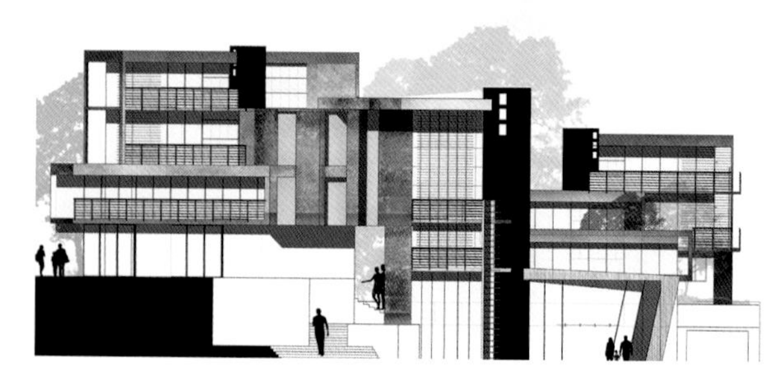

作品名称：北邦迪公寓设计
North Bondi Apartment Design

院校名：清华大学
设计人：徐菊杰
指导教师：Bruce Yaxley
课程名称：Arch B01 studio（本科生三年级）
作业完成日期：2014 年 06 月
对外交流对象：澳大利亚新南威尔士大学

NORTH BONDI APARTMENT

Context / Site Generated Design for a Living Environment

The apartment contains 17 flats and a coffee store with a dancing floor.As the location is near the famous bondi beach, the view becomes a very important factor. Moreover,the steep slope of the site make it a necessity to control privacy on the basis of a fluent transportation.And it's at a turning point,which means the street-scape should be carefully designed, so people from both sides can have a good sense of the commercial part.

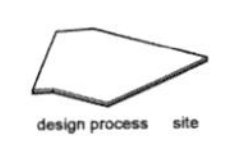
design process site

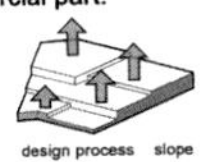
design process slope

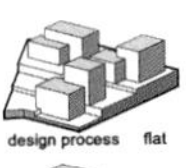
design process flat

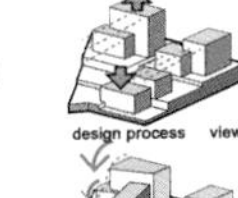
design process view

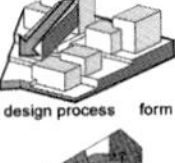
design process form
design process rotate

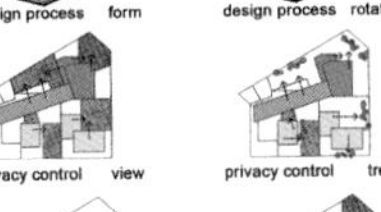
privacy control view
privacy control tree

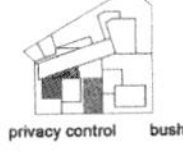
privacy control bush

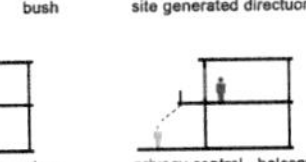
site generated direction
privacy control view
privacy control balcony

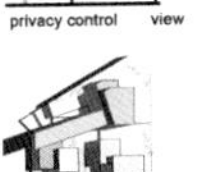
sunlight analysis 9am

sunlight analysis 9am

sunlight analysis 12am
sunlight analysis 12am

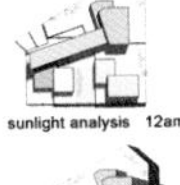
sunlight analysis 3pm

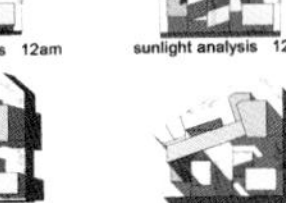
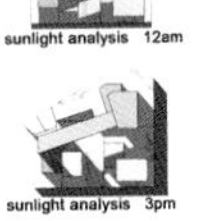
sunlight analysis 3pm

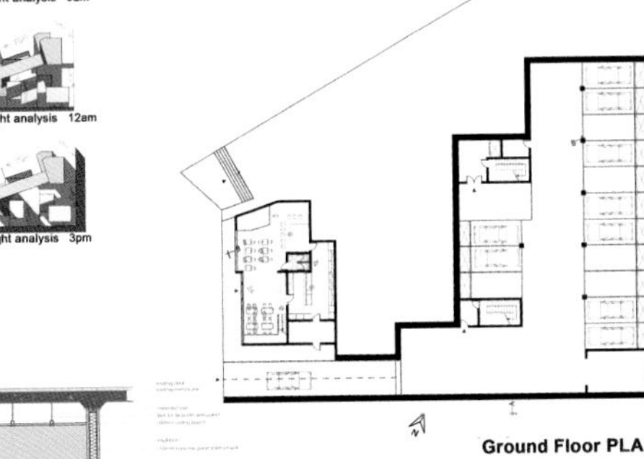
Ground Floor PLAN

First Floor PLAN

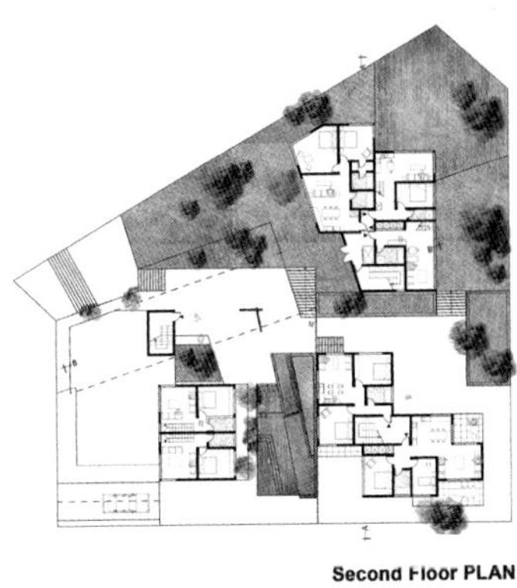
Second Floor PLAN

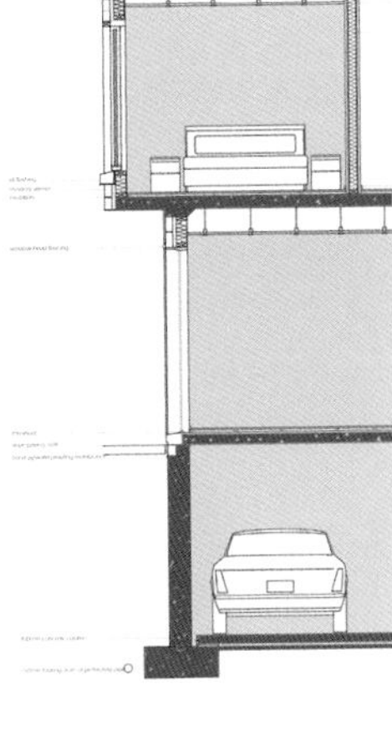

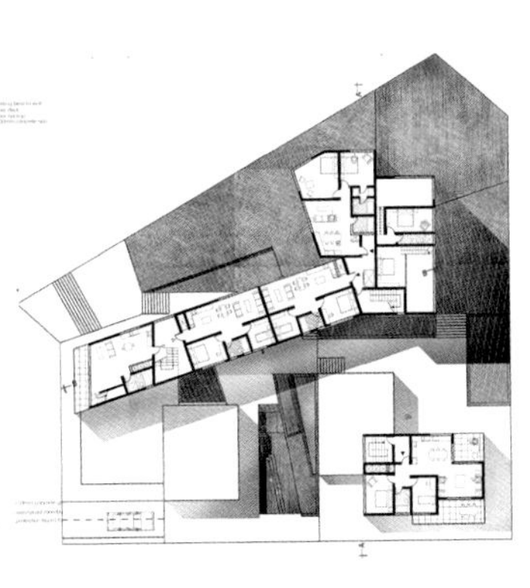
Third Floor PLAN

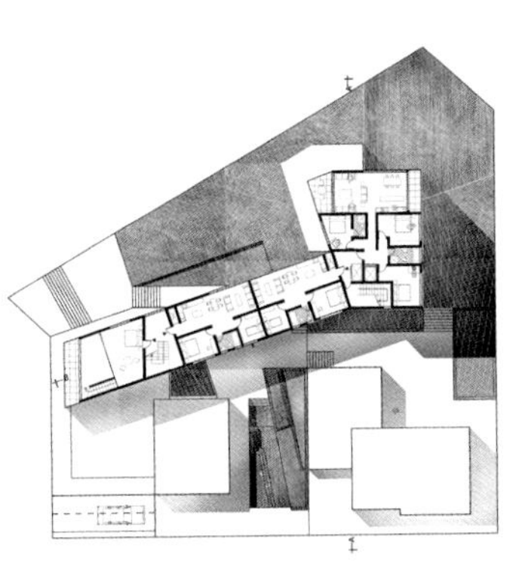
Forth Floor PLAN

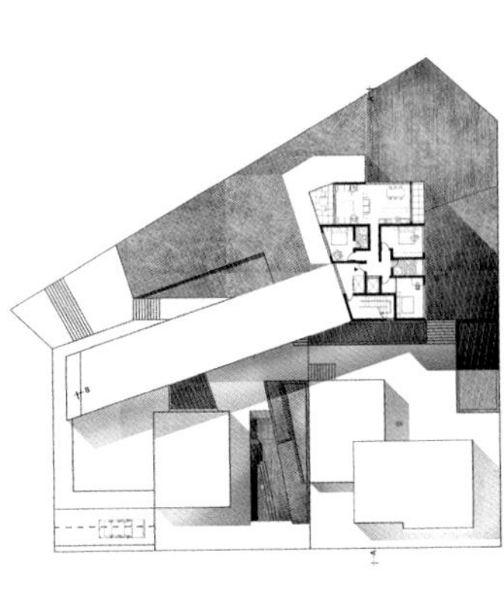
Fifth Floor PLAN

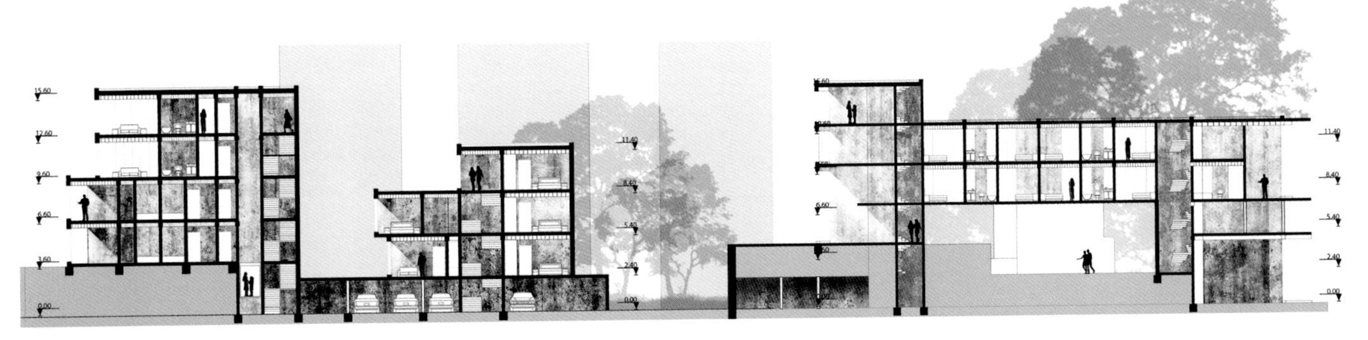

University: Tsinghua University
Designer: Xu Jujie
Tutor: Bruce Yaxley
Course Name: Arch B01 Studio 5
Finished Time: Jun. 2014
Exchange Institute: University of New South Wales

街巷 泉州历史城区文化剧场设计

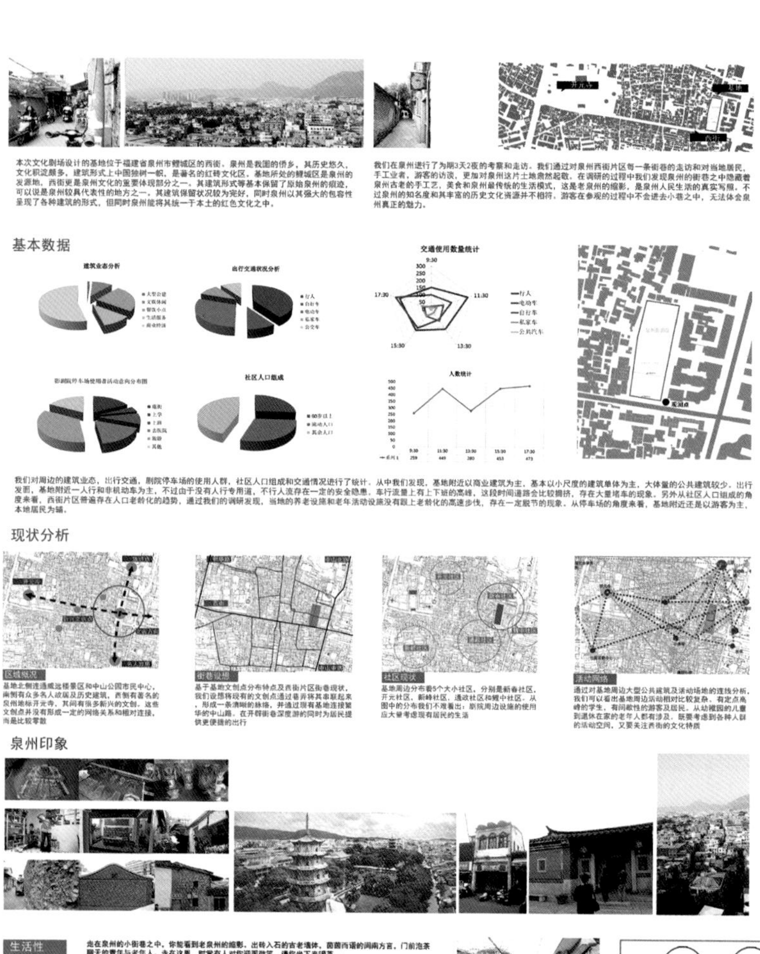

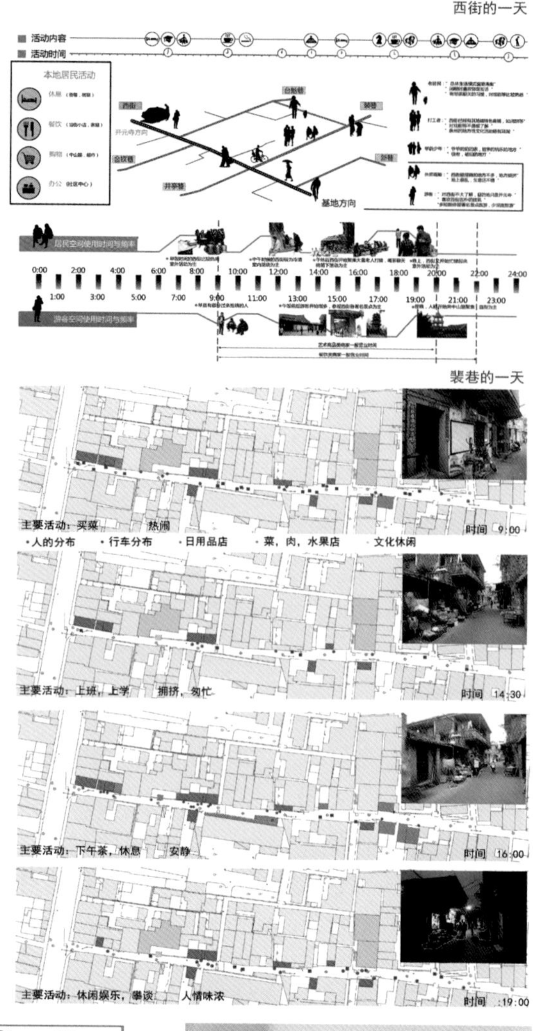

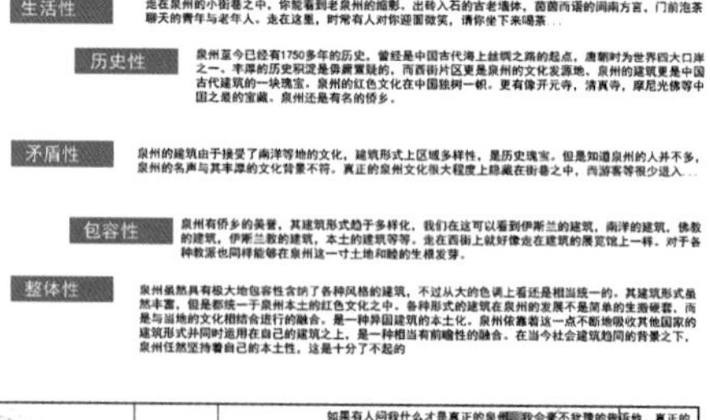

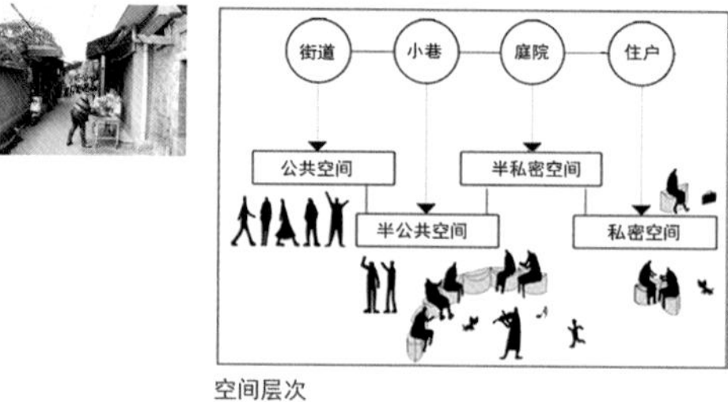

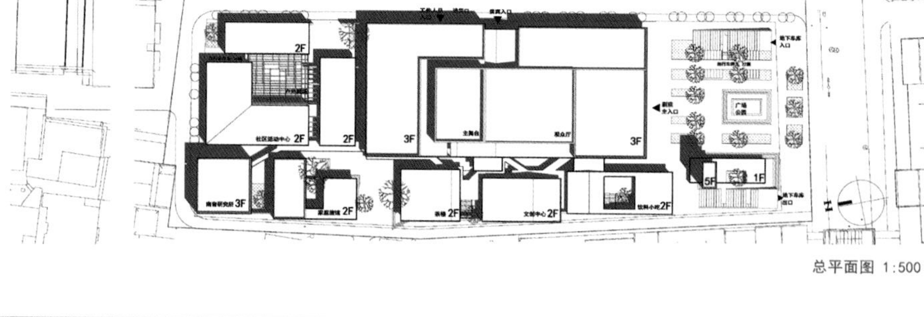

总平面图 1:500

作品名称：街巷
Streets Alleys

院校名：华侨大学
设计人：徐骏
指导教师：连旭，吴少峰
课程名称：泉州历史城区文化剧场共题设计（本科四年级）
作业完成日期：2014 年 07 月
对外交流对象：中国文化大学

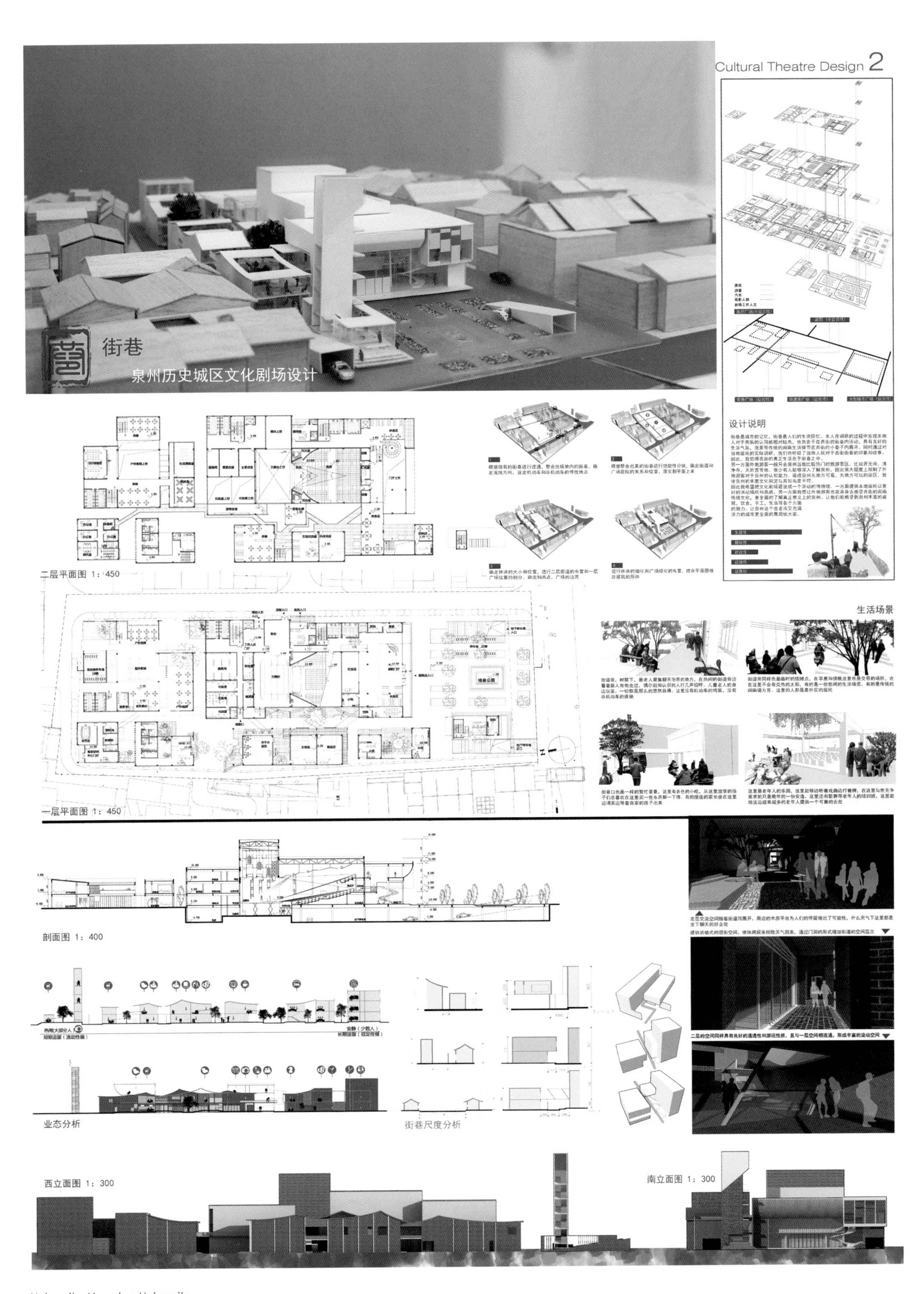

University: Huaqiao University
Designer: Xu Jun
Tutor: LianXu, Wu Shaofeng
Course Name: Quanzhou Historical and Cultural Theatre Joint Design Studio
Finished Time: Jul. 2014
Exchange Institute: Chinese Culture University

CORNER DANCING

DESIGN SPECIFICATION 设计说明

Located in the heart of the area, this design further determine the characteristics of the people living in the space -- government officer, buisiness related visitor and sightseeing visitor, by discussing the effection of the changes at the corner spcae to the centralized space in the middle of a plane as well as the changes of the space pattens. We also discussed the the new pattern of the government working space by creatively exploring completely open space, semi-open working space and the relatively closed interactive space.

本设计基地处于整个正方形地段的中心，任务书又给定了一个高层办公建筑的性质，再结合周边建筑已有的功能，我们选定市民可参与的行政办公建筑作为出发点。基于此，考虑到地段的特殊核心位置以及特殊的服务人群，我们想通过对角落和中心空间的操作，达到办公与市民参与的融合与平衡，即通过探讨角落空间的变化对于中心空间的影响，以及因此带来的空间模式的变化，试图为使用者提供一个参与度高的场所。

CONCEPTUAL DIAGRAMS 概念生成图表

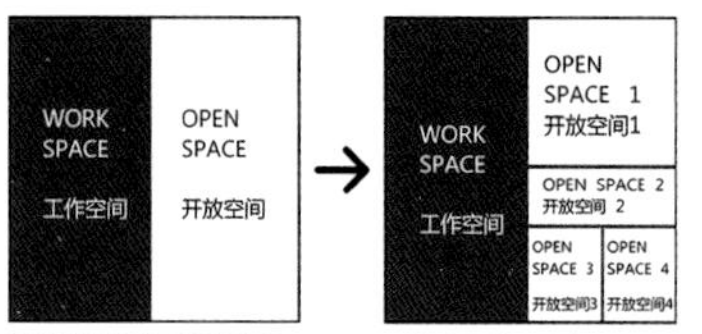

DIVIDE THE SPACE INTO 2 DIFFERENT PARTS
将空间划分成工作和开放空间。

DIVIDE THE OPEN SPACE INTO 4 PARTS
将开放空间分成四个不同的空间。

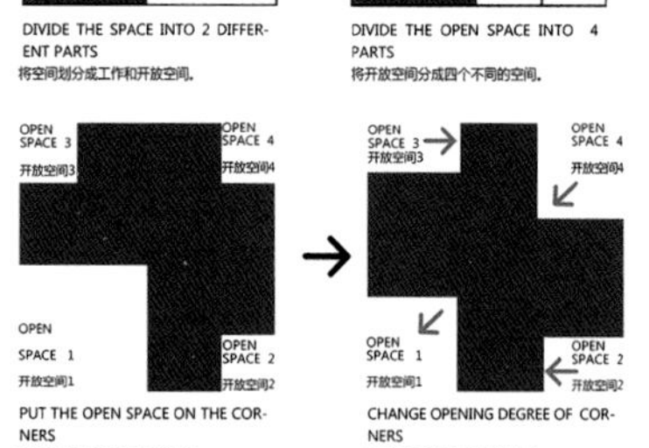

PUT THE OPEN SPACE ON THE CORNERS
不同的开放空间置于空间的四角。

CHANGE OPENING DEGREE OF CORNERS
调整四角开放空间的开放程度。

CHARACTER ANALYSIS 人群特性分析

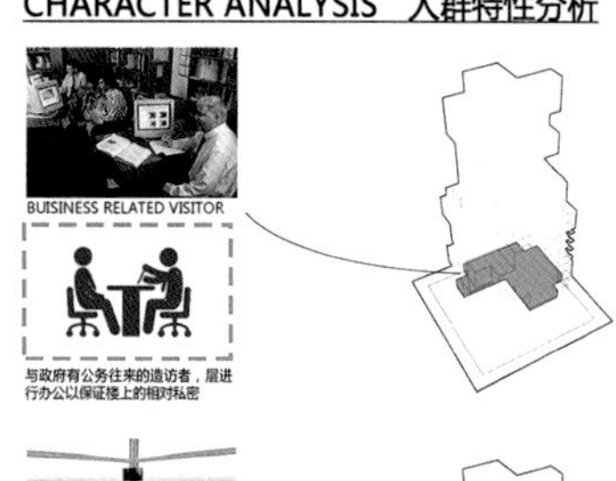

BUISINESS RELATED VISITOR

与政府有公务往来的造访者，层进行办公以保证楼上的相对私密

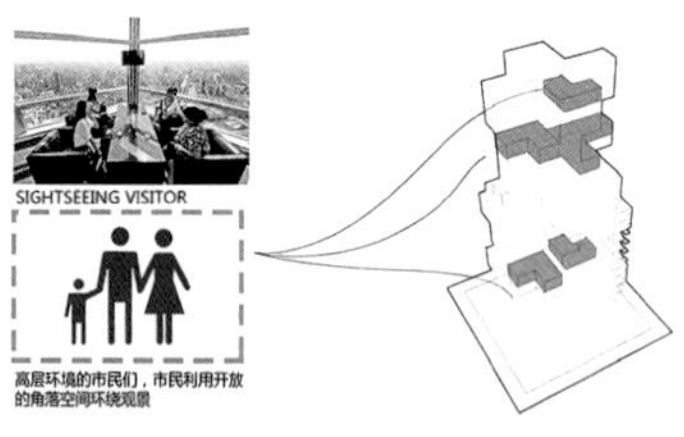

SIGHTSEEING VISITOR

高层环境的市民们，市民利用开放的角落空间环绕观景

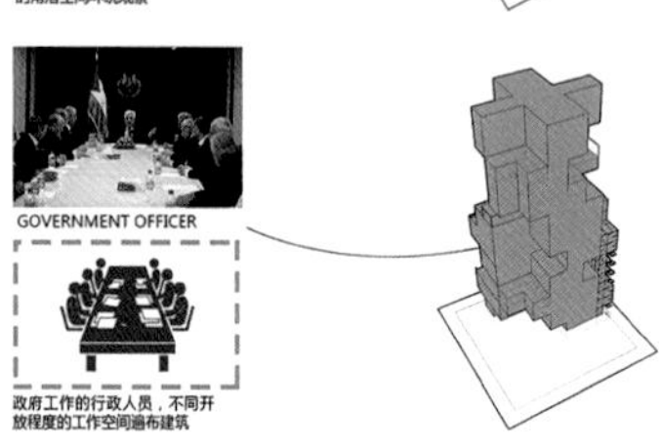

GOVERNMENT OFFICER

政府工作的行政人员，不同开放程度的工作空间遍布建筑

SITE ANALYSIS 基地分析

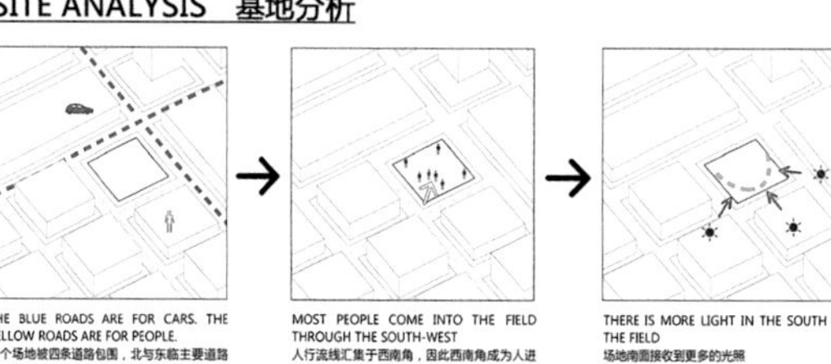

THE BLUE ROADS ARE FOR CARS. THE YELLOW ROADS ARE FOR PEOPLE.
整个场地被四条道路包围，北与东临主要道路车行为主，南与西临次要道路人行为主

MOST PEOPLE COME INTO THE FIELD THROUGH THE SOUTH-WEST
人行流线汇集于西南角，因此西南角成为人进入场地的主要方向

THERE IS MORE LIGHT IN THE SOUTH OF THE FIELD
场地南面接收到更多的光照

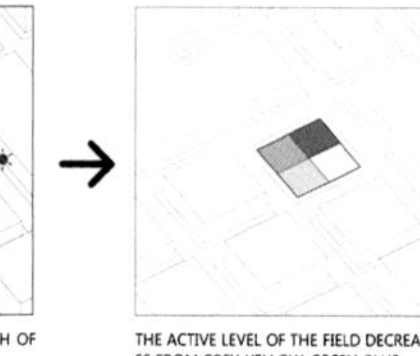

THE ACTIVE LEVEL OF THE FIELD DECREASES FROM GREY-YELLOW-GREEN-BLUE
因此场地的活跃程度可被划分为四部分，灰-黄-绿-蓝依次减弱

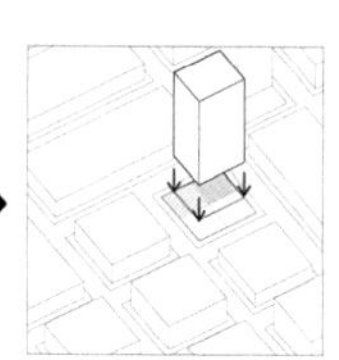

WE PUT OUR BUILDING ON THE BLUE AREA AND MAEK THE ACTIVE AREA SQUARE
我们将建筑的主体置于不活跃的蓝色区域，将其余活跃部分作为场地让出来

SPACE GROWTH 立体生成

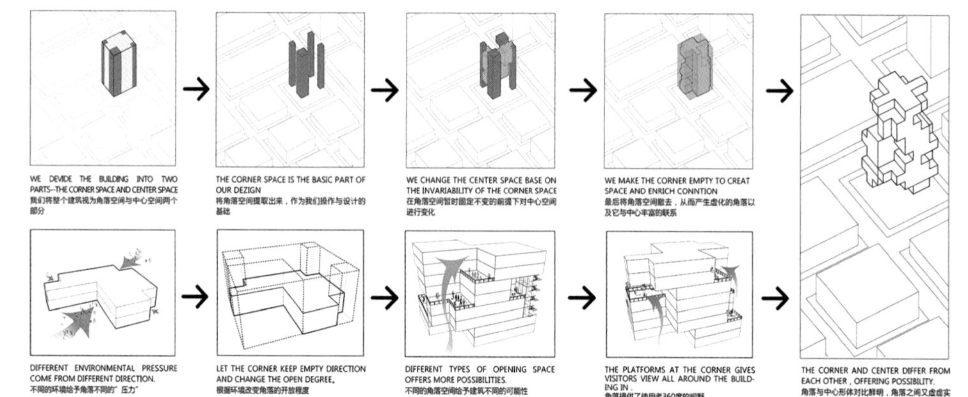

WE DEVIDE THE BUILDING INTO TWO PARTS--THE CORNER SPACE AND CENTER SPACE
我们将整个建筑视为角落空间与中心空间两个部分

THE CORNER SPACE IS THE BASIC PART OF OUR DEZIGN
将角落空间提取出来，作为我们操作与设计的基础

WE CHANGE THE CENTER SPACE BASE ON THE INVARIABILITY OF THE CORNER SPACE
在角落空间暂时固定不变的前提下对中心空间进行变化

WE MAKE THE CORNER EMPTY TO CREAT SPACE AND ENRICH CONNTION
最后将角落空间掏空，从而产生虚化的角落以及它与中心丰富的联系

DIFFERENT ENVIRONMENTAL PRESSURE COME FROM DIFFERENT DIRECTION.
不同的环境给予角落不同的"压力"

LET THE CORNER KEEP EMPTY DIRECTION AND CHANGE THE OPEN DEGREE.
根据环境改变角落的开放程度

DIFFERENT TYPES OF OPENING SPACE OFFERS MORE POSSIBILITIES.
不同的角落空间给予建筑不同的可能性

THE PLATFORMS AT THE CORNER GIVES VISITORS VIEW ALL AROUND THE BUILDING IN
角落提供了使用者360度的视野

THE CORNER AND CENTER DIFFER FROM EACH OTHER, OFFERING POSSIBILITY.
角落与中心形体对比鲜明，角落之间又虚虚实实

CONCEPT PRODUCTION 概念生成

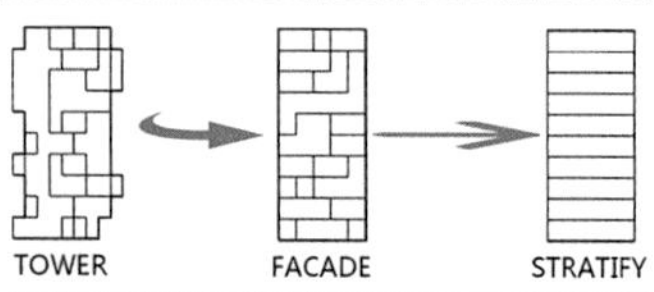

THE TOWER IS THE SYNTHESIS OF A GREAT DEAL OF DIFFERENT PEOPLE FROM DIFFERENT COMPANIES WORING IN DIFFERENT OFFICIAL PATTERNS. DESPITE THE UPPER AND LOWER STORIES, THE MIDDLE OF THIS TOWER IS ALL ABOUT OFFICIAL SPACE. TO FIND THE MOST ADAPTIVE SPACE FOR A FEATURED COMANY, WE TRY TO VARIFY PLANS TO CREATE DIVERSE WORKING SPACES. THIS IS AN EXPRIMENT TO SEE HOW TO MAKE THE RIGHT SPACE AND ESTEEM THE VARIETY WE ARE WORKING AND LIVING.

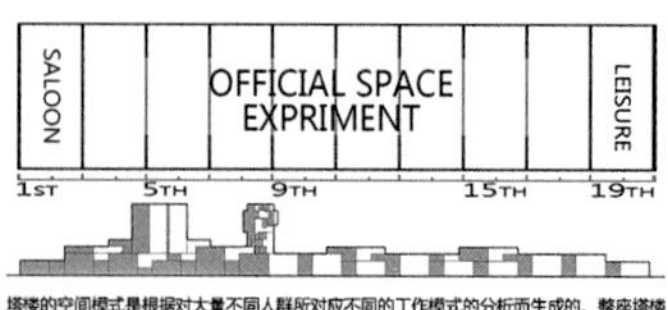

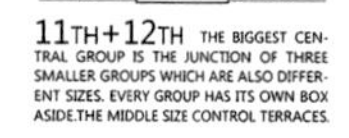

塔楼的空间模式是根据对大量不同人群所对应不同的工作模式的分析而生成的。整座塔楼除底层、顶层外皆为办公空间，而为了为不同的人群、工作模式创造不同的空间，我们对塔楼中间部分的办公空间做了各种变化。这也是一种对空间形式与人的行为对应关系的探讨，变化中充满了各种可能性。

CONCEPTUAL PLAN 概念平面

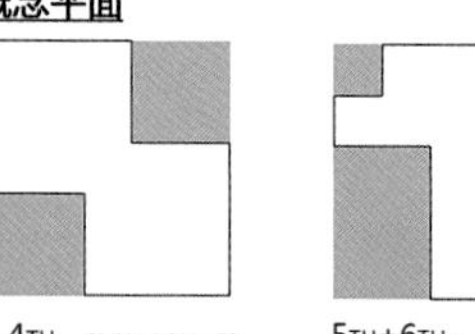

1ST+2ND PUBLIC PEOPLE IN IRREGULAR ORDER FULL OF MOST OF SPACE BOTH INSIDE AND OUTSIDE WHICH CONNECTED BY MANY DOORS. CASUAL GROUPS ARE ON THE EDGE WHILE INQURY OFFICE IS AT THE CORNER.

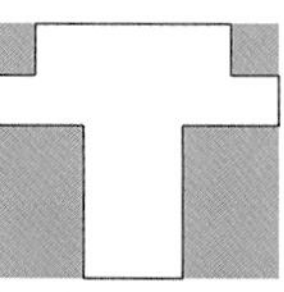

3RD+4TH DIVIDE A FLOOR INTO TWO PARTS WITH A CONNECTION BETWEEN THEM. HUMEN RAPID POURS INTO THIS CONNECTION.OFFICIAL REGULAR TURNS INTO CASUAL RANDOM THROUGH A PASS.

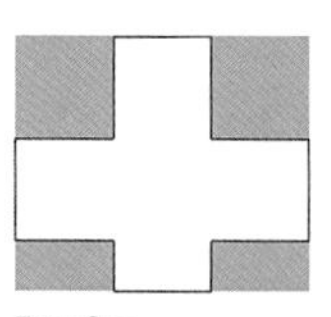

5TH+6TH A GROUP WITH SEVERAL LINES IS IN THE NORTH, WHILE A CLOSED GROUP WITH SOME BOXS IN THE SOUTH. LEISURE WORKING GROUPS ARE ON THE SIDE. THREE DIFFERENT TERRACE FOR RESTERS.

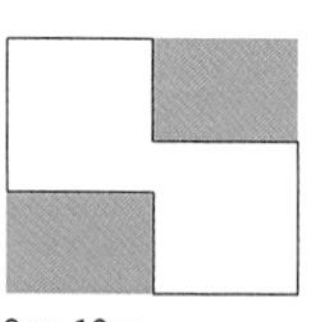

7TH+8TH A COUPLE OF CLOSED BOXES TEARS A FLOOR INTO THREE PARTS.TWO BIG GROUPS WATCH EACHOTHER AND A SMALL GROUP HOLDS MORE TERRACES. A STATIC NARROW TERRACE FOLLOWS A DYNAMIC NARROW CORRIDOR.

9TH+10TH THE BIGGER GROUP IS VERY FAR AWAY FROM THE SMALLER, WHICH MAKES THE FLOOR INTO TWO COMPLETELY DETACHED PARTS WITH LITTLE CONNECTION. ONLY A TERRACE IMBIBE THEM TOGETHER.

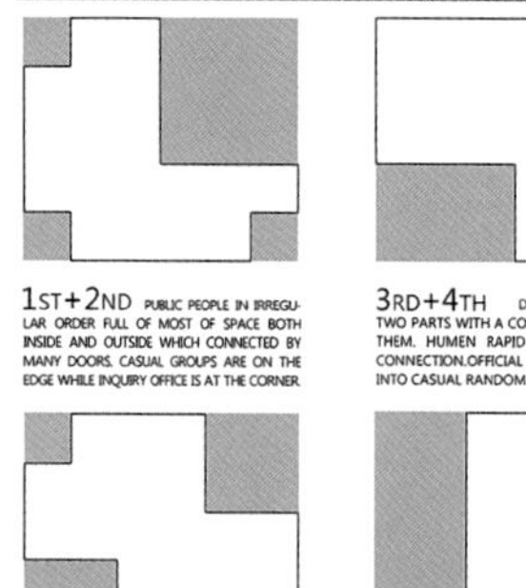

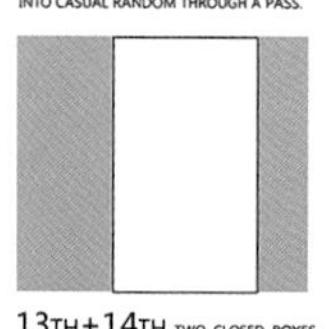

11TH+12TH THE BIGGEST CENTRAL GROUP IS THE JUNCTION OF THREE SMALLER GROUPS WHICH ARE ALSO DIFFERENT SIZES. EVERY GROUP HAS ITS OWN BOX ASIDE.THE MIDDLE SIZE CONTROL TERRACES.

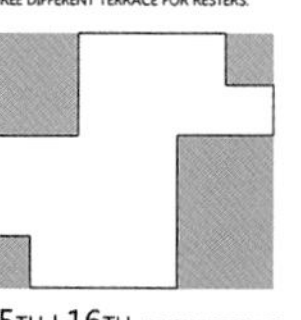

13TH+14TH TWO CLOSED BOXES DIVIDE THIS FLOOR INTO THREE GROUPS. THE FURTHER FROM THE ENTRANCE THE POSITION IS, THE SMALLER THE GROUP IS. TWO TERRACES ARE IN THE AIR BESIDE THE FLOOR.

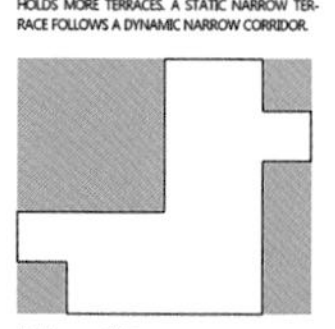

15TH+16TH THE CONNECTION OF THREE GROUPS IS A CLOSED CIRCULATION WHICH LOOKS LIKE A TRIANGLE SURROUNDING A INDEPENDENT BOX. THIS ORGANIZATION IS APART FROM AN EASTERN GROUP BY AN CORRIDOR.

17TH+18TH A ANNULAR ROUTE ALONG THE EDGE LIMITS A NARROW SPACE. A COUPLE OF BOXES AT THE CORNER DIVIDES THIS SPACE INTO TWO PARTS WHICH ARE VERY SIMILAR. A SMALL GROUP ENDS EACH PART.

19TH+20TH ONE CLOSED BOXES DIVIDE THIS FLOOR INTO THREE GROUPS. ONE GROUP SURROUNDS ANOTHER BOX IN THE SOUTH. TWO NARROW GROUPS WITH DIFFERENT SIZES HAVE DIFFERENT TERRACES.

作品名称：空角实心
Corner Dancing

院校名：哈尔滨工业大学
设计人：徐淼，王宏宇，付豪
指导教师：陈旸，邢凯，唐康硕，Giorgio Ponzo
课程名称：国际联合设计（本科三年级）
作业完成日期：2014 年 07 月
对外交流对象：荷兰代尔夫特大学

THE VARIETY OF OFFICAL PATTERN IN SPACIAL DIVERSIFICATION

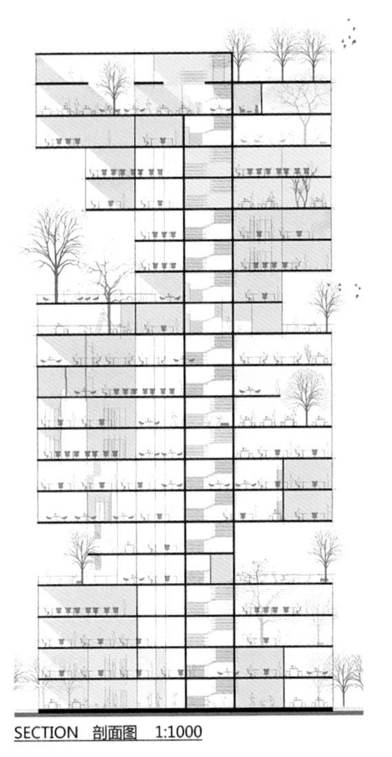

SECTION 剖面图 1:1000

MODEL PICTURES 模型照片

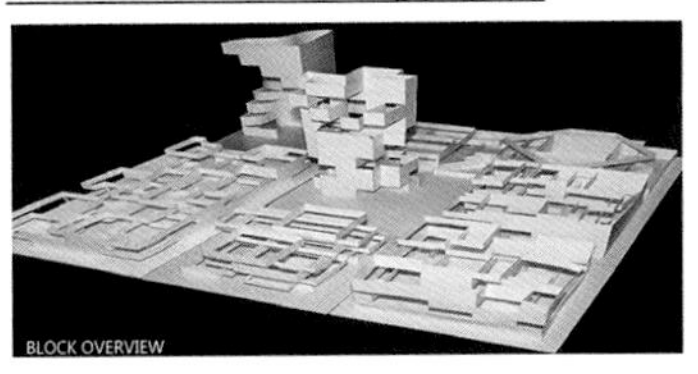

MONOCASE OVERVIEW

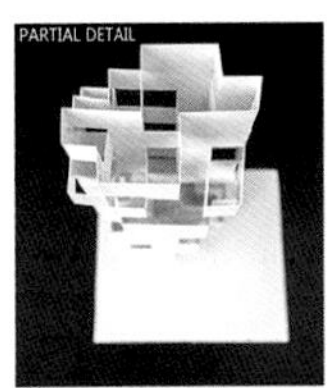

PLANS ANALYSIS 各层平面生成

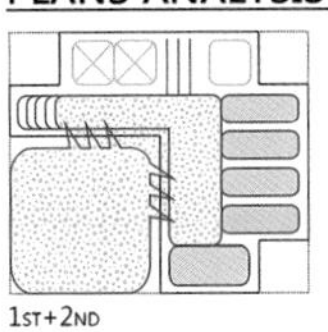

1ST+2ND
入口广场和门厅是无序空间。无序空间的尽头和分支是大厅的休闲空间。

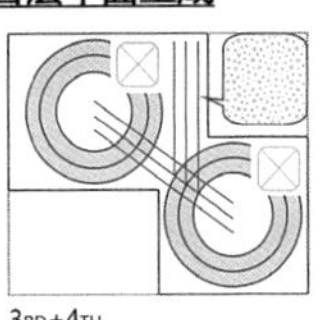

3RD+4TH
平面分成两个对等部分，。人流大部分涌入两部分的中间，少部分流入室外平台。

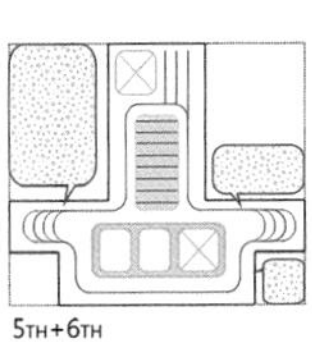

5TH+6TH
桌椅暗示北部严格办公模式。南侧封闭私密空间，两翼休闲空间。环形交通。

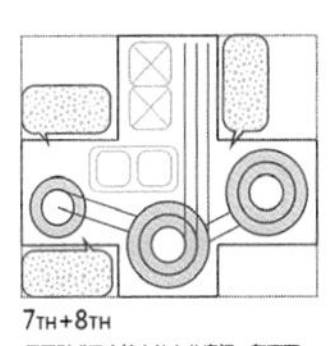

7TH+8TH
平面形成三个核心的办公空间。东南两组并置；西侧一组被私密空间分割。

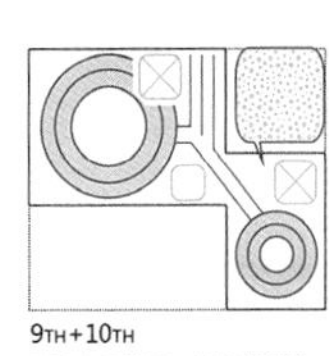

9TH+10TH
平面有大小两组组成，以私密空间遮挡，较少联系，共享一个室外平台。

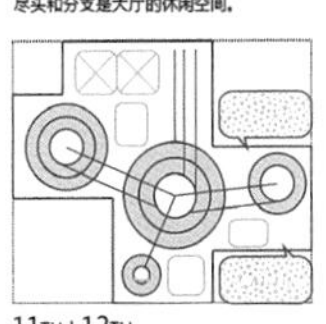

11TH+12TH
平面提供三组办公空间，不同组享有各自的私密空间。各组依靠中央空间联络。

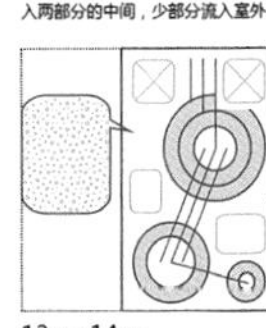

13TH+14TH
私密空间的遮挡形成了逐级传递的折线交通。在条形平面两侧各有一室外平台。

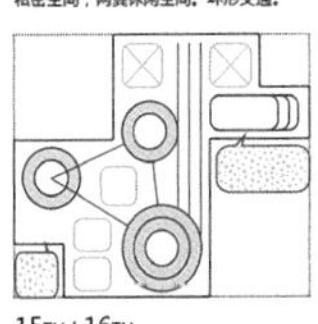

15TH+16TH
西侧形成三组联动的办公模式，并依靠中央的私密空间进行交流。东侧休闲空间。

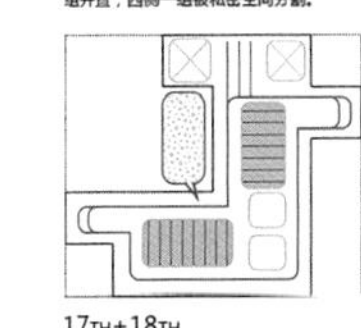
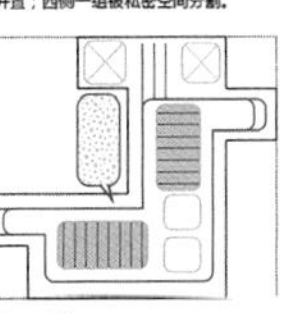

17TH+18TH
平面由两个办公单元组成，单元有严格的秩序，享有一个私密空间。

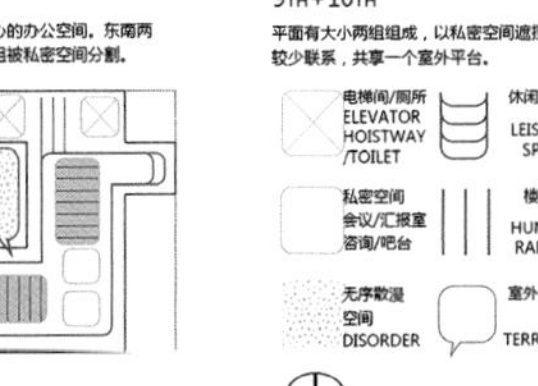

PROPOSAL

ACTUAL PLAN 实际平面

2ND FLOOR 二层平面图 1:300

3RD FLOOR 三层平面图 1:300

For the first floor, more influences of the surrounding roads and other buildings is consisdered. There' re two blocks in the west and south of the site which are relatively active, while in the northwest there is another relatively independent tower. Thus, we make a large area of the site give way out of the building to creat interaction with the two blocks.

首层平面的设计中，更多的考虑了周边道路以及其他建筑对于场地的影响。场地西面及南面相邻两个相对活跃的低层单元式建筑，而东北角与另一个相对独立的高层相邻。由此我们将建筑置于东北角，将大面积的场地退让出来，创造与其他活跃建筑交往的可能。

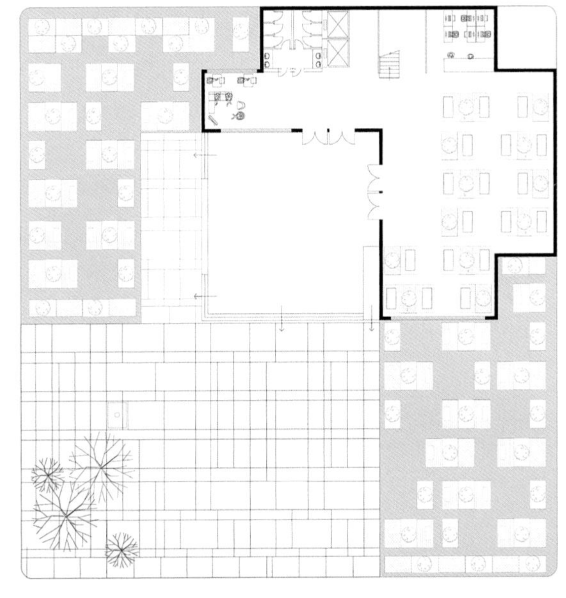

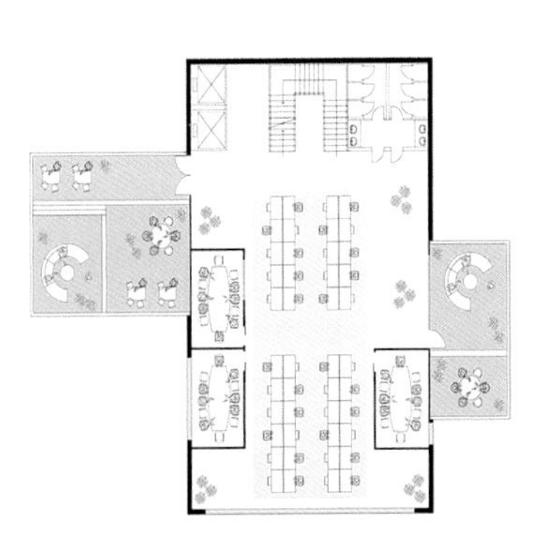

University: Harbin Institute of Technology
Designer: Xu Miao, Wang Hongyu, Fu Hao
Tutor: Chen Yang, Xing Kai, Tang Kangshuo, Giorgio Ponzo
Course Name: International Joint Design Studio
Finished Time: Jul. 2014
Exchange Institute: Delft University of Technology

ECO-MORPHOLOGY REORDERING

生态 - 形态的重组

Interfusion Canal-transportation Museum and urban redevelopment of Fengbei Relief Granary area

Master Plan 01

Conception of its future

基地未来定义

In June 2013, the Suzhou Municipal Government issued a scheme along the Pingjiang District to establish, develop and promote the development of urban tourism within the region by integrating demonstrative clutural activities. This scheme took the Pingjiang and Zhuozhen Yuan historical and cultural neighborhood as a starting point and the Fengbei Relief Granary is situated at the intersection of these two neighborhoods. However, due to the lack of relation to this scheme, the goverment has planned to revitalize the urban fabric whilst providing the opportunity for both locals and tourists to interact and create a lively social atmosphere for the city.

water urban system

major and minor connection

Reordering of Eco-Morphology

生态 - 形态的重构

This project proposes the reordering of ecomorphology which was based on the context of Suzhou. Using this site as a model for future intergration to the urban context, this proposal aims to express the ecological and morphological adaptations via its treatment of water canal system and restructuring the spatial configuration. In addition, this proposal also reinterprets the direct and indirect influences in order to reflect the local architectural atmosphere. Spaces like courtyards, laneways and building fabrics form an important element in organizing the surrounding spaces. By doing so, this proposal aims to define its relationship to the broader urban context in relation to economy and socio-political.

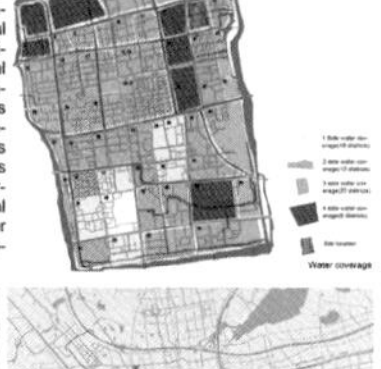

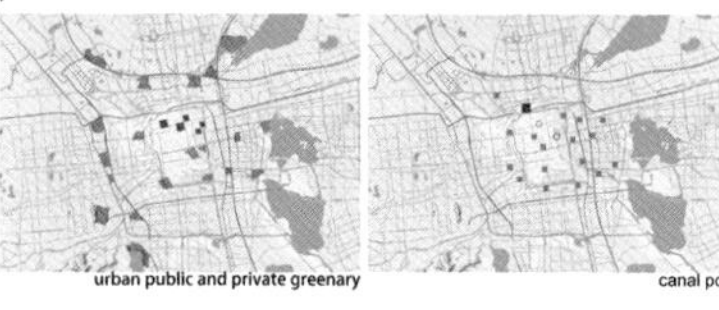

urban public and private greenary

canal port

History and characteristics of Fengbei

丰备义仓的历史与特色

The site for the project is the Fengbei Relief Granary area, which is bounded by Dongbei Street on the north, Zhang Family Estate on the west and Pingjiang River along with Pingjiang Street on the east. It was constructed in 1835 under the supervision of Lin Zexu. The graints were purchased from Wuxi and then stored in this granary in order to relieve famine. However, over the years, due to wars and social transformation, the Relief Granary gradually occupied as residential quarters and lost its original function. During the 1980s and 1990s, due to rapid economic growth and drastic urban changes, large amount of extension has been constructed, resulting in the loss of initial building envelope. Therefore, this project presents the opportunity to create a revitalized building which would celebrate the long history.

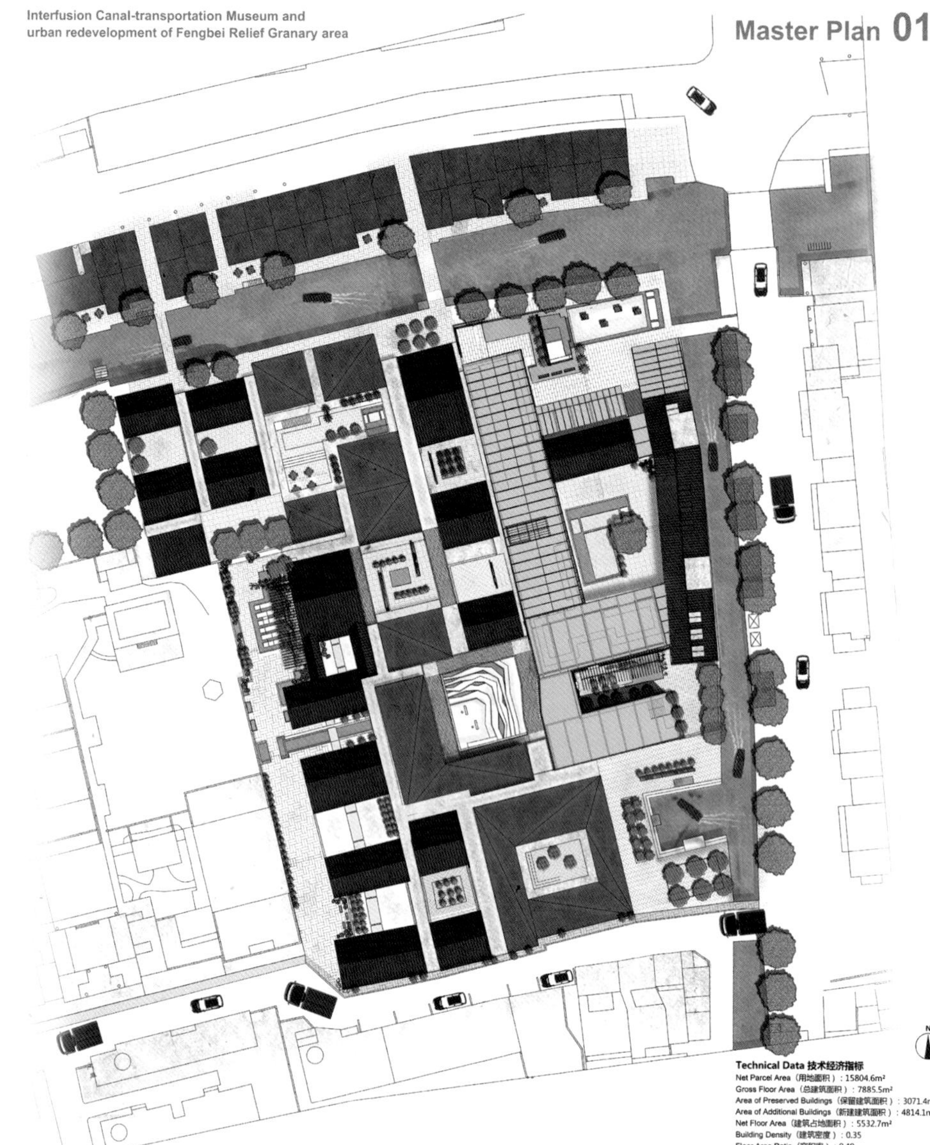

Technical Data 技术经济指标
Net Parcel Area（用地面积）：15804.6m²
Gross Floor Area（总建筑面积）：7885.5m²
Area of Preserved Buildings（保留建筑面积）：3071.4m²
Area of Additional Buildings（新建建筑面积）：4814.1m²
Net Floor Area（建筑占地面积）：5532.7m²
Building Density（建筑密度）：0.35
Floor Area Ratio（容积率）：0.49
Ratio of Green Space（绿地率）：9.53%
N

SITE PLAN 1:200

Landscape

景观

Proposed Pavement

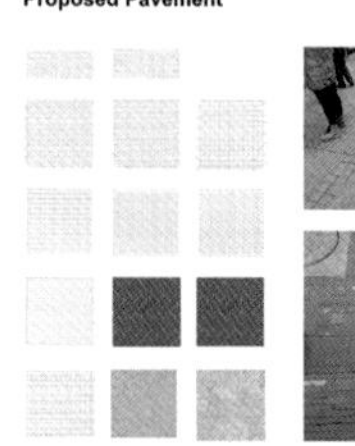

Proposed Lighting Strategy

Proposed Water Path

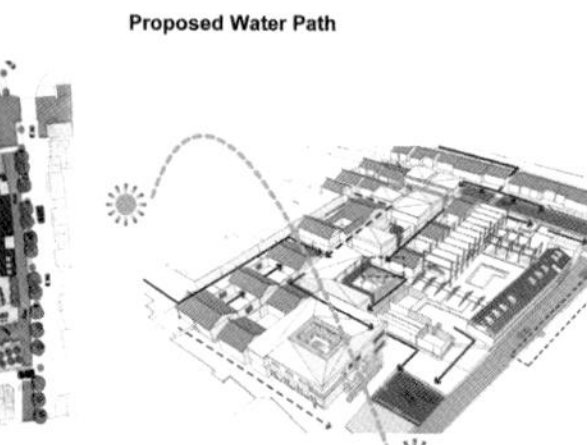

Landscape Concept

Suzhou celebrates an entensive canal network, abundance of water and a fairly flat surface. This gives Suzhou the potential to be a model of a water filtering system center for the surrounding gardens and mege water system. However, the site faces two major hurdles: under-utilized efficiency of water organization and a stagnant and polluted canal system. The desgn challenges the notion of typical rural-to-urban land transformation to instead create a model for integrated new city development. The dsign takes advantage of the site's agrarian economy, proximity to transit and urban amenities to create a village hat is compact, walkable and tightly-knit, yet has easy access to open space and maintains an agricultural character.

The success of the project depends heavily on addressing existing degraded environmental conditions by introducing treatment wetlands and reorganizing a historical canal network that will allow residents and visitors to directly engage with the water. As it exists now, the on-site canal network includes many dead-end segments and disconnected waterways, causing water to stagnant and depreciate in quality.

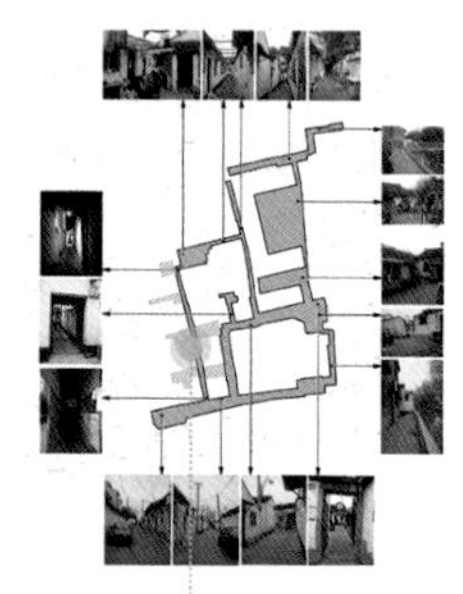

Urban Design

城市设计

The approach taken to design the urban fabric is to restructure the spatial configuration on the proposed site. By introducing a well-defined spatial arrangement, the proposal is able to provide the psychological and physical comfort in terms of visual connectivity and comfortable atmosphere as compared to the existing cluttered and poorly constructed spaces. The scheme also propose to integrate cultural related activities which would benefit Suzhou in terms of social, economic and political aspects, thereby revitalizing the urban fabric.

Contrast and Courts

The idea of contrasting spaces (e.g light, dark, narrow and wide) arises from the existing site condition. The proposal interpreted and morphosized the contrasting spaces by utilizing different materials, proportion and volumetric configuration to allow visitors to experience varying atmosphere. In order to reflect the varying atmosphere and celebrate the cultural importance of the city, the proposal proposed 4 main public courtyards and several private courtyards. The atmosphere of main courtyards are determined by the function of their immediate buildings whilst the private courtyards provide a place for contemplation and reflection.

Building Alteration

The intervention on the retained Wu Residence is to remove several existing walls to establish more connections between Zhang Family Estate and the proposed design. Also, the lane adjacent to site is transformed into a lively and interactive atmosphere by inserting featured landscapes to create a serene and calm promenade for visitors.

Canal Edge Revitalization

The canal edge is revitalized and activated by creating several pocket spaces amongst buildings. By doing so, this establishes semi-private enclosure which would activate the canal edge in order to create a lively atmosphere.

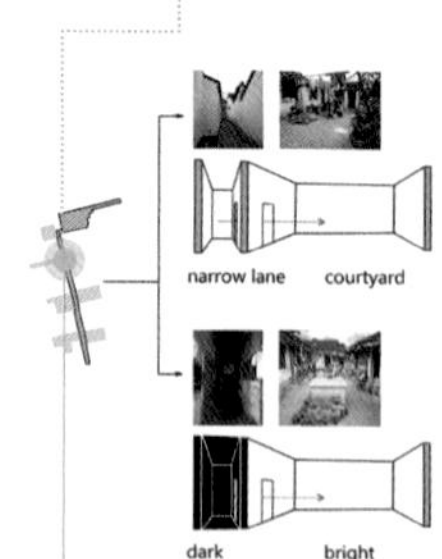

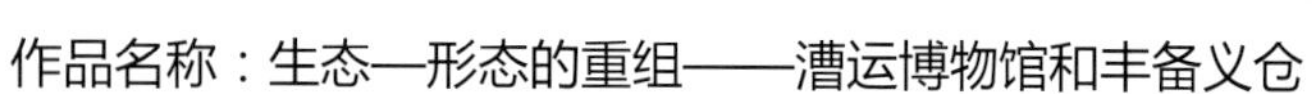

作品名称：生态—形态的重组——漕运博物馆和丰备义仓地区城市设计

Eco-Morphology Reordering

院校名：苏州科技学院

设计人：徐钰超，吕思扬，Esmonde Yap，Sarah Fayad

指导教师：胡莹，夏健，马骏华，许亦农，龚恺，王路，谢鸿权

课程名称：澳大利亚新南威尔士大学—东南大学—苏州科技学院—清华大学联合跨文化设计课

作业完成日期：2014 年 05 月

对外交流对象：澳大利亚新南威尔士大学

ECO-MORPHOLOGY REORDERING
生态 - 形态的重组

Interfusion Canal-transportation Museum and urban redevelopment of Fengbei Relief Granary area

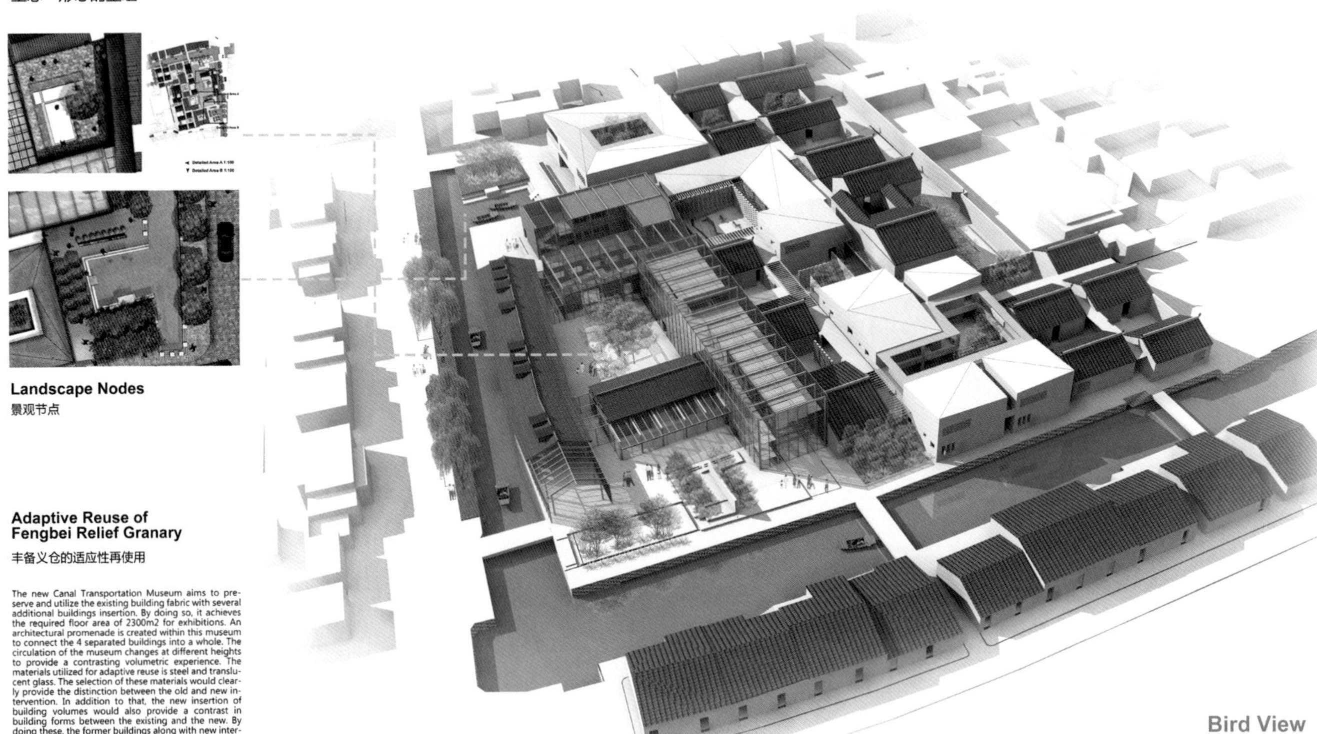

Landscape Nodes
景观节点

Bird View

Adaptive Reuse of Fengbei Relief Granary
丰备义仓的适应性再使用

The new Canal Transportation Museum aims to preserve and utilize the existing building fabric with several additional buildings insertion. By doing so, it achieves the required floor area of 2300m2 for exhibitions. An architectural promenade is created within this museum to connect the 4 separated buildings into a whole. The circulation of the museum changes at different heights to provide a contrasting volumetric experience. The materials utilized for adaptive reuse is steel and translucent glass. The selection of these materials would clearly provide the distinction between the old and new intervention. In addition to that, the new insertion of building volumes would also provide a contrast in building forms between the existing and the new. By doing these, the former buildings along with new intervention is morphosized together to provide a distinctive and ideal architectural language.

Circulation
流线

University: Suzhou University of Science and Technology
Designer: Xu Yuchao, Lv Siyang, Esmonde Yap, Sarah Fayad
Tutor: Hu Ying, Xia Jian,Ma Junhua, Xu Yinong, Gong Kai, Wang Lu, Xie Hongquan
Course Name: UNSW-SEU-SUST-Tsinghua Joint Cross-Cultural Design Studio
Finished Time: May 2014
Exchange Institute: University of New South Wales

木“扇”亭——当代“传统木榫卯”设计

木材的使用在中国曾是一种发展完善并且复杂的建筑语言，在包括住宅与文化建筑等各式各样类型的建筑中都有运用。近几十年来，中国的建筑材料文化在各种原因作用下发生了改变，人口的增长与现代化的急速扩张导致了大量的森林砍伐，得到木材的途径大大减少。

这种历史与技术发展轨迹的联系，以及当今越来越多可用的木材，使得我们的课程思考这样一个问题：什么是中国木构设计的新语言。此时，中国建筑师被摆在了一个特别的位子上，需要重新思考木材在建筑中的用途，并且为这种新的可持续材料发展一种革新的建筑语言。

案例学习
Case Study

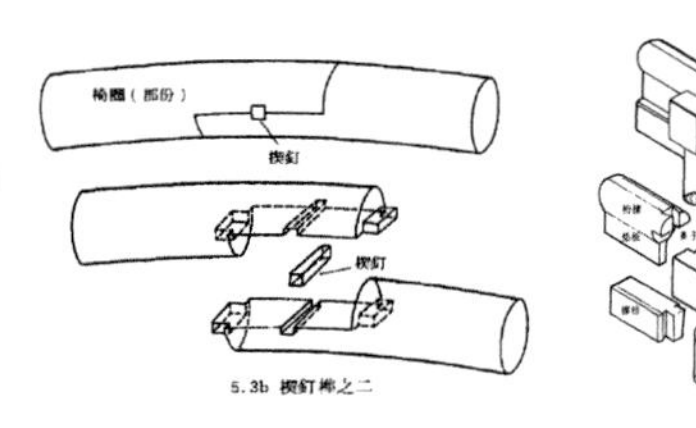
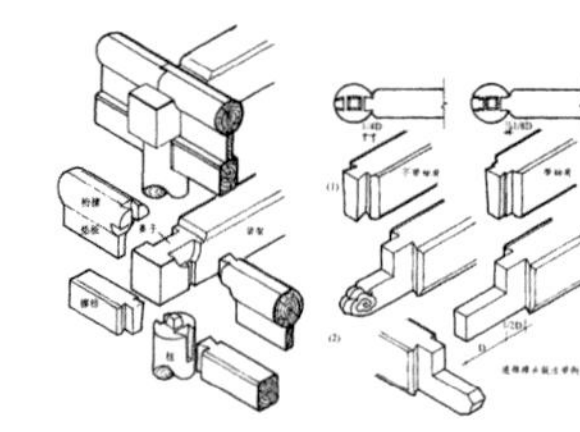

节点设计
Joints Design

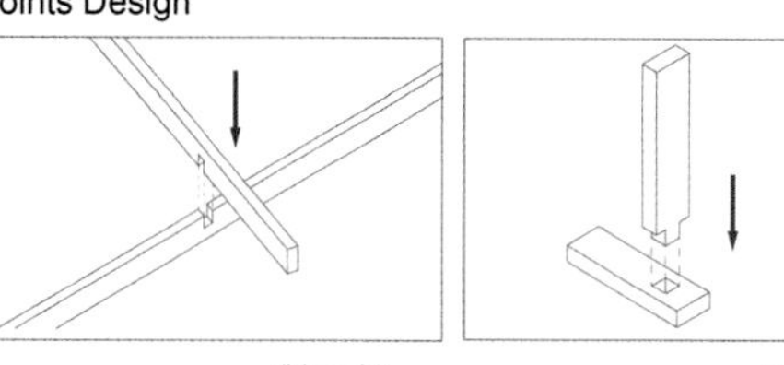
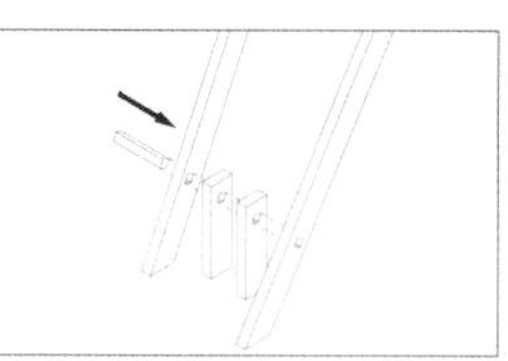
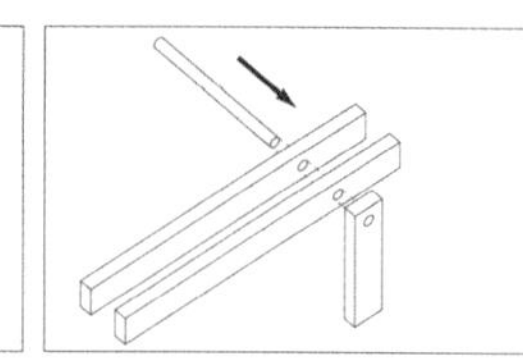

节点一：卡扣　节点二：榫头插接　节点三：方形串接　节点四：圆形串接

形态生成
Shape Generation

局部形体
typical perspective

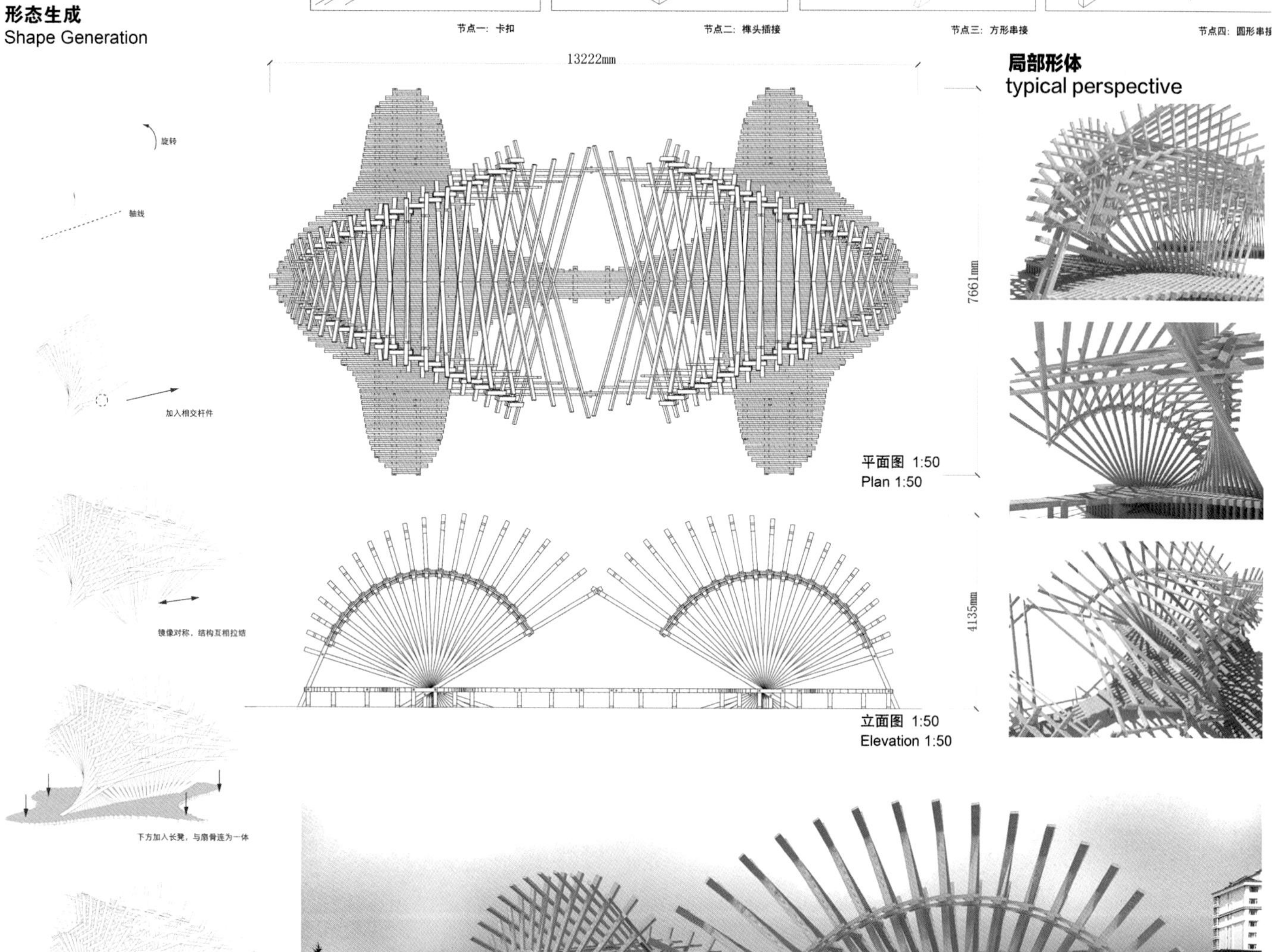

平面图 1:50
Plan 1:50

立面图 1:50
Elevation 1:50

作品名称：木“扇”亭——当代“传统木榫卯”设计
Wooden "Fan" Pavilion

院校名：东南大学
设计人：杨梦溪，王佳玲，Sara Maria
指导教师：韩晓峰，朱雷，Anna Lisa Meyboom，Blair Satterfield

课程名称：东南大学—加拿大不列颠哥伦比亚大学—加拿大木业联合设计（本科四年级）
作业完成日期：2014 年 08 月
对外交流对象：加拿大不列颠哥伦比亚大学，加拿大木业

Timber 'Fan' Pavillion——Contemporary 'traditional wood joint's design

Timber in China was once part of a developed and sophisticated architectural language. Timber was used in the construction of a variety of building types, including housing and cultural buildings. In recent decades, China has seen a dramatic decline in the use of timber. This change in China's built material culture has occurred for many reasons. Population growth combined with rapid modernization and expansion has resulted in deforestation and reduced access to wood. This nexus of historical and technological trajectories combined with the growing availability of wood, prompts Assemblage studio to ask the following questions: What is the new language of timber design in China? Chinese architects are uniquely positioned to re-consider the use of timber in architecture and to develop innovative languages for a new and sustainable material culture.

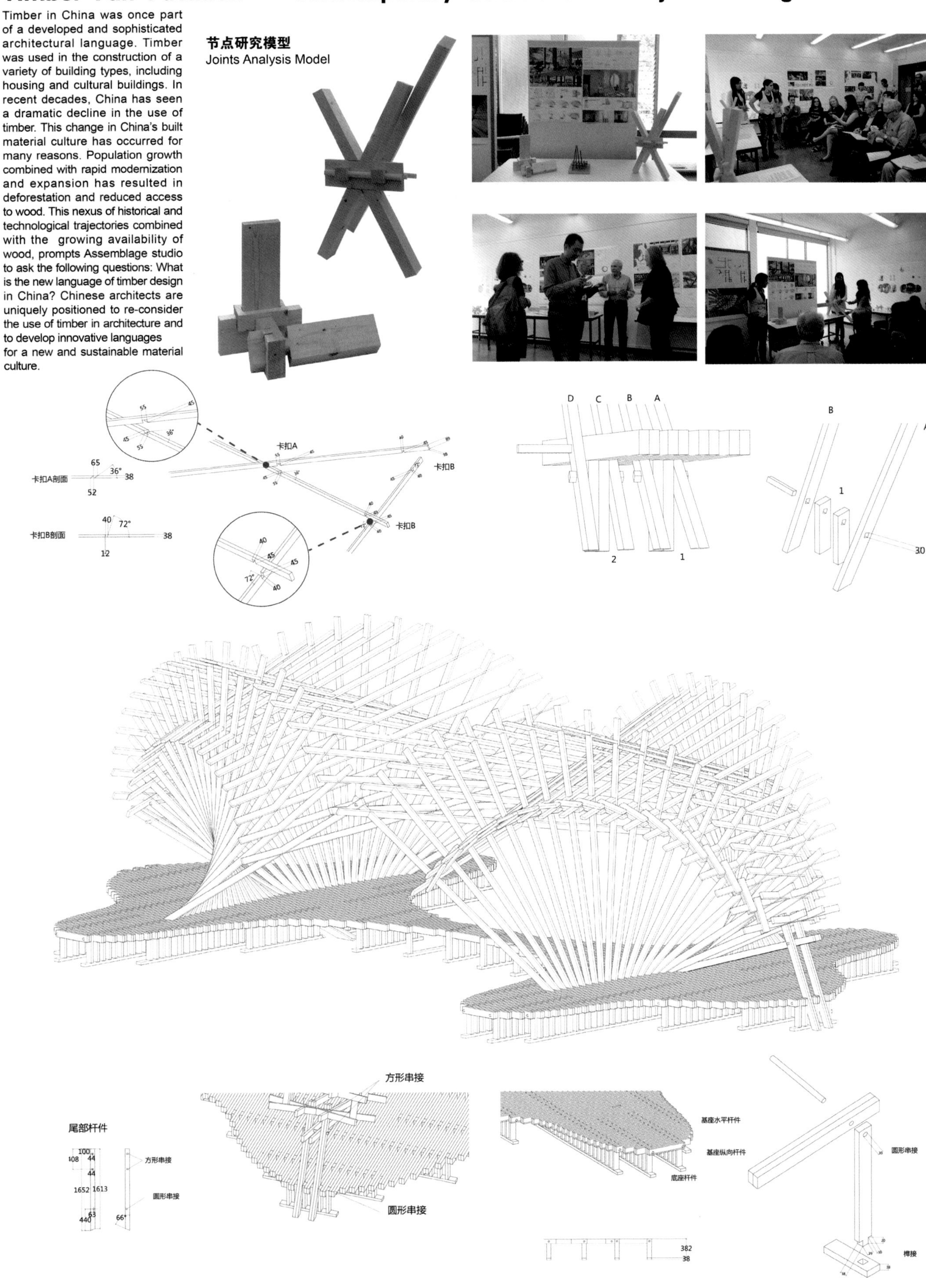

University: Southeast University
Designer: Yang Mengxi, Wang Jialing, Sara Maria
Tutor: Han Xiaofeng, Zhu Lei, AnnaLisa Meyboom, Blair Satterfield
Course Name: SEU-UBC-Canada Woods Joint Design and Construction Studio
Finished Time: Aug. 2014
Exchange Institute: University of British Columbia, Canada Woods

BRUNNEN HOTEL

Focus on structure

Lobby Rendering
大厅效果图

MODELS
模型

Model making is undoubtedly the best way to research for space. Structure from the previous discussion, to determine the type of structure and space gradual way. Most of the ideas are refined and produced in the model production process. Starting with triangle prototype structure, divergent a new manner of organic structure. Finally, combain transform triangle logic and cube logic to create a new structure form.

模型无疑是推荐空间的最好的方式。从前期结构形式的探讨，到逐步确定结构类型和空间方式，大部分思路都产生于模型的推敲和制作中。结构以三角形为原型出发，发散为有机的结构方式。将变形的三角形关系和立方体的结构逻辑结合衍生一种新的结构形式。

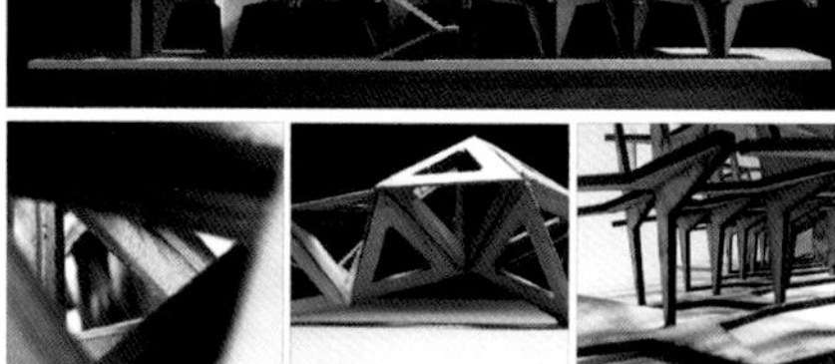

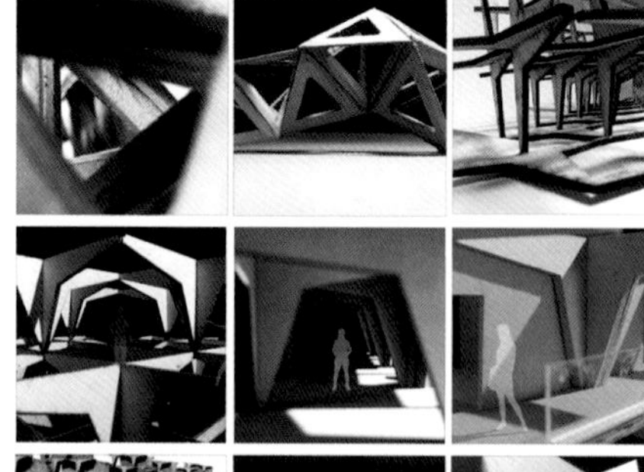

Fuction Parts　Corridor Parts

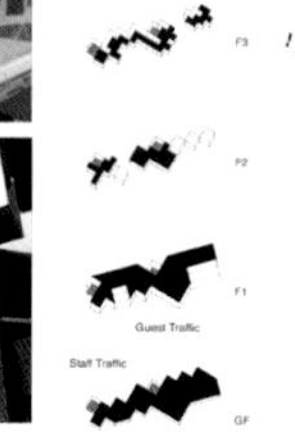

INSPIRED
灵感

Old Traditional Chinese Towns
古老的中国传统村落

Port
船港

Port is a traditional Chinese fishermen important traffic locations. It tandem with people's daily lives. Convenient transportation between the building has a better reachability.

港口是传统中国渔民重要的交通地点，它串联着人们的日常生活，便捷的交通关系，使建筑拥有更好的可达性。

TRIFFIC
交通

The whole village was built in the rock cliff hole, and then in the vertical transportation system built on cliffs. It runs through all the traffic behavior. This spatial relationship particularly large impact on me. I tried to experience this kind of traffic references to this scenario.

整个村落都建在悬崖的岩石洞中，然后在垂直的峭壁上建立交通系统，它贯穿所有人的交通行为，这种空间关系对我影响特别大，我试图将这种交通体验引用到这次方案中。

STRUCTURE
结构

Monastery in Shanxi still stands today on steep cliffs. Wooden structure system through the vertical and horizontal directions fixed at 90 degrees on a cliff. Feeling like the whole building is filled with the vitality of life, as if growing out from the rock. It also gave me a lot of notice.

在山西的悬空寺如今仍然屹立在陡峭的悬崖上，木结构系统通过垂直和水平两个方向固定在90度的悬崖上，感觉整个建筑充满了生命般的活力，好像从岩石中生长出来，这也给了我很大的启事。

CORRIDOR
走廊

Located in Chongqing's air suspension also built on a cliff above the road. Climbing the mountain throughout the corridors and on. Visitors can very close to experience the charm of nature.

位于重庆的空中悬道同样建在公路上面，整个走廊架设在山壁面上，游客可以很近距离体验到大自然的魅力。

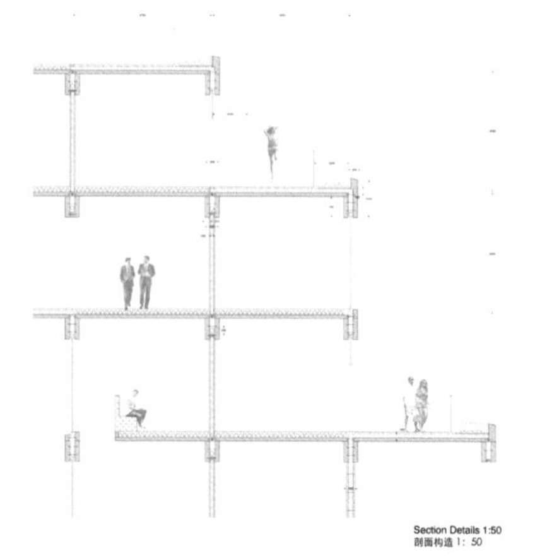

Section Details 1:50
剖面构造 1：50

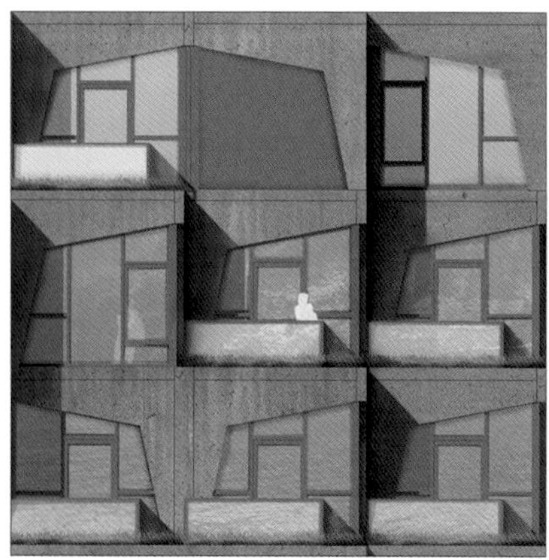

Facade 1:50
立面构造 1：50

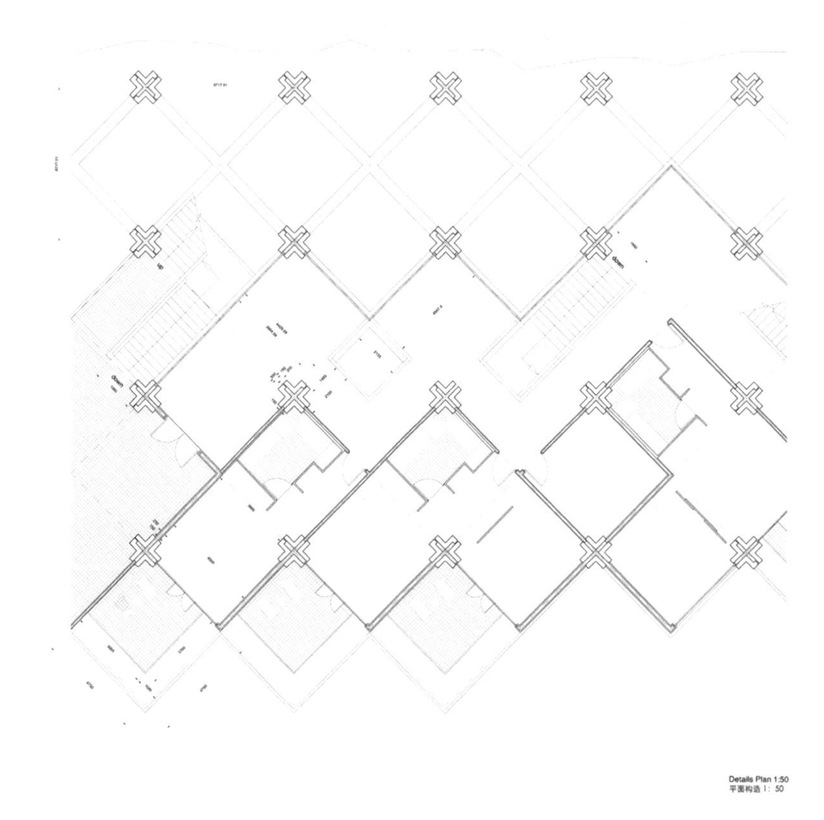

Details Plan 1:50
平面构造 1：50

作品名称：结构与空间探究
Focus on Structure-Hotel

院校名：中央美术学院
设计人：袁智敏
指导教师：Niklaus Graber，Bernhard Maurer
课程名称：酒店设计——结构与空间
作业完成日期：2013 年 07 月
对外交流对象：瑞士卢塞恩应用技术与艺术大学

BRUNNEN HOTEL

Site Sketch
基地速写

CONCEPT
概念

SITE
基地

Project is located in a small town in Switzerland, Brunnen. The city is dominated by tourism, there is a small amount of light manufacturing. Base is located in a suburban foot of the mountain, just across a road from the famous Brunnen Lake, has a good vision. Base was previously quarry, so backing surface with a large area was rocky. The project is between the rock and the lake.

项目坐落于瑞士一座小城Brunnen。该城以旅游业为主，有少量轻工制造业。基地位于城市郊区的一座山脚下，仅隔着一条公路便是著名的Brunnen湖水，有很好的视野。基地之前是采石厂，故靠山面有一片很大面积的岩石。项目位于岩石于湖水的中间。

STRUCTURE
结构

Conceived within a large area from the base sloping rock. This architecture system is supported entirely by ramp. Not only does each one structural unit and a relationship with the mountain, but also the entire building forms have a positive interaction with the environment. Let the structure grow out from the rock.

构想从基地内大面积倾斜的岩石出发。建立一个完全靠斜坡支持的结构体系。不仅每一个结构单元和山体发生关系，而且整个建筑形态也与环境产生积极的互动。让结构从岩石中生长出来。

FUNCTION
功能

To take advantages of the structure, so that the room scattered generate backward-style balcony, to get an organic facade effect. Between the users and the lake, there is an interesting conversation. Between the building and the mountain is abundant traffic space. Through the gap between the building and the mountain, it improves lighting conditions inside. In order to make users experience the corridor space well.

充分发挥结构的优势，让房间错落产生退步式的阳台，得到一个有机的立面效果。让使用者之间，湖水，发生有趣的对话。建筑与山体之间是丰富的交通空间，通过建筑与山体间的空隙，改善内部采光条件，使用者能更好体验山体与建筑之间的关系。

MATERIALS
材料

The whole structure system is made of concrete materials and structures exposed. So use warm wood materials to offset the concrete's sense of apathy. Facade is consist of entire walls, to ensure the overall sense of organic architecture.

整个建筑结构采用混凝土材料，且结构外露。以木材这种暖材料来中和混凝土的冷漠感。立面以整墙的处理手法，保证建筑有机的整体感。

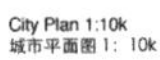

City Plan 1:10k
城市平面图 1：10k

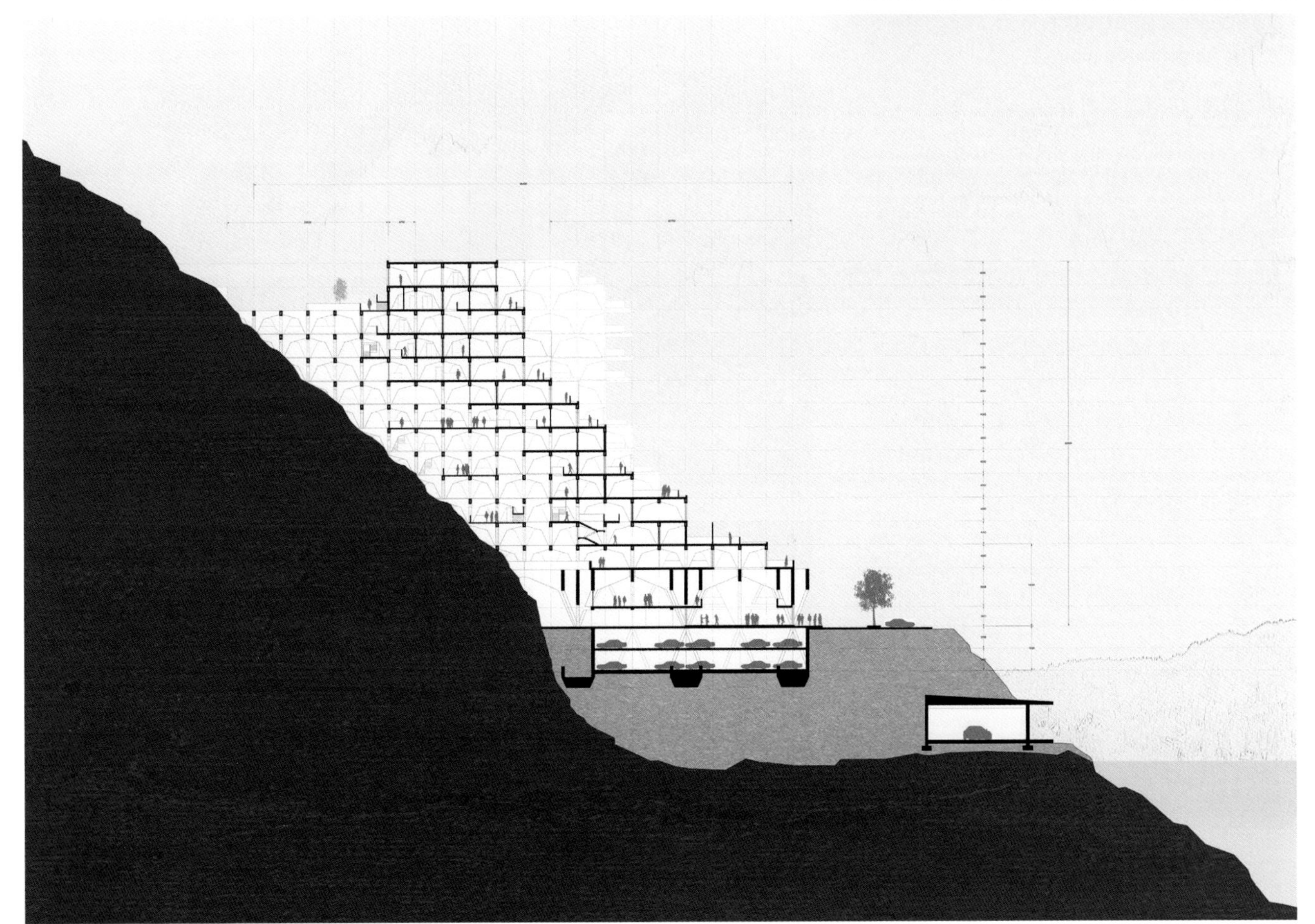

University: Central Academy of Fine Art
Designer: Yuan Zhimin
Tutor: Niklaus Graber, Bernhard Maurer
Course Name: Focus on structure-hotel
Finished Time: Jul. 2013
Exchange Institute: Hochschule Luzern

Evolution of The Block

SITE STUDY

EVOLUTION PROCESS

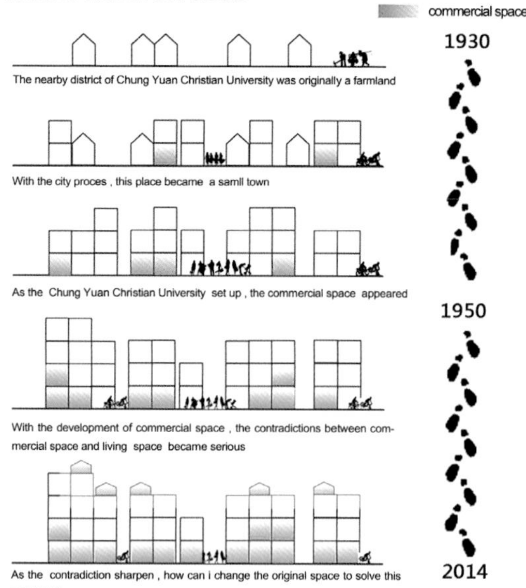

commercial space
1930
The nearby district of Chung Yuan Christian University was originally a farmland
With the city proces , this place became a samll town
As the Chung Yuan Christian University set up , the commercial space appeared
1950
With the development of commercial space , the contradictions between commercial space and living space became serious
As the contradiction sharpen , how can i change the original space to solve this problem ?
2014

RECORD OF THE BLOCK

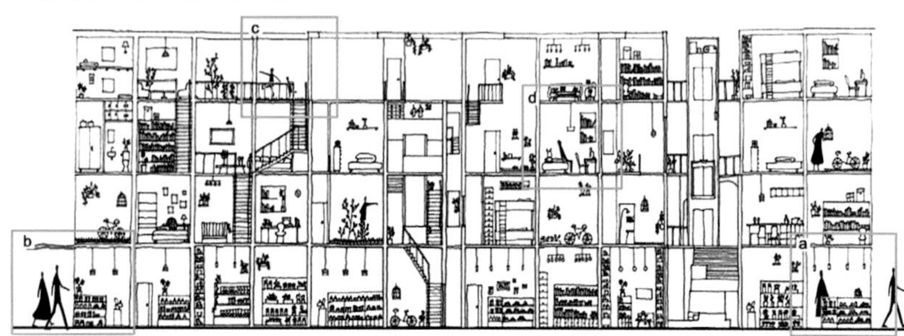

WHAT SHALL WE DO ?

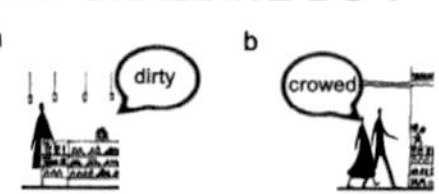

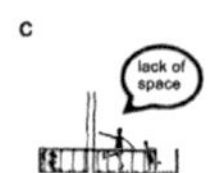

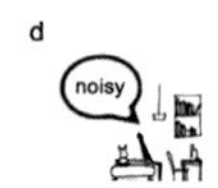

The research object is the nearby regions of Chung Yuan Christian University in the city of Chung Li of TaiWan, China. The nearby district of Chung Yuan Christian University was originally a farmland, and it is gradually taking shape with the foundation of Chung Yuan Christian University.

The earliest block gives priority to residence, and gradually produces a large amount of business because of the nearby college students' requirements. With the development of commercial space, the intermixing mode of commerce and residence becomes the dominant mode of the district. In the process of block evolution, many problems were exposed, such as, environmental deterioration, night market noises, crowded living space, and lack of public communication space, and so on. I hope to make the blocks become more reasonable in the intermixing mode of residence and commerce by further design in the process of researching the block evolution from the mode of residence to the intermixing mode of residence and commerce.

CONCEPT & RESEARCH

A Buildings on the roof

1. To increase living space and alleviate the living pressure of the blocks;

2. To increase the living space on the roof and move the living space up to reduce the surrounding noises of the commerce ;

3. To increase space on the roof for storage and alleviate the crowded living space pressures and reduce the disadvantages of storage in the old house;

4. To increase space on the roof to meets people's requirements of communication, create the roof platform and set the teahouse, pergola as rest space to meets people's requirements of making friends and having parties in the night market area with high density;

5. Increasing space on the roof meets the surrounding students' requirements of renting house and alleviates the pressures of renting house, and it makes preparation for students far away to contact with the local people and melt themselves into the life of central plain.

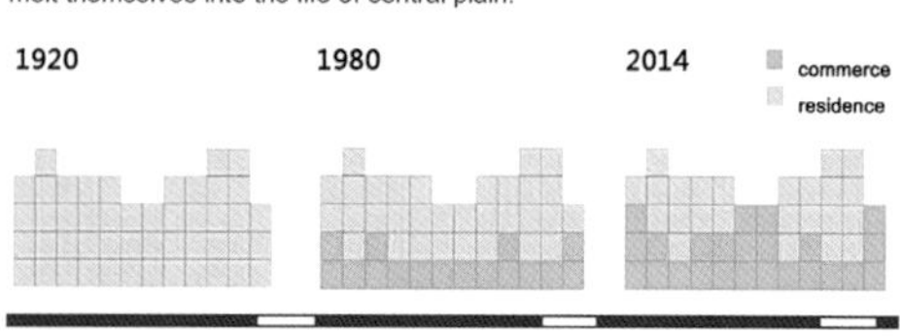

Construction process

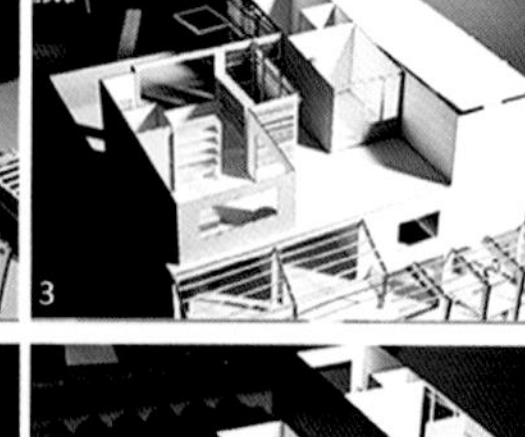

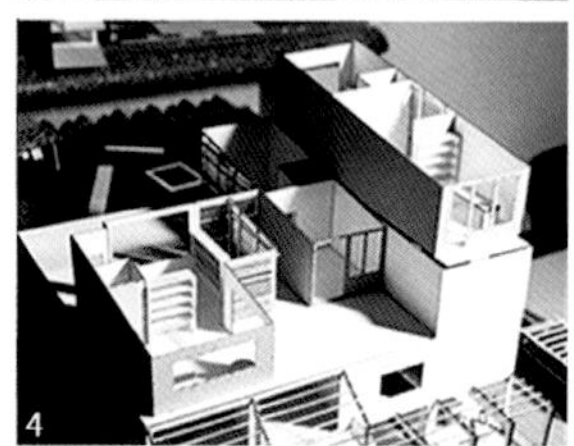

B Expansion on the exterior wall

1. To shield off wind, rain and sunlight and produce a cool environment.

2. The changing space of the new arcade meets people's different requirements and enriches the night market with the combination of the signature.

3. Break the bondage of a single isolated family and produces a shared space for the customers, it presents the modern consciousness of mutual respect. Walking under the arcade, you feel free and comfortable, cozy and close, neat and tidy with no dust on your feet. And it shows a harmonious and interactive interpersonal relationship.

4. There is a rich flavor of life in the arcade; and it becomes a place for drinking tea, chatting, enjoying the cool, meeting the customers, exchanging information and sleeping in the cool night.

Expansion type

作品名称：街区进化论——基于中坜市中原校区周边街区的演变研究
Block Evolution Theory

院校名：哈尔滨工业大学
设计人：张睿南，董奕兰，陈聪
指导教师：董宇，薛名辉，邵郁，张姗姗，杨秋煜，黄承令，黄俊铭
课程名称：开放式研究型建筑设计（本科四年级）
作业完成日期：2014 年 03 月
对外交流对象：中原大学

Evolution of The Block

A Buildings on the roof

Model A

Communication mode of roof space provides communication space for the residents in the old house.

Model B

Living mode of roof space opens up new residential places with quiet environment on the roof and alleviates the pressures of crowded living space;

Model C

Storage mode of roof space, in the meantime, set the pergola, pet house, and dove nest, etc.

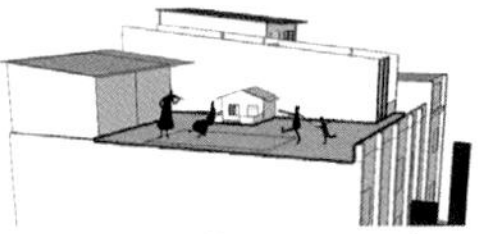

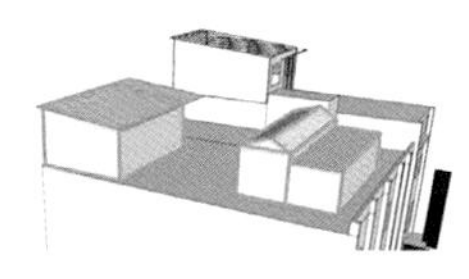

Model A

Model B

Model C

Design for different people (I design three model , as the users' need ,they can combine them at will)

a. For children

Playing space + Quite space

As for children ,they need spacious and quite space to play and sleep well . So I combine the Model A and Model B to make children have a happy life in the special home on the roof .

b. For housewife

Communication space + Storage space

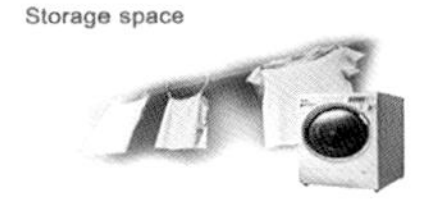

As for housewife ,they need storage space and communication space to do the housework and make full use of spare time . In view of their spare time , I use the Model A to give them spacious space to communicate with others.

c. For the young

Quite space + Communication space

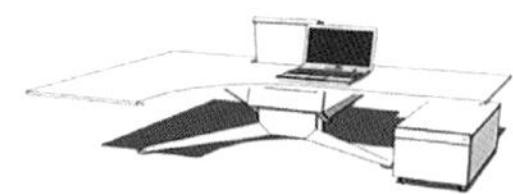

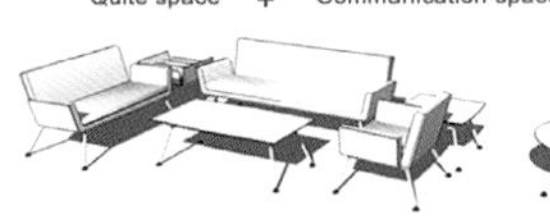

As for the yong , maybe they have a lot of work to do after work in home ,so they need quiet space to work .However , they also need communicate with the same age ,then I design the Model A and Model B for them .

B Expansion on the exterior wall

Policy A

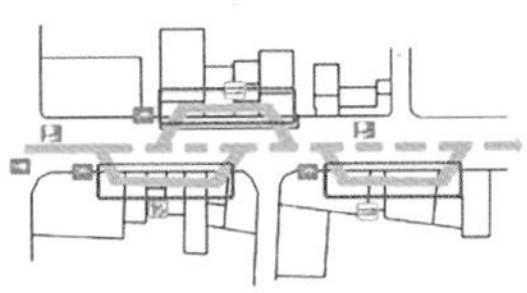

Making people walk under the arcade and reserving appropriate places for the roadside vendors to direct traffic.

Policy B

Expanding outdoor space for having rest on the exterior wall to extend space for teahouse and restaurant.

Policy C

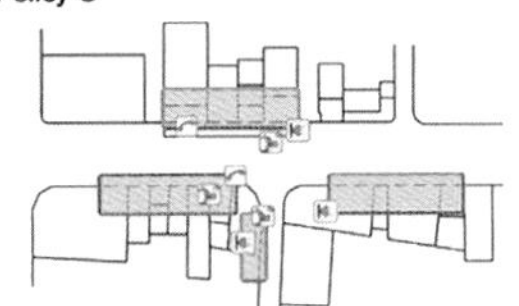

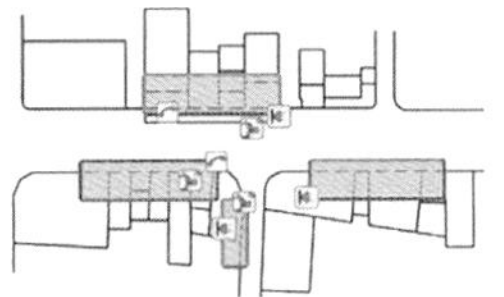

To combine space expansion with the signature, and it enriched facade wall without affecting the sight.

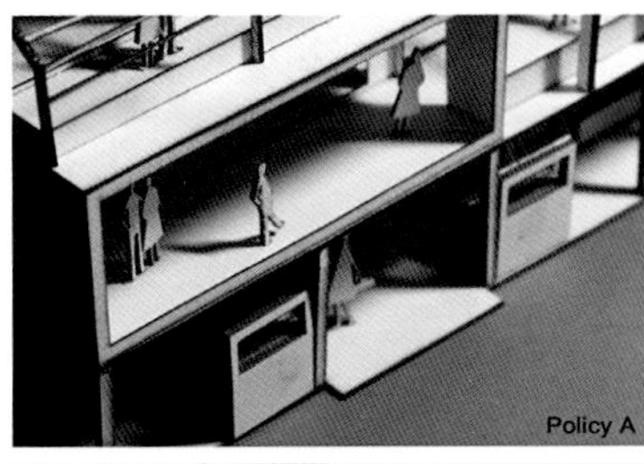

Policy A

Solve different problems (I put forward original policies , in viewof the different problems , I combine them on purpose to solve problems)

Problem a

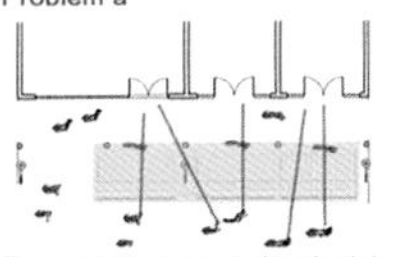

The roadside vendors randomly putting their goods in front of the arcade in traditional night market seriously affected the business of the fixed stores behind the arcade, and it is inconvenient for the pedestrians to go into the arcade.

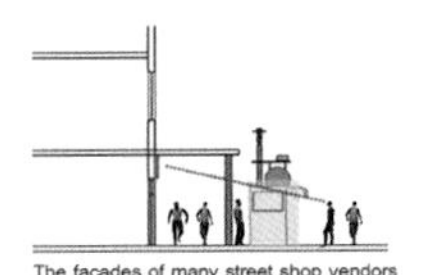

The facades of many street shop vendors block people's sight to see the signboards behind, the garbage produced by the snack vendor made a bad influence to the arcade.

Solution : policy A +C

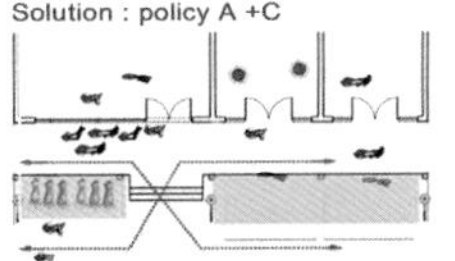

To plan the positions of street vendors to a corresponding fixed position, and reserve space for parking haulage motor.

To raise the height of the arcade and seperate street shop vendors from the arcade space and make the pedestrians see the signboard of the fixed shops behind.

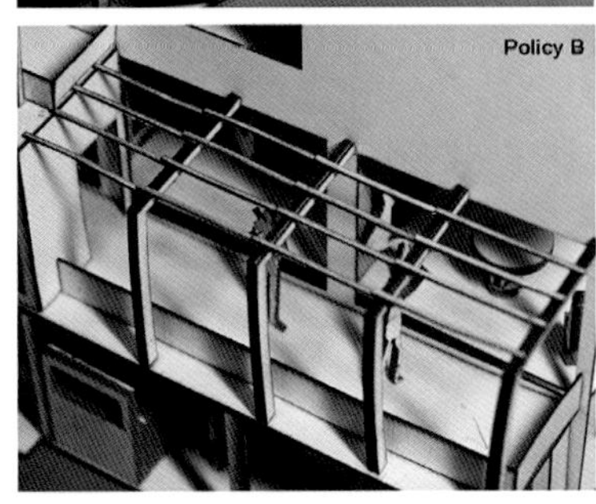

Policy B

Problem b

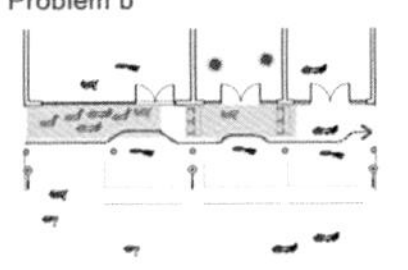

The space under the arcade is occupied by the temporary stalls and it is inconvenient for pedestrians to walk through the arcade.

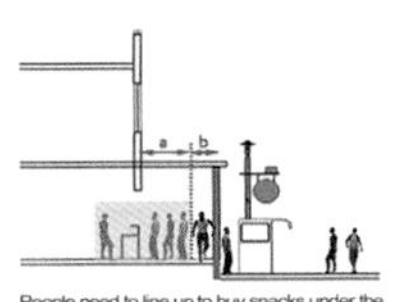

People need to line up to buy snacks under the arcade, but the crowd who buy the snacks seriously affects the traffic under the arcade.

Solution : policy A

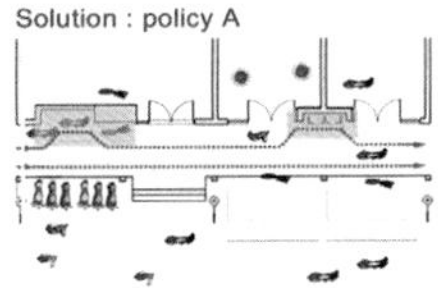

To reserve space for the temporary stalls and also for the people who line up to buy the snacks.

There is a change in the width of the arcade space after the adjustment, and it alleviate the chaotic stream of people under the arcade.

Problem c

Some of the signatures on the second-floor are ignored by the pedestrians because of the restrictions of the arcade.

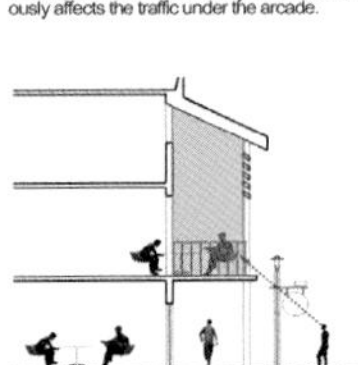

To utilize upper space of the arcade to expand exhibition space for second-floor commerce and make it become a live signature.

Solution : policy B+C

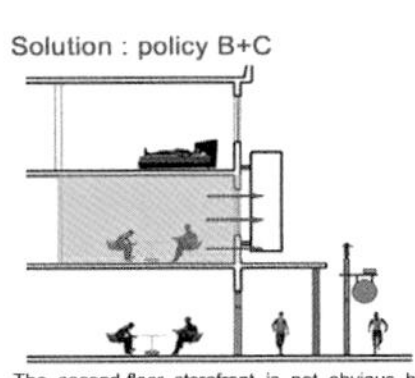

The second-floor storefront is not obvious because of lack of commercial space.

However, to use the expanding space over the arcade, the storefront can be obvious.

Policy C

CONCLUSION

University: Harbin Institute of Technology
Designer: Zhang Ruinan, Dong Yilan, Chen Cong
Tutor: Dong Yu, Xue Minghui, Shao Yu, Zhang Shanshan, Yang Qiuyu, Huang Chengling, Huang Junming
Course Name: International Joint Design Studio
Finished Time: Mar. 2014
Exchange Institute: Chung Yuan Christian University

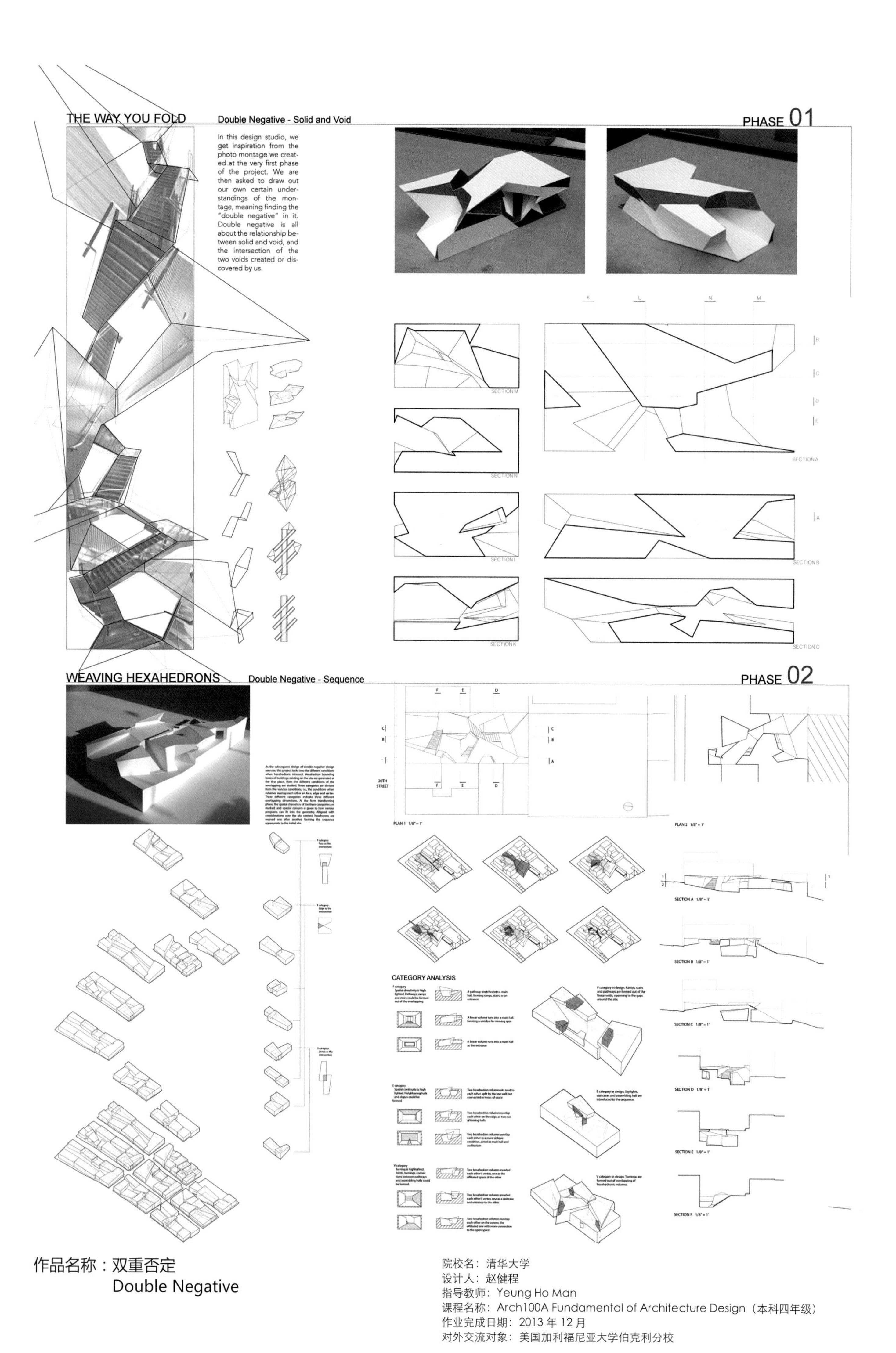

作品名称：双重否定

Double Negative

院校名：清华大学

设计人：赵健程

指导教师：Yeung Ho Man

课程名称：Arch100A Fundamental of Architecture Design（本科四年级）

作业完成日期：2013 年 12 月

对外交流对象：美国加利福尼亚大学伯克利分校

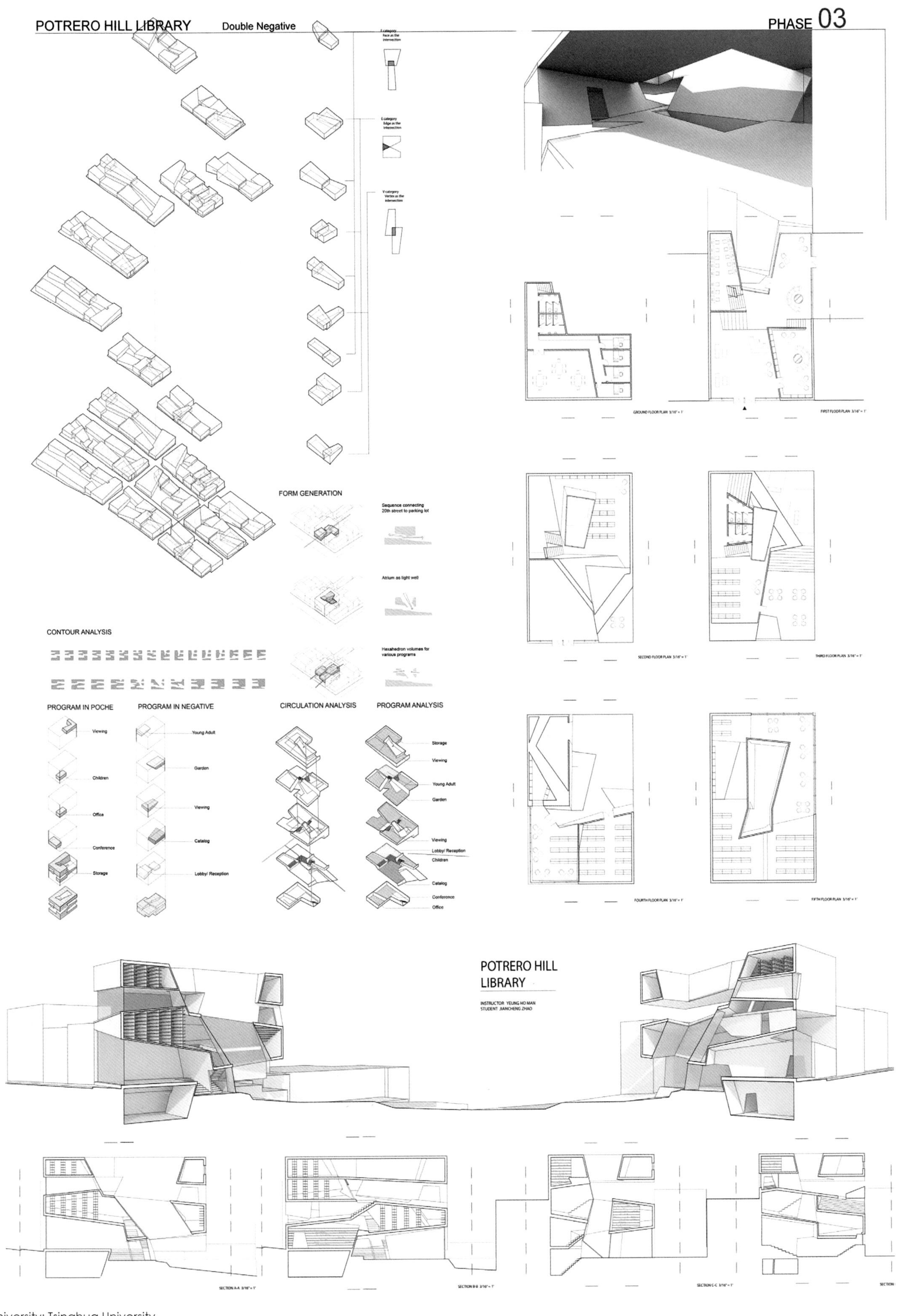

University: Tsinghua University
Designer: Zhao Jiancheng
Tutor: Yeung Ho Man
Course Name: Arch100A Fundamental of Architecture Design
Finished Time: Dec. 2013
Exchange Institute: UC Berkeley

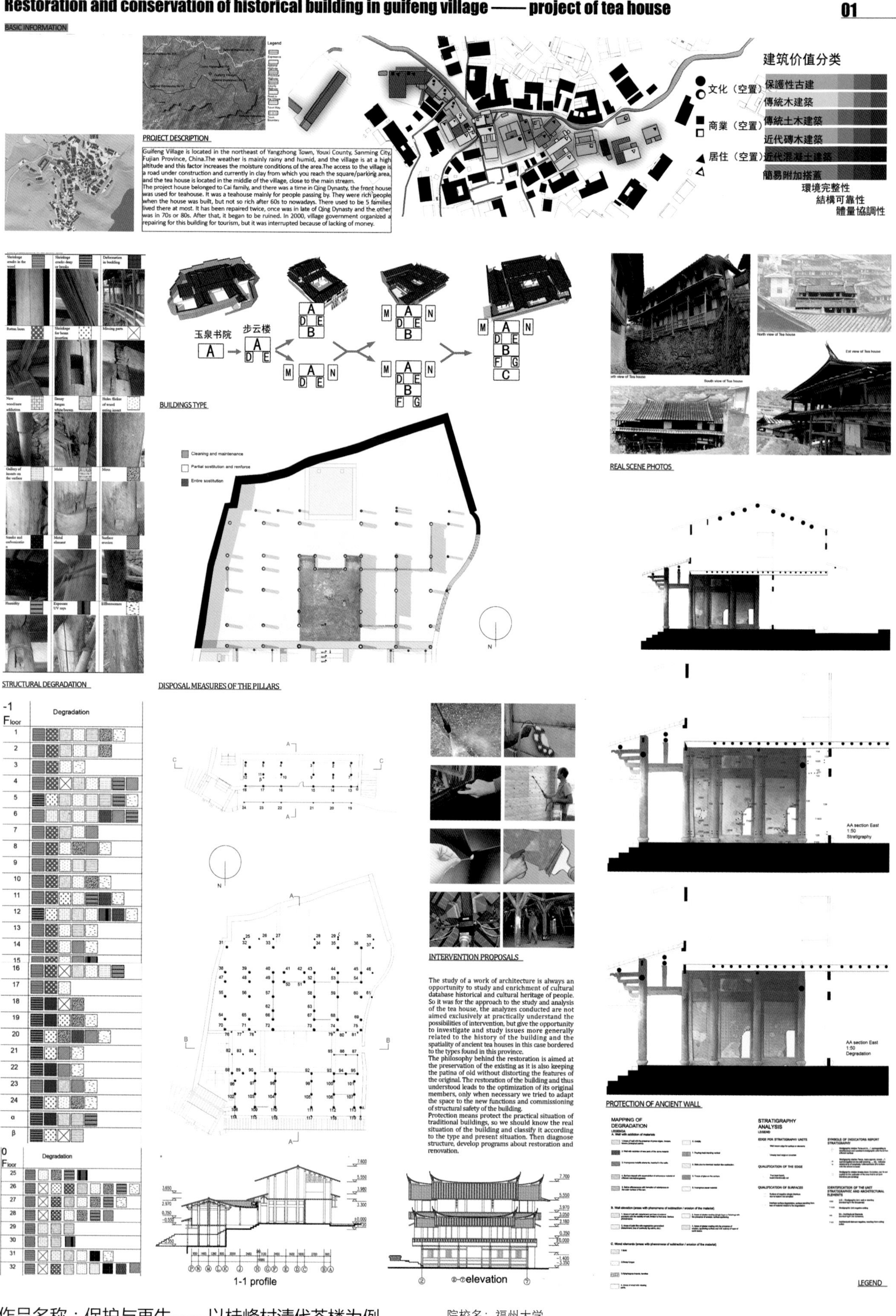

作品名称：保护与再生——以桂峰村清代茶楼为例

The Protection and Regeneration of Traditional Architecture — Teahouse in Guangxi Village in Qing Dynasty as an Example

院校名：福州大学
设计人：Claudia Stancanelli，柏苏玲
指导教师：张鹰
课程名称：福州大学—威尼斯建筑大学联合设计课程教学
作业完成日期：2014 年 05 月
对外交流对象：意大利威尼斯建筑大学

Restoration and conservation of historical building in guifeng village —— project of tea house

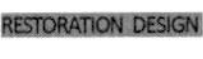

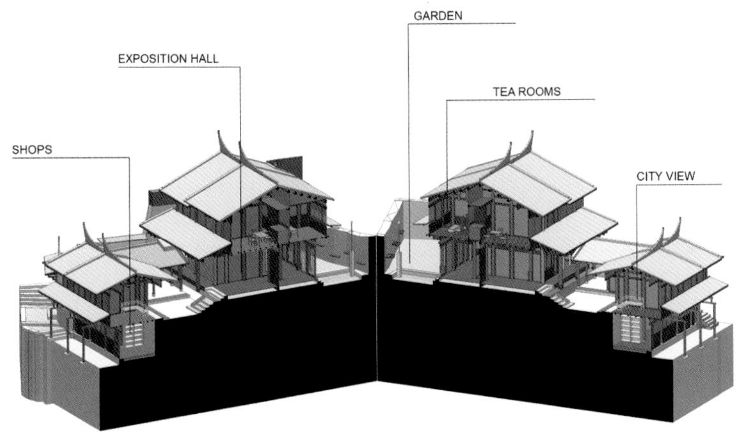

FUNCTIONAL DIVISION

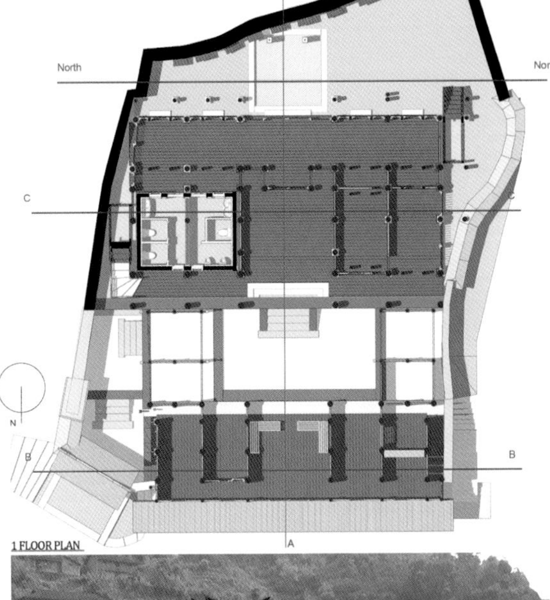

1 FLOOR PLAN

VIRTUAL SCENE OF TEAROOM

RENDERING

PHOTOS

The purpose of the restoration of the tea house is due to the importance and strategic location of this within the village for tourism purposes and for the promotion of local products of the village. The intent is to make this house a home manifesto for the restoration and restauration for the good of the old buildings of wood around the village . Being a predominantly agricultural village and having experienced firsthand the dishes of the place there is no doubt that the tea house should have the function of restaurant but most of all it should regain its original function as a tea house and a meeting place of passage. On the first floor there will then function as the rooms are more and divisions intact.

On the second floor there will then function as the tea rooms are more and divisions intact. On the ground floor underside of the tea rooms will be used as exhibition space, the topics will vary from the history of Chinese tea to the beautiful landscape paintings of the personalities in the history of the village from the same village, the space in front is clear of furniture and seating to observe the village overlooking the Shen Gui creek. The floor below ground will be used as shops selling local crafts and tea.

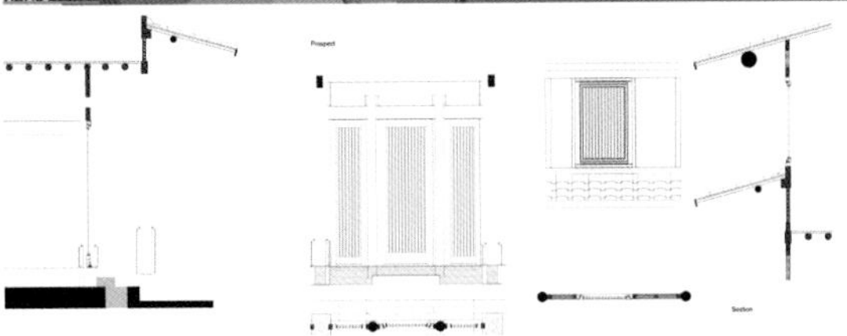

STRUCTURE NODE

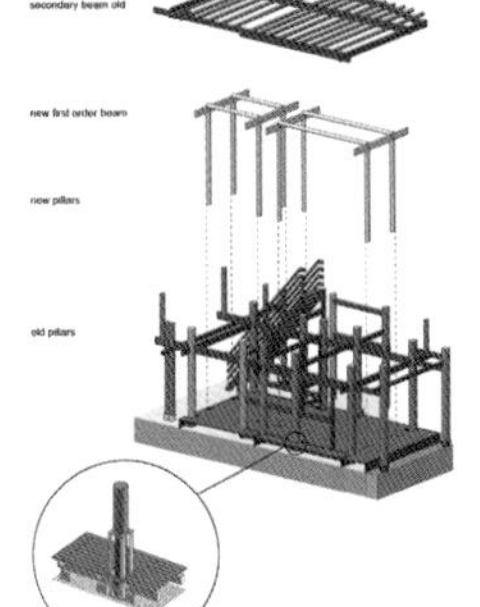

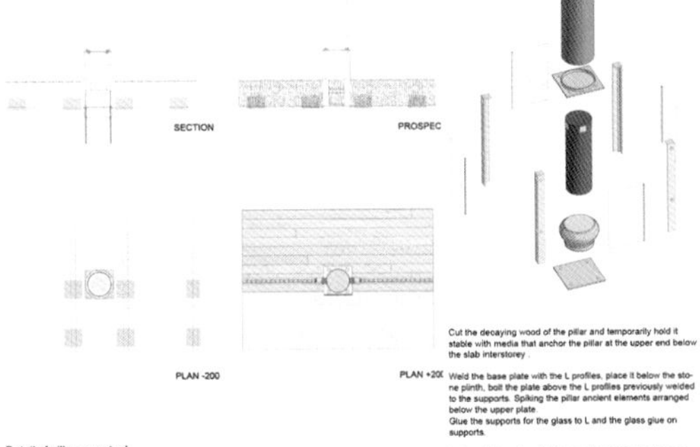

NEW STEEL OF PILLAR

A-A SECTION

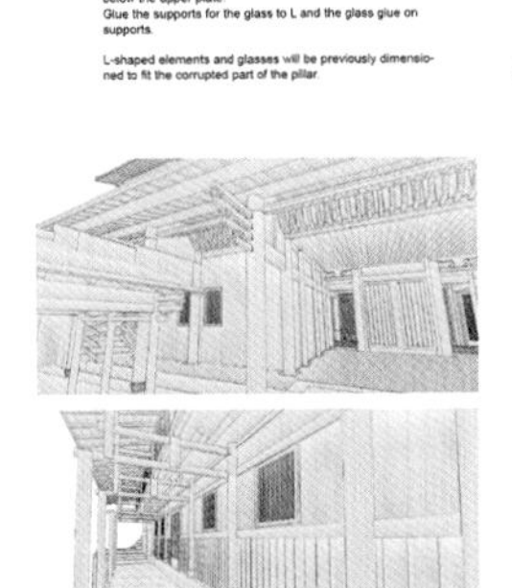

INTERIOR SCENE MODEL

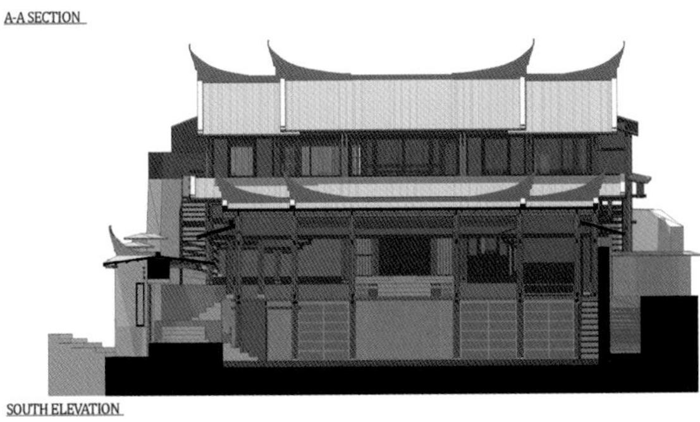

SOUTH ELEVATION

University: Fuzhou University
Designer: Claudia Stancanelli, Bai Suling
Tutor: Zhang Ying
Course Name: Fu Zhou - IUAV University Joint Design Studio
Finished Time: May 2014
Exchange Institute: IUAV University

MAPPING THE VOID II

映射无形 II--2013

A study of the "Edge" and the Port of Aarhus

Taking as our artistic expression and method the walks along the edge of the port. The walks were to be considered as works and narratives,which thematise the linear motif in the intersection of the horizontal and the vertical. Simultaneously they were to function as mapping devices and measuring tools;performative instruments with the ability to archive and reconstruct spatial experiences and sensations.

The first walk-entitled"guided walk drift"-Took place on 23 September
The second walk-entitled"walk in one continuous line"-Took place on 3 October
The third walk-entitled"walking montage"-Took place on 14 October

The assignment was inspired by works, methods and techniques applied in Performance art, Minimal art and Land art.

G2: CABINET OF CURIOSITIES
"I AM SO HUNGRY....."EVERYTHING SEEMS DIFFERENT""ARE YOU SCIENTISTS?"

奥胡斯港口与边界研究

这次课题将行走作为一种艺术表现和方法，沿奥胡斯的港口边缘进行了一系列的集体行走实验。将行走看作是一种艺术作品和记述过程，它将线性的主题归结为一种横向与纵向的交错和延伸。同时行走也是映射的手段和丈量的工具，是一种演示性的参照，附带有重组空间的经验和感受。

行走一：引导与漂流（2013.09.23）；
行走二：线之旅一编舞的行走（2013.10.03）；
行走三：行走蒙太奇（2013.10.14）

行走的想法来源于表演艺术，极简艺术和大地艺术的作品、方法和技巧。

第三组：分类学
Olga Sigporsdottir;Rosemary Jeremy; 王金璐；段邦禹

SEVEN INTEREST POINTS
七个兴趣点

About Bondray
关于边界

The Way of WALK 行走方式

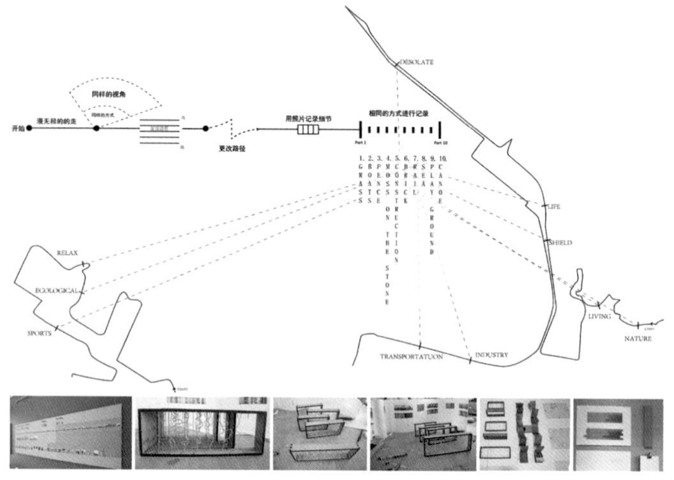

FIRST WALK--Free Walk 第一次行走 — 自由行走

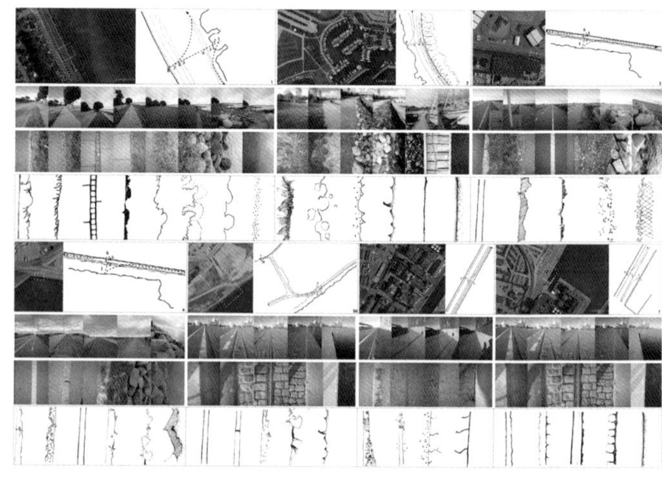

SECOND WALK--Special Walk--Walk on line 第二次行走 — 排队行走

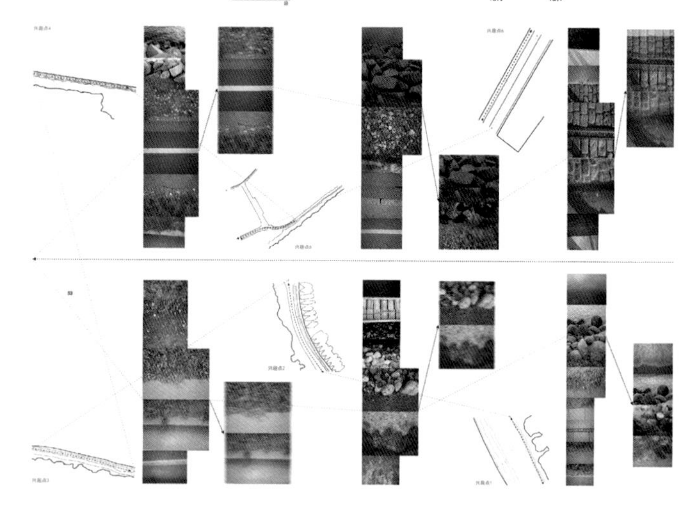

THIRD WALK--Team Walk 第三次行走 — 团队行走

作品名称：分类学——边界重构
Taxonomy

院校名：中央美术学院
设计人：段邦禹，王金璐，Olga Sigporsdottir，Rosemary Jeremy
指导教师：何可人，王威，Anne Elisabeth Toft，Claudia Carbone
课程名称：2013 中央美术学院与奥胡斯建筑学院联合课题
作业完成日期：2013 年 10 月
对外交流对象：丹麦奥胡斯建筑学院

Detail of textures

Wang Jinlu

exploration of textures and the emotions they provoke. the relationship between time and these textures, influenced by the ratio and extent of one texture to another.

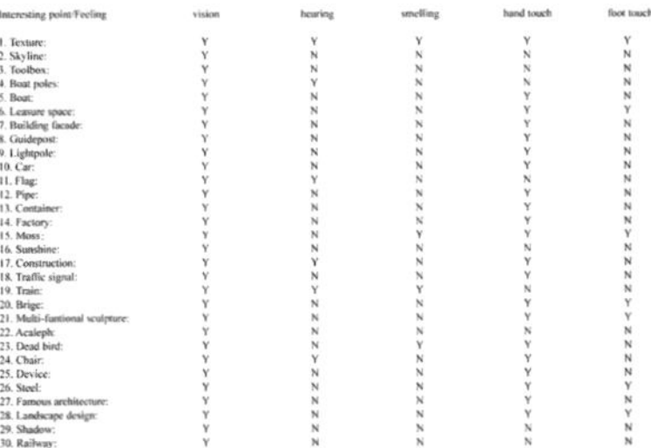

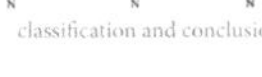

cooperation

1. Texture:
There are so many different textures from the land to the sea, from the manmade to nature. The different colors and gradual change attracts me so much, and all kinds of feeling of my foot, my hand, my nose, my ear and my eye.
2. Skyline:
The skyline on the sea is so flat, and it is seems that the cloud is too low to touch it by hand.
3. Toolbox:
It is heard that these boxes is used to carry the tools for fishing and sailing.
4. Boat poles:
When I see the vertical poles which break the horizontal skyline, the different direction and the white-blue colors are interesting for me.
5. Boat:
This is my first time to meet so many personal boat, there are inverted reflection in sea, many seagulls fly around, the boat poles make a silivery sound.
6. Leasure space:
I don not think of the table and chair be emerged nearby the wooden trestle, it is so harmnization.
7. Building facade:
The building facade nearby the dock are colorful and warm, compared with the cool sea.
8. Guidepost:
The color blendent, position, form, layout are all different from chinese.
9. Lightpole:
The open space in the harbor make the lightpole prominent- high and thin, the design of it is simple and brief.
10. Car:
The yellow color of the car here is so prominent in this kind of enviroment.
11. Flag:
I love these flags not cause the color but the cute design when I see them first sight.
12. Pipe:
Huge scale pipes on the grass are shocking.
13. Container:
Many different color containers are superposition together,like a series of architecture.
14. Factory:
The gray surface building looks like a super mueum.
15. Moss:
A lot of circle moss are green and soft.
16. Sunshine:
The sunshine on the building form the abundant shadow which are changing according to the time.
17. Construction:
Membrane structure construction is so cool, which is placed beside the house and sea.
18. Traffic signal:
The color and design are different from chinese.
19. Train:
The train is beautiful.
20. Brige:
The wooden brige is made so delicate.
21. Multi-funtional sculpture:
Many people are attracted by the color and the form of the sculpture, it can be sitted and standed carried something.
22. Acaleph:
Light and transparent body is floating in the sea.
23. Dead bird:
The bird which has white feather is laying on the ground, it is so hard when we touch it.
24. Chair:
Orderless wooden chairs stands in a sand hill.
25. Device: A huge size glass is putting faced the sea, and the pink-black color and the form attract many people to take photo with it.
26. Steel: I find them beside the area which is under construction.
27. Famous architecture: It is hoared that here is a famous architecture. When I get approach of it, the triangle roof, window and blue glass are interesting to this enviroment, the white pure space and the delicate public service facilities attract me so much.
28. Landscape design: The brief design around the buildings creat a pure and peace space for people.
29. Shadow: The shadow on the steel stairs which has a lot of holes is so amazing .
30. Railway: It is my first time to see the railway above it, and the color and the trace are beautiful.

discribing the feeling during the walk

Interesting point/Feeling	vision	hearing	smelling	hand touch	foot touch
1. Texture:	Y	Y	Y	Y	Y
2. Skyline:	Y	N	N	N	N
3. Toolbox:	Y	N	N	N	N
4. Boat poles:	Y	Y	N	N	N
5. Boat:	Y	N	N	Y	N
6. Leasure space:	Y	N	N	Y	Y
7. Building facade:	Y	N	N	Y	N
8. Guidepost:	Y	N	N	Y	N
9. Lightpole:	Y	N	N	Y	N
10. Car:	Y	N	N	Y	N
11. Flag:	Y	Y	N	N	N
12. Pipe:	Y	N	N	Y	N
13. Container:	Y	N	N	Y	N
14. Factory:	Y	N	N	Y	N
15. Moss:	Y	N	Y	Y	Y
16. Sunshine:	Y	N	N	N	N
17. Construction:	Y	Y	N	Y	N
18. Traffic signal:	Y	N	N	Y	N
19. Train:	Y	Y	Y	N	N
20. Brige:	Y	N	N	Y	Y
21. Multi-funtional sculpture:	Y	N	N	Y	Y
22. Acaleph:	Y	N	N	N	N
23. Dead bird:	Y	N	Y	Y	N
24. Chair:	Y	Y	N	Y	N
25. Device:	Y	N	N	Y	N
26. Steel:	Y	N	N	Y	Y
27. Famous architecture:	Y	N	N	Y	N
28. Landscape design:	Y	N	N	Y	Y
29. Shadow:	Y	N	N	N	N
30. Railway:	Y	N	N	N	N

classification and conclusion

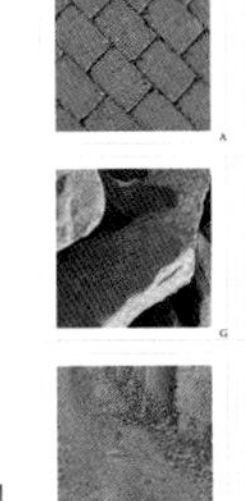

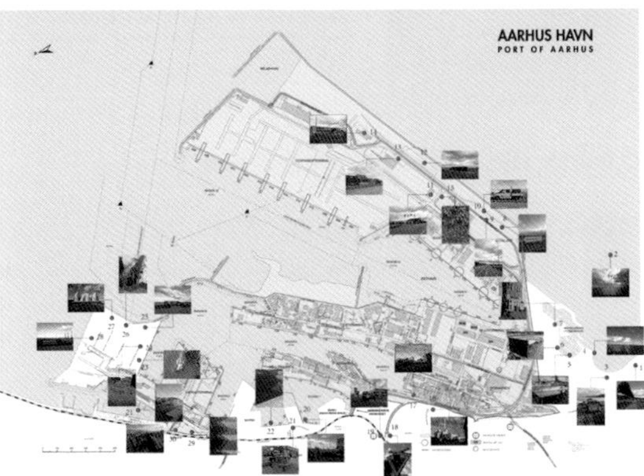

interesting places

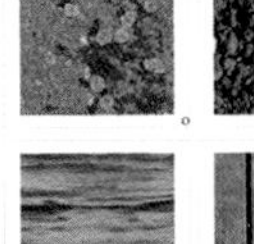

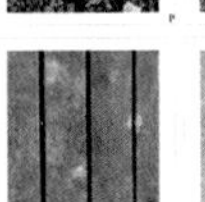

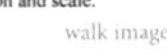

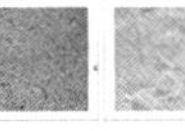

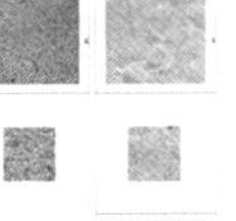

detailed textures abstracted from the section

1th walk aimlessly

This is my first time to come Europe, and the first time to see the real sea. Everything to me is so beautiful and I am so exciting to visit the harbor in Aarhus. According these conclusion, only the textures contains five feelings of mine, and I like the different and abundant feelings of them.

2th walk in a row

When I walk along the coastline with no speaking, no laughing, just seeing, hearing, smelling, touching and feeling with my heart,

I can find other different points as rising tide of the sea, relic of the construction, the shadow of our line, the strange flower, and so on.

I smell the air, touch the wind, hear all the sound I have never heard, feel all kinds of feeling of different spaces.

3th walk-Change

Our members change the interesting point to investigate!

This time, I focus on the plants and their reference substances.

I am thinking about when plants beside different reference substances, it is showed many condition and scale.

walk image

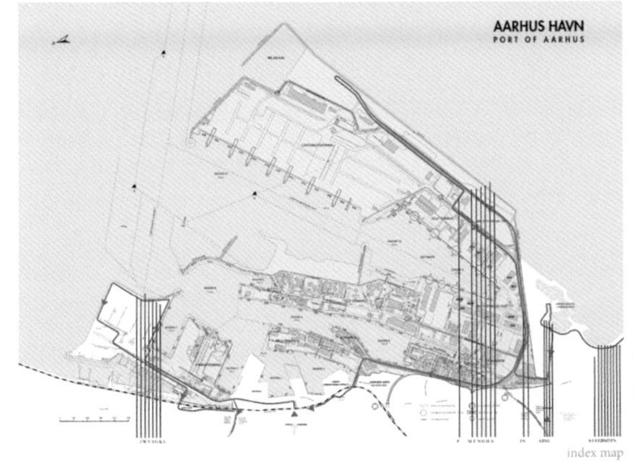

index map

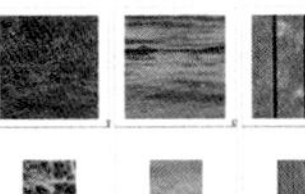

classification and conclusion

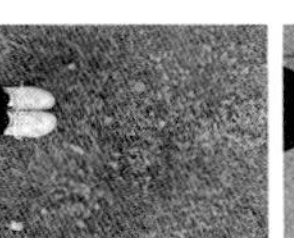

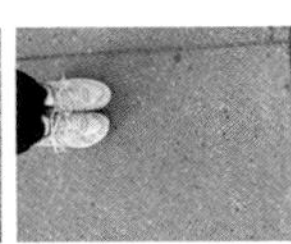

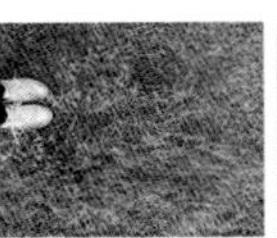

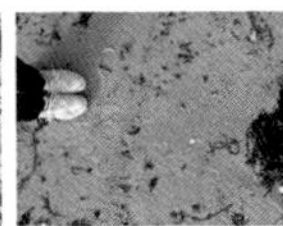

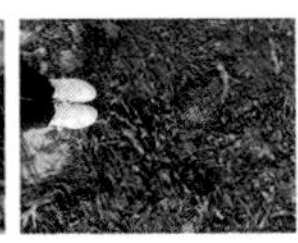

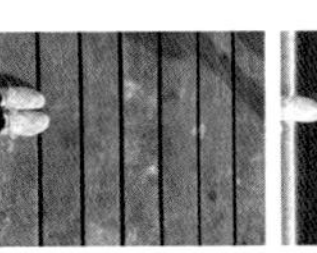

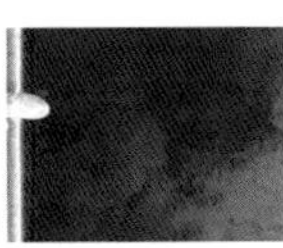

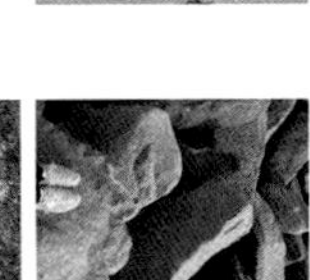

walk with ground textures

University: Central Academy of Fine Art

Designer: Duan Bangyu,Wang Jinlu,Olga Sigporsdottir,Rosemary Jeremy

Tutor: He Keren,Wang Wei,Anne Elisabeth Toft, Claudia Carbone

Course Name: Works from the CAFA I AAA Joint Studio Programme 2013

Finished Time: Oct. 2013

Exchange Institute: Aarhus Architecture School

CLAIM THE TERRITORY BY CHAOS
— WORK & WORKPLACE

EACH SINGLE LAYER OF STRUCTURE
IS UNSTABLE, INPERFECT, WEAK, EASY TO READ.
THEY DEPEND ON EACH OTHER,
COMBINE TOGETHER TO CHECK THE BLANCE OF EACH ELEMENT,
FINALLY, FIND THE BLANCE OF THE WHOLE PROJECT.

THE DESIGN CHALLENGE

This is a program when I studied as an exchange student in B3 Studio, AHO, Oslo.

The task is to draw a workplace related to your understanding of work and your attitude towards the new collective. you are to make your own program.
Scale: from 200 ㎡ to 10.000 ㎡, including complimentary activities related to the workplace spatial idea.
Site/Context: urban, rural or a constructed context.

Your project as work/ workplace shall serve as inspiration towards the new collective as an idea.
To sense an architectural structure that has a spatial power to stimulate human life and work is this semester, and the new collectives challenge.

THINKING

WHAT IS WORK BEYOUD JOB ?
WHAT IS WORKPLACE BEYOUD OFFICE ?

You go to work by the way of you go to the workplace. Then you turn yourself from the relax mood to a working mood. It's both the spacial property and the sign in your mind lead you to concerntrate on your work.

You have a desk or maybe have a room in your workplace. That space is yours. you mark the space by put your works on the desk, somebooks which you are reading, a photograph of you and your family, a coffee cup which you received as a present from your best friend. even a bunch of flowers you picked on your way to work. Finally, "a desk" become "the desk", "a space" become "the space". you marked your territory and be comfortable and happy working inside.
The desk become much more important as your workplace than the building which your office is in.

THINKING

WHAT IS THE RELATIONSHIP BETWEEN
INDIVIDUAL AND COLLECTIVE ?

INDIVIDUAL:
What makes you individual in your work colleutive is that your special work capacity and your own work desk/room. With the development of industry. Everything become commercial process. the individuals are dying. One is always have a desire to keep the individual in the collective.

COLLECTIVE:
Everyone is certainly belong to a collective. It can be small as two, or big as thousands. The small or big is not only according to the scale of a company but also how you define the scope of " the collective".
The people collective in work is the people you work with/ for. Some of them are people you quite familaer with. And some of them you know them detailly but not familar them at all.
The spacial collective in workplace is the similar working atmosphere you share with others.

INSPIRATION

The RELATIONSHIP between INDIVIDUAL & COLLECTIVE inspired by nature photographys.

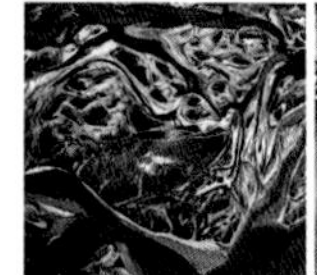

Tunhná river, Iceland

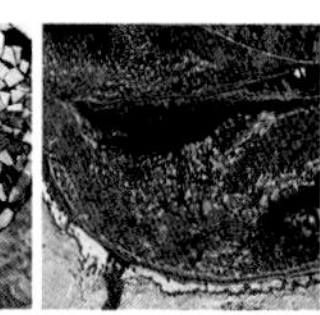

Þingvallavatn Lake, Iceland

Þjórsá river delta, Iceland

Skeiðarársandur coastal sands, Iceland

Photographer: Sigurgeir Sigurjónsson

PROCESS OF THE PROJECTS BY MODELS

A -- COLLECTIVE OF SOLID WOOD

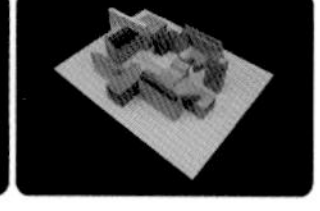

- no scale
Pick some small piece of wood from garbage box in workshop. Cut two or three sides, maintain the charactor of the original woods. Place them, try to find the right position of each one and the collective.

B -- HOW TO IMPLY A SHAPE BY LESS -- 1

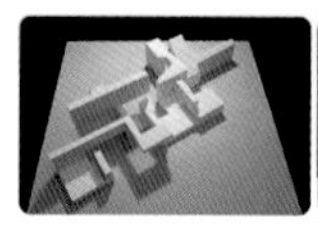
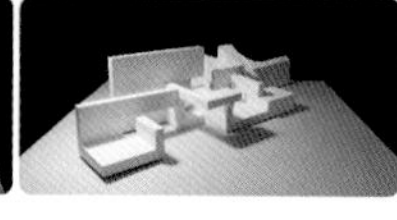

- no scale
Form a part of the "model stage 1" in a spacial way. to find the space can be used.

B -- HOW TO IMPLY A SHAPE BY LESS -- 2

- no scale
Turning 90 degress on each corner. Try different shapes
Try to cut a part of the frame, place it in different ways to see the change of space.

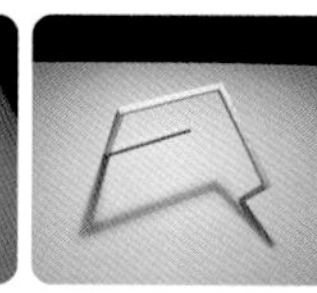

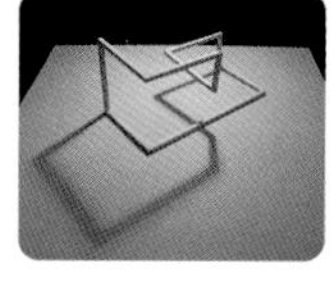
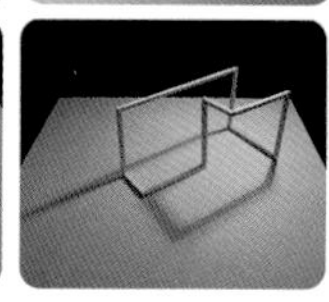
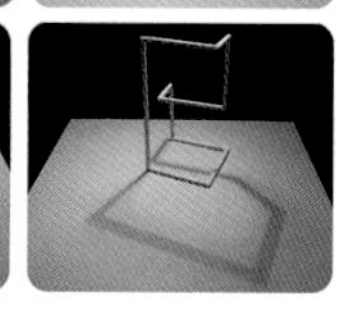
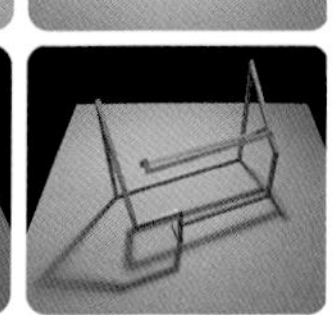

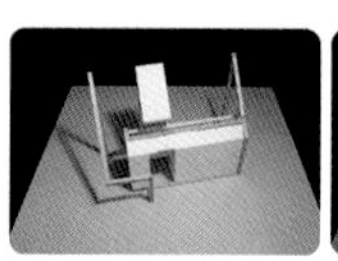
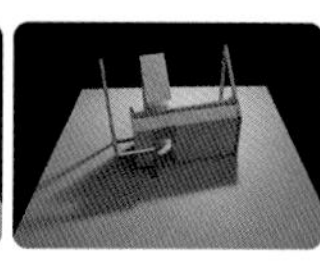

- no scale
try a specific shape to see the method of turning 90 degree at each corner can work. combain the two simplify ways together to see the complexity potencial.

C -- HOW TO IMPLY THE EXTENTION OF A SPACE

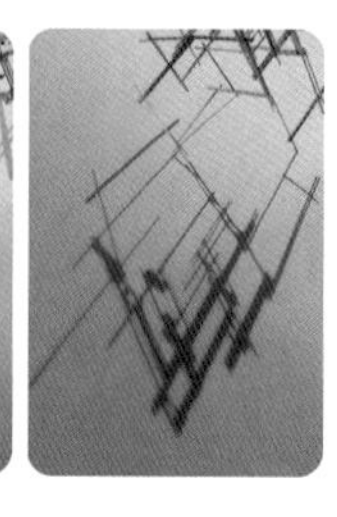

- no scale
how the frame connected. the construction of the node. learned from B2. the way it stands change the space a lot. each node has a foot make it can stand in different ways. But the result of side-effect of complexity goes to the different way of mode B

D -- HOW FRAMES COMBINE TOGETHER

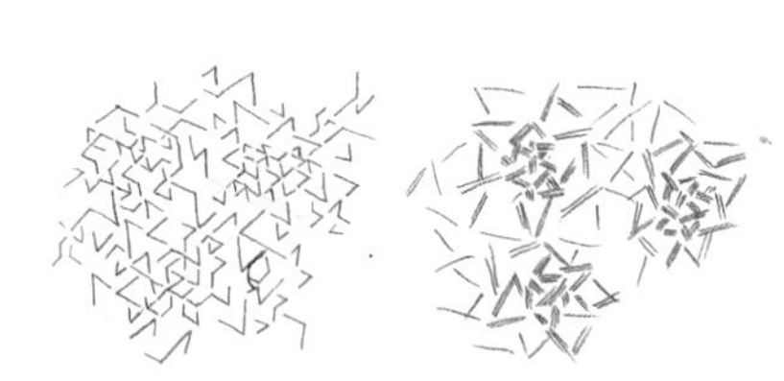

SKETCHES - no scale
two kinds of way to make individual frames into collective.
With center or wothout center

MODEL scale - 1:50
Give the project a context. A lot of prefabricate frames in two hierarchies. Try to find it's own match with another one to make a small collective. and the position of this small collective unit in the whole context.
The context comes to be a place where comparely rural. kind of unartificial.

PROCESS OF THE PROJECTS BY SKETCHES

Study about LIGHT AND SHADOW; how people USE them; how frames HIT THE GROUND; and the relationship between every element.

A: concrete vs earth

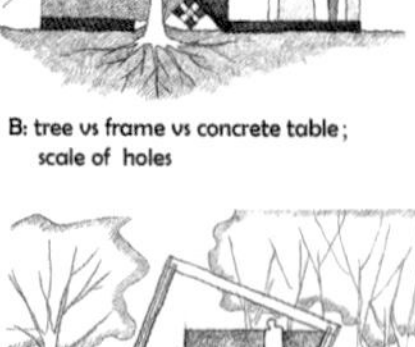

B: tree vs frame vs concrete table;
scale of holes

C: tree vs frame vs concrete table

D:concrete vs rammed earth vs rural earth;
scale of the holes; light and shadow.

作品名称：模糊场域的界定
Claim the Territory by Chaos

院校名：中央美术学院
设计人：范劼
指导教师：Per Olaf Fjeld，Rolf Gerstlauer，Lisbeth Funck
课程名称：工作空间
作业完成日期：2013 年 12 月
对外交流对象：挪威奥斯陆建筑与设计学院

CLAIM THE TERRITORY BY CHAOS
— WORK & WORKPLACE

FINAL DESIGN

MY UNDERSTANDING OF
GENERAL / COLLECTIVE / CONTEXT

GENERAL:

The project is working on how architecture through different structures in related to light and shadow, open and close to claim their territories.
each single layer of structure is unstable, inperfect, weak, easy to read. They depend on earch other, combine together to check the blance of each element, finally find the blance of the whole project.
The result came to be like chaos. But compare to a complet form, chaos is more fair, dynamic and creative potencial.

COLLECTIVE:

every two individual desk make a small collective. the four layers of rammed earth, wood frame, concret space and shadow are connected. they share one concret close dark space. I call this an unit. The four units makes another collective. this collective connected by rammed earth path. here, the unit become individual. plants in between them. isolated them and combine them at the same time. Certainly, this collective will act as individual in another bigger scale.

CONTEXT:

The project is located on a rural area. an area where earth on the ground. not very flat. Grasses/ mushrooms/ flowers/ bushes/ trees are growing. Not very density but in it's own order. "rural" is a relative speaking. It's not necessary to be rural as really far from the city or really wild. It can be a national park/ a beach in the city/ a small island or peninsula, etc.

FINAL DRAWINGS AND MODEL PICTURES

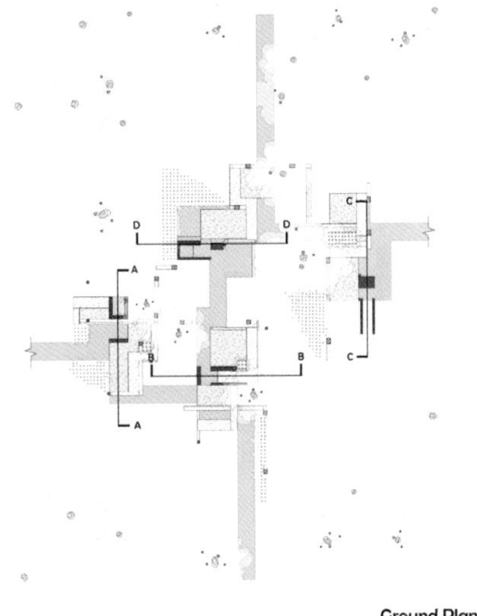

Ground Plan

Ground Plan

Section A-A

Section C-C

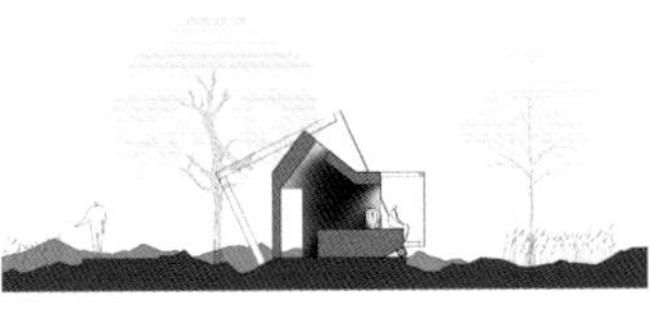

Section B-B

Section D-D

FINAL TEXT

MY EXPLAINATION OF
ORDER / RAMMED EARTH / SHADOW / ROJECTWOOD FRAME / CONCRST SPACE IN THE PROJECT

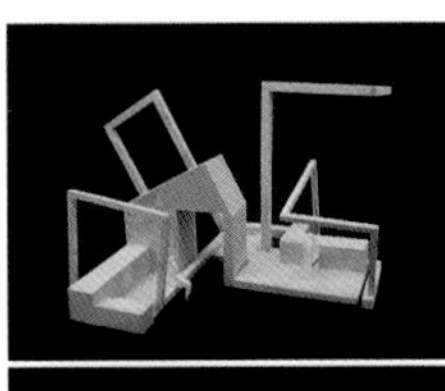

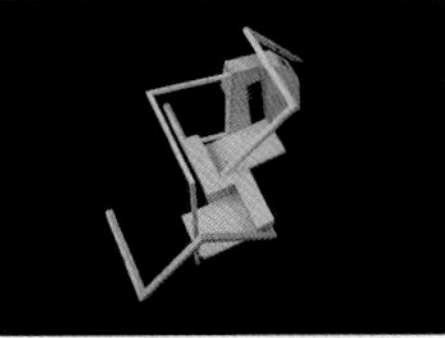

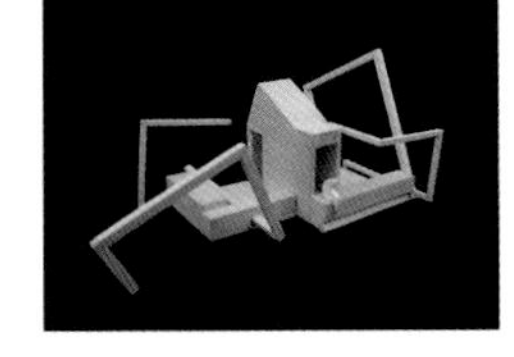

ORDER :

It looks chaos, but actually, has it's own order/ sequence.
It has 4 layers. rammed earth- wood frame- concret space- shadow.

RAMMED EARTH :

Earth is a part of the context. The earth can have different textures. It can be soft, sticky, wet, rough, hard, etc. Here, I rammed the earth and make the texture closest to an architecture flat.
The border which the rammed earth flat made is blured because of the same material. And the edge of the flat will be less sharp as time goes by.

SHADOW :

The shadow of the project is dynamic driven by the sun.
It's keep changing because of the weather but the quality maintines.
the shadow of frame circles spaces together with the physical structue, they claim the dynamic space territory.

WOOD FRAME :

Each frame is a part of one whole shape.
The frame is a continuous line turned 90 degree at each corner. The frame is made of hollow wood. the construction of the node of corner is hidden inside the wood. in order to make the feeling of "line" more obvious and abstruct.
They are prefabricate before they located on the base. It can be inching during the process of the project but have to maintain the general way of the frame. It's the individual/charactor of the frame itself.
It stand on it's own feet. The way they standing is different from the way that original shape stands. But they imply the space of the original one and occupy it.

CONCRET SPACE:

A parallel structure of the frame. They two structures support each other. They really match in shapes partly.
It's not only a wall or a room. It's a desk/ a seat/ a window/ a door/ a wall/ a floor/ a border/ a space/ a light container/ a view spot etc.
A view spot where you sitting/lie in a narrow, dark space, seeing outside nature as a bystander comparing as a participator outside in the nature.

University: Central Academy of Fine Art
Designer: Fan Jie
Tutor: Per Olaf Fjeld, Rolf Gerstlauer, Lisbeth FunckCourse
Course Name: Workplace
Finished Time: Dec. 2013
Exchange Institute: Oslo School of Architecture and Design

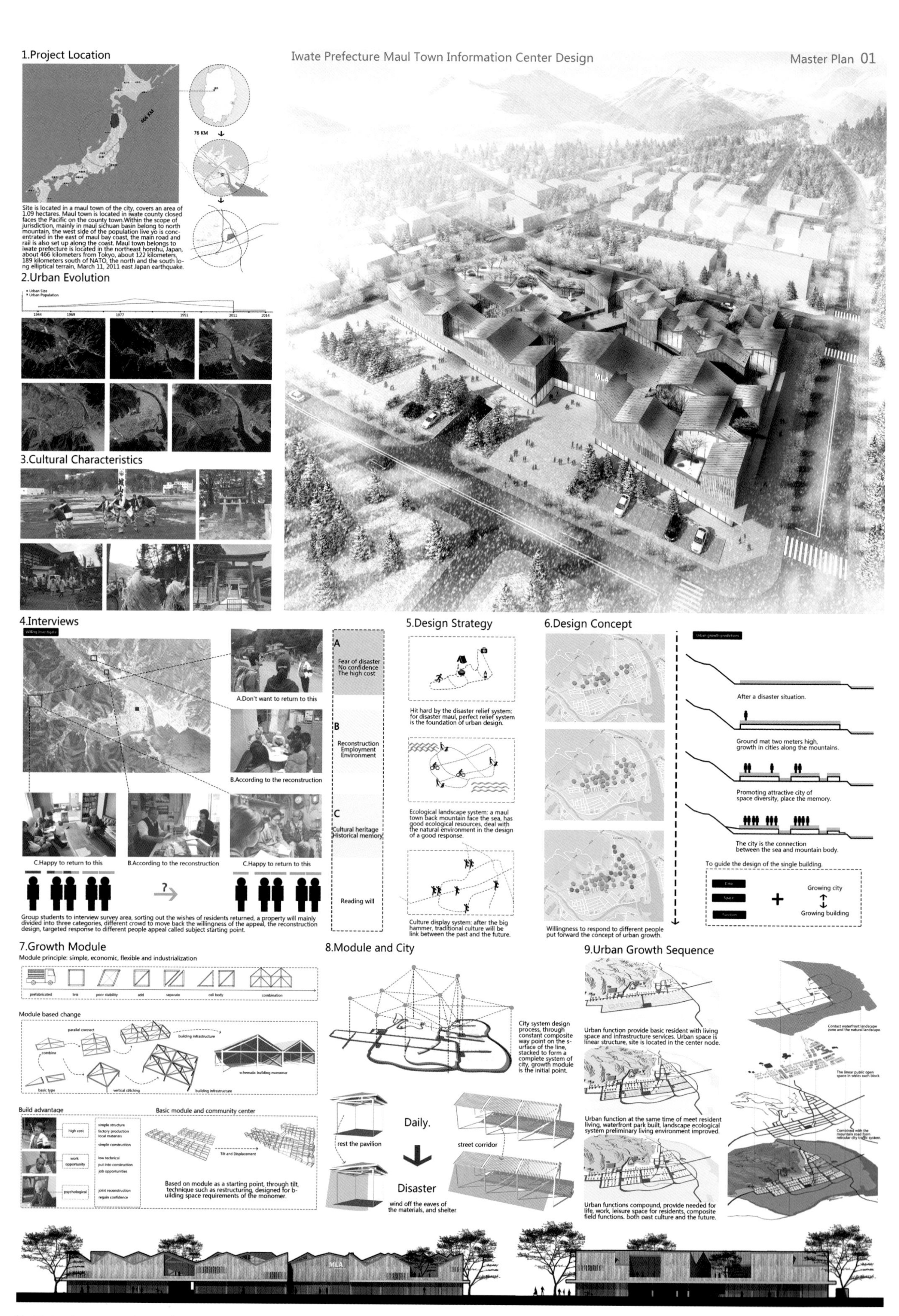

作品名称：3.11东日本震灾复兴计划——岩手县大槌町社区信息交流中心设计
3.11 East Japan Earthquake Recovery Plan—Design of Iwate Tsuchi Community Information Exchange Center

院校名：重庆大学
设计人：傅悦，胡瑶婷
指导教师：邓属阳，翁季，蒋家龙
课程名称：重庆大学—清华大学—早稻田大学 2014 年中日联合设计
作业完成日期：2014 年 07 月
对外交流对象：日本早稻田大学

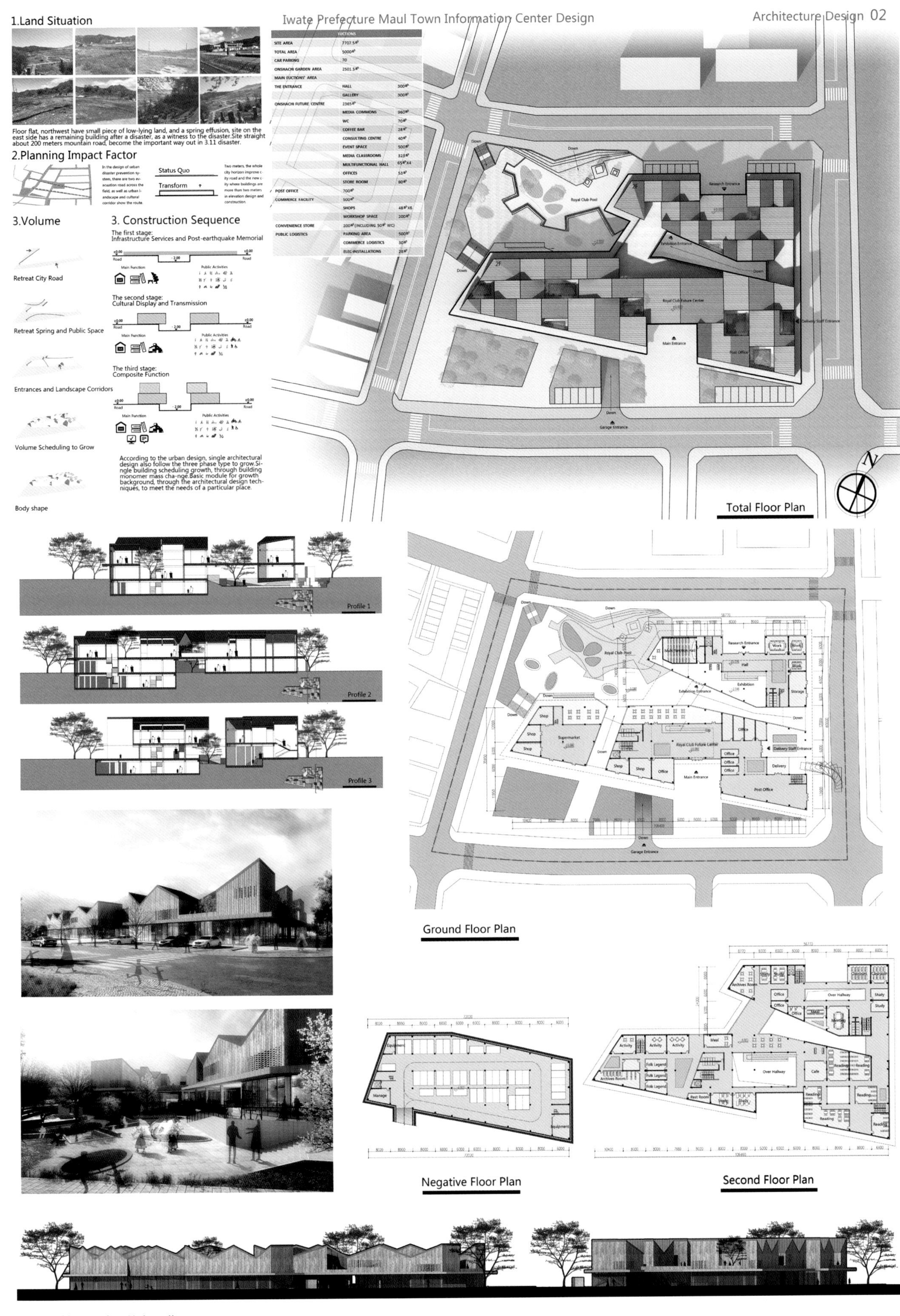

University: Chongqing University

Designer: Fu Yue, Hu Yaoting

Tutor: Deng Shuyang, Weng Ji, Jiang Jialong

Course Name: Chongqing University-Tsinghua University-Waseda University, 2014 China-Japanese joint design studio

Finished Time: July 2014

Exchange Institute: Waseda University

工業遺産 & 社區復興

INDUSTRIAL HERITAGE & COMMUNITY REVIVAL

设计概念 Conception

钢的云 Rion Cloud

奇观建筑 Building Spectacle

如同库哈斯描绘的广谱城市精神内核，建筑奇观带来人们对未来的狂热期待，对技术和力量的崇拜。建筑外表的简洁集合形态最大程度地服务于城市空间，而内部则应对功能复杂多变的现代生活。

As the spirit of Generic City described by Rem Koolhaas, building spectacle brings zealous expectation of future, worship of technology and strength. The building appearance of simple geometry extremely provides service for urban space, while the internal space response to complicated and changeable modern life.

区位及用地分析 Location and Property

此次中德联合设计课题的选址位于德国杜塞尔多夫市莱茵河西岸 Oberssel 地区，在此处旧城区进行相关城市更新研究与城市设计，探讨老城区在新的城市发展中的可能性。

The location of this China-Germany joint design is located in Dusseldorf of Germany, Oberkassel in the west bank of Rhine river. The subject aims at the study on relevant urban renewal and urban design of the site, and discusses the possibility of old town in the development of city.

空间组织关系 Spatial Structure Analysis

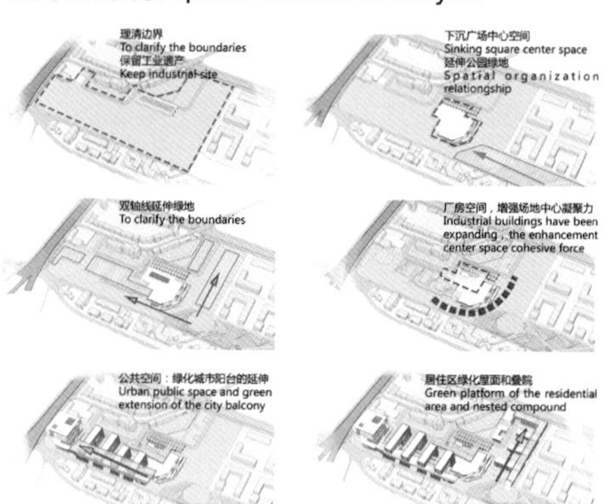

交通组织关系 Traffic Organization

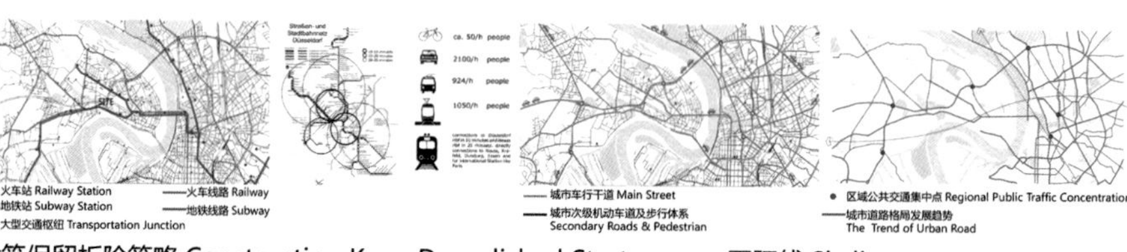

建筑保留拆除策略 Construction Keep Demolished Strategy

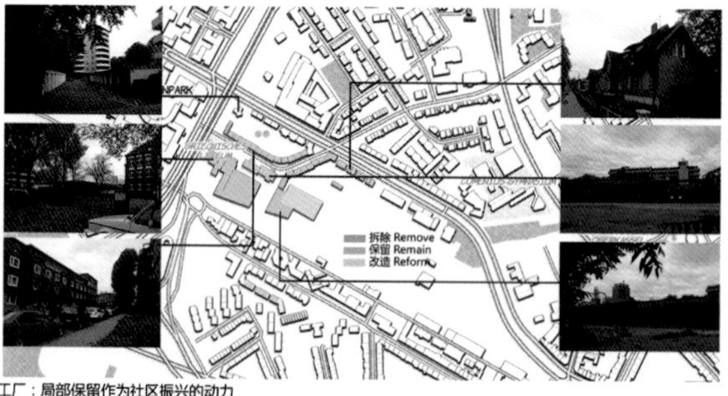

工厂：局部保留作为社区振兴的动力
Factory:Remain parts of it to be the power of community
沿街小住宅：年代久，条件差，建议拆除
Small houses:Poor construction conditons,it is recommended that the demolition
沿街居民楼：建筑形式完整，使用状况良好，保留
Residential building:Suggested to remain those in good condition
学校和办公楼：场所记忆的重要组成部分，合理利用
Schools and Office building:Important parts of the memory of the site to be used rationally

天际线 Skyline

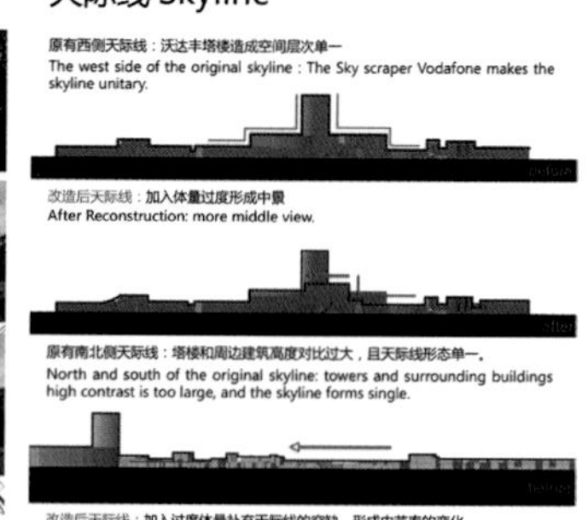

原有西侧天际线：沃达丰塔楼造成空间层次单一
The west side of the original skyline : The Sky scraper Vodafone makes the skyline unitary.
改造后天际线：加入体量过度形成中景
After Reconstruction: more middle view.
原有南北侧天际线：塔楼和周边建筑高度对比过大，且天际线形态单一。
North and south of the original skyline: towers and surrounding buildings high contrast is too large, and the skyline forms single.
改造后天际线：加入过度体量补充天际线的空缺，形成由节奏的变化
After transformation: Add excessive volume supplement vacancy skyline, forming a rhythmic changes.

空间要素 Space Element

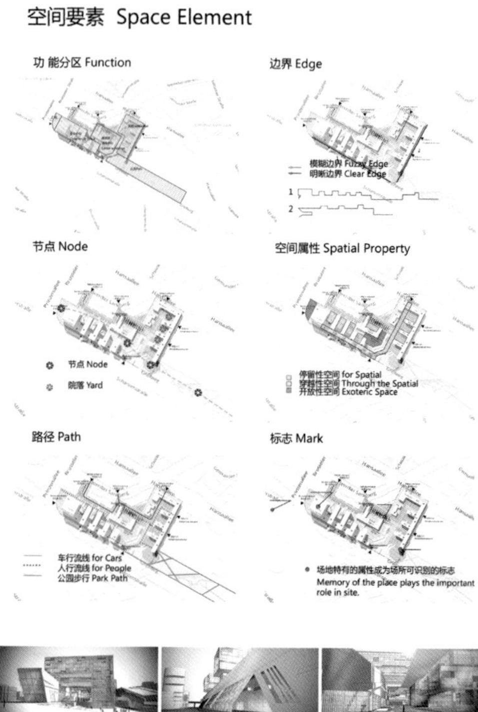

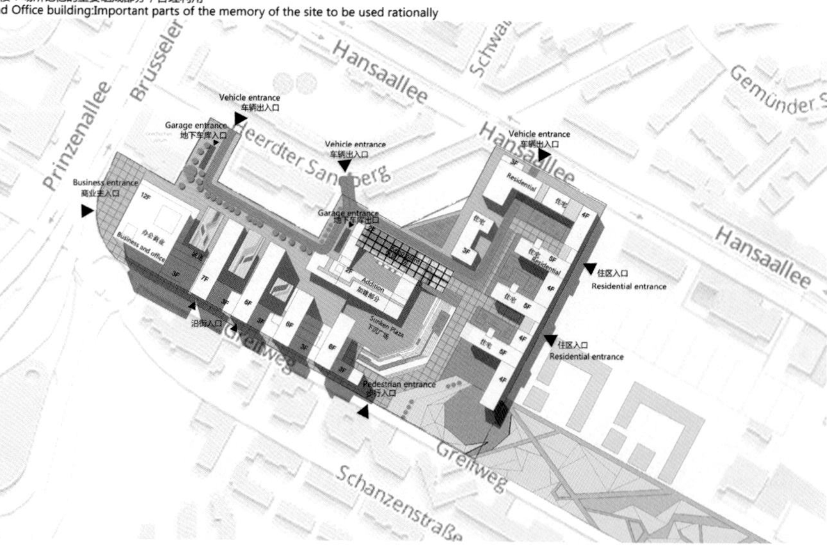

作品名称：工业遗产 & 社区复兴
Industrial Heritage & Community Renewal

院校名：重庆大学
设计人：高澍，曾柳银，程文楷
指导教师：龙灏，褚冬竹，Jorg Leeser，Juan-Pablo Molestina
课程名称：重庆大学—杜塞尔多夫应用科学大学研究生联合设计课程教学
作业完成日期：2014 年 01 月
对外交流对象：德国杜塞尔多夫应用科学大学

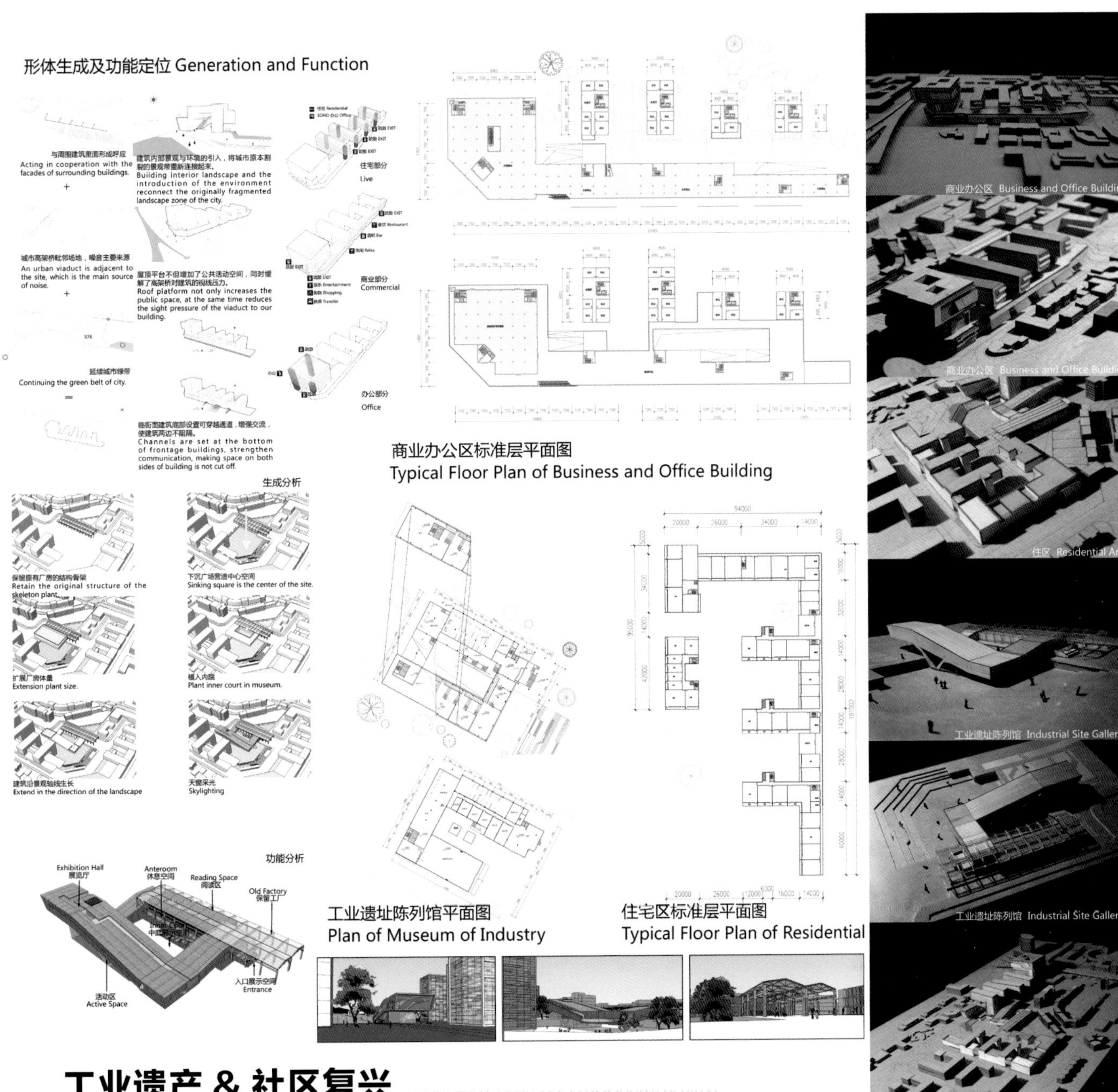

工业遗产 & 社区复兴

University: Chongqing University
Designer: Gao Shu, Zeng Liuyin, Cheng Wenkai
Tutor: Long Hao, Chu Dongzhu, Jorg Leeser, Juan-Pablo Molestina
Course Name: Chongqing University - Dusseldorf University of Applied Sciences Joint Design Studio
Finished Time: Jan. 2014
Exchange Institute: Dusseldorf University of Applied Sciences

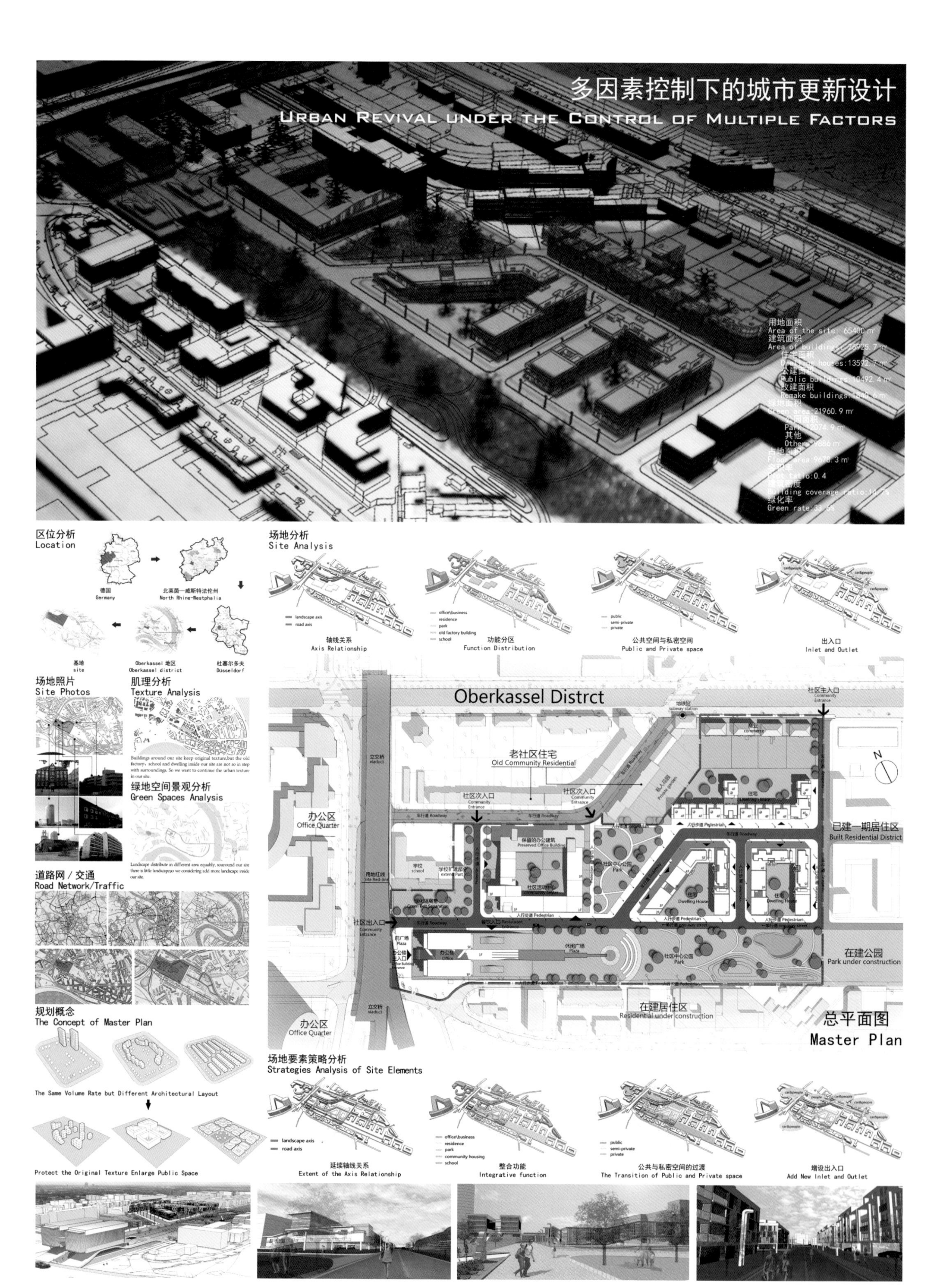

作品名称：多因素控制下的城市更新设计

Urban Renewal Design Under Multi-Factor Control

院校名：重庆大学

设计人：黄一夫，况毅，杨涵

指导教师：龙灏，褚冬竹，Jorg Leeser，Juan-Pablo Molestina

课程名称：重庆大学—杜塞尔多夫应用科学大学联合设计课程教学

作业完成日期：2014 年 01 月

对外交流对象：德国杜塞尔多夫应用科技大学

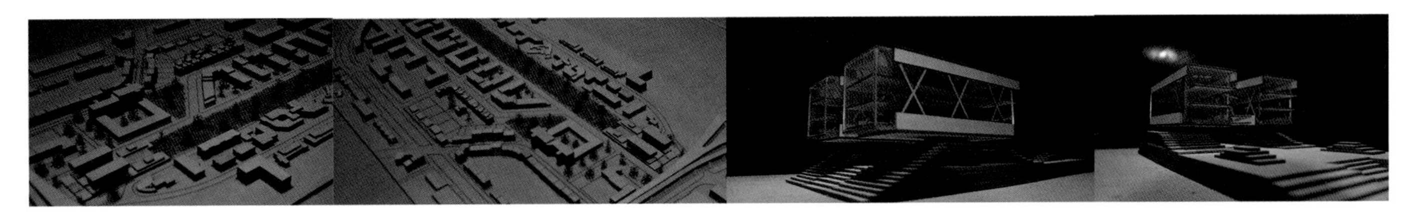

多因素控制下的城市更新设计

URBAN REVIVAL UNDER THE CONTROL OF MULTIPLE FACTORS

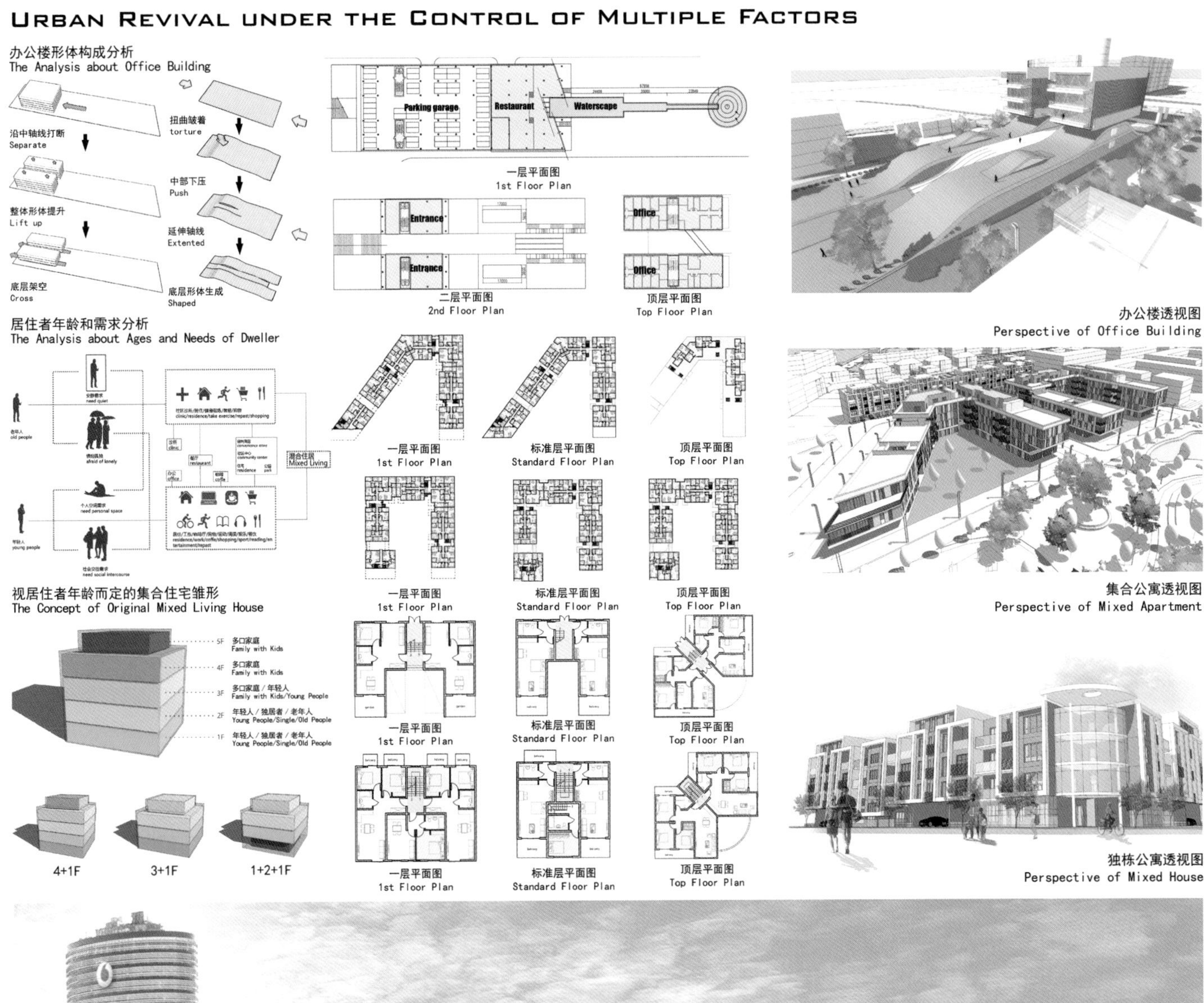

University: Chongqing University
Designer: Huang Yifu, Kuang Yi,Yang Han
Tutor: Long Hao, Chu Dongzhu，Jorg Leeser，Juan-Pablo Molestina
Course Name: Chongqing University-Düsseldorf University of Applied Sciences Joint Design Studio
Finished Time: Jan. 2014
Exchange Institute: Dusseldorf University of Applied Sciences

概念阐释 CONCEPT

本案地块位于日本日本大槌町，背山面海，具有良好的自然景观，海啸后，居民由于心理阴影惧怕回到这个区域，分散到海拔较高的山腰上。
让人们愿意回到这个地方，消除的海啸的恐惧，自由的散步，同时恢复原来的邻里关系和生活场景，是设计面临的关键问题，进过调研分析，人们的生活、传统文化以及原来城市建筑的肌理和“山与海”具有紧密的联系，因此本案提出“山与海的舞台”的概念，在城市设计中，在与山与海有重要关系的节点处设置不同的舞台，满足日常生活、散步、及非日常逃跑的需要，同时具有纪念性意义。
山与水的舞台，它可以是日常生活中的庭院，也可以是非日常时期的避难所；可以是对逃跑路径的指示，也可以是灾害发生的纪念场所。舞台点选择在具有历史意义的节点处、遗留的重要建筑处、同山与海相关的肌理转折处、海啸前形成的邻里庭院处……

This project is located in Otsuchi, Japan with mountains and the sea natural landscape, after the tsunami, residents fear of the psychological shadow and dare not go back to this area, spread across the higher elevations of the mountain.
So that people are willing to return to this place, eliminating the fear of tsunami, freedom of walking, while restoring the original neighborhood relationships and scenes of life, is the key issue facing the design, research and analysis over into people's lives, culture and tradition of the original city building texture and "Mountains and sea" has a close relationship, and therefore the case put forward the concept of "mountains and Sea stage" in urban design, set a different stage in an important relationship with the mountains and sea node, to meet the daily life , walking, and non-routine needs to escape, while the monumental significance.

地段调研 SITE RESERCH

震前村镇肌理

地段背山面水，地震前形成的地段肌理自然灵活，根据山脊山谷线舒展开，人造环境也遵从山水自然的联系

震后重建保障房肌理

重建房屋为保证建造速度和使用效率，忽略了地段原有的自然肌理，方格网打破了山水的联系

山水自然肌理 NATURAL

文化舞蹈舞台 CULTURAL

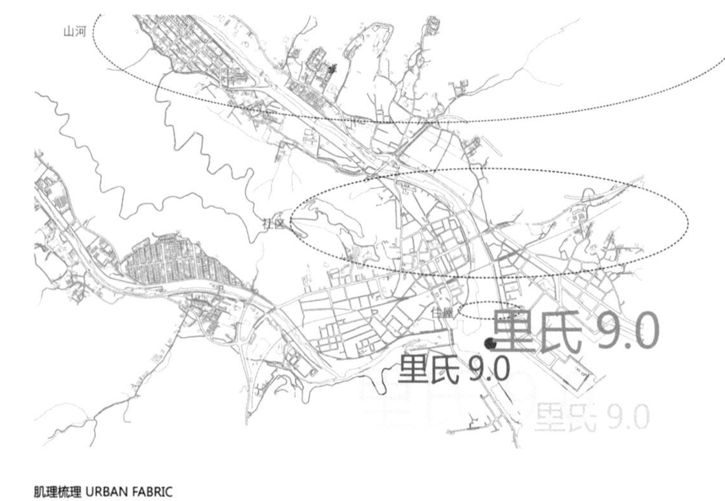

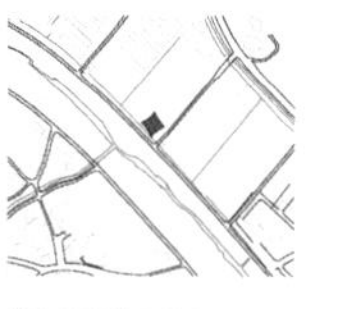

类型一 田间场地 TYPE I FARM
地区文化的连接，配合大槌町地区的主要产业，将文化活动空间设置在田间地头

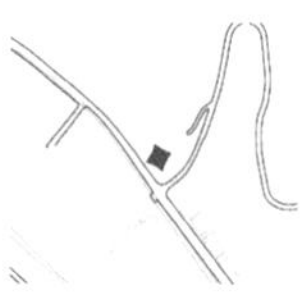

类型二 林间场地 TYPE II FOREST
林地之中，具有旅游资源采用木材作为中心场地的材质，体现了当地的文化

肌理梳理 URBAN FABRIC

传统肌理转折处
THE OLD AND NEW

传统肌理的转折，印证了传统居民在自发营造时考虑“山、海、人”之间的联系

新旧道路相交处
ACROSS THE ROAD

新老道路的交汇点是传统与现代的连接点，应营造一种标志舞台

邻里交流空间处
COURTYARD

传统邻里交流空间是以前居民生活的印记，记忆着人们生活的场景

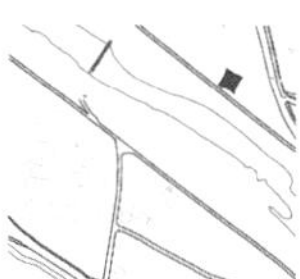

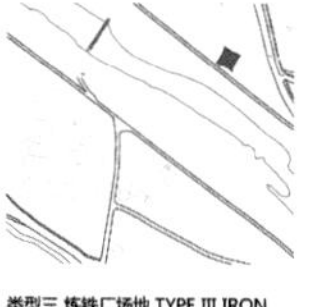

类型三 炼铁厂场地 TYPE III IRON
就地取材，采用钢材作为中心场地的材质，体现了当地的文化，住宅较多，居民交流功用

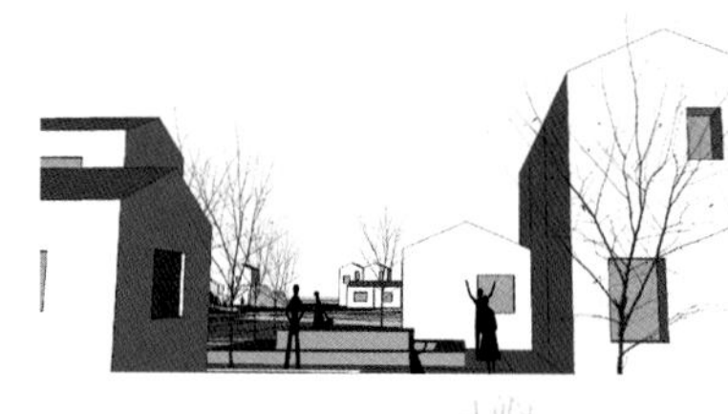

宅基地梳理 OLD FABRIC

穿插公共空间 PUBLIC SPACE

新的城市肌理 NEW FABRIC

类型四 山地场地 TYPE IV MOUNT
按照地形走势，建于高地平台，作为登山人群的休息平台，同时也是非日常时期的避难所

作品名称：山水舞台——岩手县大槌町社区信息交流中心设计
Landscape Stage — Design of Iwate Tsuchi Community Information Exchange Center

院校名：清华大学
设计人：吉亚君，蔡长泽
指导教师：许懋彦，罗德胤
课程名称：清华大学—早稻田大学—重庆大学联合设计课程教学
作业完成时间：2014 年 7 月
对外交流对象：日本早稻田大学

University: Tsinghua University
Designer: Ji Yajun, Cai Changze
Tutor: Xu Maoyan, Luo Deyin
Course Name: Tsinghua-Waseda University-Chongqing University Joint Design Studio
Finished Time: Jul. 2014
Exchange Institute: Waseda University

隙 间

该联合教学工作坊由东南大学、东京工业大学、同济大学和华南理工大学共同举行，以广东老城区中居住商业文化混合功能的城市设计为主题，设计对象为荔枝湾南:广东文化艺术保育和创新基地。本次联合教学的基地选址在广州荔湾区荔枝湾以南，地属西关，以羊城八景、行商园林、西关大屋闻名于世。近年来，连接珠江内外水道的荔枝湾涌复涌工程建设以后，整个广州西关串联起来，提供了重构公共空间体系、促进城市更新的机会，但同时呼唤对空间价值的恰当理解和对城市发展的高度想象力。

本设计主题为“隙间”，通过在一到三层利用传统民居“西关大屋”和“竹筒屋”中的特殊空间——隙间，获得了丰富且充满差异性的各种小尺度公共空间。另一方面亦成功的塑造了河岸到中心公园的可达性的同时，并利用“冷巷”（亦为隙间的一种）将滨水区域的微风引入场地深处以此改善气候，使的当地传统的生态设计手法与当代城市设计相结合。

荔枝湾南——广东文化艺术保育和创新基地设计

现状 / Main Problem

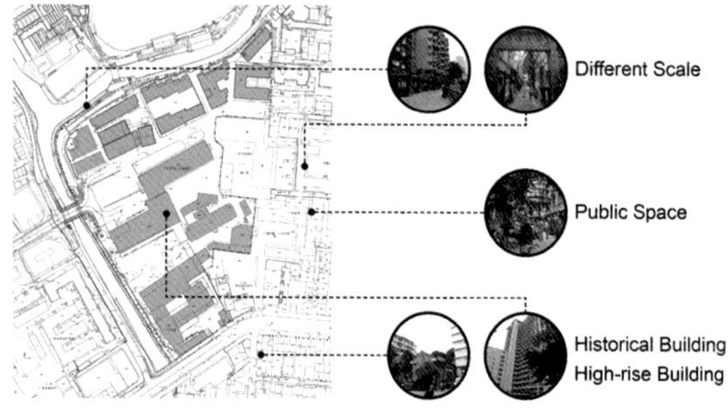

关键词 / Keyword&Destination

西关大屋研究 / Study in Xiguan House

重组“院屋巷” Reform Patio Room&Road

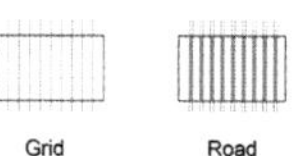

形体生成 / Form Growing

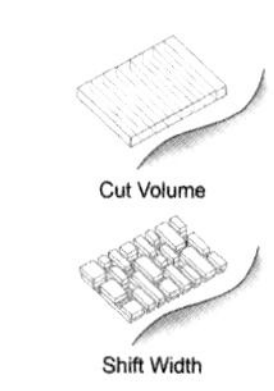
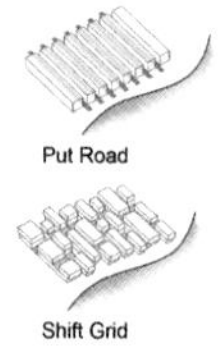
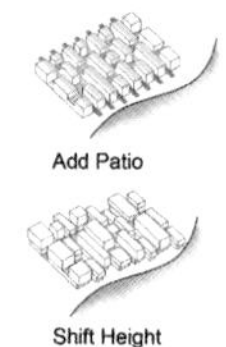
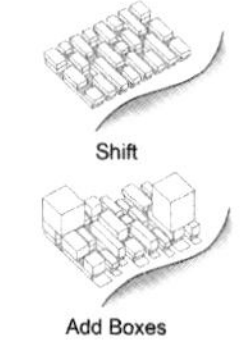

平面生成 / Plan Growing

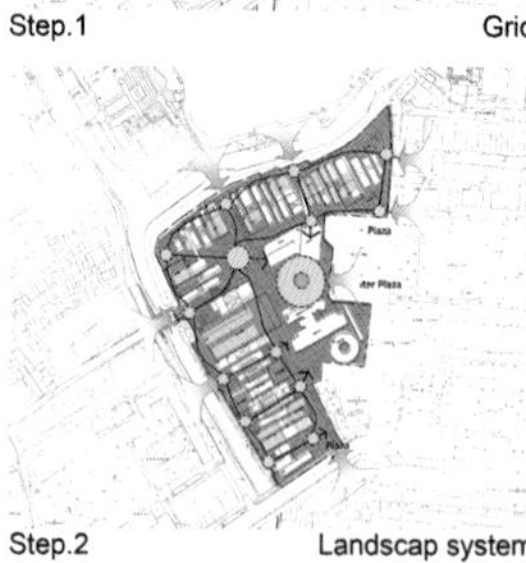

首层平面 / First Floor Plan

总平面 / Master Plan

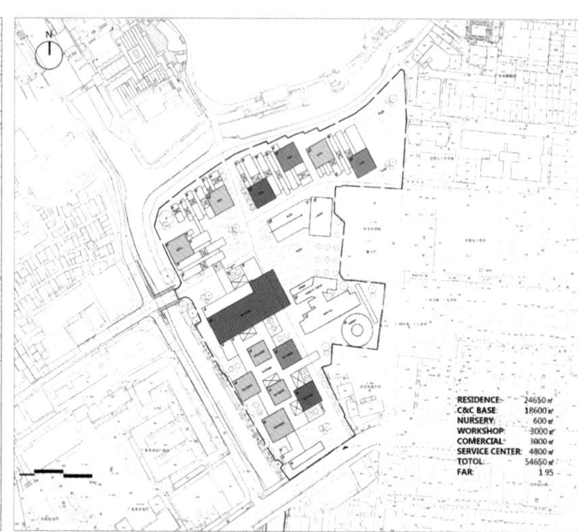

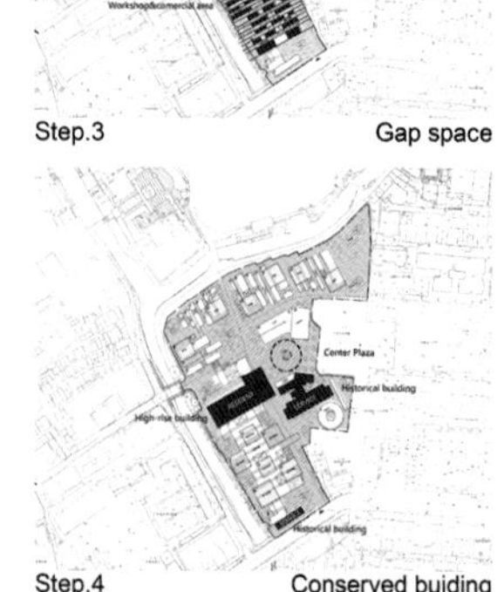

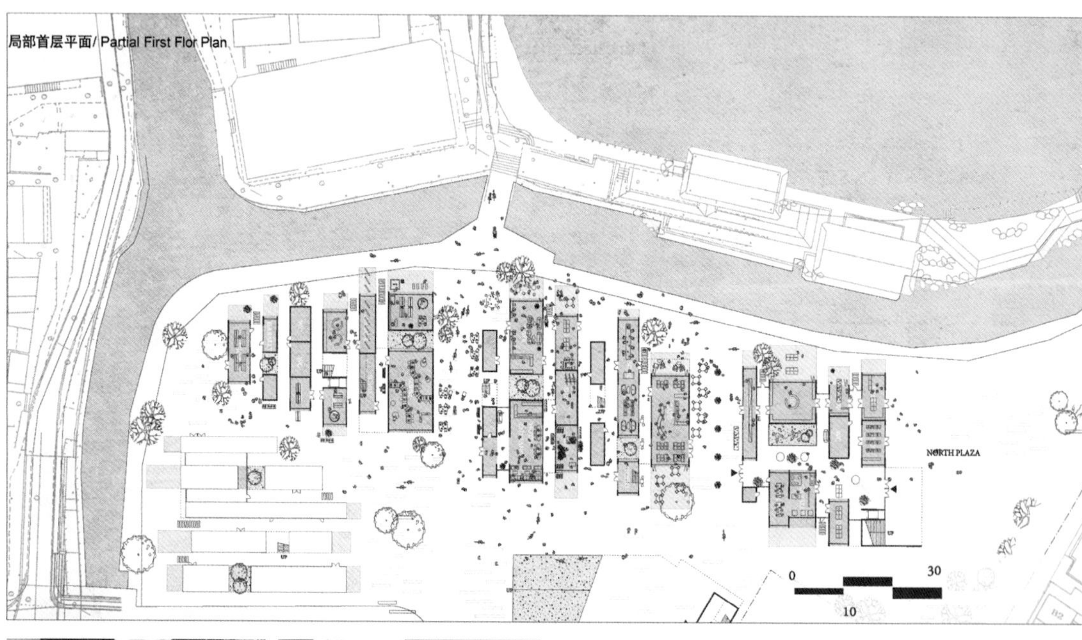

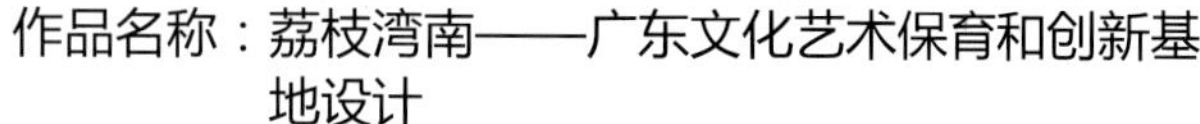

作品名称：荔枝湾南——广东文化艺术保育和创新基地设计

South of Litchi Gulf — Design of Cultural and Art Conservation and Innovation Base in Guangdong Province

院校名：东南大学
设计人：金海波，洪梦扬，许铎，南泽智规
指导教师：唐芃，葛明，王伯伟，孙一民，王方戟，冯江，徐好好，奥山信一
课程名称：东南大学—东京工业大学—同济大学—华南理工大学联合设计工作坊
作业完成日期：2014年02月
对外交流对象：日本东京工业大学

GAP SPACE

Cultural conservation and creation Base for Cantonese traditional and contemporary Culture and Arts

The site of the workshop that locates in the south of Lychee Bay is well known of beautiful sceneries, private gardens and Xiguan housing. In recent years, thanks to the recovery of the Lychee Bay stream, the whole Xiguanconnects, providing chances to restructure the public spaces and the historic city. In the meantime, it calls for a new understanding of spatial value and a fantasy imagination for the future urban development.

By using "gap space" and the methods of traditional Guangzhou residence(Xiguang house&Bamboo house) in 1F to 3F, we obtained a kind of urban space full of diversity and funny images of our comunity. We also used gap space to lead people into to central plaza. And Gap space is also a kind of cold lain which could be used for improving climate of this site.

分解轴侧 / Explosion

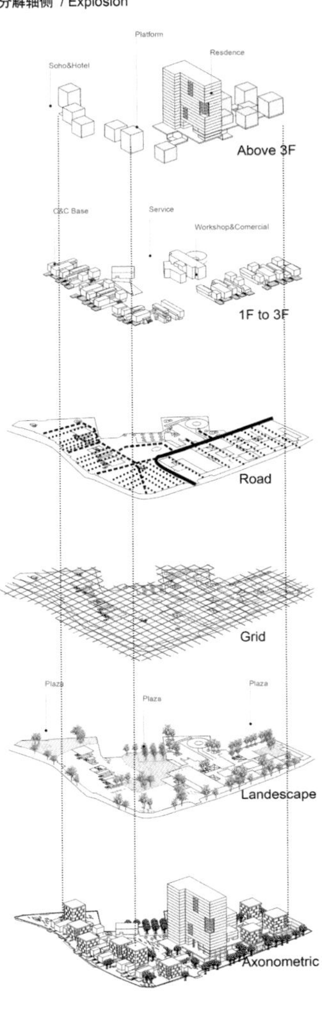

隙间透视 / Gap Space Perspective

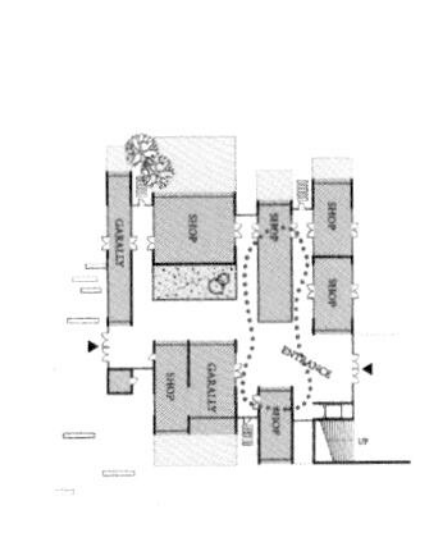

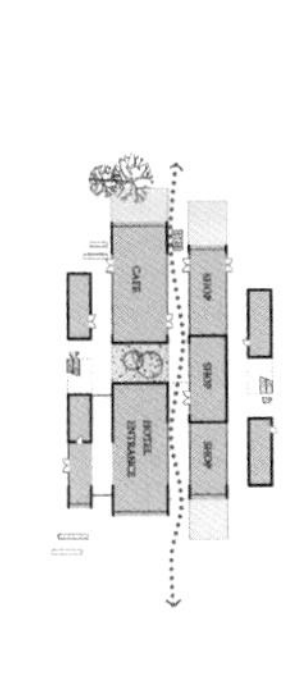

材质拼贴 / Material Pasteup

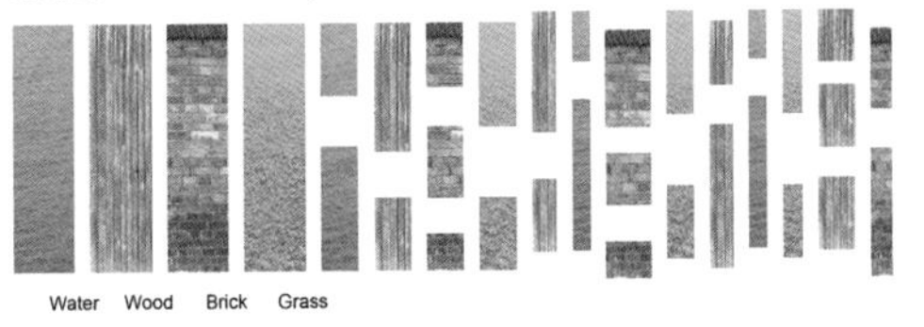

局部剖面 / Part Section

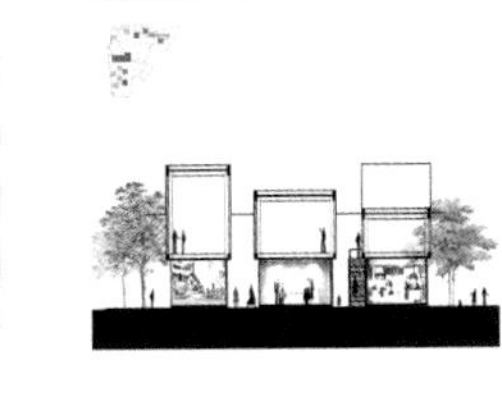

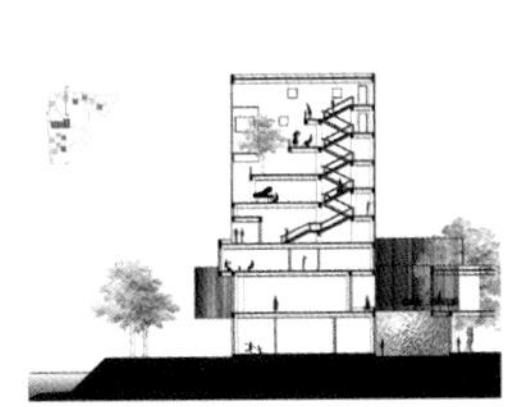

局部剖面 / Part Section

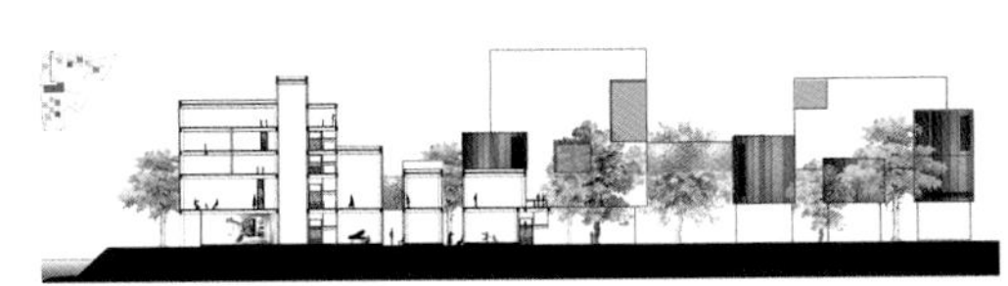

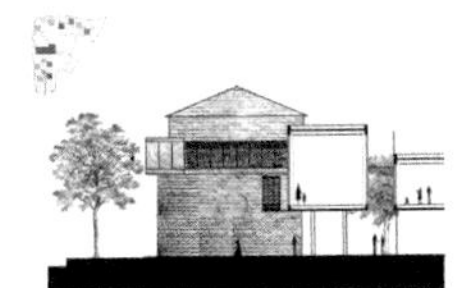

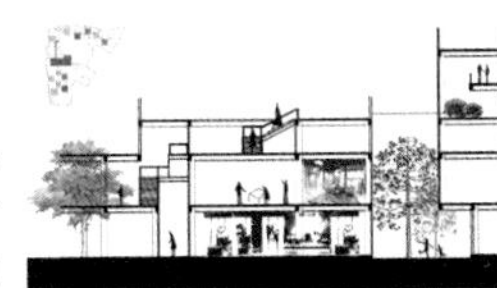

长剖面 / Long Section

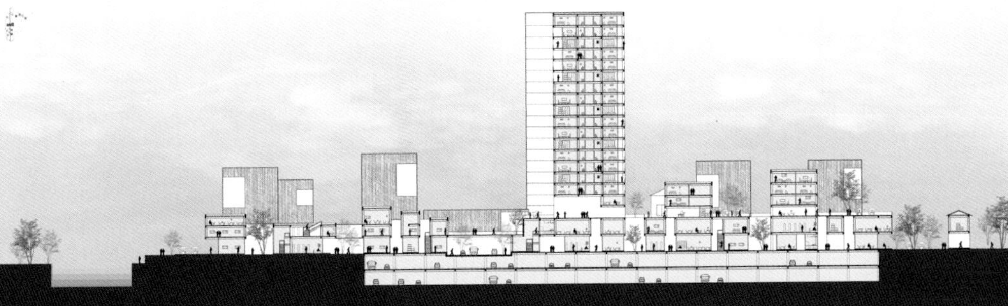

评图照片 / Photos of Presentation

University: Southeast University
Designer: Jin Haibo, Hong Mengyang, Xu Duo, Minamisawa Tomonoli
Tuto: Tang Peng, Ge Ming,Wang Bowei, Sun Yimin, Wang Fangji, Feng Jiang,Xu Haohao,Okuyama Shinichi
Course Name: SEU - TIT - TJU-SCUT Joint Design Studio
Finished Time: Feb. 2014
Exchange Institute: Tokyo Institute of Technology

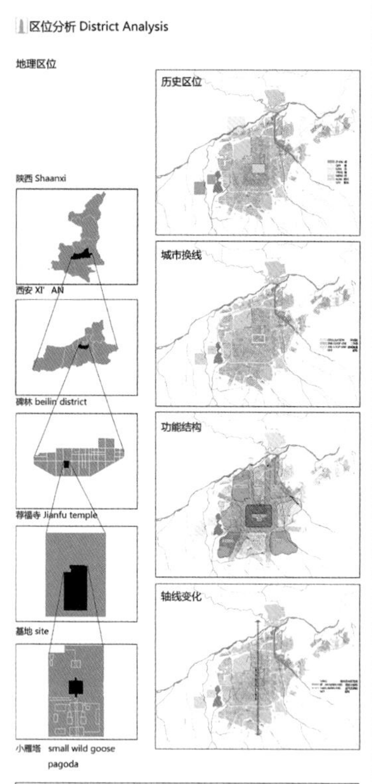

设计说明：

1 以区域功能分区为基础：历史文化区，景观过度区，当代生活区。
2 基地现状分析：主要人群分类，基地内功能分区，道路系统分层。
3 建筑形体设计：重塑轴线，退让景观，视线引导，辐射范围。

Design Notes:

1 functional area to area based on: the historical district, landscape excessive area, contemporary living area.
2 base situation analysis: the main population classified within the base zoning, road system hierarchy.
3 architectural form design: remodeling axis, concessions landscape, sight guidance, radiation range.

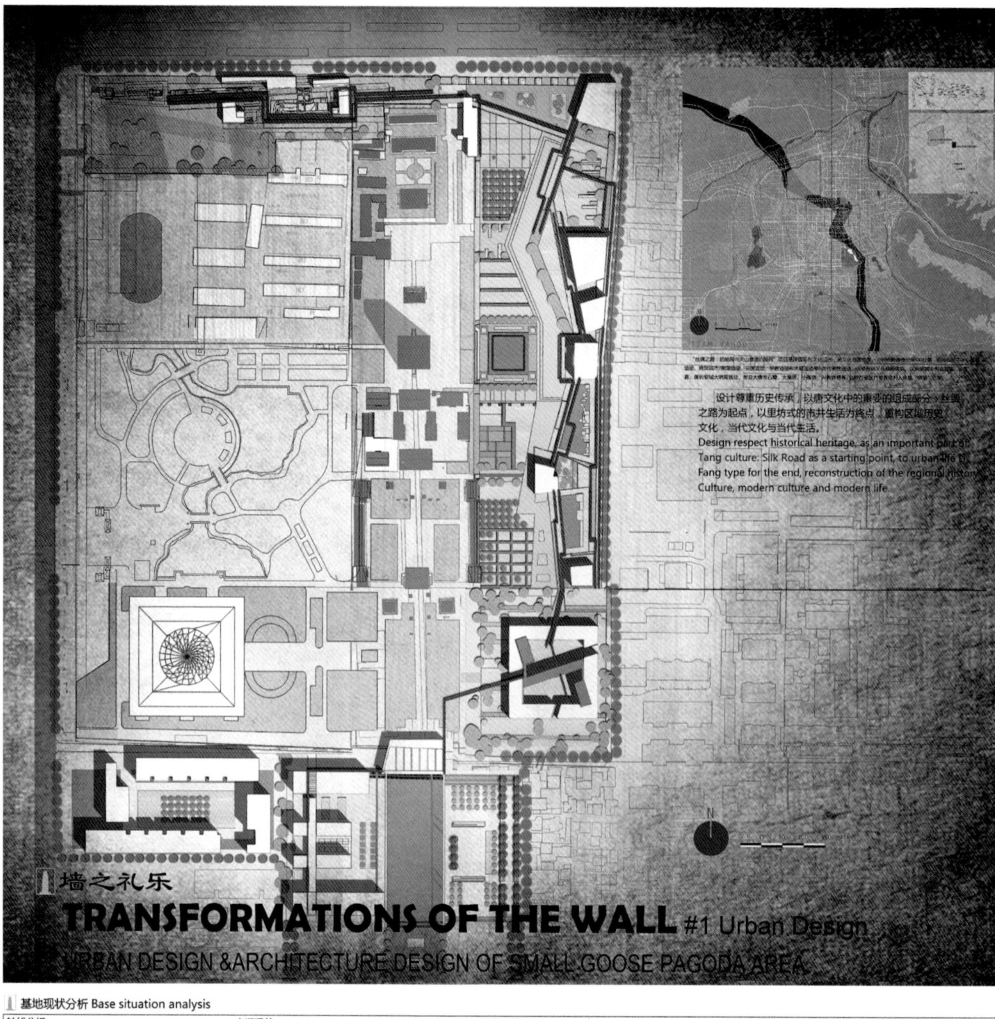

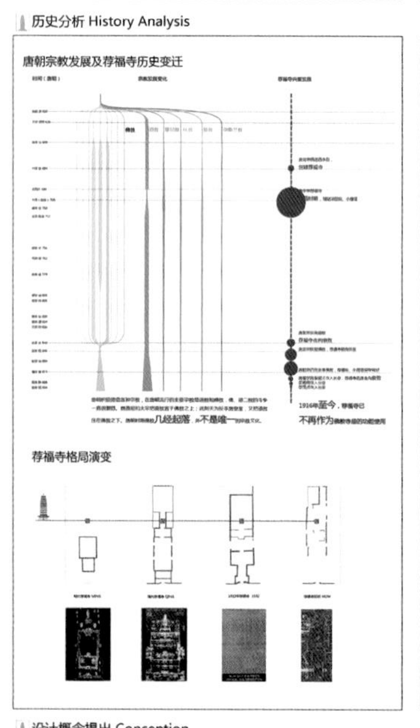

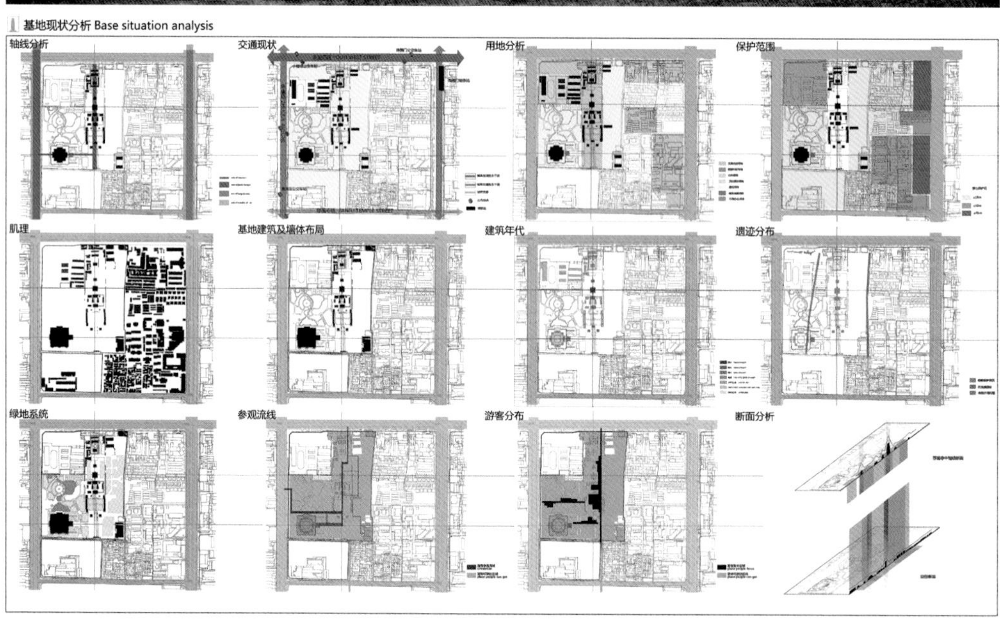

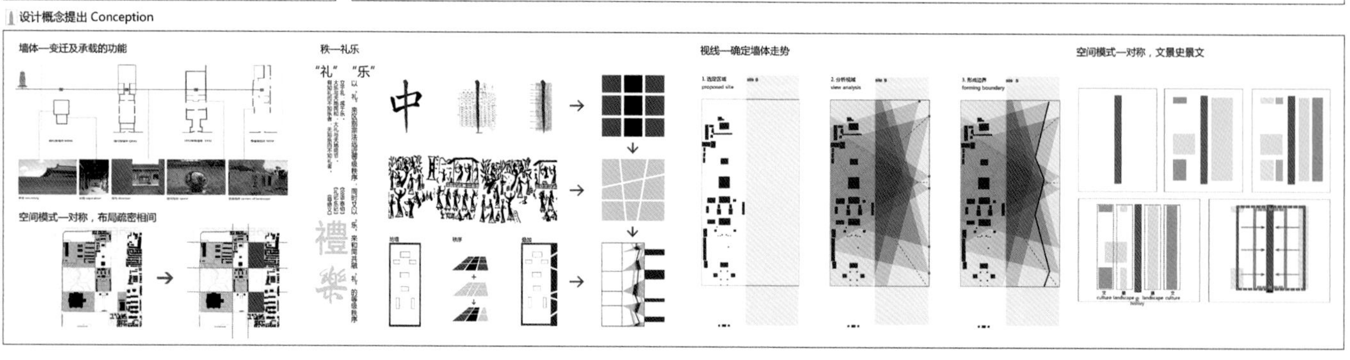

作品名称：墙之礼乐——西安博物院城市设计及重点建筑设计
The Wall of the Ritual — Xi'an City Museum Design

院校名：西安建筑科技大学
设计人：刘思源，方异辰，邹宜彤，张娜，Jared Lambright，Quang Nguyen，MingJin Hong
指导教师：肖莉，常海青，苏静，Albertus Wang，鲁旭，任中琦
课程名称：西安建筑科技大学—佛罗里达大学联合课程设计教学
作业完成日期：2014 年 08 月
对外交流对象：美国佛罗里达大学

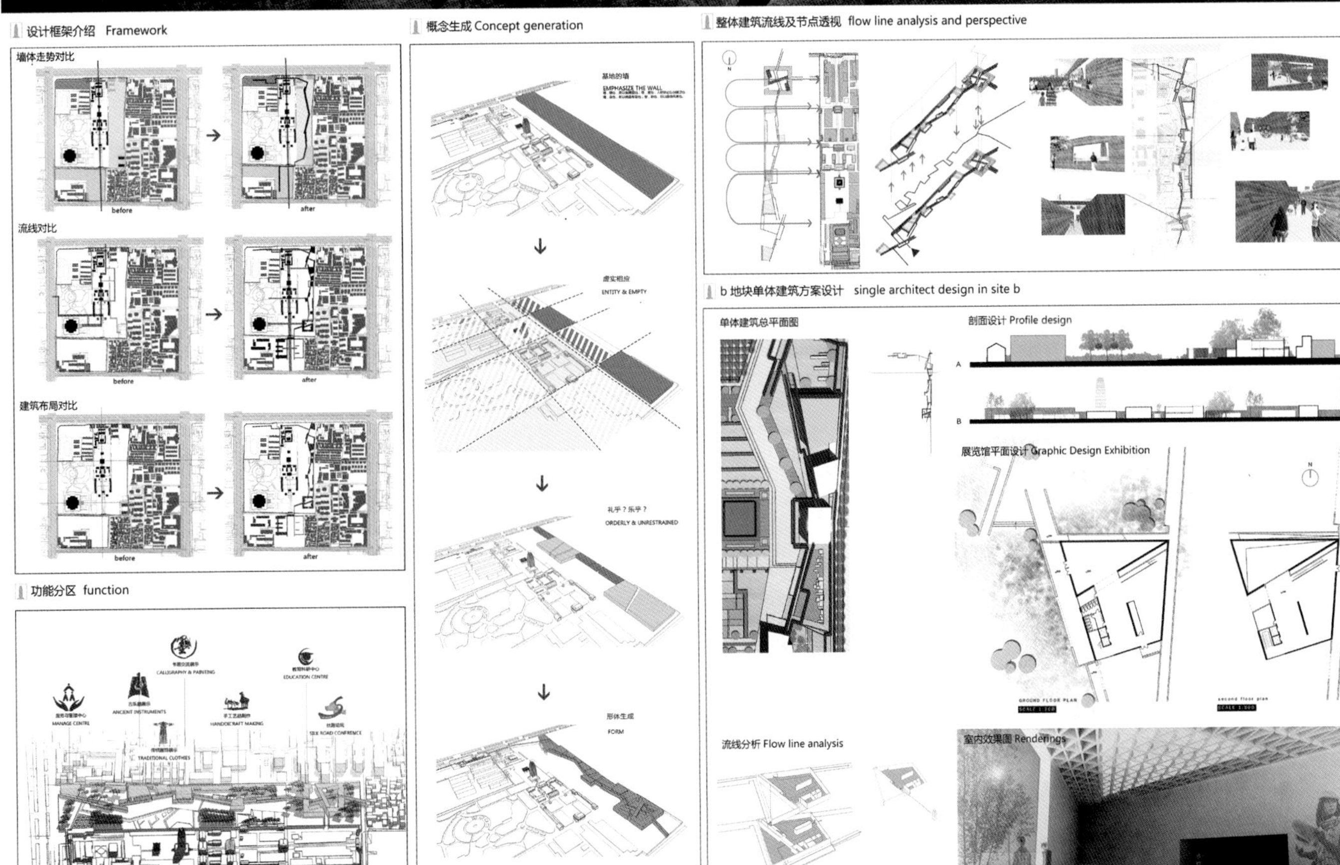

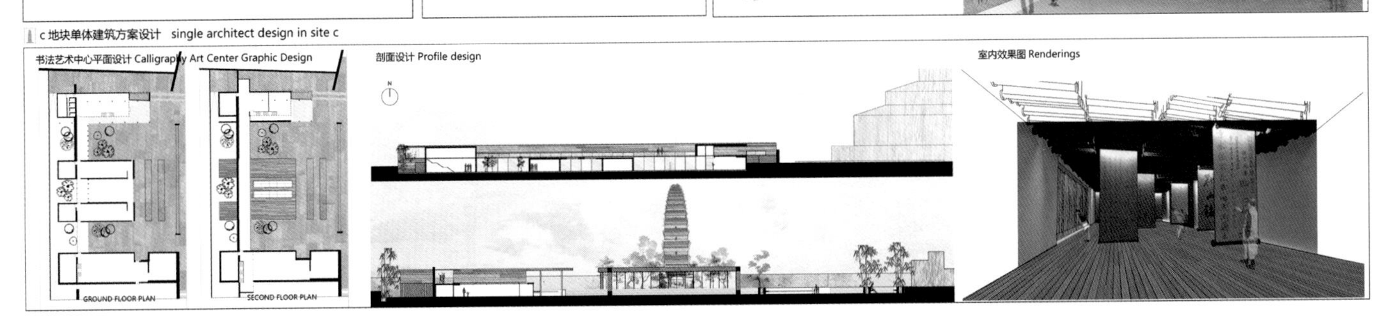

University: Xi'an University of Architecture and Technology
Designer: Liu Siyuan, Fang Yichen, Zou Yitong, Zhang Na, Jared Lambright, Quang Nguyen, MingJin Hong
Tutor: Xiao Li, Chang Haiqing, Su Jing, Albertus Wang, Lu Xu, Ren zhongqi
Course Name: Xi'an University of Architecture and Technology-University of Florida Joint Design Studio
Finished Time: Aug. 2014
Exchange Institute: University of Florida

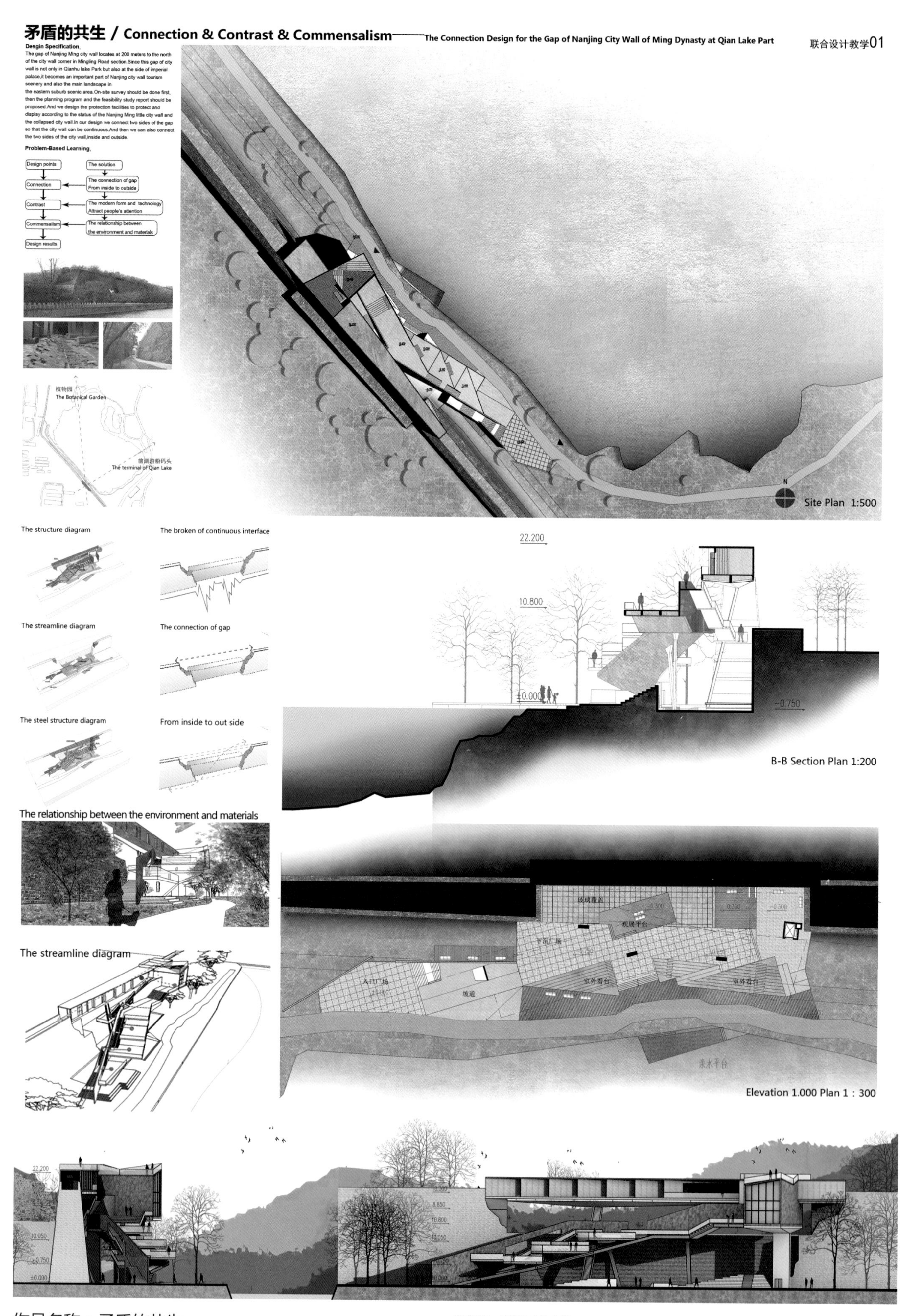

作品名称：矛盾的共生

The Contradiction of Symbiosis

院校名：南京工业大学

设计人：潘江海，黄豪，Linda Schmidt，蒋碧冰

指导教师：郭华瑜，Marina Stankovic，张蕾，李国华，胡振宇，姚刚，段忠诚

课程名称：南京明城墙前湖段缺口连接体设计

作业完成日期：2014 年 8 月

对外交流对象：德国莱比锡应用科学大学

矛盾的共生 / Connection & Contrast & Commensalism——The Connection Design for the Gap of Nanjing City Wall of Ming Dynasty at Qian Lake Part

Description of project:

The gap of Nanjing Ming city wall locates at 200 meters to the north of the city wall corner in Mingling Road section. This part was collapsed due to a long period of heavy rains in 1991.The ruins wasremoved and cleaned by Nanjing Cultural Relics Bureau in late 90s. This collapse reve
aled the inner short wall (51m long, 10m high, 2.5m wide) which was built in early Ming dynasty (1368-1644). Some part of the wall was even built by bricks made in Six Dynasties and Song Dynasty. This provides an important basis for the research of architectural history of Ming city wall, and has a great historical value.

The block analysis

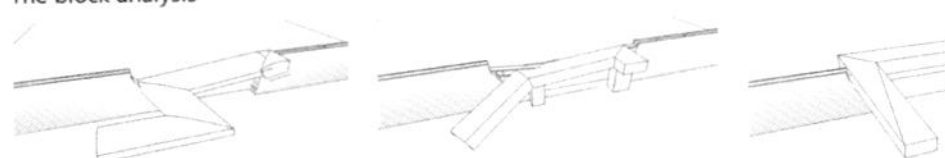

The distortion of the single pipe　The connection between two citywall　The integration of the blocks

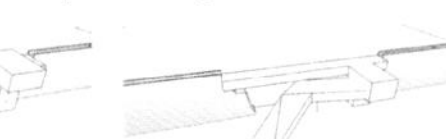

Make it simple　Make the streamline clear　Confirm the details

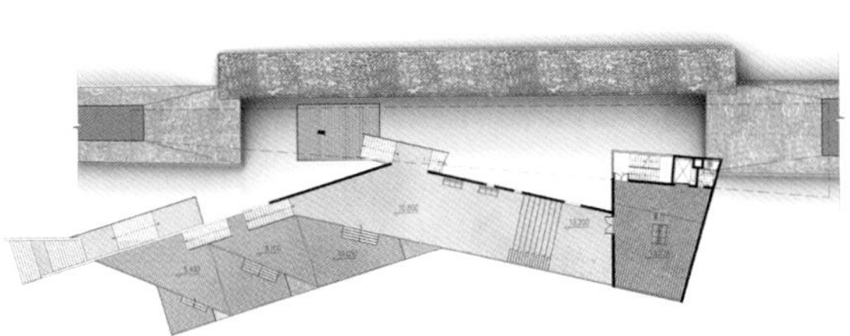

Elevation 14.200 Plan 1：700

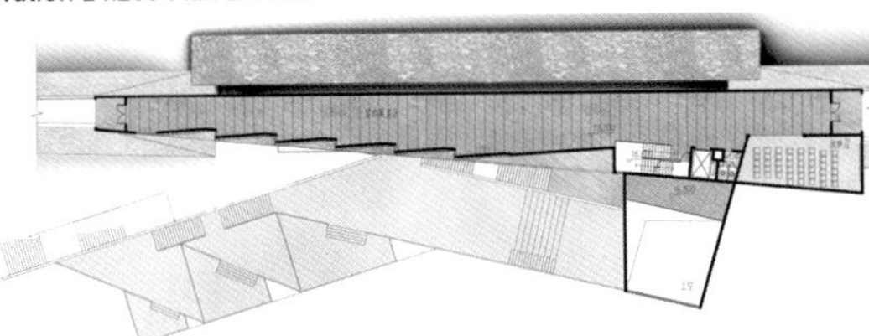

Elevation 19.300 Plan 1:700

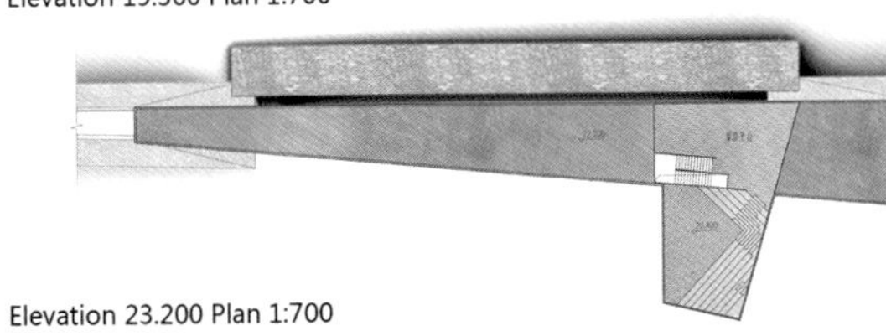

Elevation 23.200 Plan 1:700

A-A Section Plan 1:200

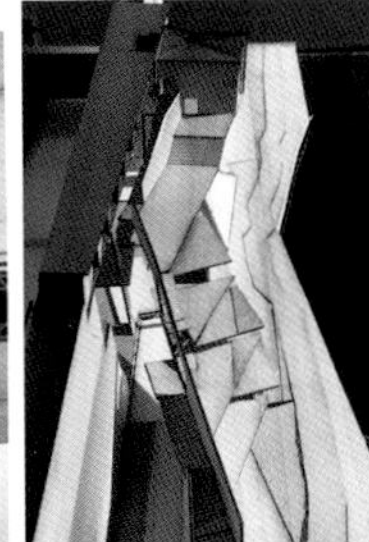

University: Nanjing Tech University
Designer: Pan Jianghai, Huang Hao, Linda Schmidt, Jiang Bibing
Tutor: Guo Huayu, Marina Stankovic, Zhang Lei, Li Guohua,Hu Zhenyu,Yao Gang,Duan Zhongcheng
Course Name: The Connection Design for the Gap of Nanjing City Wall of Ming Dynasty at Qian Lake Part
Finished Time: Aug. 2014
Exchange Institute: CUMT University

MAPPING THE VOID II

映射无形 II

A study of the "Edge" and the Port of Aarhus

Taking as our artistic expression and method the walks along the edge of the port. The walks were to be considered as works and narratives,which thematise the linear motif in the intersection of the horizontal and the vertical. Simultaneously they were to function as mapping devices and measuring tools;performative instruments with the ability to archive and reconstruct spatial experences and sensations.

The first walk-entitled"guided walk drift"-Took place on 23 September

The second walk-entitled"walk in one continuous line"-Took place on 3 October

The third walk-entitled"walking montage"-Took place on 14 October

The assignment was inspired by works, methods and techniques applied in Performance art, Minimal art and Land art.

G2: CABINET OF CURIOSITIES
"I AM SO HUNGRY....." "EVERYTHING SEEMS DIFFERENT" "ARE YOU SCIENTISTS?"

Thi Duy An Tran; May Damagaard Sorensen; Wang Qi;Shen Lu

奥胡斯港口与边界研究

这次课题将行走作为一种艺术表现和方法，沿奥胡斯的港口边缘进行了一系列的集体行走实验。将行走看作是一种艺术作品和记述过程，它将线性的主题归结为一种横向与纵向的交错和延伸。同时行走也是映射的手段和丈量的工具，是一种演示性的参照，附带有重组空间的经验和感受。

行走一：引导与漂流（2013. 09. 23）；
行走二：线之旅 -- 编舞的行走（2013. 10. 03）；
行走三：行走蒙太奇（2013. 10. 14）

行走的想法来源于表演艺术，极简艺术和大地艺术的作品、方法和技巧。

第二组：好奇的格子
Thi Duy An Tran; May Damagaard Sorensen;
王琪；沈璐

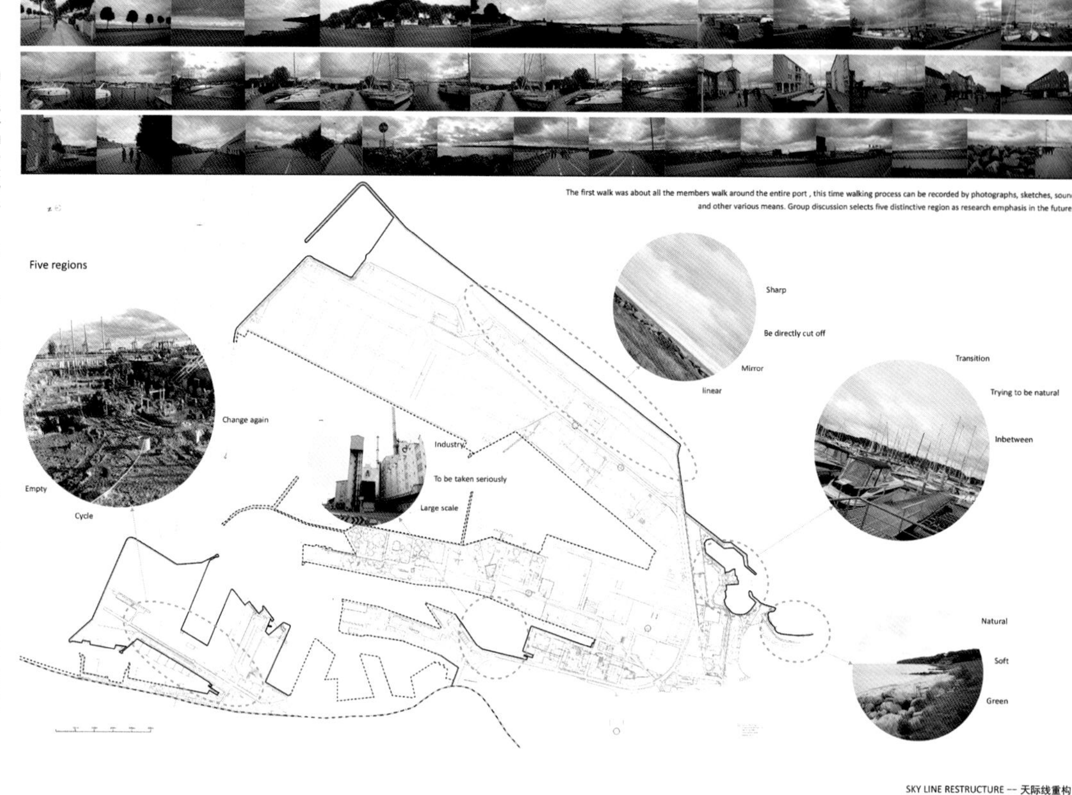

The first walk was about all the members walk around the entire port , this time walking process can be recorded by photographs, sketches, sound and other various means. Group discussion selects five distinctive region as research emphasis in the future

SKY LINE RESTRUCTURE -- 天际线重构

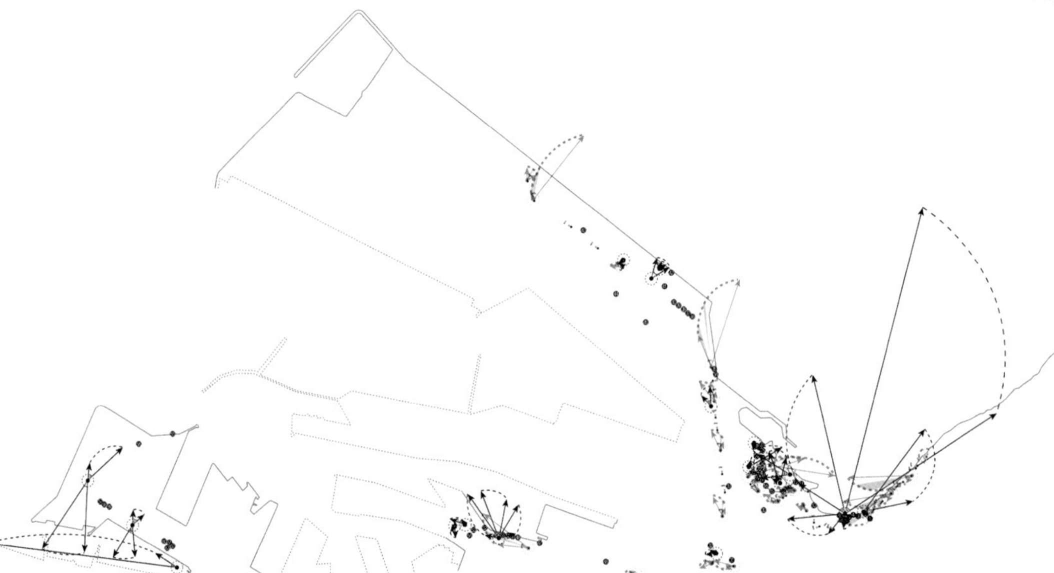

作品名称：好奇的格子
Cabinet of Curiosity

院校名：中央美术学院
设计人：沈璐，王琪，Thi Duy An Tran,May Damagard Sorensen
指导教师：何可人，王威，Anne Elisabeth Toft, Claudia Carbone
课程名称：2013 中央美术学院与奥胡斯建筑学院联合课题
作业完成日期：2013 年 10 月
对外交流对象：丹麦奥胡斯建筑学院

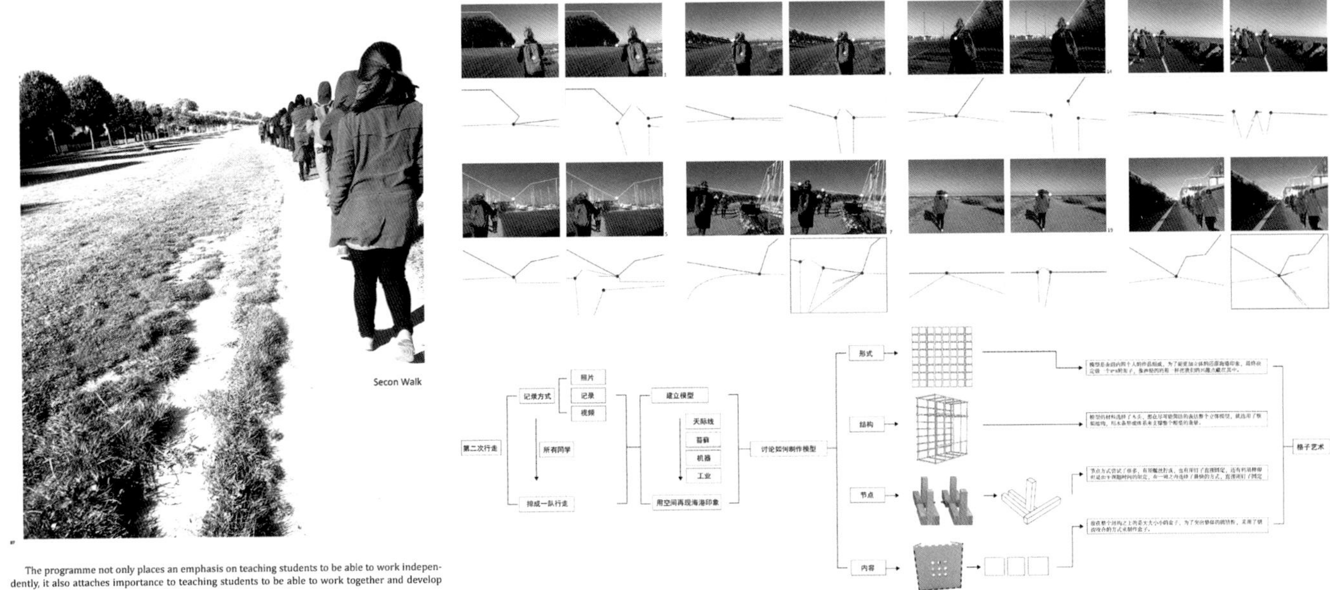

The programme not only places an emphasis on teaching students to be able to work independently, it also attaches importance to teaching students to be able to work together and develop projects in teams.

Teaching is process and method-oriented and is based on student supervision at their drafting tables and on situations involving joint critiques.

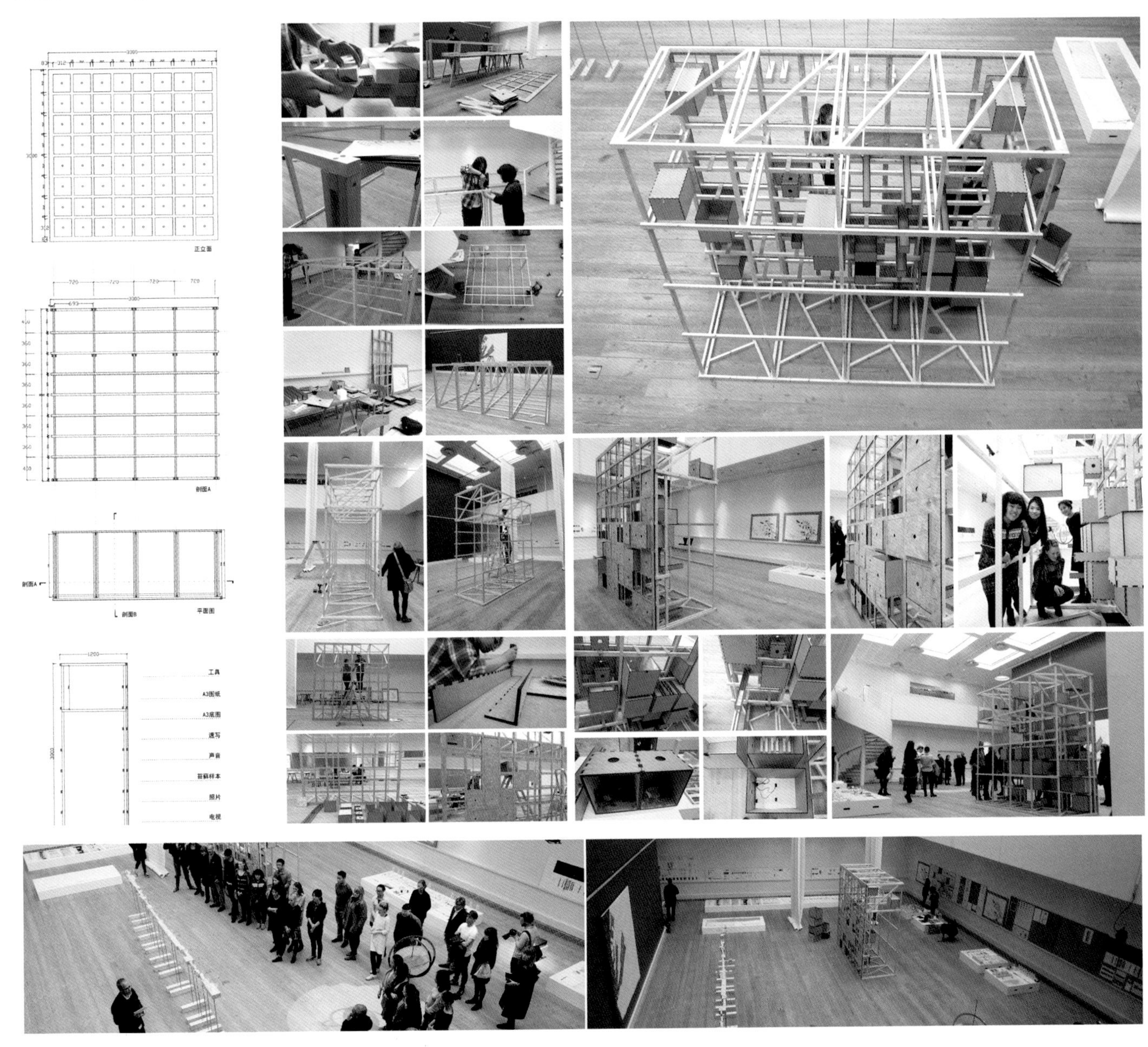

University: Central Academy of Fine Art
Designer: Shen Lu, Wang Qi, Thi Duy An Tran, May Damagaard Sorensen
Tutor: He Keren,Wang Wei,Anne Elisabeth Toft, Claudia Carbone
Course Name: Works from the CAFA I AAA Joint Studio Programme 2013
Finished Time: Oct. 2013
Exchange Institute: Aarhus Architecture School

行与市

本次国际联合设计的主题是关于城市交通基础设施的更新设计。深圳近年快速发展的多条地铁线路，一方面解决了深圳市小汽车过多造成的城市拥挤问题，另外一方面由于地铁等公共交通系统等规划与城市发展规划未能整合，造成地铁站点与站点周边城市建筑、城市空间未能高效复合，这造成了地铁站点本身所蕴含等商业潜力未能发挥，最典型的情况就是，地铁出站口与最近的商业综合体距离过远，导致地铁站点周边土地闲置、人烟稀少。本课题选择了深圳市地铁1号线车公庙站、大剧院站两个站点进行城市设计，旨在通过新的站点周边500米范围内城市空间、建筑功能的重新整合，激发地铁站点应有的活力。

交通与换乘的交集

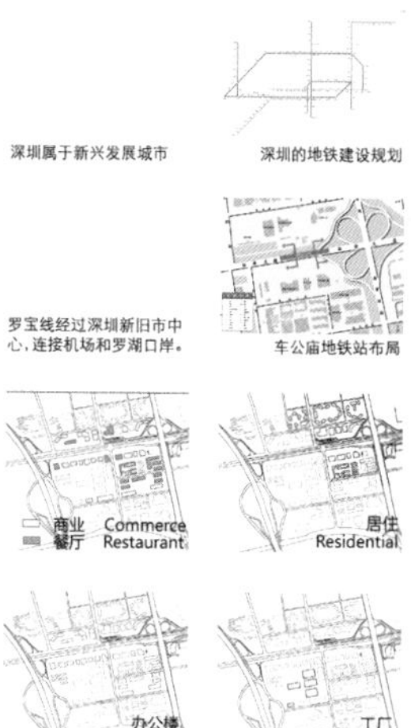

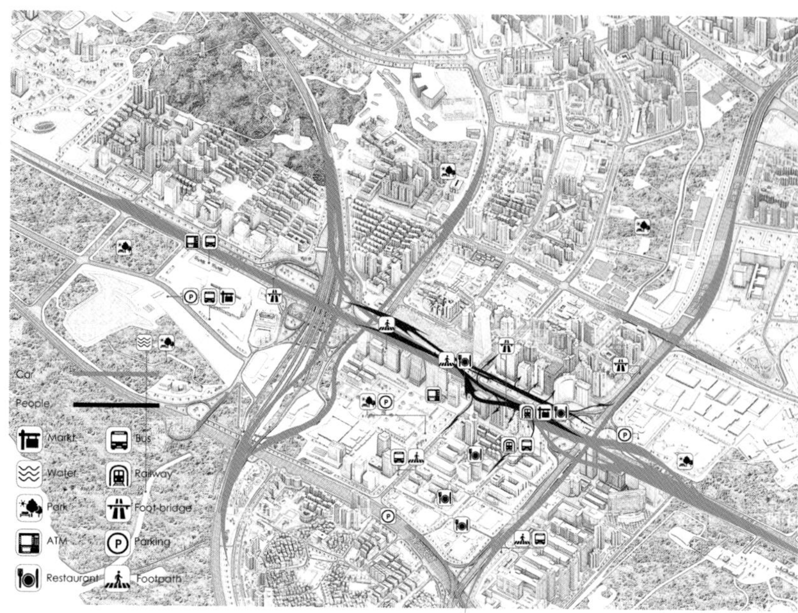

车公庙丰盛町商业街现存业态与问题分析

位于车公庙深南大道南北两侧，两线之间设有ABCD四区，布局为线性。

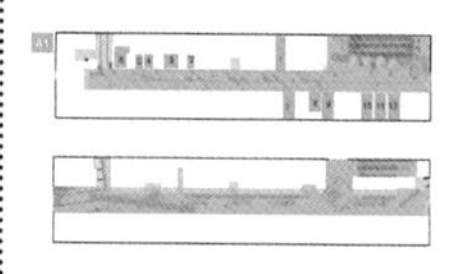
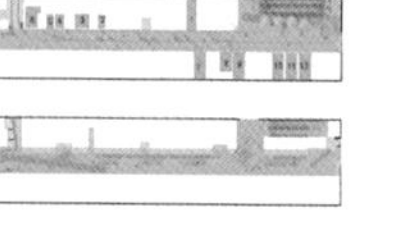

A区域距离地铁站较远，开业率仅为24.5%。

A1 accessibility is poor .

B区尚未正式开始招商。

B area has not yet been opened to the public .

C区与地铁站C出口连接，人流最多区域。

Area C are closely linked to MTR station (exit C).

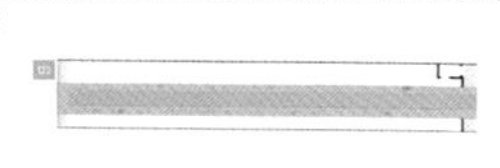

D区为一条贯通深南大道南北两侧的步行通道。

D area connect north and south on both sides of theShennan Road.

问题

建筑、景观、交通被宽阔的道路阻隔，整体性的缺乏使得地铁站点的交通、商业及景观枢纽价值未能得到充分发挥。

W ithin

站点只与地下商业街相连与周边建筑缺乏联系，天桥与周边隔绝使用率低，商业街内功能基本以餐饮为主，也体现地段商业外延性的匮乏。

各种城市交通方式间缺乏有效衔接，建筑的私密性与城市交通的开放性之间缺乏过渡空间，导致到达与离开的不便利。

策略

建筑、景观、城市交通通过地上、地下关联串通，形成直接与立体的联系体系，建立立体化空间，形成人车分流体系，提升交通系统的通行能力

在建立立体、便捷、整体的联系体系基础上，进行功能拓展，强化交通功能与生活功能的融合，复合多种功能，激发地段的城市活力。

地铁站点与建筑内部空间进行更新与改良，将室内中庭空间与城市交通转换场所相结合，形成多重换乘节点空间，减少城市交通接驳时间。

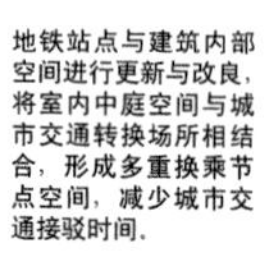

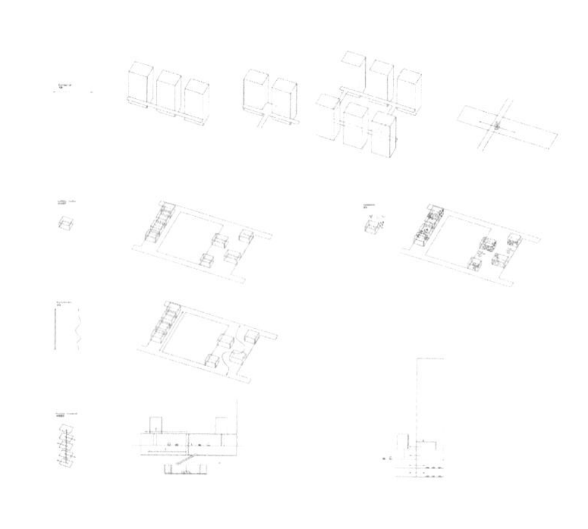

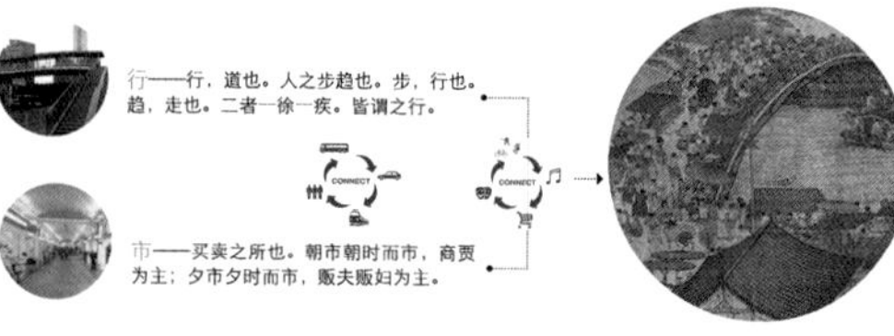

行——行，道也。人之步趋也。步，行也。趋，走也。二者一徐一疾，皆谓之行。
市——买卖之所也。朝市朝时而市，商贾为主；夕市夕时而市，贩夫贩妇为主。
CONNECT

行，说的是地铁站是是城市交通系统的一部分，它快速、高效，在这里产生了大量而集中的人口流动。

市，说的是随着人流聚集而发生的商业行为。商业功能和交通功能的融合与冲突，在各种人流交叠中展开。

我们希望将这两种矛盾又统一的特性结合起来，解决车公庙现状中地面空间利用不足、地下流线长且混乱的问题，在组织立体换乘枢纽空间的同时，为市民提供便捷轻松的购物以及提高的生活品质，视实现城市综合体与城市交通系统融合。

作品名称：行与市

Traffic and City

院校名：东南大学

设计人：时楠，张倩倩，陈凤娇，克里斯托夫

指导教师：韩晓峰，葛明，龚恺，莫拉德·亚德力奇

课程名称：东南大学—同济大学—深圳大学—维也纳理工联合设计课程教学

作业完成日期：2013年10月

对外交流对象：奥地利维也纳理工大学

By And Buy

This is the theme of the international joint design of urban transportation infrastructure renewal design.In recent years,Shenzhen had undergone a rapid development of metro line,the operation of many metro lines effectively solved Shenzhen urban congestionproblems caused by too much cars.However on the other hand,Public transit system planning such as metro travel were failed to integrate with Urban development planning,metro stations had unefficient composition with surrounding buildings and urban space ,the result is metro station could not bring its commercial potential into full play .One typical problem is metro station exits complex are too distant to commercial complexes, even the nearest that leaves surrounding land of metro stations fallow and sparsely populated .This urban design project chosen Che Kung Temple and Grand Theater Station on No. 1 metro line as the study area,aimed at inspiring the due vigor of metro stations by renewal design of integration of urban space and building function around 500 m range of metro stations.

Transit infrastructure：Hybrid Transitivity

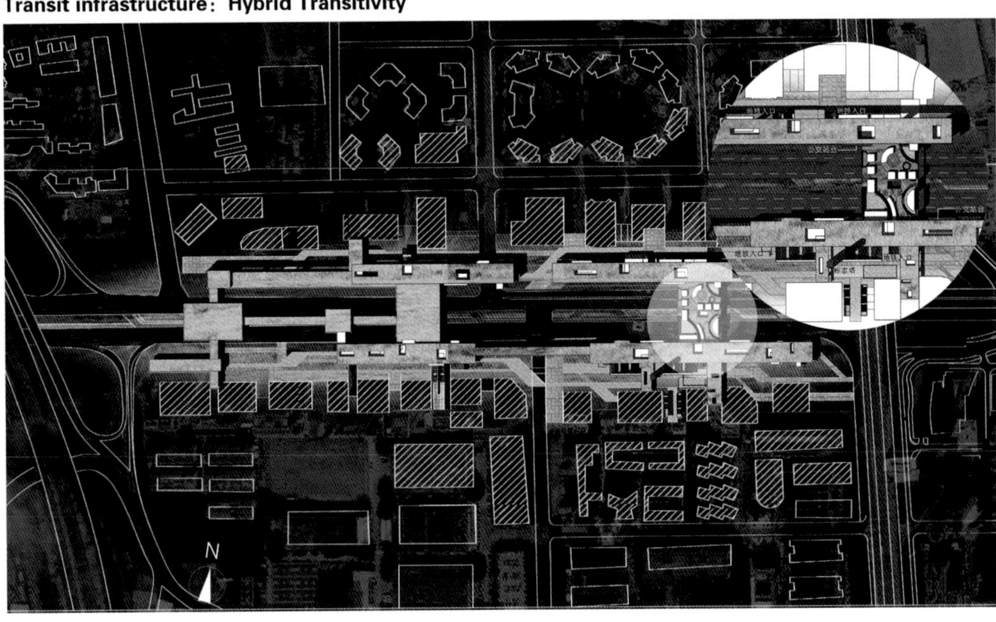

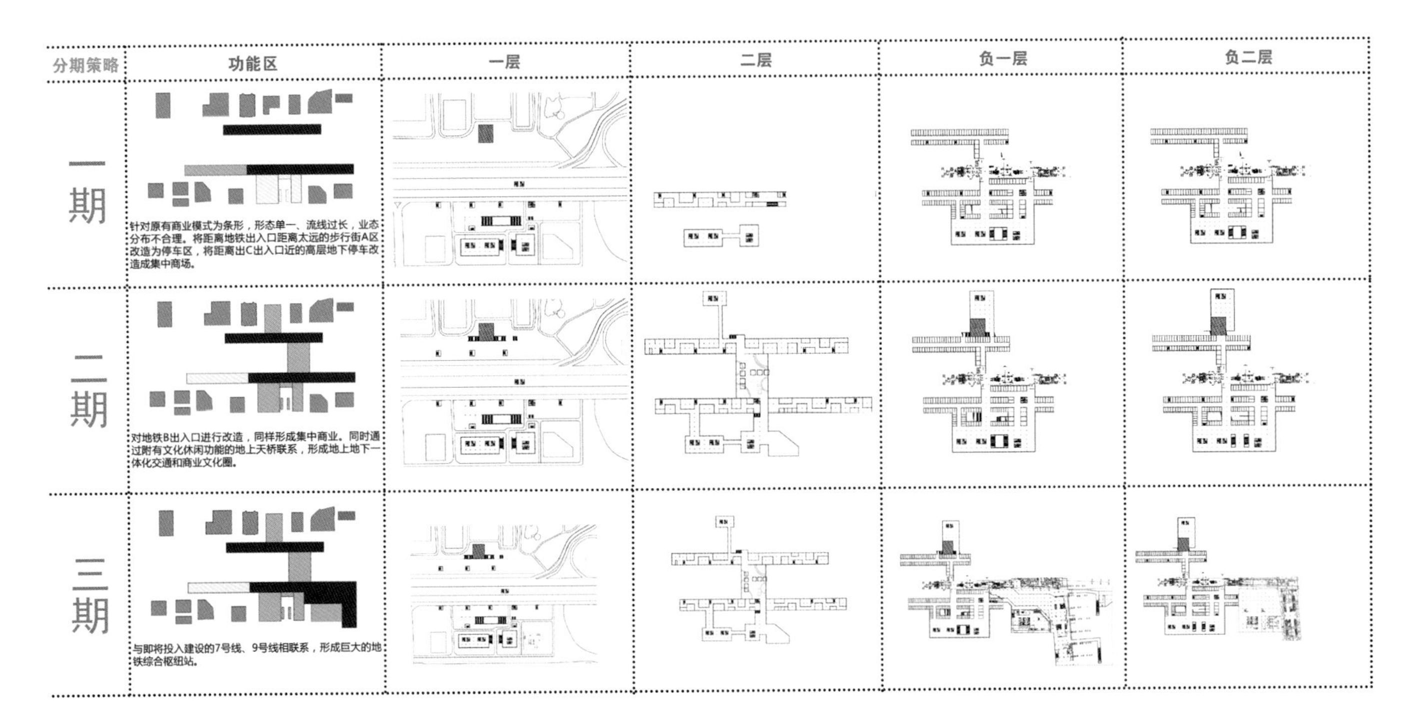

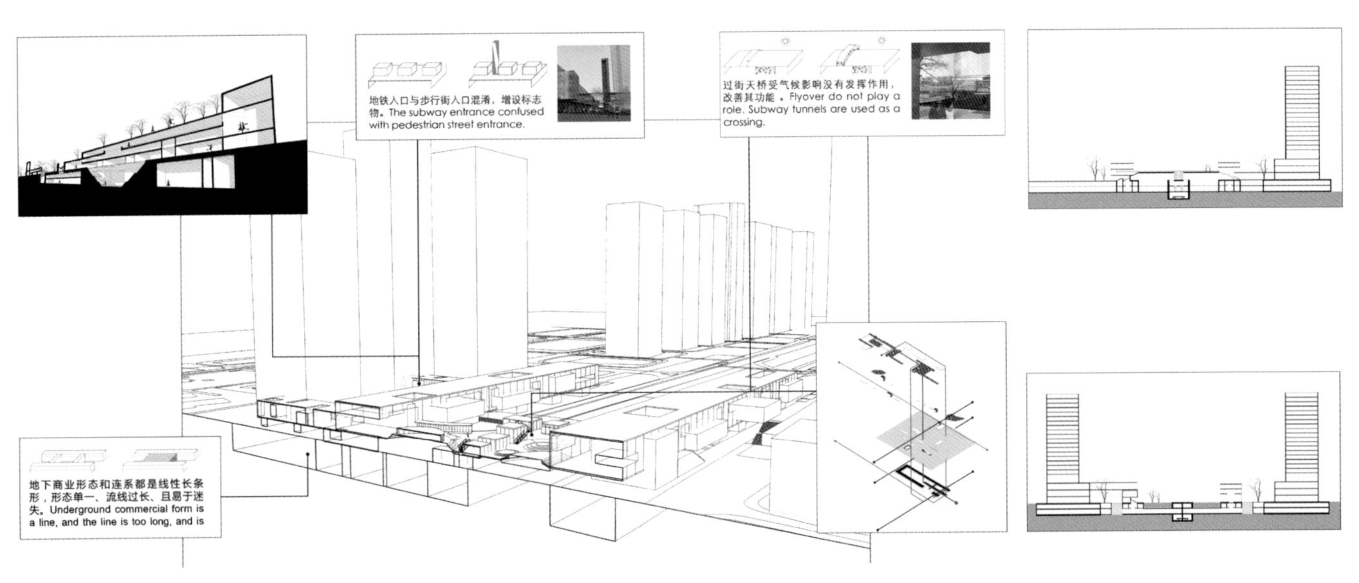

University: Southeast University
Designer: Shi Nan, Zhang Qianqian, Chen Fengjiao, Christoph
Tutor: Han Xiaofeng,Ge Ming, Gong Kai, Mladen Jadric

Course Name: SouthEast University - TongJi University - Shenzhen University - Vienna Tech University Joint Design Studio
Finished Time: Oct. 2013
Exchange Institute: Vienna Tech University

POST-WASTE CITIES Reprotocolising Bio-mass

CONCEPT

This Urban Codebook focuses on using bio-mass waste and features one specific geographic area to the north of Tucson, Arizona close to the case study site. The built landscape exhibits a marked contrast in both the scale of development as well as bio-mass volumes produced. Catalina, is a rural suburb with large development plots and a low population density, whereas the neighbouring suburb of Saddle Brooke is more modern and has high density, high value developments with associated shopping and entertainment facilities, and produces significantly more bio-mass per resident.

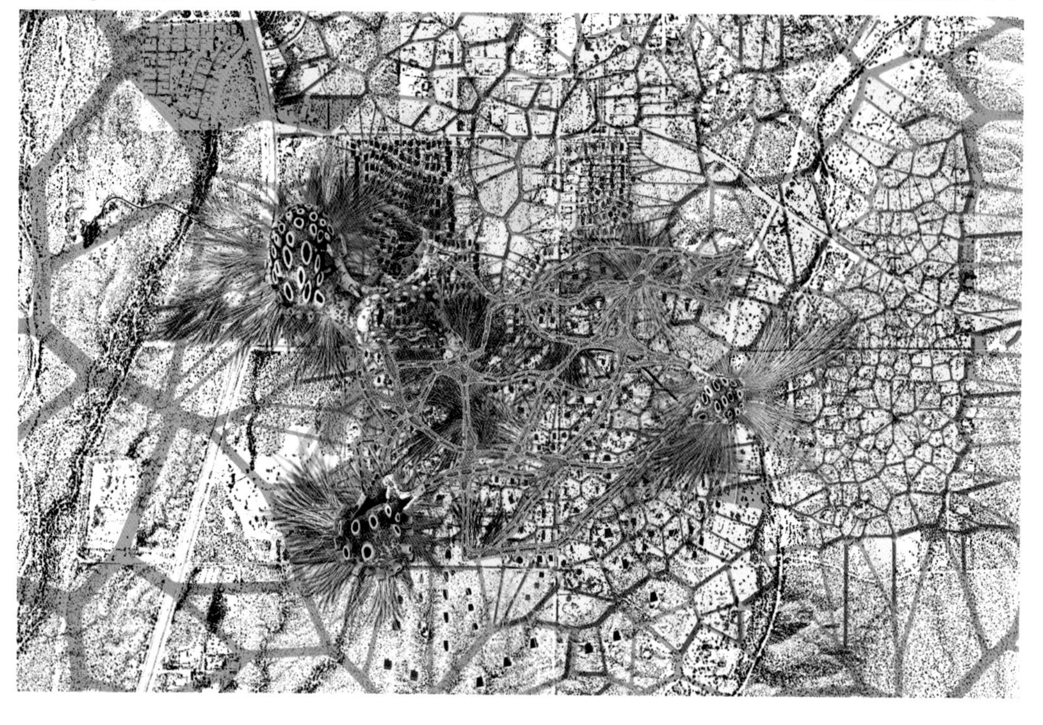

These waste streams are produced within urban and suburban areas and follow an iterative protocol revealed by the robotic mechanism, which demonstrates how the biomass/waste can be plugged into an existing scheme and generate new divergent dynamic systems and new urban scenarios. The volumes of waste produced in these areas, which are expressed in bio waste values, would allow for community energy demand to be supplemented by small generation projects. The application of a open bio-technological system suggests possible urban morphologies that operate by using real time data to stream reused waste.

SECTION

GROWING TRACE

Path system stage 1

Path system stage 2

Path system stage 3

DEFINE BIO-UNITS

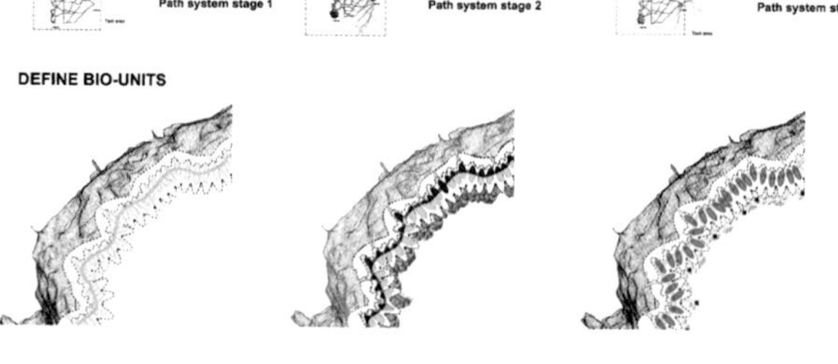

MINIMAL NETWORK

Based on the optimised network of the Attini colony and the analog experiment inspired by Frei Otto an efficient minimal network was developed.
Multiple simulations related to the micro-climate and the shortest routes were run to explore and determine what optimum structure would connect six separate bio-units.

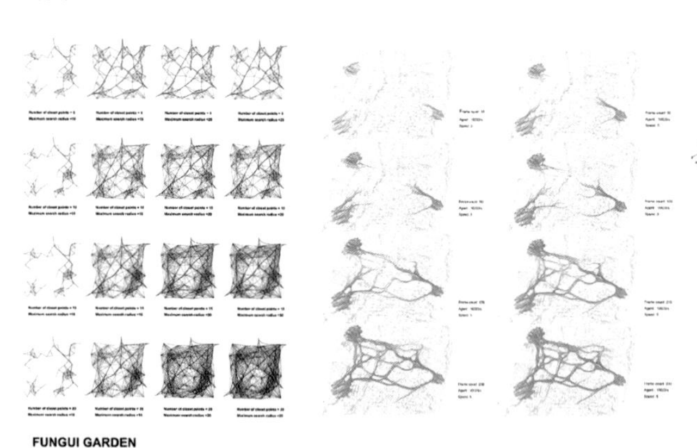

3D MASTER MAPPING

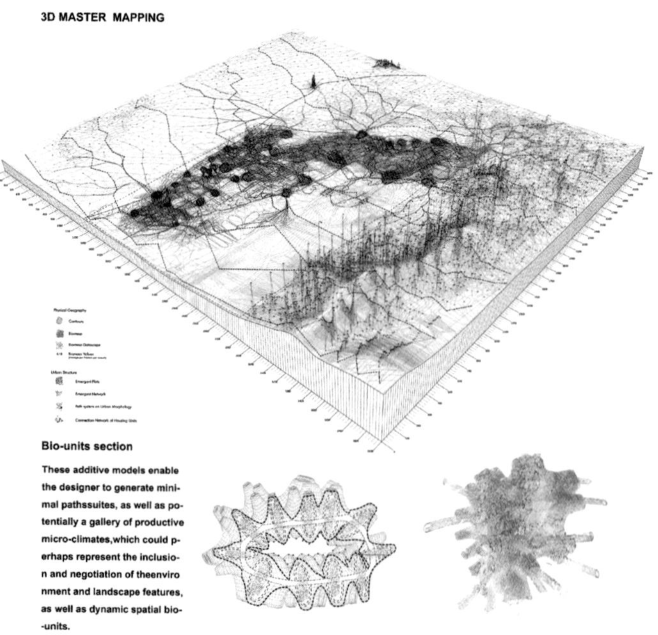

FUNGUI GARDEN

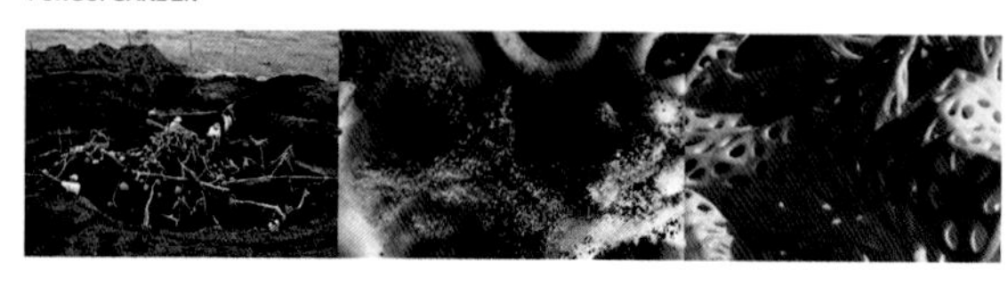

Bio-units section

These additive models enable the designer to generate minimal pathssuites, as well as potentially a gallery of productive micro-climates,which could perhaps represent the inclusion and negotiation of theenvironment and landscape features, as well as dynamic spatial bio-units.

作品名称：变废为宝之城
Post-Waste Cities

院校名：合肥工业大学
设计人：唐洪亚，任国乾，贾宁
指导教师：Claudia pasquero，Marco Poletto
课程名称：伦敦大学学院毕业设计
作业完成日期：2014 年 08 月
对外交流对象：英国伦敦大学学院

POST-WASTE CITIES Reprotocolising Bio-mass

MICRO CLIMATE

Climatic data input from the counties of Pinal and Pima in Arizona. The time and frequency of data input aid with the biodigesting of bio-mass. The grey coloured sections indicate peak performance, where as the colored sections show reduced output of bio-digester.

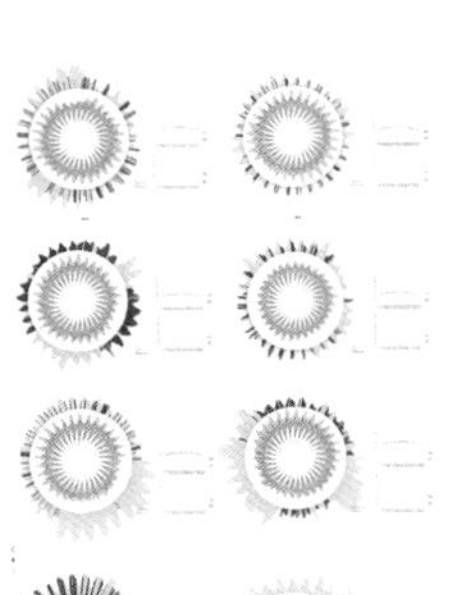

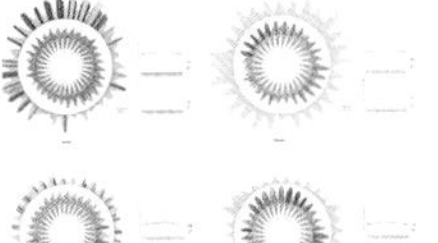

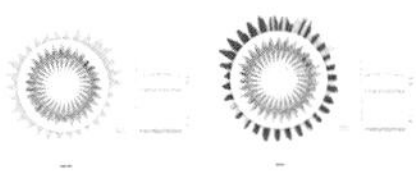

NETWORK DISTANCE CALCULATION

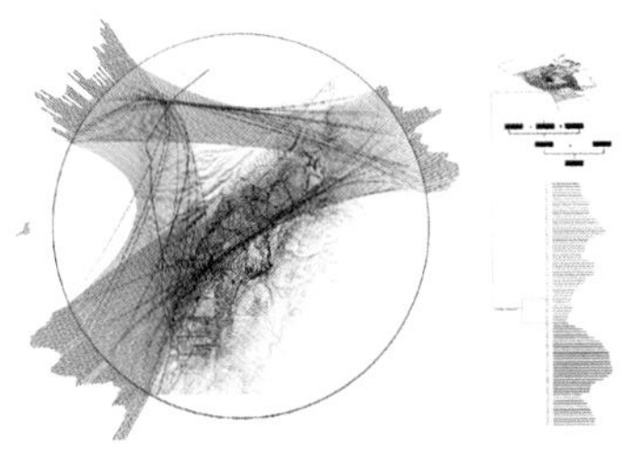

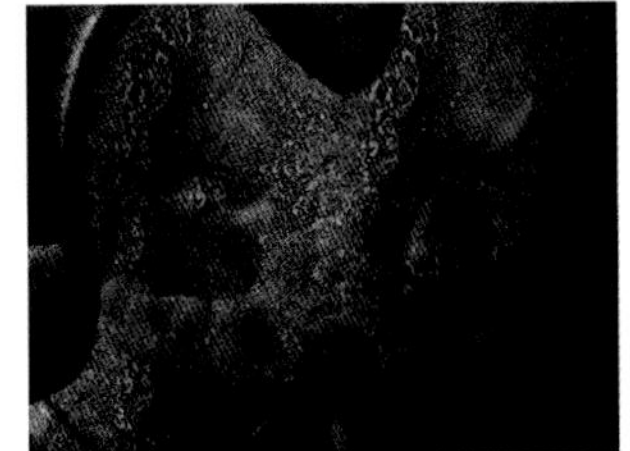

GROWING MODEL

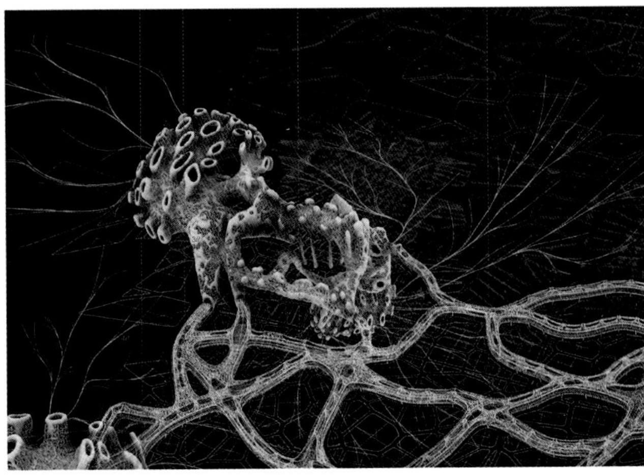

Bio-mass is biological material derived from living, or recently living organisms. It most often refers to plants or plant-based materials which are specifically called lignocellulos bio-mass.

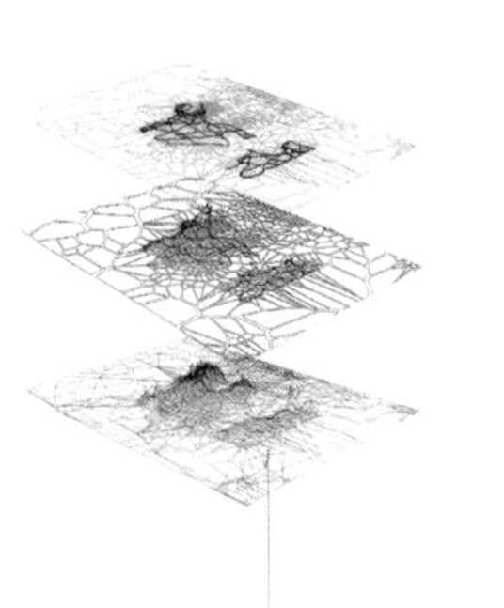

Through the use of a mathematical modelling and experimental data it was possible to efficiently place bio-reactor nodes in an urban setting, redesigning the citified landscape around tomorrow' s energy requirements.

VALUE MAPPING

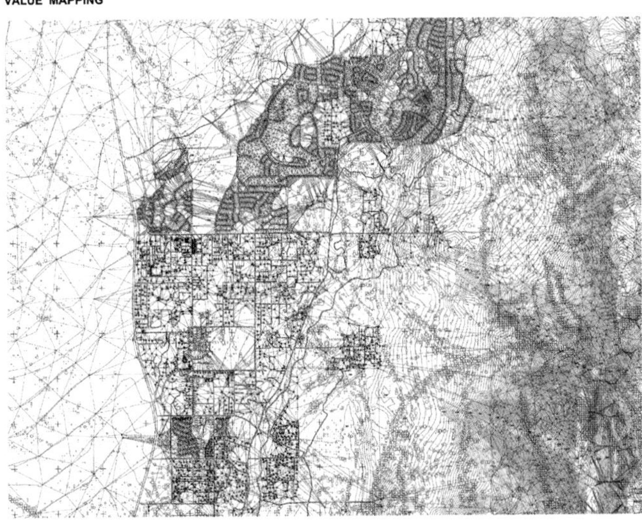

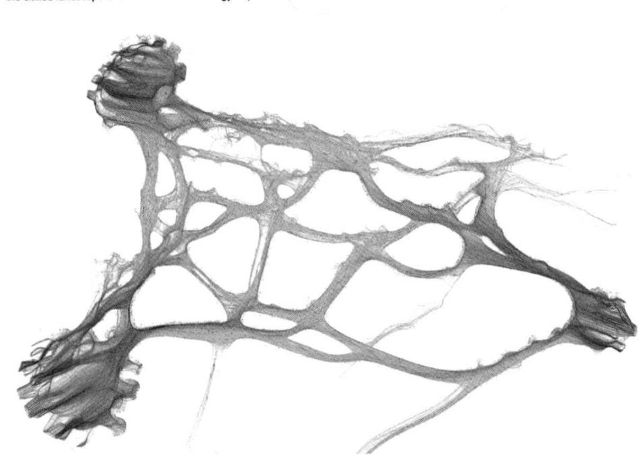

University: Hefei University of Technology
Designer: Tang Hongya, Ren Guoqian, Jianing
Tutors: Claudia pasquero, Marco Poletto
Course Name: University of London graduation design
Finished Time: Aug. 2014
Exchange Institute: College of London University

REBUILD THE CHAIN Green-Eco-Sustainable Town Design toward Glocalization

Master Plan 01

背景介绍 BACKGROUND

余市位于北海道仁木地区，以威士忌酒和水果种植业闻名。和其他日本的村庄一样，余市拥有丰富的自然资源，对于我们这种外来者来说，余市的种种风景都非常有吸引力。但实际上，余市正面临着严重的人口减少与老龄化以及经济衰退问题。

Yoichi is a small town in Hokkaido which was well-known by its NIKKA WHISKY and fruits.As the other villages of japan,Yoichi is rich in natural resources.Through the view of us outcomers,the ocean ,the rivers,the mountains, and the earth are charming. But in fact it faces deflation, an ageing and shrinking population and economy sagging problems.

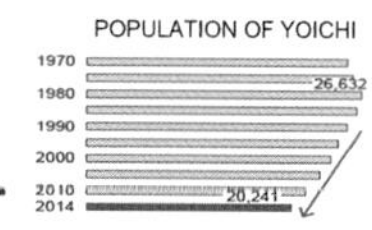

场地研究 LANDFORM STUDY

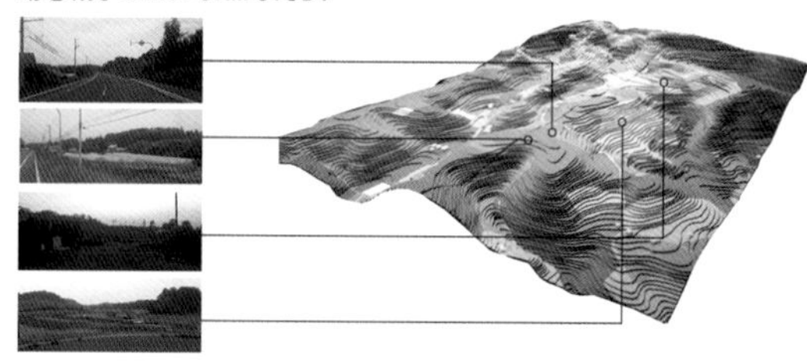

规划场地位于丘陵地带，山都不高，山谷比较平，场地内主要有三条山脉延伸，山脉中形成了两条山谷

The site located at moutainnous zone,the moutains are relatively short,the valleys are flat.There are mainly three banding muntains on the site.Two valleys are surounded by them.

空间策略 SPACE STRATEGIES

空间在这里依据可到达性的强弱分为两个层级：核心区域与周边区域。核心区域以其较高的开放度和丰富的活动向周边区域的农场形成辐射，带动整体的发展。

Area is divided into two partsMain Area and Surrounding Area.High degree of openness and variety of activities in Main Area can help the Surrounding Area develop.

核心区域内部则根据其场地自身特点及周边情况规划出具有不同的功能的空间，各空间相互联系，形成一条完整的游览路径。

There are spaces with different functions planned in the Main Area. Every space is connection to others which forms a complete sight seeing path.

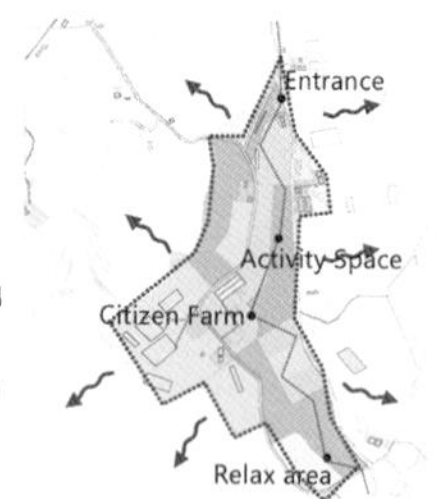

时间策略 SEASON STRATEGIES 规划结构 PLANNING STRUCTURE

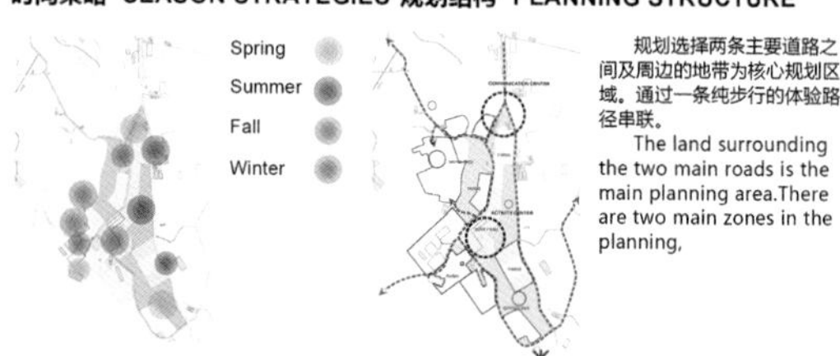

规划选择两条主要道路之间及周边的地带为核心规划区域。通过一条纯步行的体验路径串联。

The land surrounding the two main roads is the main planning area.There are two main zones in the planning.

北海道的冬季严寒且漫长，在此期间市民没有活动可以进行。针对这项情况，对各个季节进行不同的活动规划，以修复时间链。

Hokkaido has a long and boring winter ang people have nothing to do then. We rebuild the time chain by arranging different activities for different seasons.

重点区域设计 KEY AREAS DESIGN

经过现场调研，我们发现场地内缺少接待志愿者的功能用房同时，生态农场内部遭受严重的昆虫灾害，所以我们选择了两个重点区域进行设计，移动志愿者中心和小鸡培养站。针对我们提出的休闲谷所缺少的功能补充而进行的单体设计，形成一种有效的可移动的功能设施。

we choose two areas to make further design, and their funtions are mainly about volunteer' truck and chicken station.

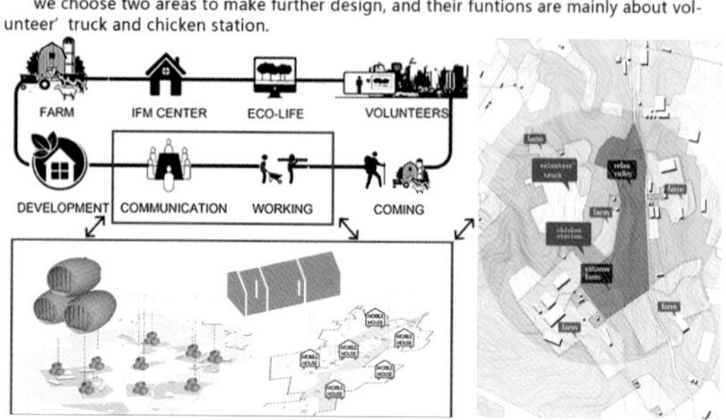

LEGEND
Existing private farm
Key area of relax valley
Forest & Oxygen bar
Key zone
green house
farm house

Master Plan

概念 concept

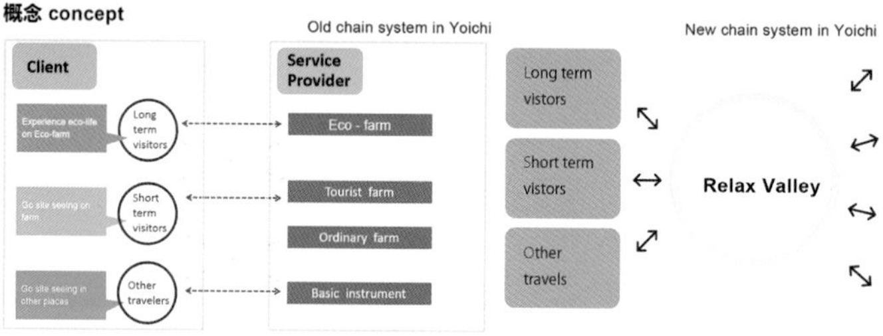

我们希望通过这次规划设计提供一种新的方法，结合目前余市已有的产业循环，通过修补余市产业链，增加吸引游客的设施，激发场地活力，刺激余市经济增长，使年轻人留在余市。

We hope this plan would provide a new way to solve the situation. Considering the present situation of Yoichi industry, improved infrastructure and guest services , recall the site activity to rebuild the industry chain in Yoichi

规划解析 PLANNING ANALYSIS

总体规划 LANDFILL SITES

流线分析 FLOW ANALYSIS

用地性质 LAND USAGE

作品名称：生态链接，产业重塑——国际化绿色生态城市规划设计
Rebuild the Chain — Green-Eco-Sustainable Town Design toward Glocalization

院校名：重庆大学
设计人：王凌云，曾渝京，尚白冰，李超
指导教师：龙灏，杨培峰，谷光灿，森傑（日）
课程名称：重庆大学—北海道大学联合设计
作业完成日期：2014 年 8 月
对外交流对象：日本北海道大学

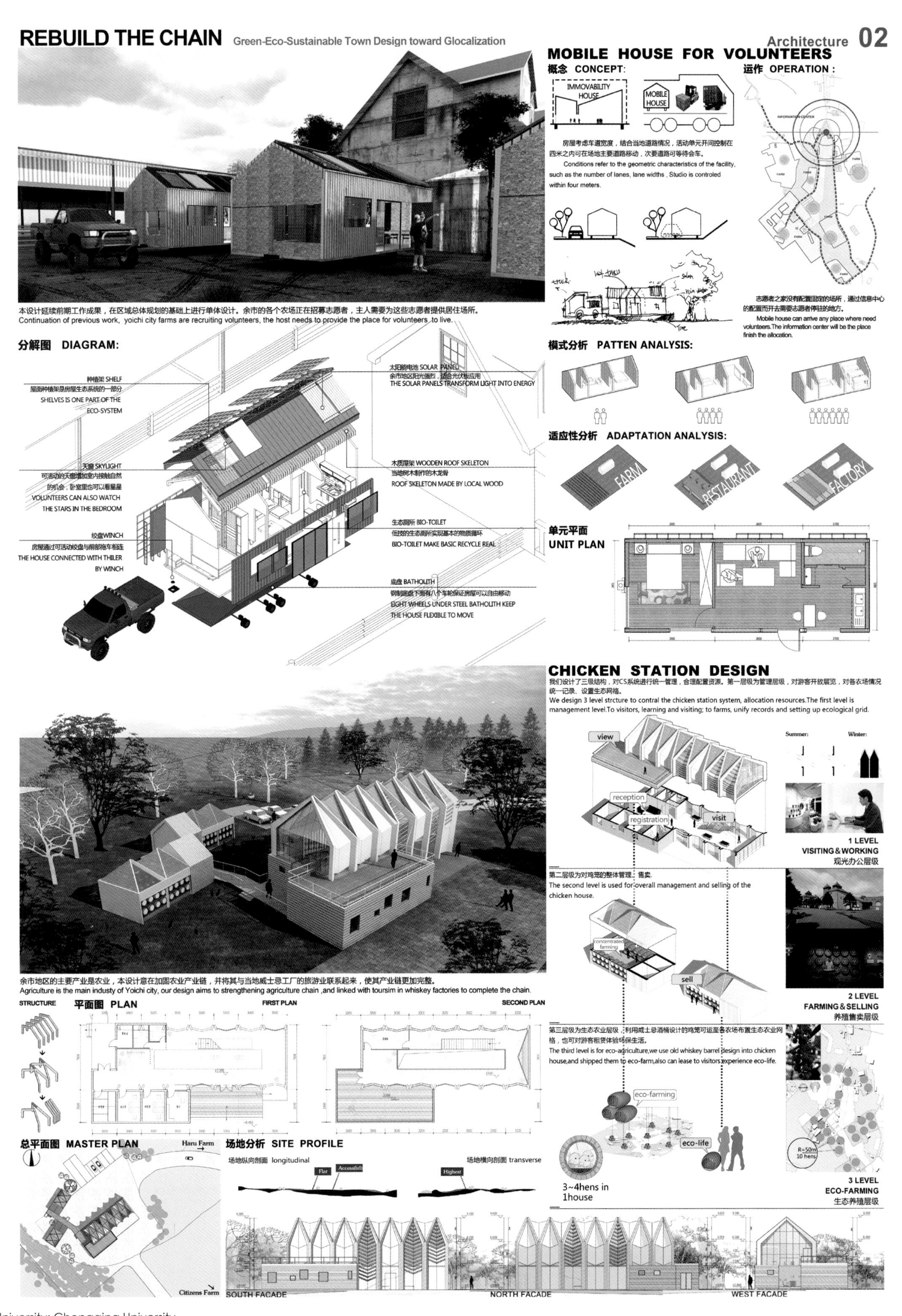

University: Chongqing University
Designer: Wang Lingyun, Zeng Yujing, Shang Baibing, Li Chao
Tutor: Long Hao, Yang Peifeng, Gu Guangcan, Moritex
Course Name: Chongqing University-Hokkaido University Joint Design Studio
Finished Time: Aug. 2014
Exchange Institute: Hokkaido University

Location - Gap

Site is located in the position of the wall near the south Mingling road in front of lake . The east side of site is close to the lake, which is Zhongshan Ling and Mingxiao Ling; South of site is Yueyahu park; In addition to the Ming imperial palace, the west side of site distributes residential areas.According to the distance of each big scenic spot on the drawing, the base is in the middle of residential area and zhongshan scenic area. The crowd activities are hold very frequently.

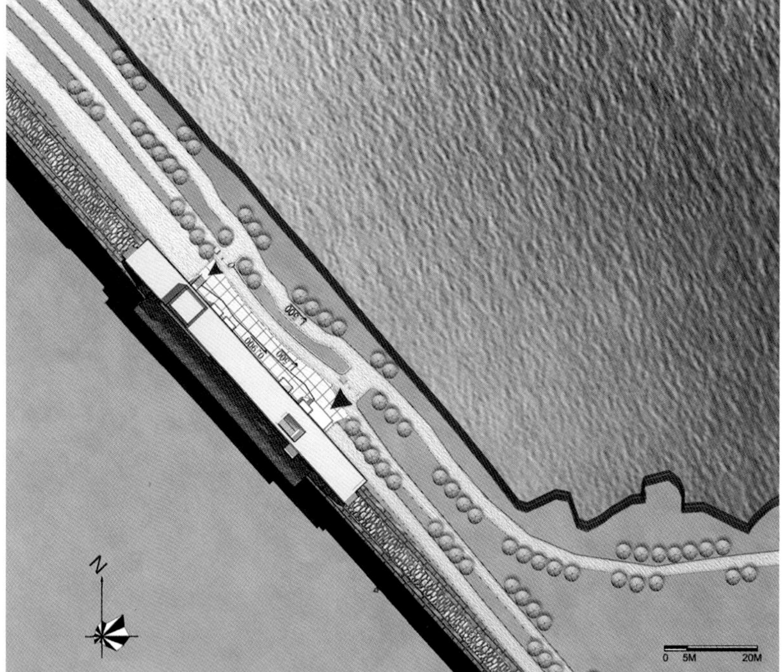

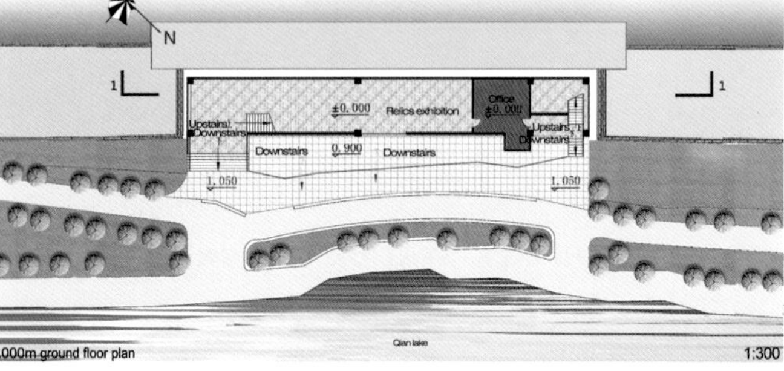

Spatial analysis

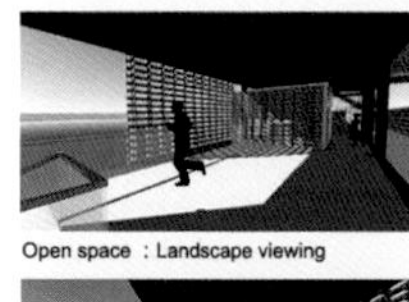
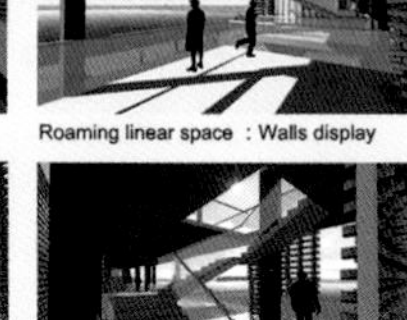
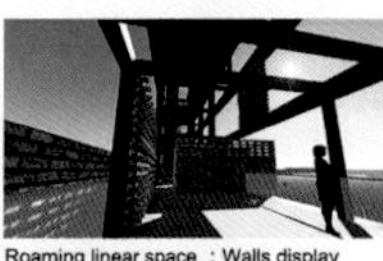

Open space : Landscape viewing

Roaming linear space : Walls display

Roaming linear space : Walls display

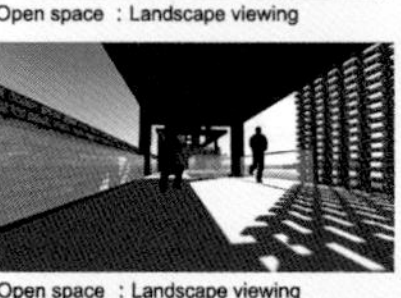
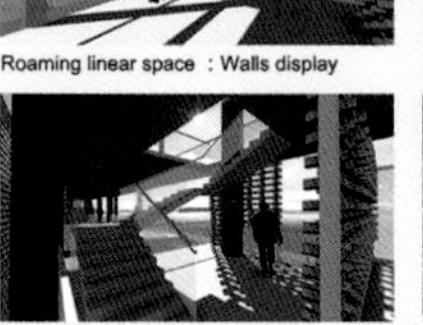

Open space : Landscape viewing

Traffic space : Vertical transportation

Enclosed space :Indoor exhibition gallery

3.600m floor plan 1:300

7.200m floor plan 1:300

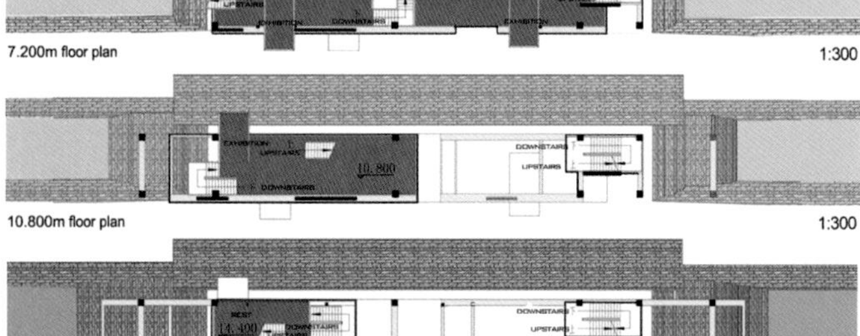

10.800m floor plan 1:300

14.400m floor plan 1:300

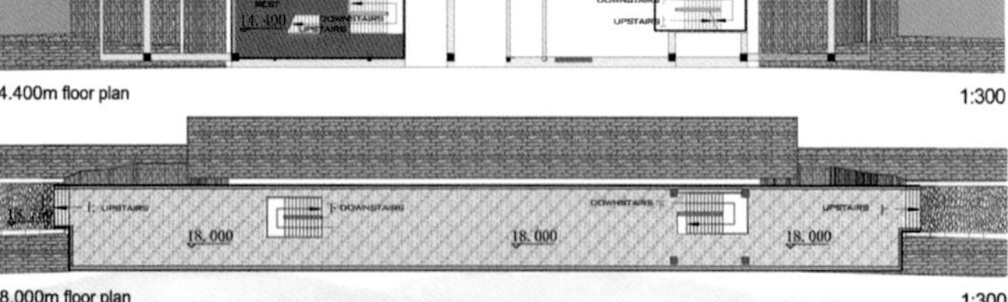

18.000m floor plan 1:300

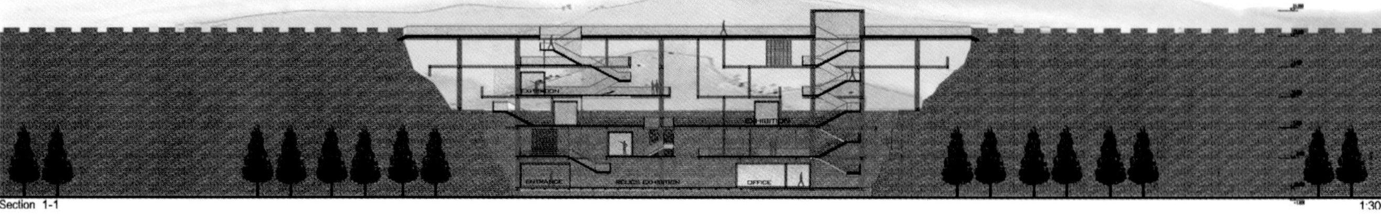

Section 1-1 1:300

作品名称：城墙博古架

Modern Frame — The Design of a Connection for the Gap of Nanjing Wall

院校名：南京工业大学
设计人：杨璐，史悦，Chantal Marschall，丁少华，冯赫
指导教师：郭华瑜，Marina Stankovic，张蕾，李国华，姚刚，朱冬冬
课程名称：南京明城墙前湖段缺口连接体设计
作业完成日期：2014 年 8 月
对外交流对象：德国莱比锡应用科学大学

Modern Frame

The Design Of A Connection For The Gap Of Nanjing Wall

Design Concept

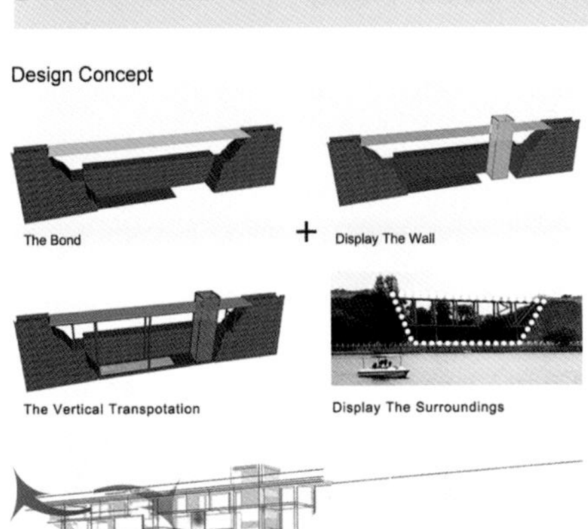

Concept of the "Ancient Frame":
"Ancient Frame" is a indoor multi-storey wooden frame which can display antique treasures, similar to the bookshelf.

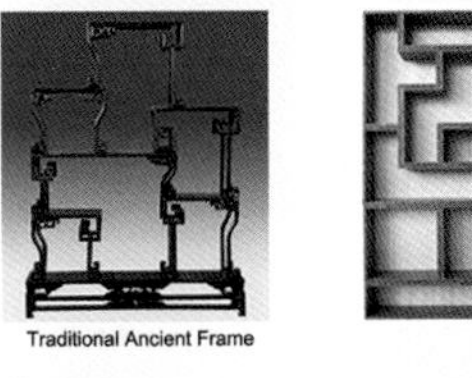

Traditional Ancient Frame

Modern Ancient Frame

Space

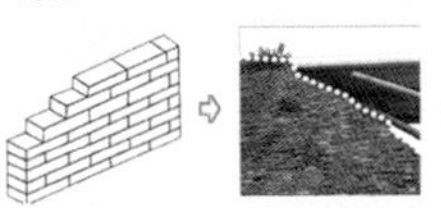

Ancient frame can be separated in different form of grids which display a variety of antiques.

Form

Current Situation

Concerning to the walls on both sides of the fracture, the late design can considered the exhibition. It is different from the whole period of the Ming city wall, low wall built in the han dynasty, is a more ancient ruins. In addition, the existing temporary structure can be used partly.

The crowd and functional requirements

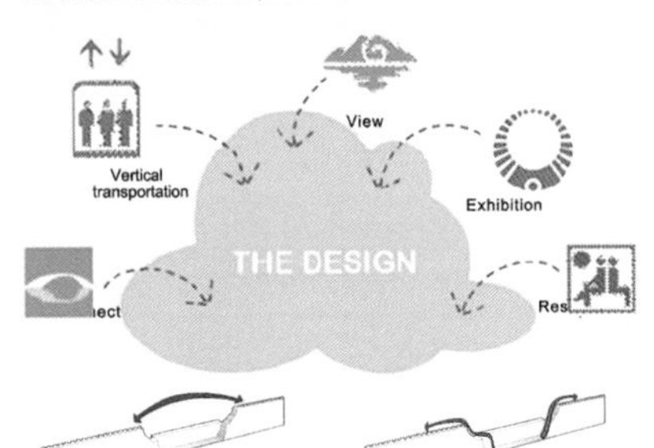

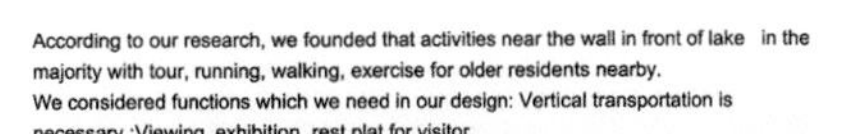

According to our research, we founded that activities near the wall in front of lake in the majority with tour, running, walking, exercise for older residents nearby.
We considered functions which we need in our design: Vertical transportation is necessary ;Viewing, exhibition, rest plat for visitor.

Analysis

Circulation organization diagram / Function analysis

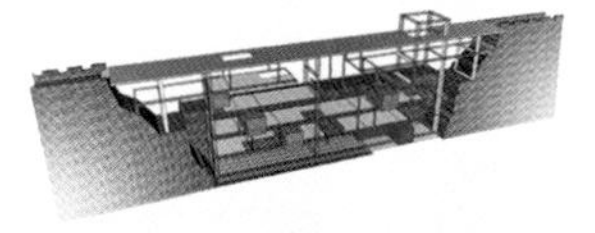

View analysis

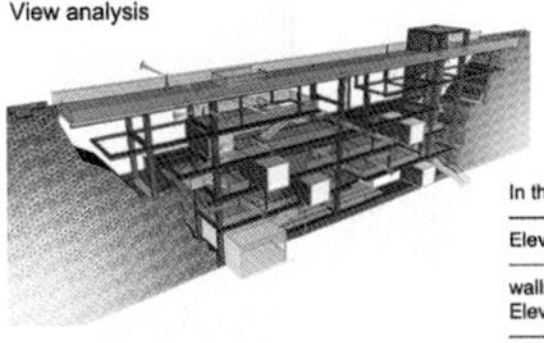

In the bottom of the crowd
—— visit ground sites;
Elevation under 10m
—— short wall and two side walls fracture;
Elevation above 10m
—— both sides of the landscape.

Section Perspective

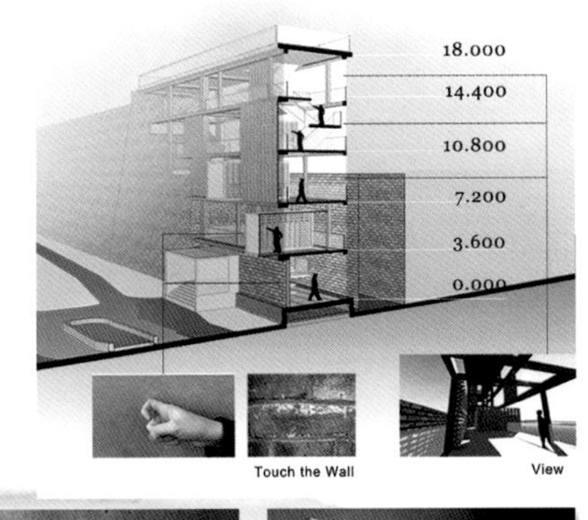

Touch the Wall

View

University: Nanjing Tech University
Designer: Yang Lu, Shi Yue, Chantal Marschall, Ding Shaohua, Feng He
Tutor: Guo Huayu, Marina Stankovic, Zhang Lei, Li Guohua, Yao Gang, Zhu Dongdong
Course Name: The Connection Design for the Gap of Nanjing City Wall of Ming Dynasty at Qian Lake Part
Finished Time: Aug. 2014
Exchange Institute: CUMT University

作品名称：共享居室——2050年的曼哈顿
Shared Living Rooms — Manhattan 2050

院校名：同济大学
设计人：张林琦，郗晓阳
指导教师：杨春侠，黄林琳
课程名称：同济大学—美国圣路易斯华盛顿大学联合城市设计课程教学
作业完成日期：2014 年 07 月
对外交流对象：美国圣路易斯华盛顿大学

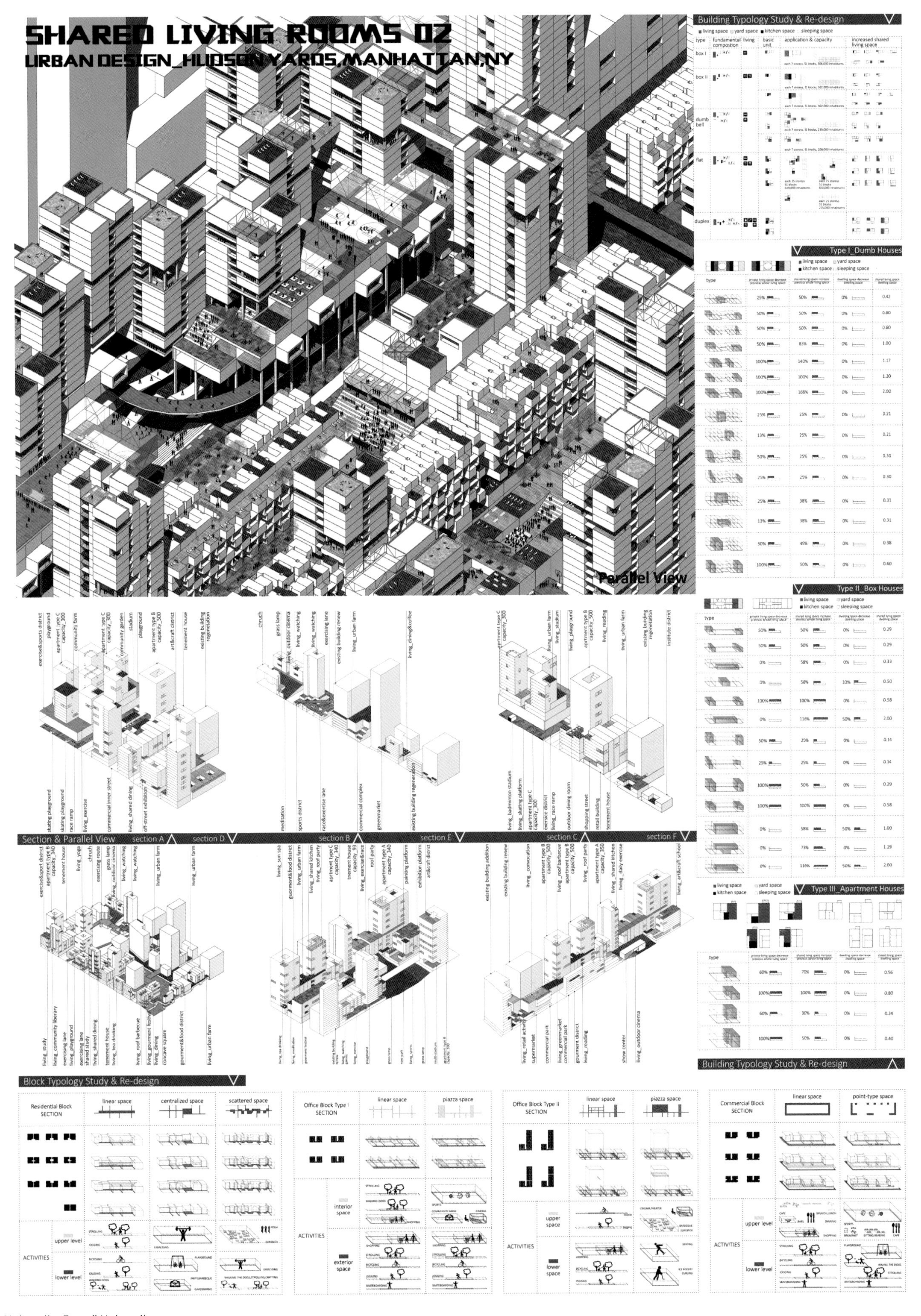

University: Tongji University
Designer: Zhang Linqi, Xi Xiaoyang
Tutor: Yang Chunxia, Huang Linlin
Course Name: Tongji University - Washington University in St.Louis Joint Urban Design Studio
Finished Time: Jul. 2014
Exchange Institute: Washington University in St.Louis

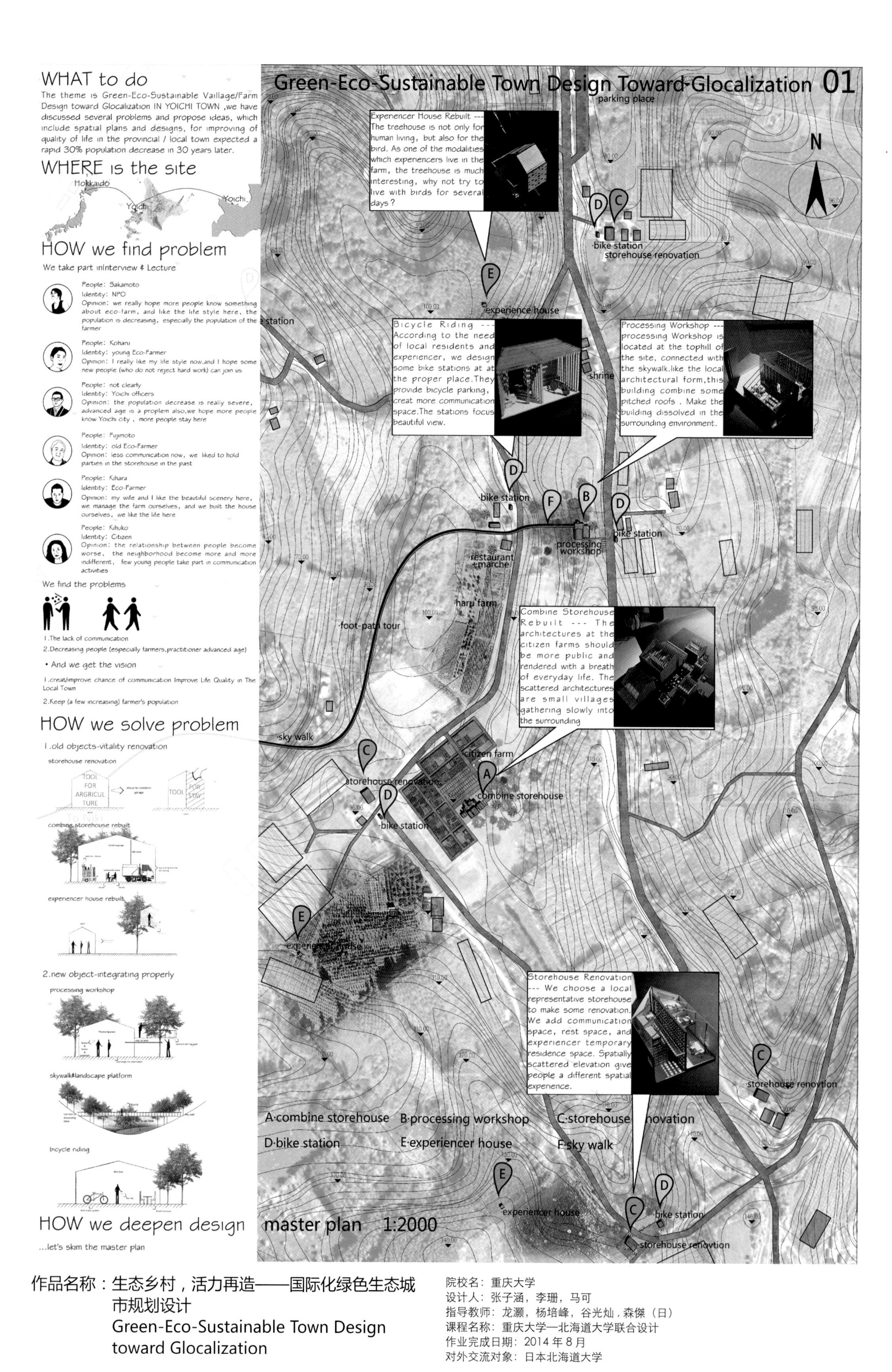

作品名称：生态乡村，活力再造——国际化绿色生态城市规划设计
Green-Eco-Sustainable Town Design toward Glocalization

院校名：重庆大学
设计人：张子涵，李珊，马可
指导教师：龙灏，杨培峰，谷光灿，森傑（日）
课程名称：重庆大学—北海道大学联合设计
作业完成日期：2014 年 8 月
对外交流对象：日本北海道大学

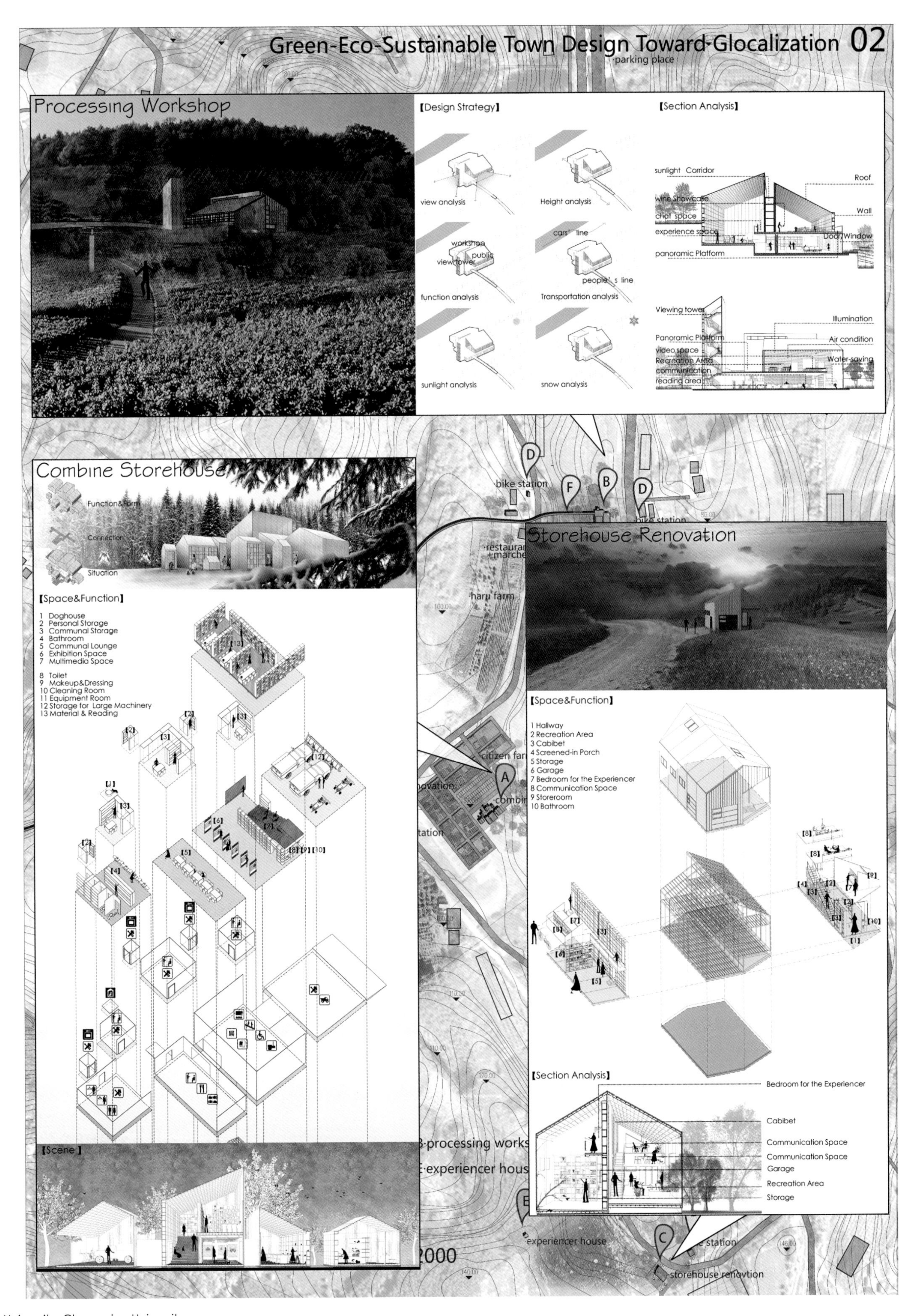

University: Chongqing University
Designer: Zhang Zihan, Li Shan, Ma Ke
Tutor: Long Hao, Yang Peifeng, Gu Guangcan, Moritex
Course Name: Chongqing University-Hokkaido University Joint Design Studio
Finished Time: Aug. 2014
Exchange Institute: Hokkaido University

7
Siping Road / Dalian Road | 四平路/大连路
multi-level connections
多层连接体
Zhao Yuling - Xia Qin - Lei Shaoying
With good connection to the city metro system and poor condition on the ground (heavy traffic and narrow pedestrian path), pedestrians are rushing to the underground compartments, but would not go for a walk at the corner of Dalian Road and Siping Road.
In our projects, we will build up multi-level connections between the existing buildings and the new complex at the clear space, and construct a sky park above the corner for leisure and cultural activities. The project is to active public activities by a 3D pedestrian system in the high-density city area.
基地位于大连路与四平路交叉口，其与上海城市地铁系统具有便捷的连接，却由于地面条件的局限（包括拥堵的机动车交通与狭窄的步行空间），使该区域内的步行者大多直奔地铁站，却不愿在地面空间停留。
这次的方案中，我们在基地转角的空地上新建了一栋城市综合体，将它与周边原有的建筑进行多层次的连接，并在此基础上于大连路与四平路机动车道上方设立了一座悬浮于空中的花园，以满足区域内的休闲及文化活动需求，并希望通过这个三维步行系统的设立，激发该高密度城市区域内的公共生活。
07

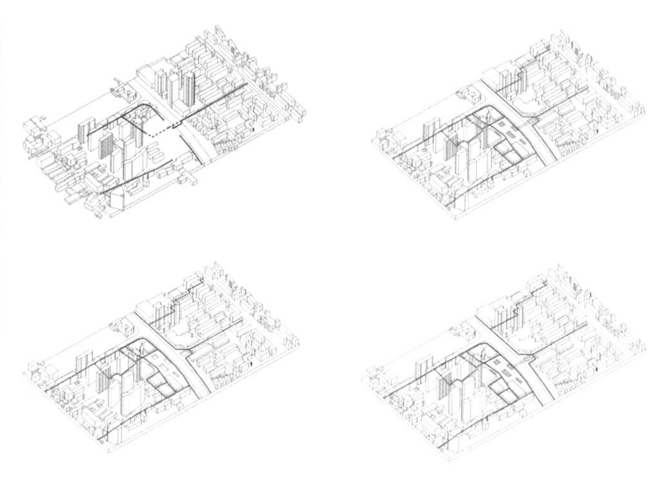

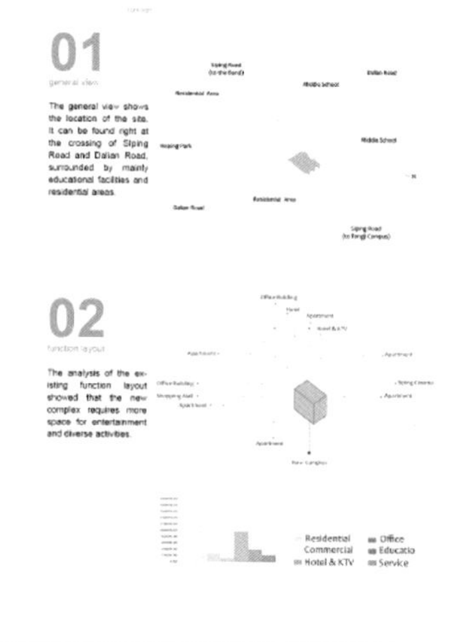

01
general view
The general view shows the location of the site. It can be found right at the crossing of Siping Road and Dalian Road, surrounded by mainly educational facilities and residential areas.
02
function layout
The analysis of the existing function layout showed that the new complex requires more space for entertainment and diverse activities.
Residential
Commercial
Hotel & KTV
Office
Educatio
Service

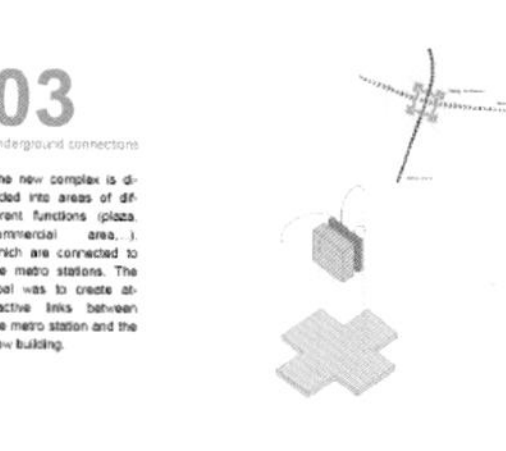

03
underground connections
The new complex is divided into areas of different functions (plaza, commercial area,...), which are connected to the metro stations. The goal was to create attractive links between the metro station and the new building.

04

connection skypark

The newly created skypark is an attractive meeting point for the public. It allows diverse activities to happen, in contrast to Heping park, where most of them are forbidden. Ranging from camping, bicycling to walking dogs - almost everything is allowed.

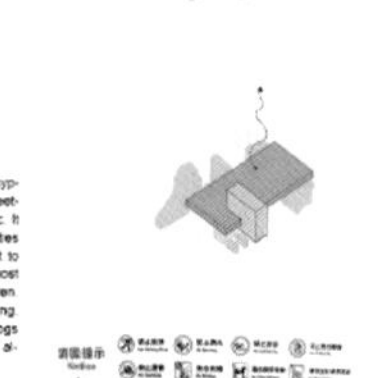

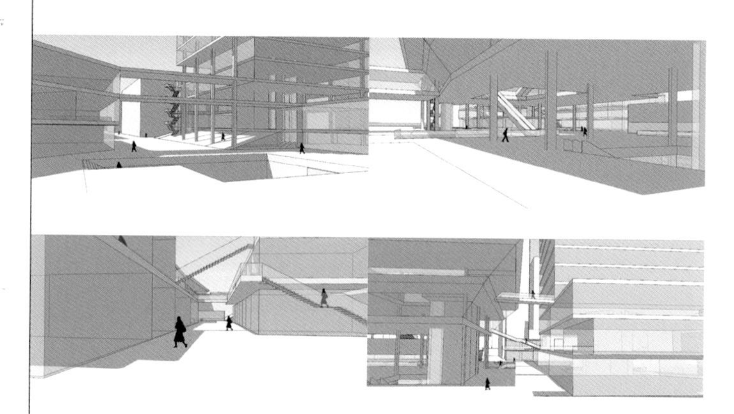

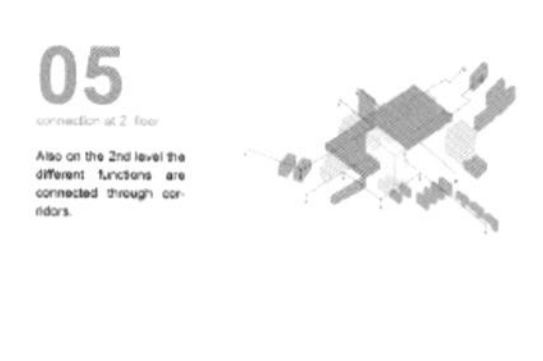

05
connection at 2. floor
Also on the 2nd level the different functions are connected through corridors.

06
connection of all levels
Finally, all levels are connected to each other - starting on the underground floor, reaching up to the skypark.

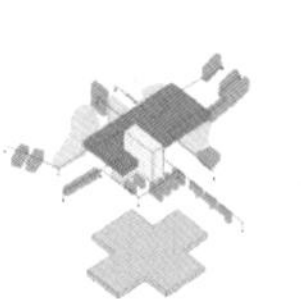

07 DALIAN ROAD
Siping Road and Dalian Road intersection
四平路大连节点
GROUP MEMBERS: Zhao Yuling - Xia Qin - Lei Shaoying

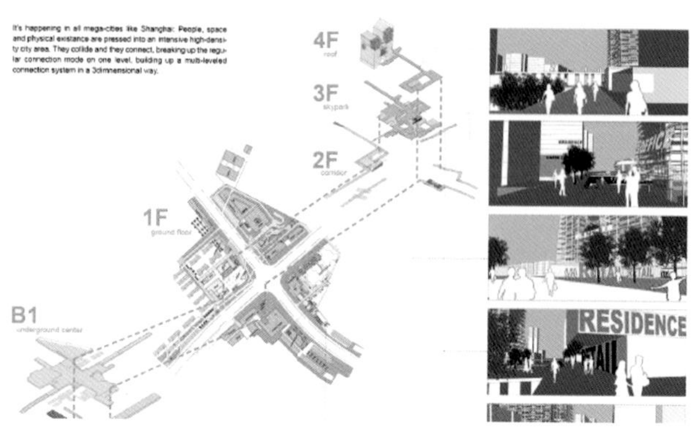

作品名称：步行者天堂——城市步行系统节点城市设计
Dalian Road — Siping Road and Dalian Road Intersection

院校名：同济大学
设计人：赵玉玲，夏琴，雷少英
指导教师：孙彤宇，许凯
课程名称：同济大学硕士双学位国际设计课程教学
作业完成日期：2014 年 05 月
对外交流对象：奥地利维也纳工业大学，法国凡尔赛大学，法国斯特拉斯堡大学，比利时布鲁塞尔自由大学

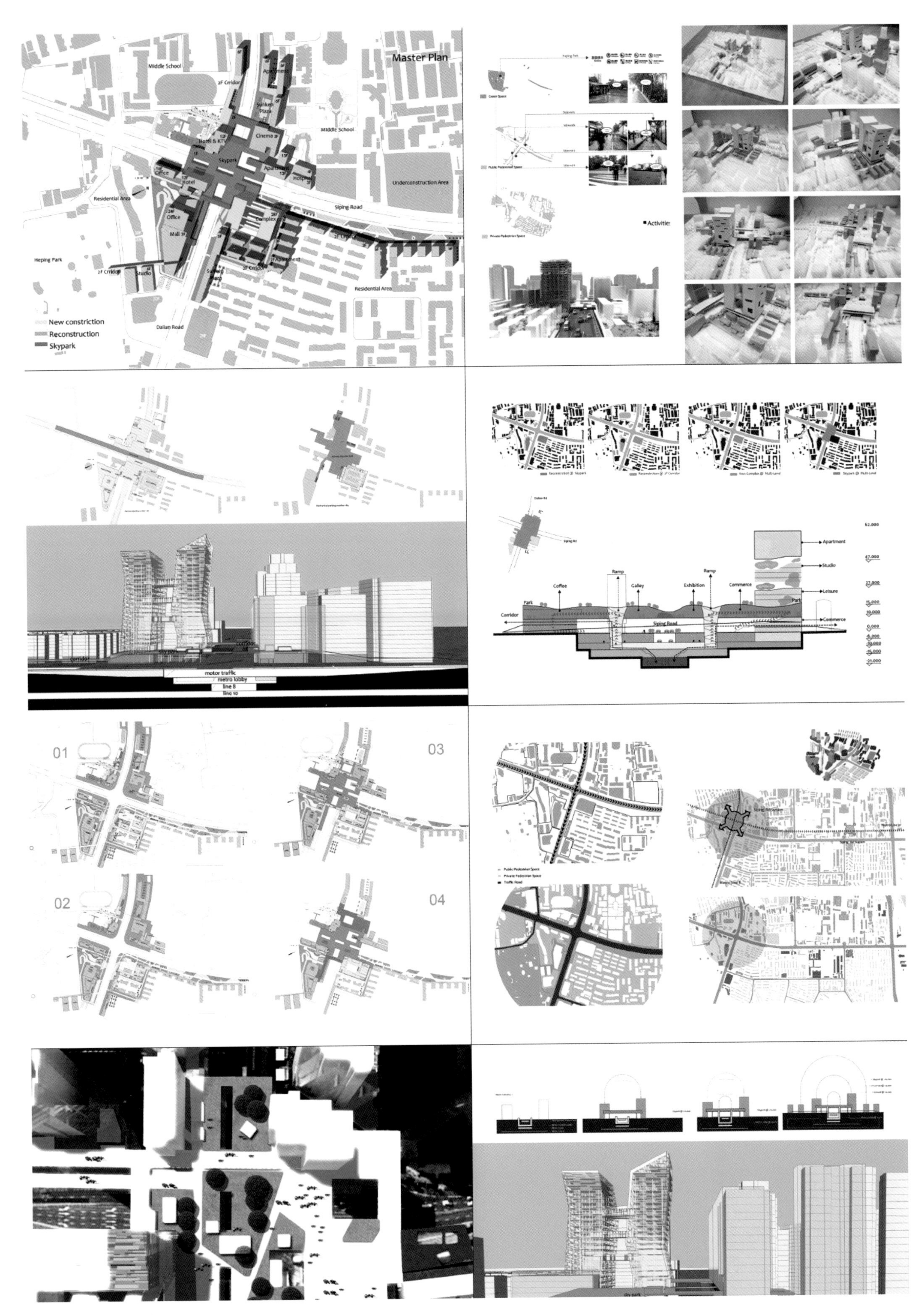

University: Tongji University
Designer: Zhao Yuling, Xia Qin, Lei Shaoying
Tutor: Sun Tongyu, Xu Kai
Course Name: International Master's Course Design Studio
Finished Time: May 2014
Exchange Institute: Vienna University of Technology，Université de Versailles，Université de Strasbourg，Université libre de Bruxelles

Mapping the Void: Consequence I
奥胡斯城市空间图解：迹

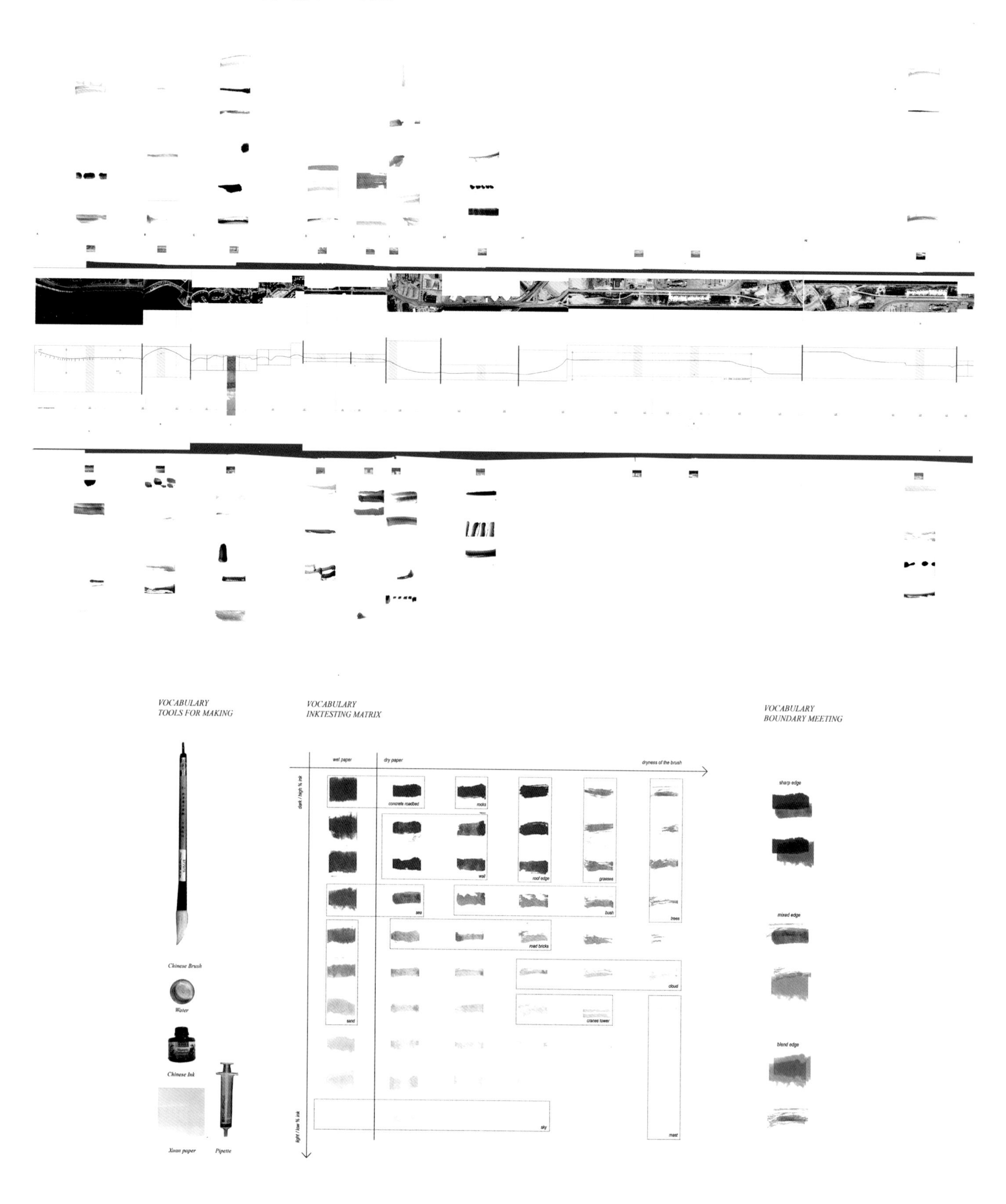

作品名称：“迹”
Unfolding Harbor Sequence

院校名：中央美术学院
设计人：周格，王羽，Siv Bøttcher
指导教师：何可人，王威，Anne Elisabeth Toft，Claudia Carbone
课程名称：2013 中央美术学院与奥胡斯建筑学院联合课题
作业完成日期：2013 年 10 月
对外交流对象：丹麦奥胡斯建筑学院

Mapping the Void: Consequence II
奥胡斯城市空间图解：迹

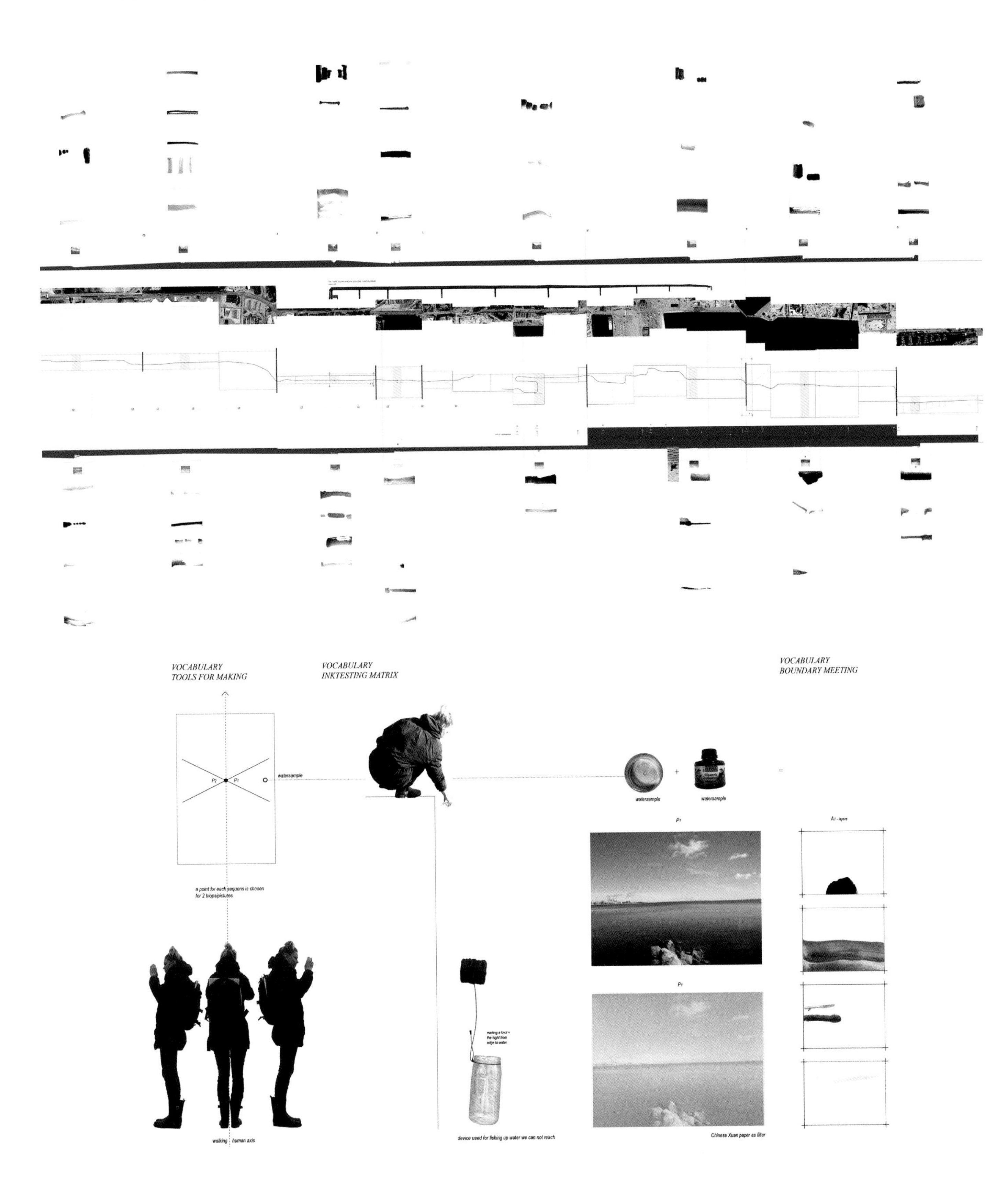

University: Central Academy of Fine Art
Designer: Zhou Ge, Wang Yu, Siv Bøttcher
Tutor: He Keren, Wang Wei, Anne Elisabeth Toft, Claudia Carbone
Course Name: Works from the CAFA I AAA Joint Studio Programme 2013
Finished Time: Oct. 2013
Exchange Institute: Aarhus Architecture School

织·补 WEAVING OF THE OLD COMMUNITY

REFRESH_LINK_ACTIVE VERTICAL CIRCULATION

WITH THE DEVELOPMENT OF CHINESE INDUSTRY IN 50'S, MANY FACTORIES AND COURTYARDS HAVE BEEN SET UP THROUGHOUT THE COUNTRY.AS TIME GOES BY, MANY HUGE FACTORY BUILDINGS AND THE HALL OF RESIDENCE FOR STAFF AND SPECIALISTS WHICH BUILT IN BRICK STRUCTURES AT THAT TIME PERIOD ARE STILL STAND IN THE ROUND. THESE STAFF GET OLD AND THE AGING OF POPULATION IN THIS COMMUNITY LEADS TO A TEDIOUS LIFE AFTER RETIREMENT. THE OPPORTUNITY NEEDS TO BE PROVIDED TO THESE PEOPLE TO CHANGE THEIR LIFE.

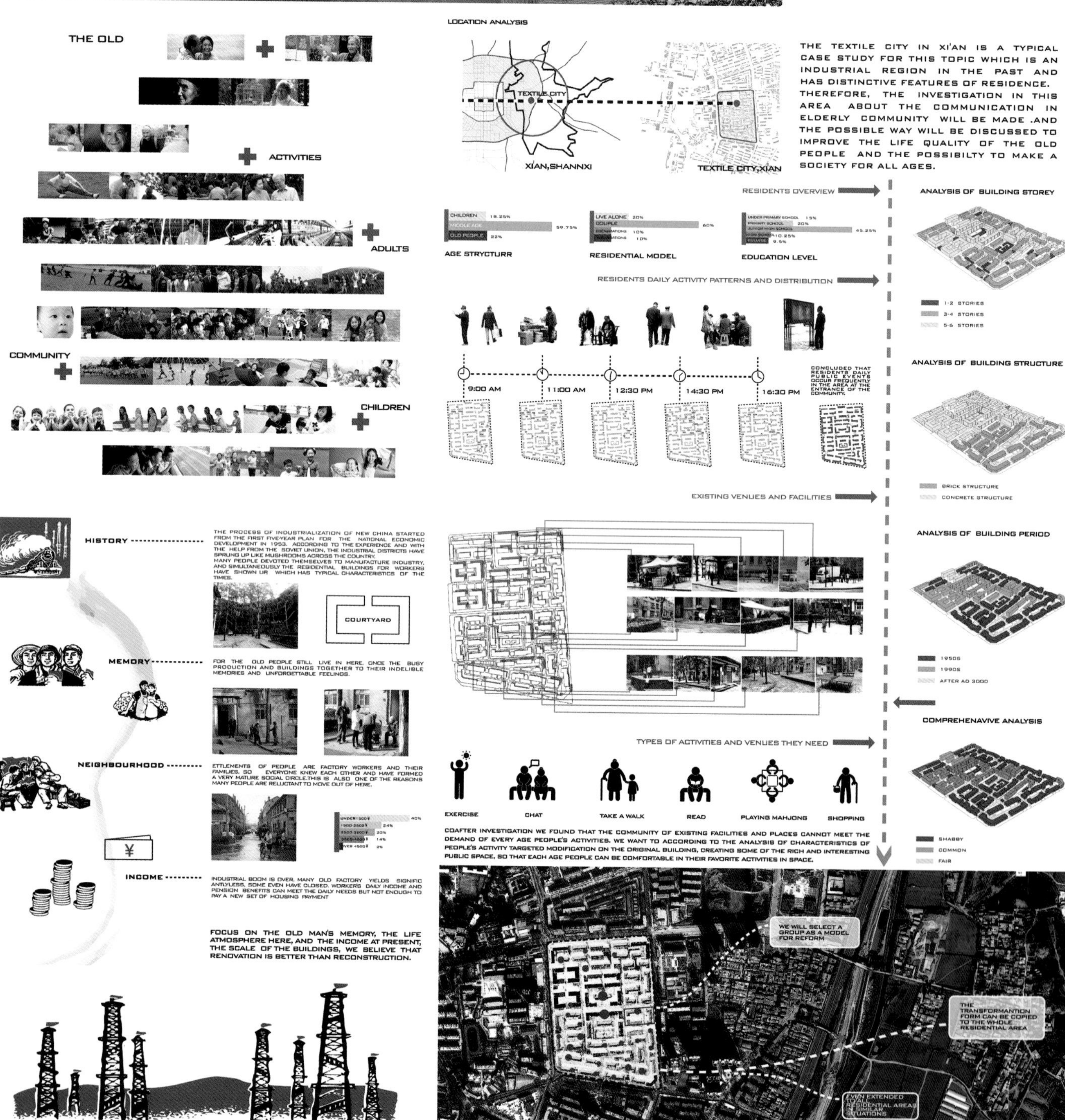

作品名称：织·补
Weaving of the Old Community

院校名：西安建筑科技大学
设计人：白纪涛，王阳，王静，王思睿
指导教师：张倩，王芳
课程名称：2014 ICCC 国际学生设计大赛（本科五年级）
作业完成日期：2014 年 03 月

织·补 WEAVING OF THE OLD COMMUNITY

REFRESH_LINK_ACTIVE VERTICAL CIRCULATION

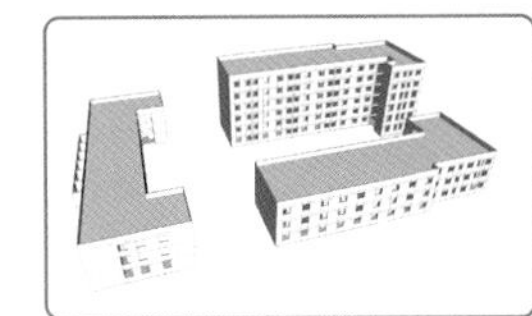

CURRENT SITUATION

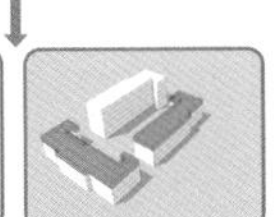

ON THE BASIS OF THE ORIGINAL HEIGHT OF THE BUILDING, THREE RESIDENTIAL OVERLAY LAYER TO HAVE A BETTER SENSE OF CONNECTION WITH THE SIX HIGHLY VISUAL.AT THE SAME TIME TO ENSURE THAT AFTER THE TRANSFORMATION THE BUILDINGS CAN ACCOMMODATE ALL THE PEOPLE WHO LIVE IN HERE BEFORE.

INCREASE THE BRISK ELEMENT OF CLUSTER TO MAKE SURE THEIR ACTIVE ATMOSPHERE. USE BRIGHT COLOUR TO MAKE BUILDINGS EASIER DISTINGUISH,SO THE OLD CAN EASIER FIND THEIR HOME.AND VIVID COLOUR CAN MAKE PEOPLE FEEL PLEASANT.

OMIT SOME ELEMENTS TO MAKE THE BODY MORE LIGHTSOME

PUBLIC SPACE OF INCREASING

EACH FUNCTION SIGNIFICANTLY

CLUSTER GREENING DESIGN

GENERAL PLAN 1: 500

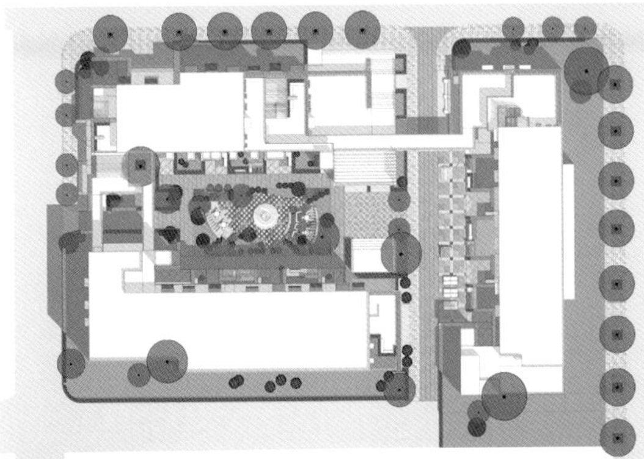

RESIDENTIAL BUILDING

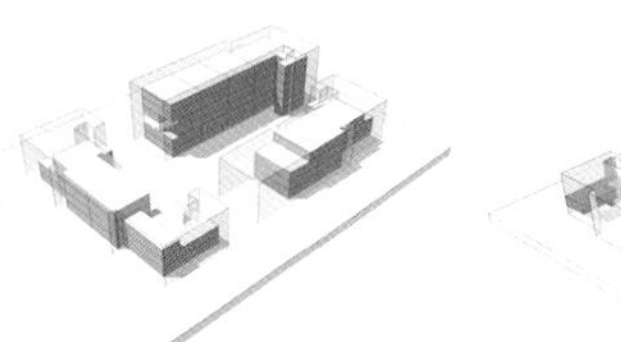

PUBLIC SPACE

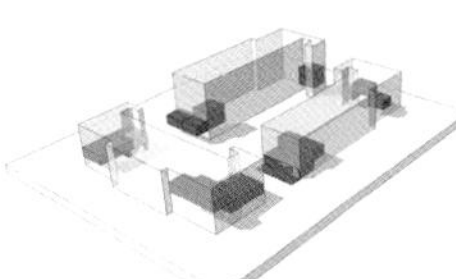

RETALL

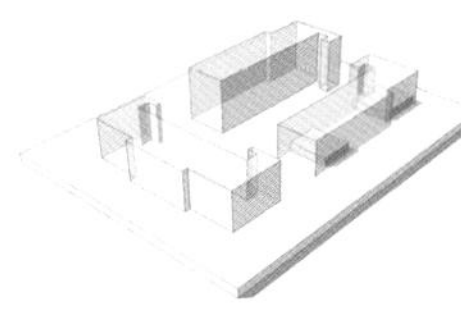

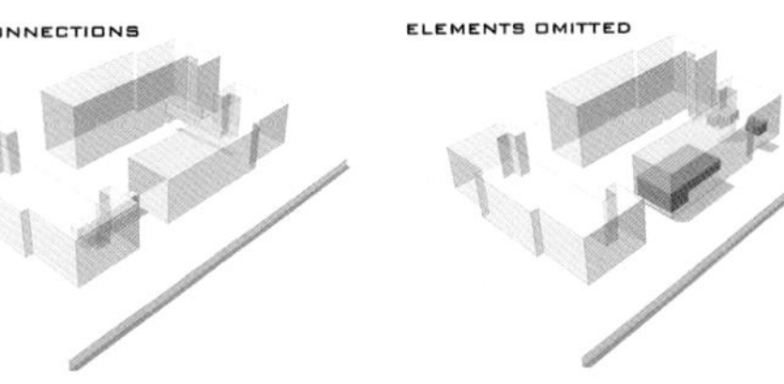

BLOCK CONNECTIONS

ELEMENTS OMITTED

ACTIVE ELEMENTS

BEFORE AFTER

THE NEW ENTWR ROAD

A STAGE FOR THE OLD TO PERFORM THIER"QIN QIANG"

SECONDARY ENTRANCE

SPORT AREA

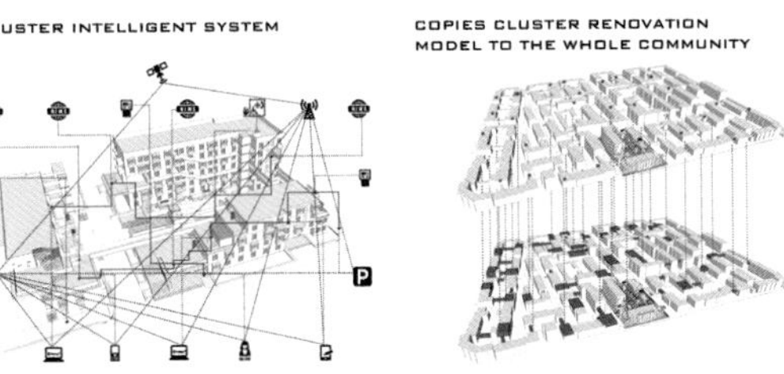

CLUSTER INTELLIGENT SYSTEM

COPIES CLUSTER RENOVATION MODEL TO THE WHOLE COMMUNITY

IN OUR DESIGN, WE WILL INTEGRATE MODERN INFORMATION TECHNOLOGY TO BUILDINGS AND PUBLIC SPACES, AND EXPECT TO ACHIEVE INFORMATION SHARING AND EFFICIENT EMERGENCY RAPID WAY. COMBINED WITH WIRELESS SIGNALS AND ARTIFICIAL ONLINE SERVICES EFFECTIVELY IMPROVE THE DAILY LIVES OF RESIDENTS OF CONVENIENCE.

University: Xi'an University of Architecture and Technology
Designer: Bai Jitao, Wang Yang, Wang Jing, Wang Sirui
Tutor: Zhang Qian, Wang Fang
Course Name: 2014 International Student Design Competition
Finished Time: Mar. 2014

THE CHILDREN'S MUSEUM

--------the regeneration of edge Louisville

对路易斯维尔当地气候特点，街道立面，居民生活习惯的调研后，结合当地气候、地形地貌特点，借用路易斯维尔郊外的溶洞、山石为语汇，营造了一个空间层次丰富的中庭与外皮，以营造探秘洞穴似的空间体验为主要手段，以满足儿童博物馆吸引儿童的基本要求。

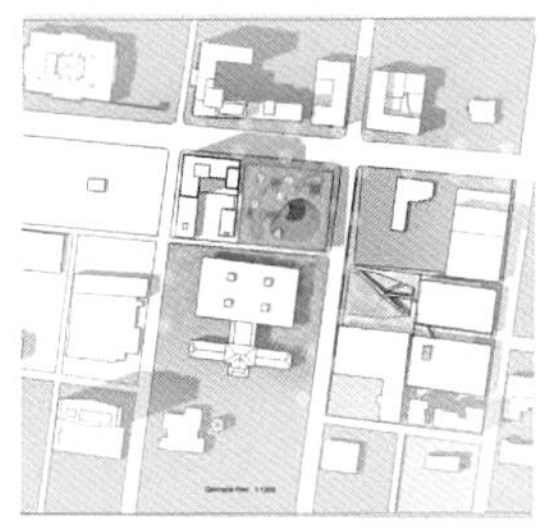

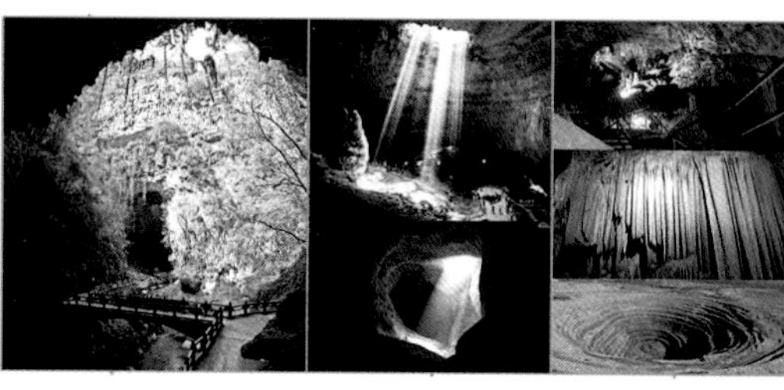

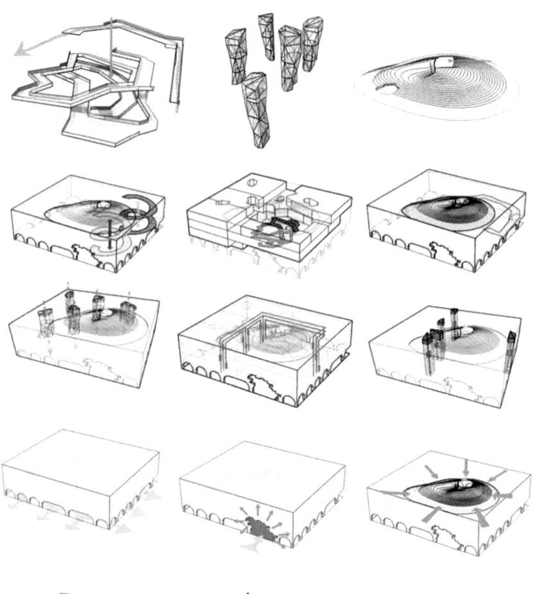

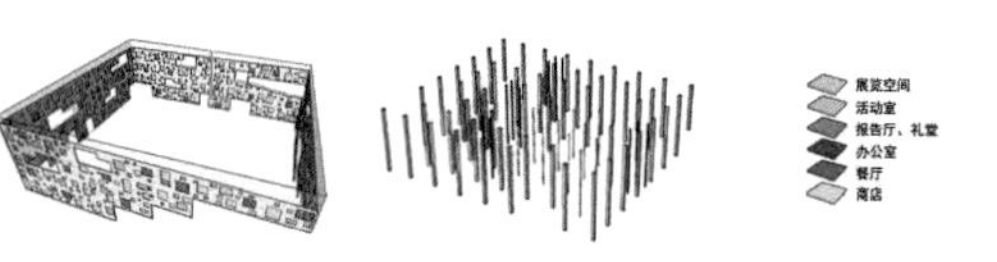

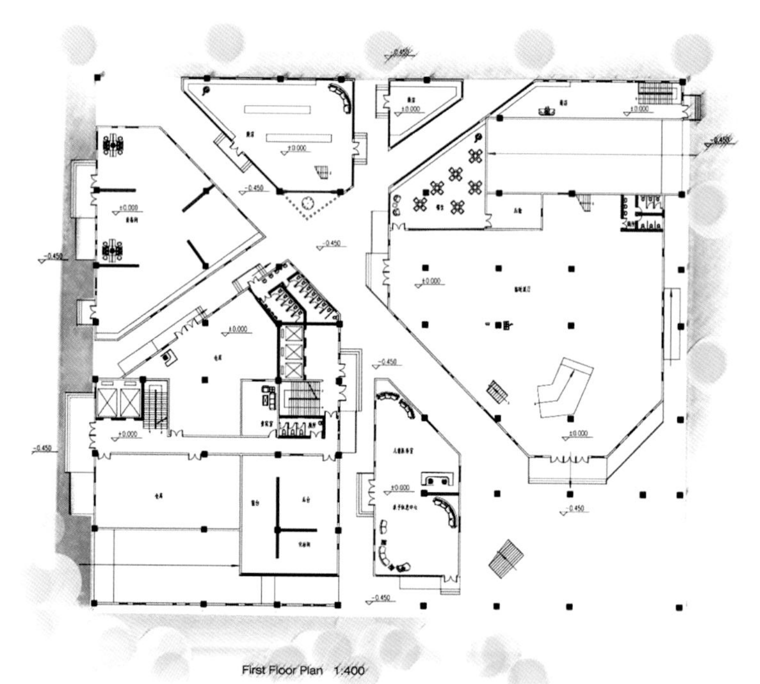

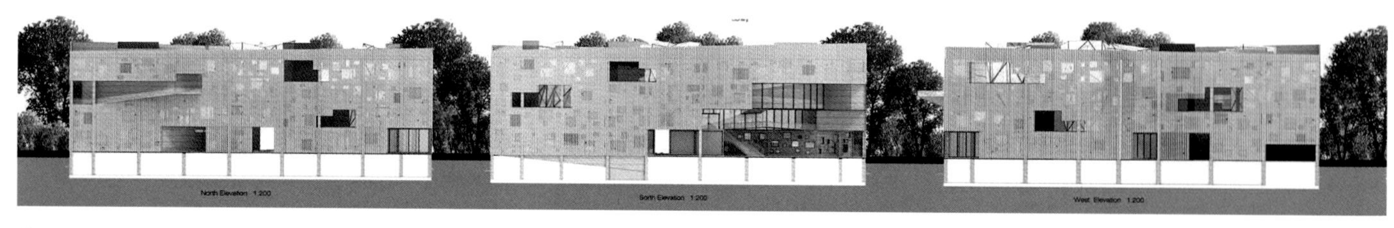

作品名称：儿童博物馆
The Children's Museum

院校名：青岛理工大学
设计人：蔡昆洋，张沛，韩梦，谭君睿
指导教师：许从宝，聂彤，徐翀
课程名称：2014AIA 路易维尔儿童博物馆建筑设计竞赛
作业完成日期：2014 年 02 月

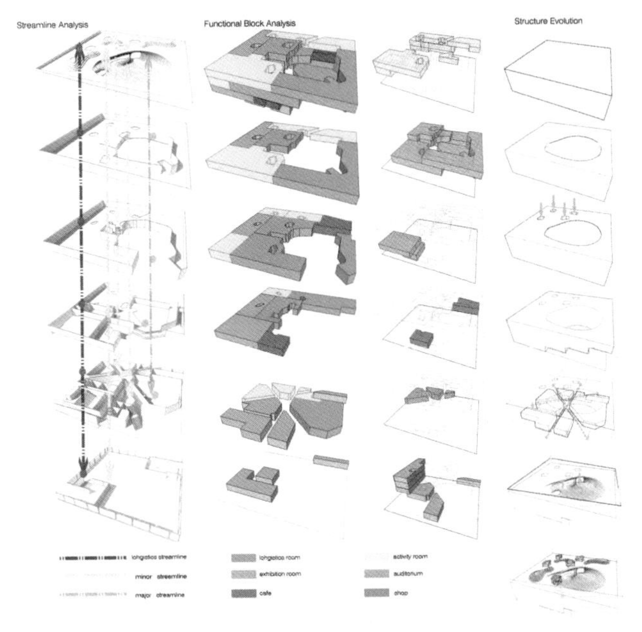

The second floor Plan 1:400

The third floor Plan 1:400

The fourth floor Plan

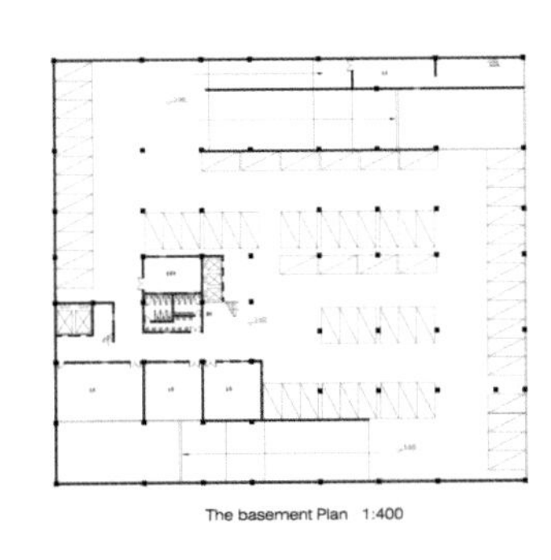
The basement Plan 1:400

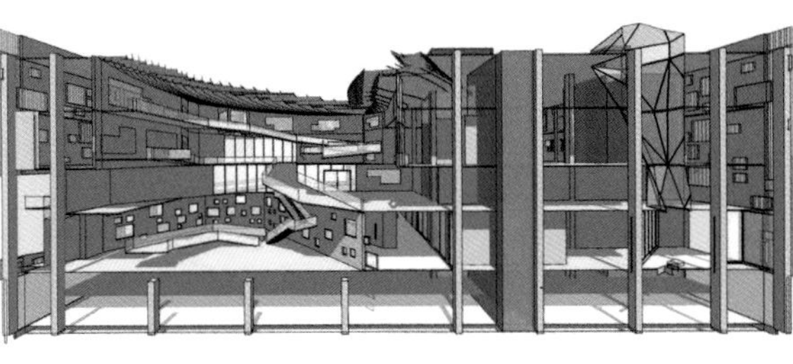

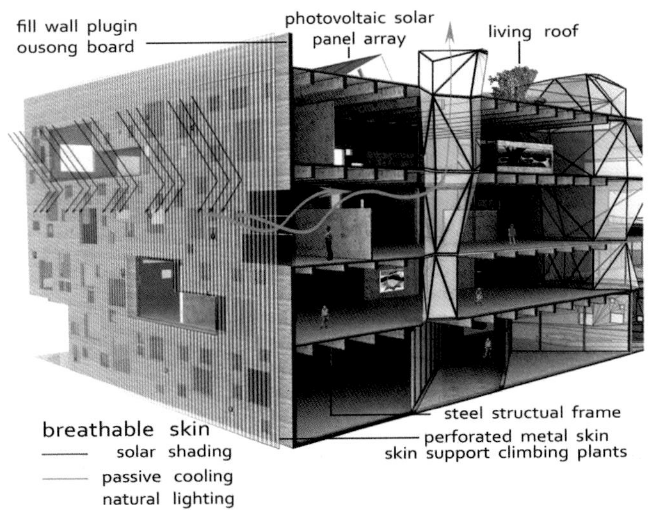

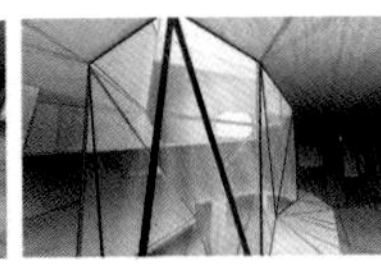

University: Qingdao Technological University
Designer: Cai Kunyang, Zhang Pei, Han Meng, Tan Junrui
Tutor: Xu Congbao, Nie Tong, Xu Chong
Course Name: AIA Children's Museum—The Regeneraiton of Edge Louisville
Finished Time: Feb. 2014

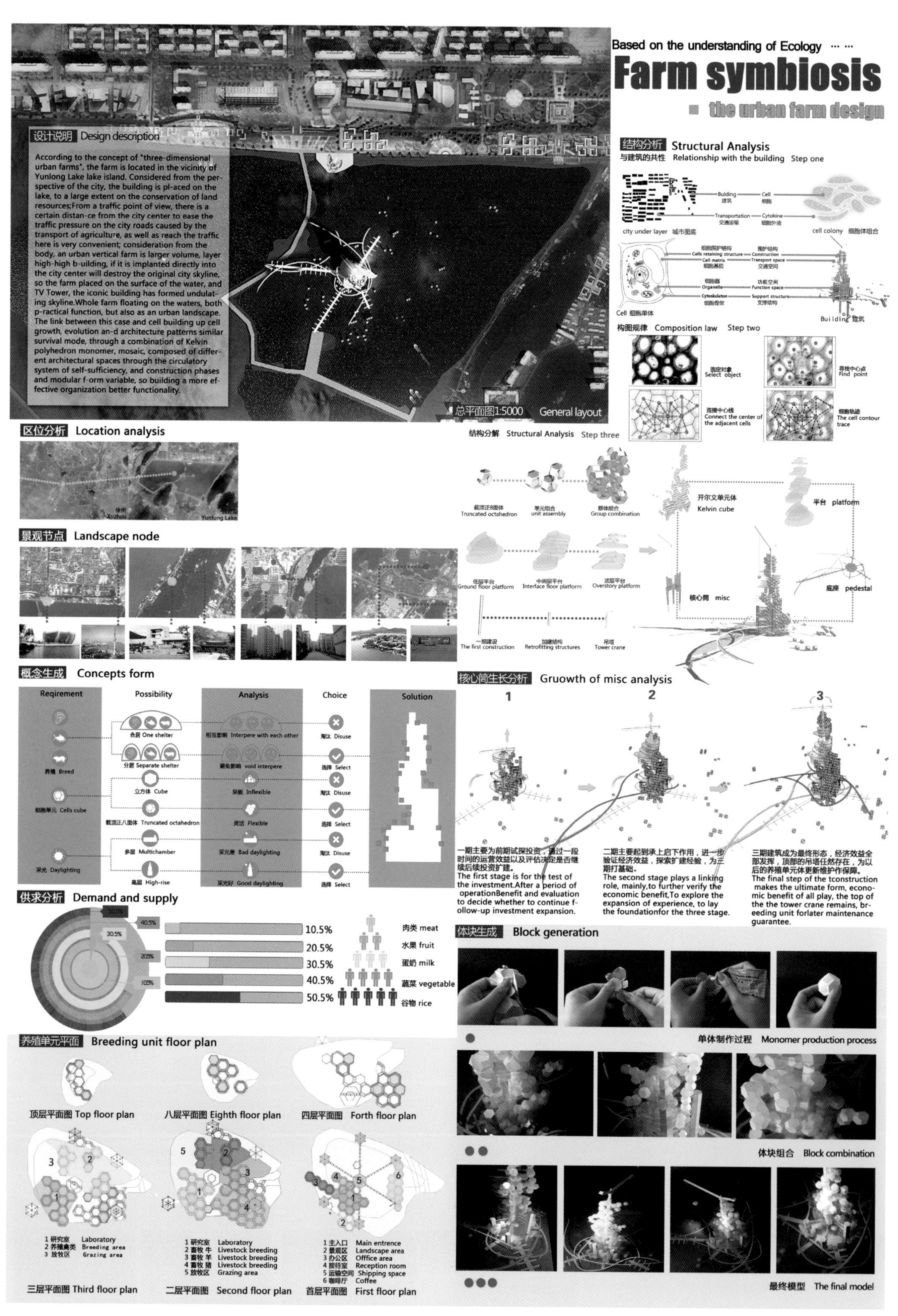

作品名称：循环农场——城市立体农场设计
Farm Symbiosis the Urban Farm Design

院校名：中国矿业大学
设计人：曹琳琳，刘彬峰，胡小华，张琪，于小果，张涛
指导教师：刘茜，林涛
课程名称：城市立体农场国际大学生建筑设计竞赛
作业完成日期：2013 年 12 月
对外交流对象：博德西奥（BDCL）国际建筑设计有限公司

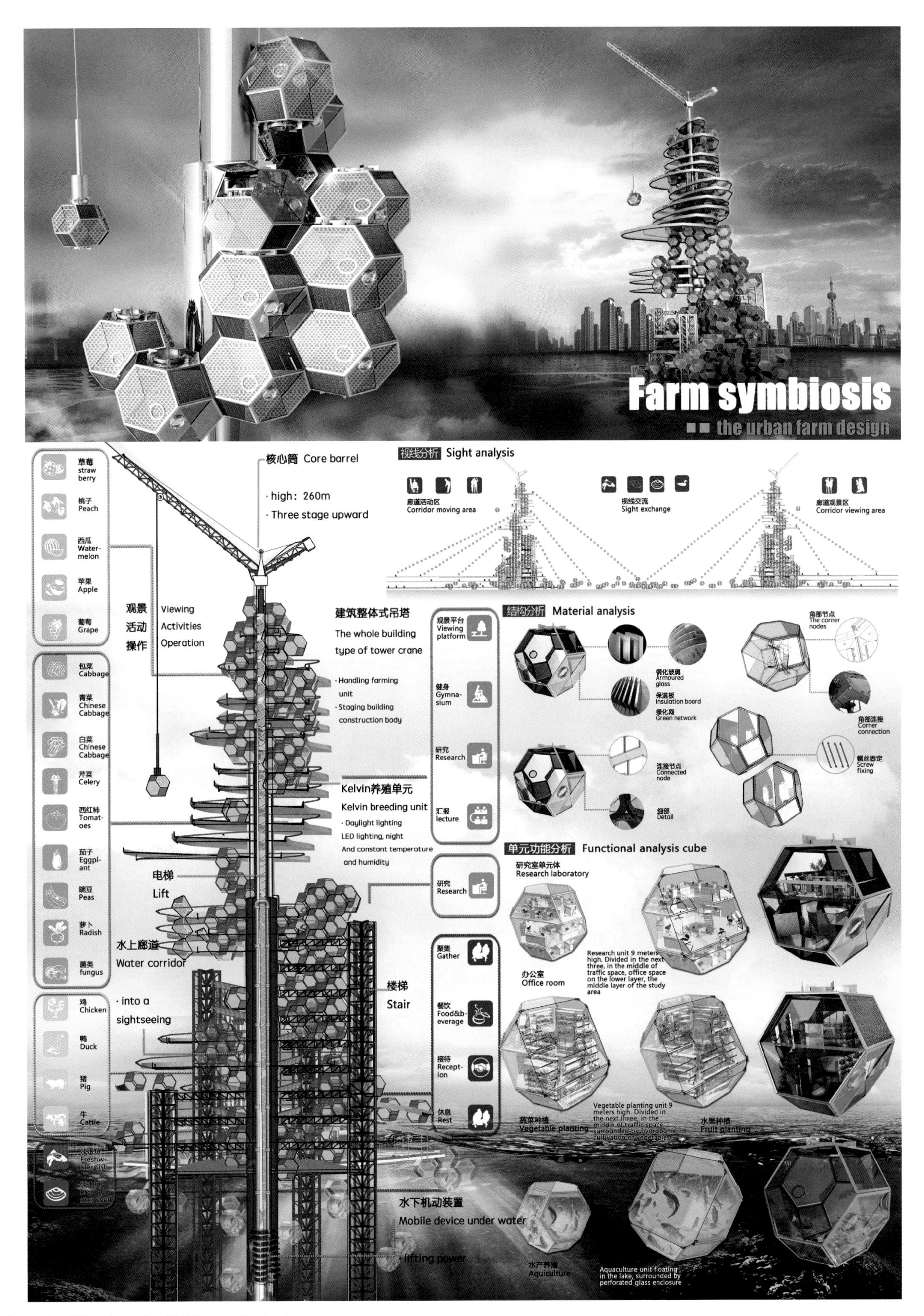

University: China University of Mining and Technology
Designer: Cao Linlin, Liu Binfeng, Hu Xiaohua, Zhang Qi, Yu Xiaoguo, Zhang Tao
Tutor: Liu Qian, Lin Tao
Course Name: Vertical Farming International College Architecture Design Competition
Finished Time: Dec. 2013
Exchange Institute: BDCL

Babel City[1] 中国香港 **巴别城**

The architecture is located in HongKong China.HongKong is the final shape of hyperaccumlation city.The plot ratio is chose to13.And it will never stop.
For very long, people always immersed the old knowing of "density" which makes us turn a blind eye to the turth unexpected city is making us to think: When the densely becomes a supernormal problem, the measure should be extraordinary as well.
Named of our design is Bable City.In the Bible,it said there were some people wants to build a tower to heavy which is called Bible.But God stopped them by giving them different language and race.So they can' t communicate.
So Babel city is a overhead bridge for communicating.And as well ,it will be regard as a masterplate,coming into everywhere of the city ,which connects all the high—rise.Then become a real city.
In the histoty HongKong presented a hot strategies in the predicament.This is what HongKong' s charm. We make use of the secondary land.There is vertical farm,overhead bussiness,zigzag jogging vertical parking and sewage treament vertically so on.
This is a life entity which can adjust and develop itself .Developping the city by cxisting unexpectly.

Babel City HongKong

Vertical Farm
Zigzag Jogging
Vertical Parking
Sewage Treament
Vertically
Overhead Bussiness

Location analysis

The position of the architecture spreads the whole of the Queen' s Road.It is situated by The XiHuan Building which is on the crowded Queen' s Road.
Queen' s Road acrosses densely inhabited Commercial area——Central.The densely inhabited district——West Point&East Point. This road is 5km

1.East of the Queen' s Road
2.Middle of the Queen' s Road
3.West of the Queen' s Road

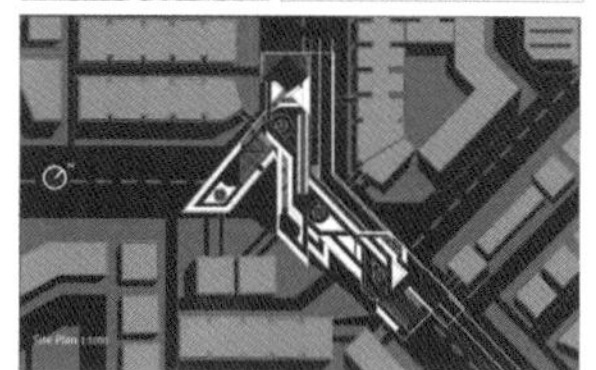

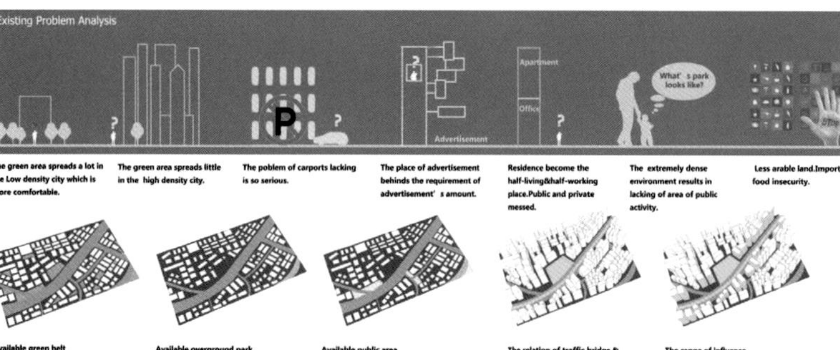

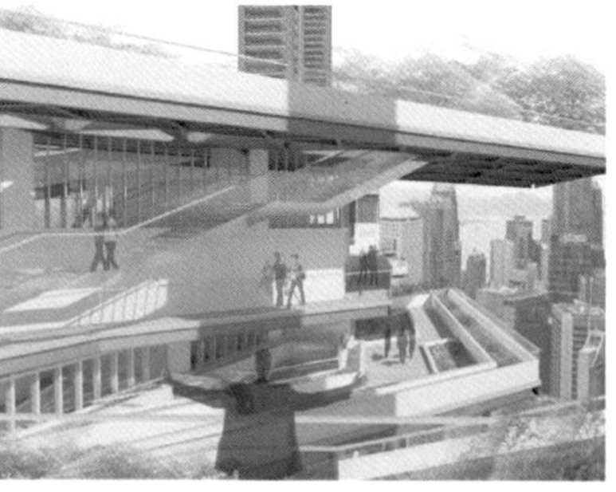

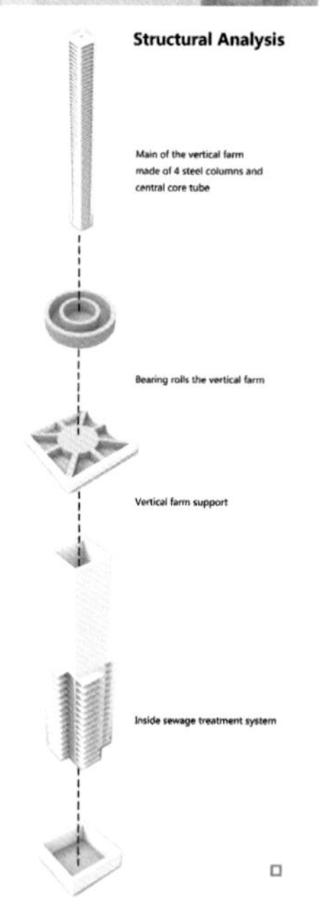

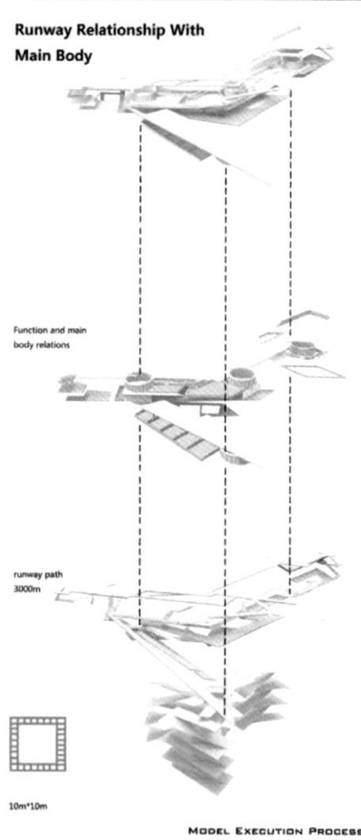

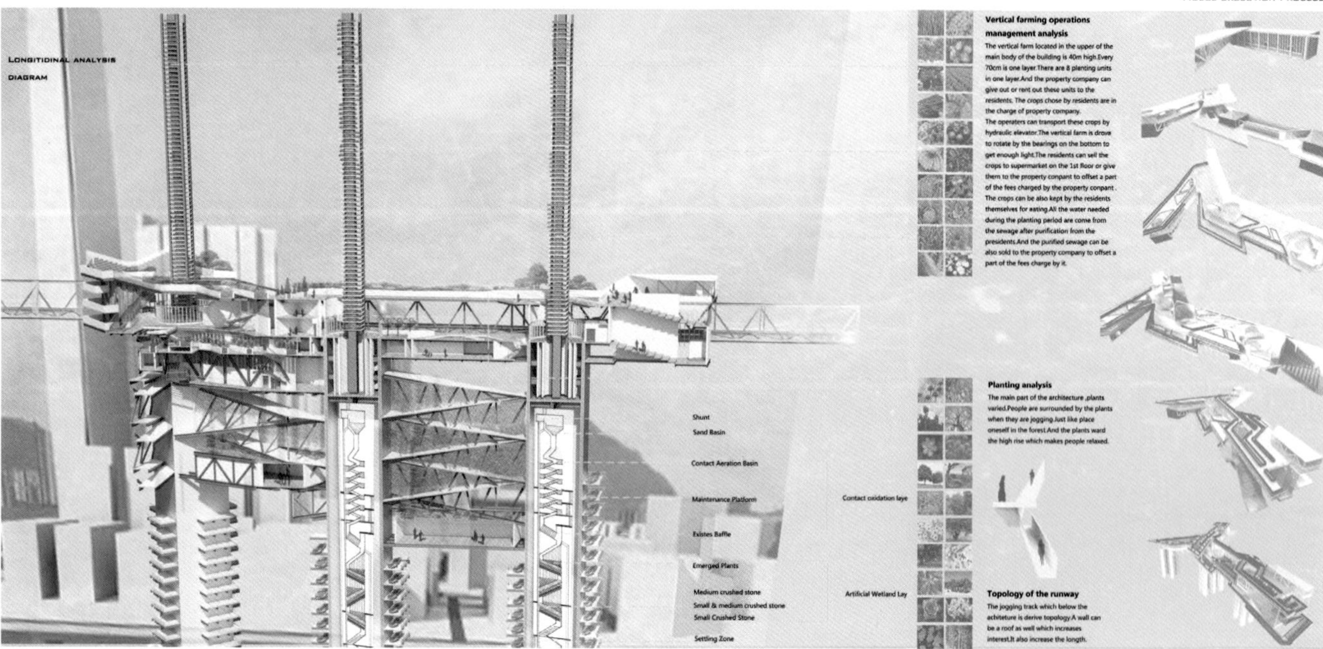

Vertical farming operations management analysis

The vertical farm located in the upper of the main body of the building is 40m high.Every 70cm is one layer.There are 8 planting units in one layer.And the property company can give out or rent out these units to the residents. The crops chose by residents are in the charge of property company.
The operaters can transport these crops by hydraulic elevator.The vertical farm is drove to rotate by the bearings on the bottom to get enough light.The residents can sell the crops to supermarket on the 1st floor or give them to the property conpant to offset a part of the fees charged by the property conpant . The crops can be also kept by the residents themselves for eating.All the water needed during the planting period are come from the sewage after purification from the presidents.And the purified sewage can be also sold to the property company to offset a part of the fees charge by it.

Planting analysis

The main part of the architecture ,plants varied.People are surrounded by the plants when they are jogging.Just like place oneself in the forest.And the plants ward the high rise which makes people relaxed.

Topology of the runway

The jogging track which below the achiteture is derive topology.A wall can be a roof as well which increases interest.It also increase the length.

作品名称：出乎意料的城市——巴别城
Unexpected City: Babel City

院校名：昆明理工大学
设计人：褚剑飞，马杰茜
指导教师：张欣雁
课程名称：昆明理工大学—研究型设计霍普杯国际竞赛
作业完成日期：2014 年 07 月

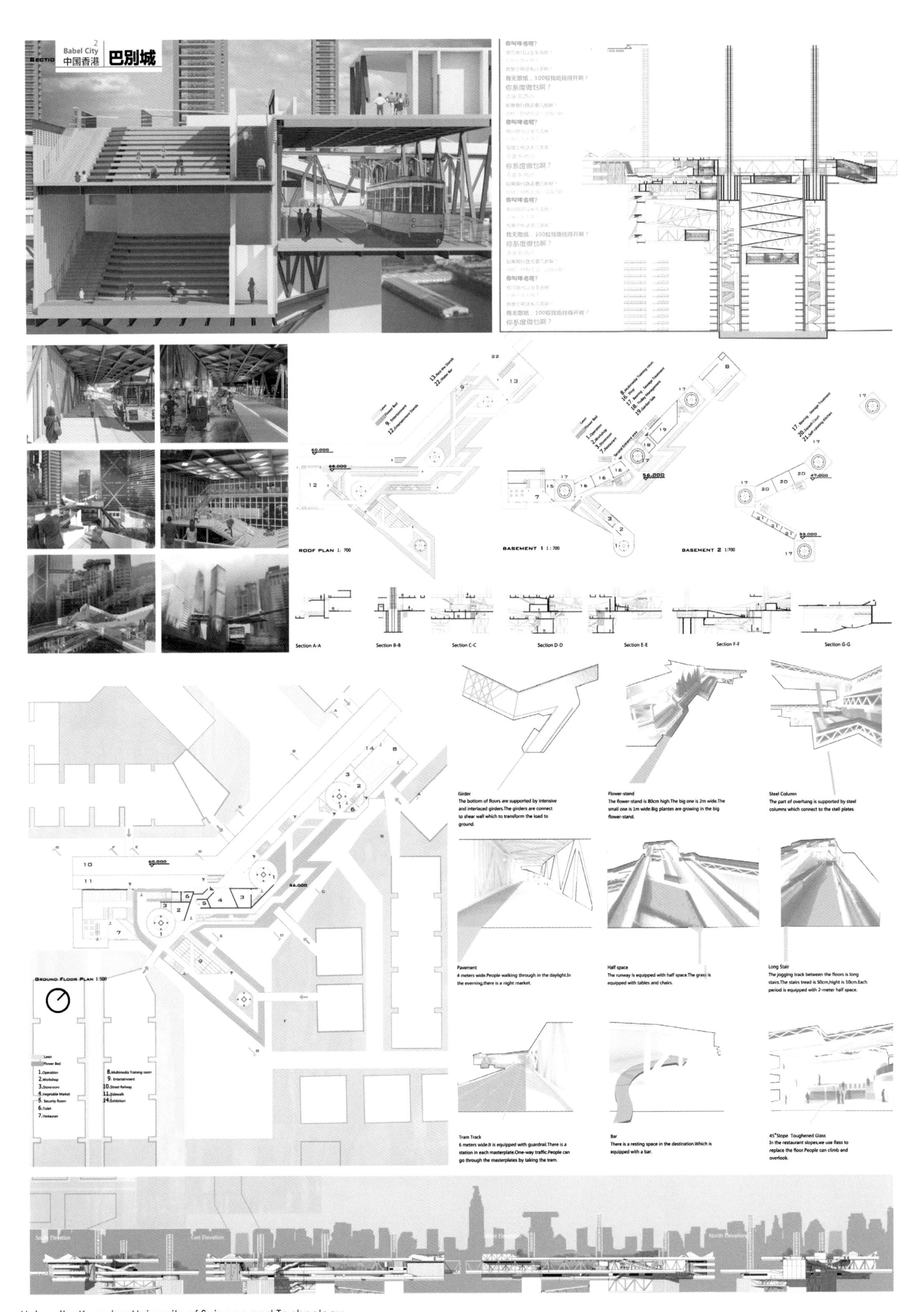

University: Kunming University of Science and Technology
Designer: Chu Jianfei, Ma Jieqian
Tutor: Zhang Xinyan
Course Name: UIA-HYP cup 2014 International student competition in architecture design
Finished Time: Jul. 2014

Nodes & Activation: the village reformation and architectural design in Jicun village

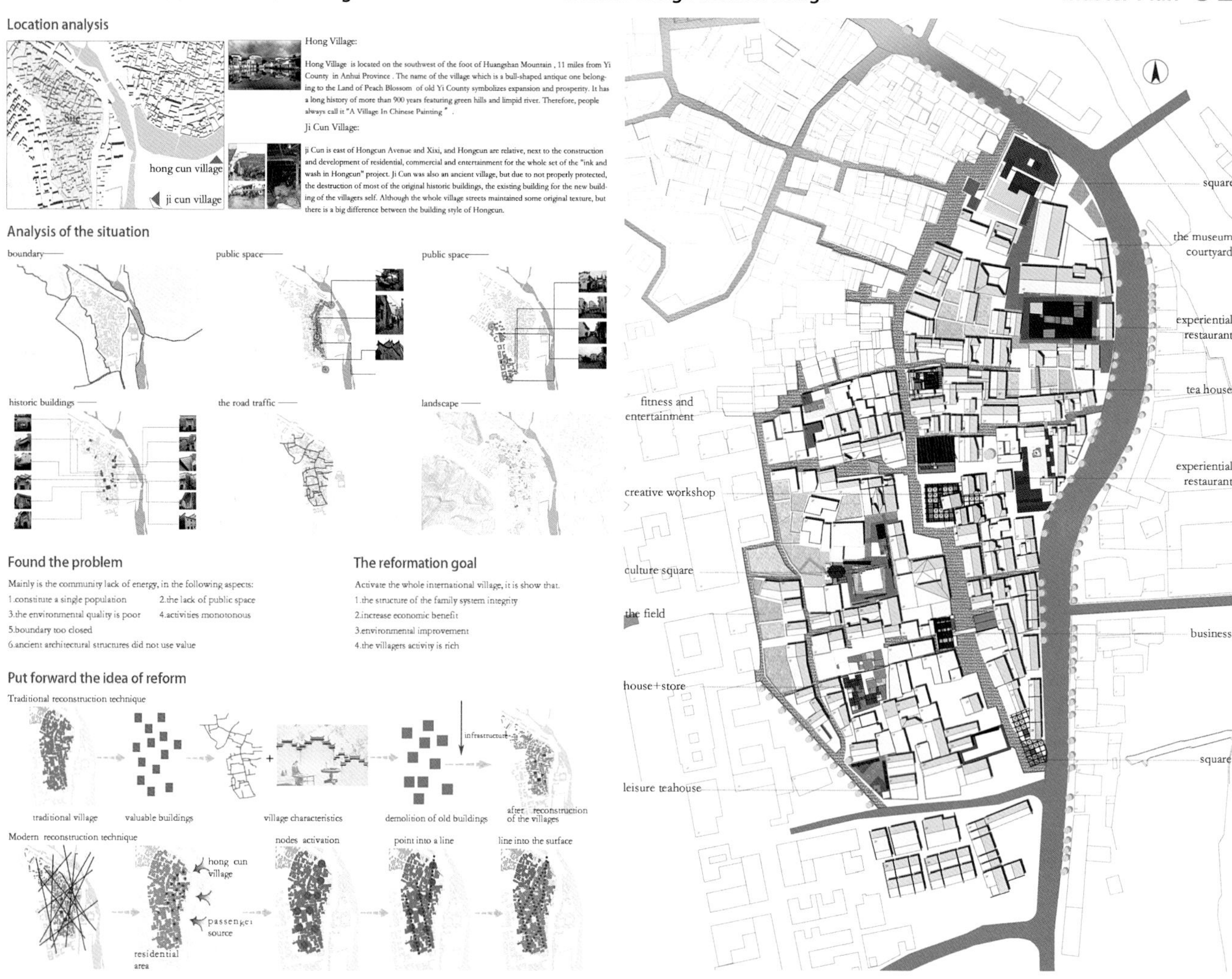

The design process

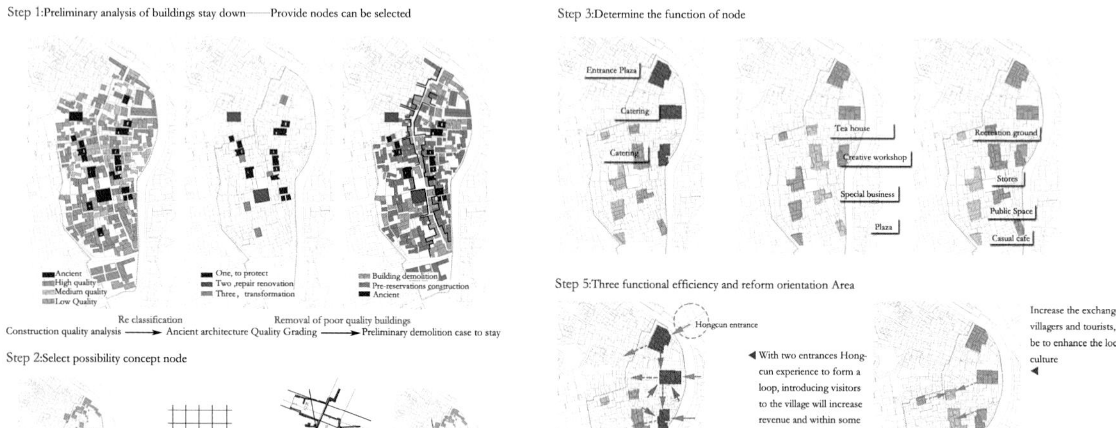

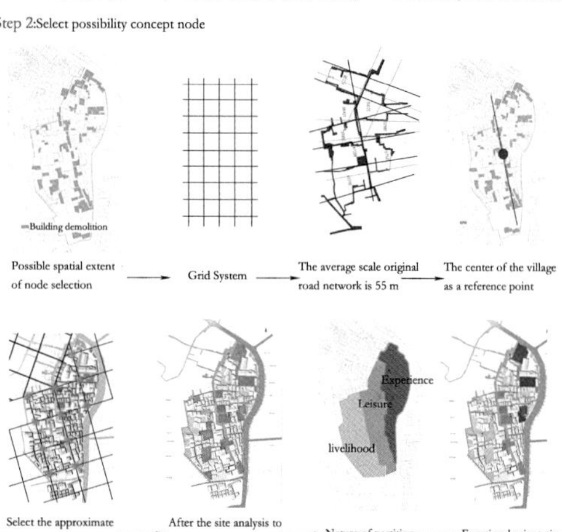

Model show

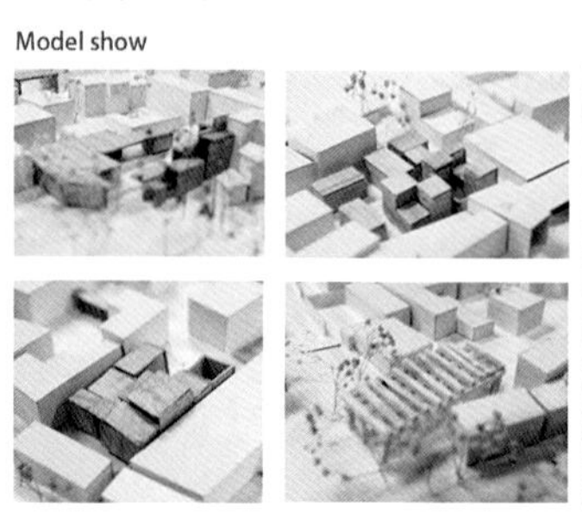

作品名称：节点激活——黟县际村村落改造与建筑设计
Nodes & Activation: The Village Reformation and Architectural Design in Jicun

院校名：合肥工业大学
设计人：邓慧丽，杨三瑶，汪宇宸
指导教师：苏剑鸣，李早，刘阳，任舒雅
课程名称：全球毕业设计大赛（本科五年级）
作业完成日期：2014 年 09 月

Nodes & Activation: the village reformation and architectural design in Jicun village

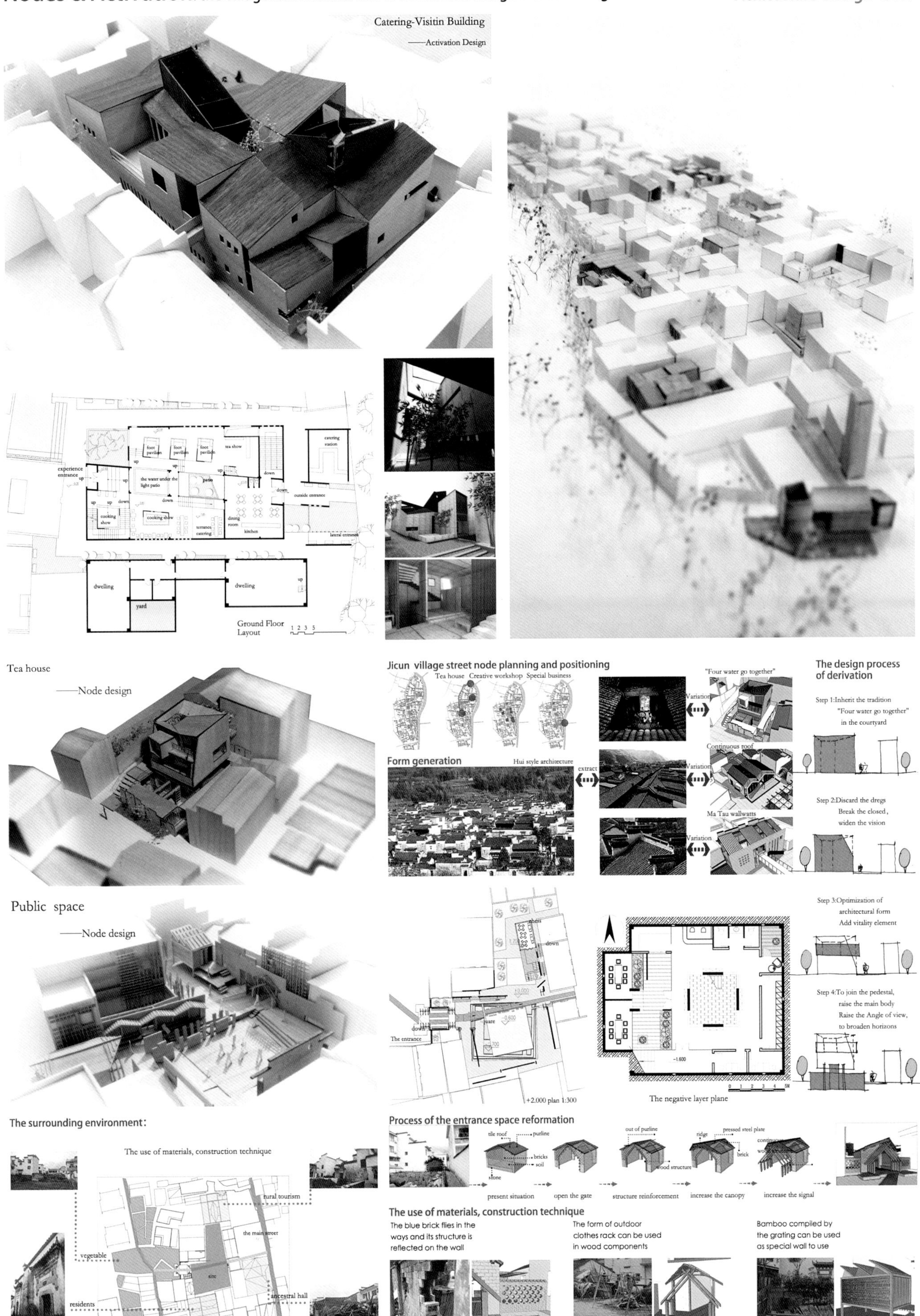

University: Hefei University of Technology
Designer: Deng Huili, Yang Sanyao, Wang Yuchen
Tutor: Su Jianming, Li Zao, Liu Yang, Ren Shuya
Course Name: Archiprix International
Finished Time: Sep. 2014

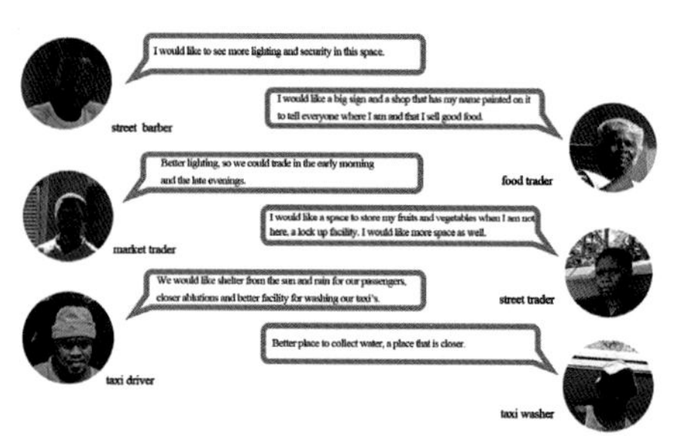

Reorganization of the Traffic Streamline

Before

After

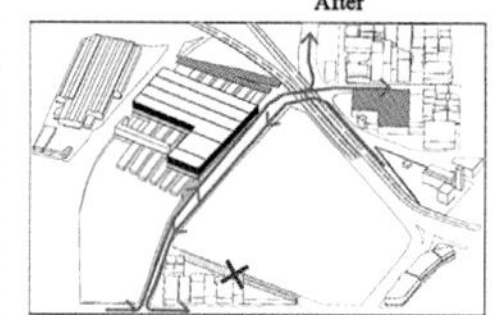

road in jam / market / one-way road

The current state of traffic is too complicated, especially at the intersection of Brook Street and Theatre Lane.

road in jam / market / road

We open up Brook Street and turn it into a two-way road, then isolate Theatre Lane from vehicles to become a pedestrian-only street, relocating the vendors here.

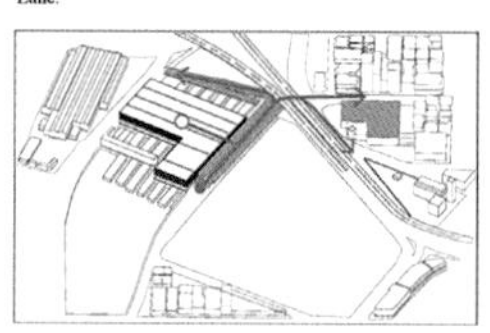

market / passenger streamline

Currently, the passengers who exit the station either take the minibus on Brook Street or walk away from the North side.

market / passenger streamline

The vendors sell their goods on the edges of Theatre Lane, Joseph Nduli, and Monty Naicker.

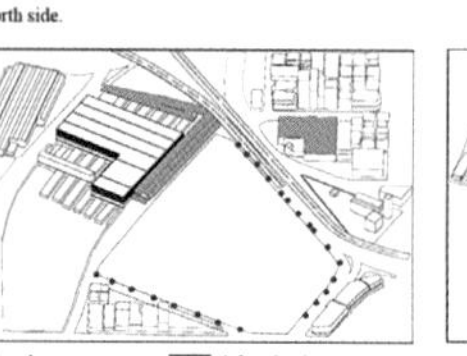

market / informal trader

The vendors sell their goods on the edges of Theatre Lane, Joseph Nduli, and Monty Naicker.

market / triangle plaza / pedestrian street

We should first attract the vendors to move, using a catalyst, to the pedestrian-only Theatre Lane, which can be easily supervised.

The Brook Street

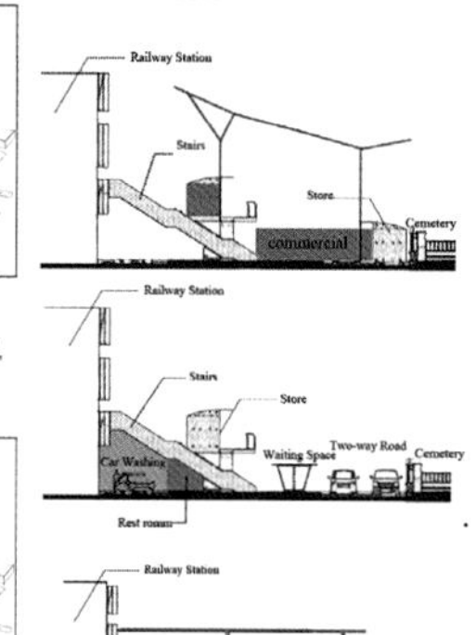

The Theatre Lane

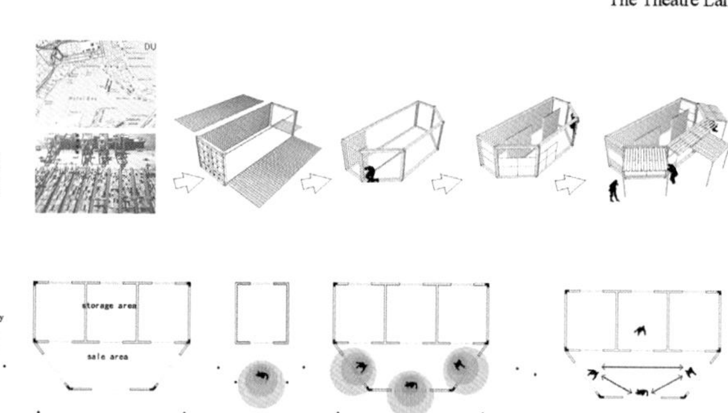

1.Containers are transformed into the catalyst which provides storage and commercial space for the vengors.

2.Gather the vendors together in number of 3 and 5 to sell together

3.Since the goods is stored together,vendors can take care of each other's table and goods to keep everying safe.

Situation and Expectation

Market distribution mart

Vendor distribution

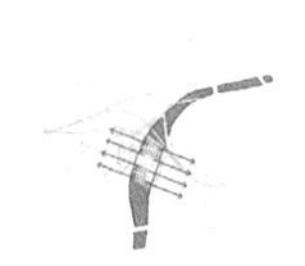

Railway divided the city into two parts.

Rail station is the tranportation junction.

The train station acts as a transportation junction between the two parts of the city;it sends people to many different detionations.

CITY FIBER OF DURBAN DISTRBUTARY ACCOMMODATION EXPANSION

作品名称：德班城市光纤
City Fiber of Durban

院校名：合肥工业大学
设计人：高翔，沙成鑫，刘梦柳，封瑞牧，黄宇超，林夏冰，张亚伟
指导教师：徐晓燕
课程名称：UIA 国际建筑竞赛（本科四年级）
作业完成日期：2014 年 04 月

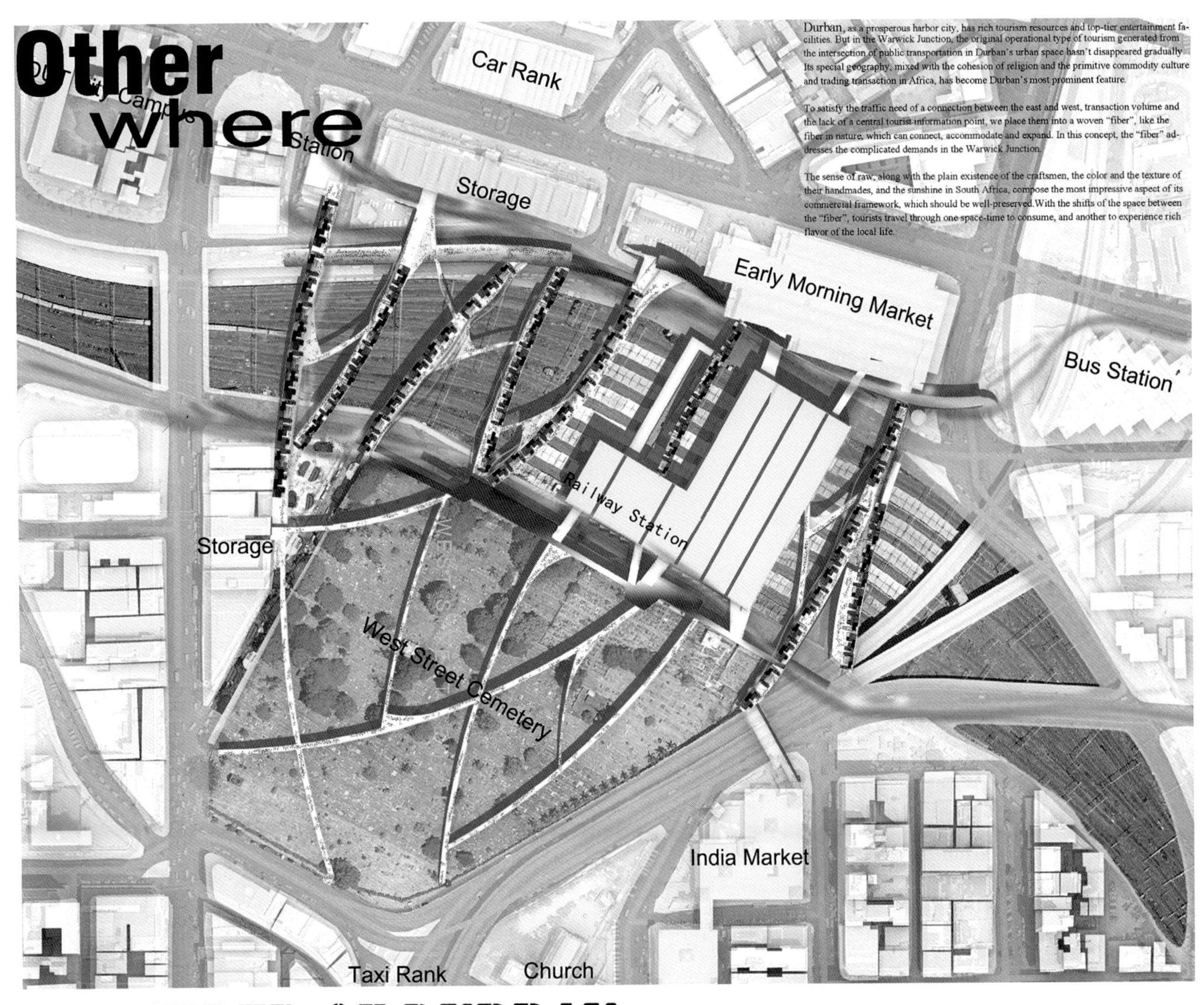

CITY FIBER OF DURBAN DISTRBUTARY ACCOMMODATION EXPANSION

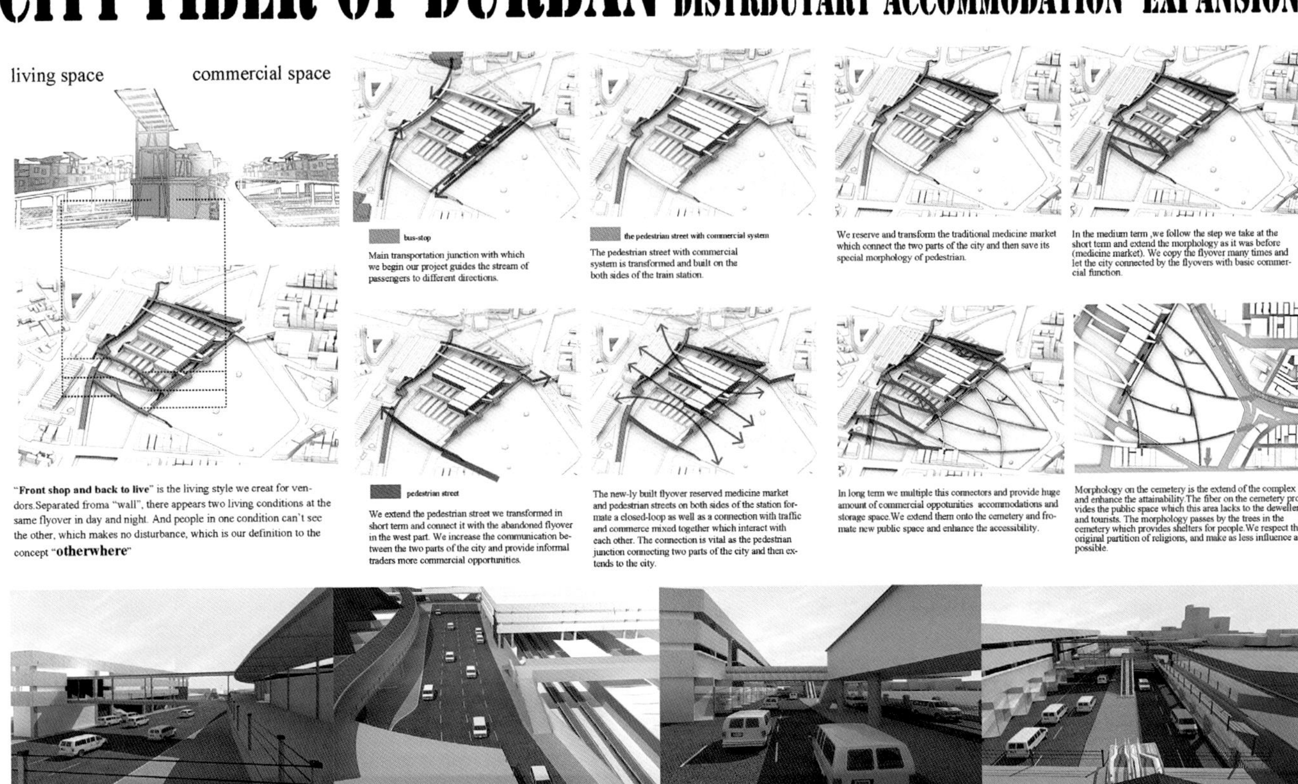

University: Hefei University of Technology
Designer: Gao Xiang, Sha Chengxin, Liu Mengliu, Feng Ruimu, Huang Yuchao, Lin Xiabing, Zhang Yawei
Tutor: Xu Xiaoyan
Course Name: UIA International Student Architecture Design Competition
Finished Time: Apr. 2014

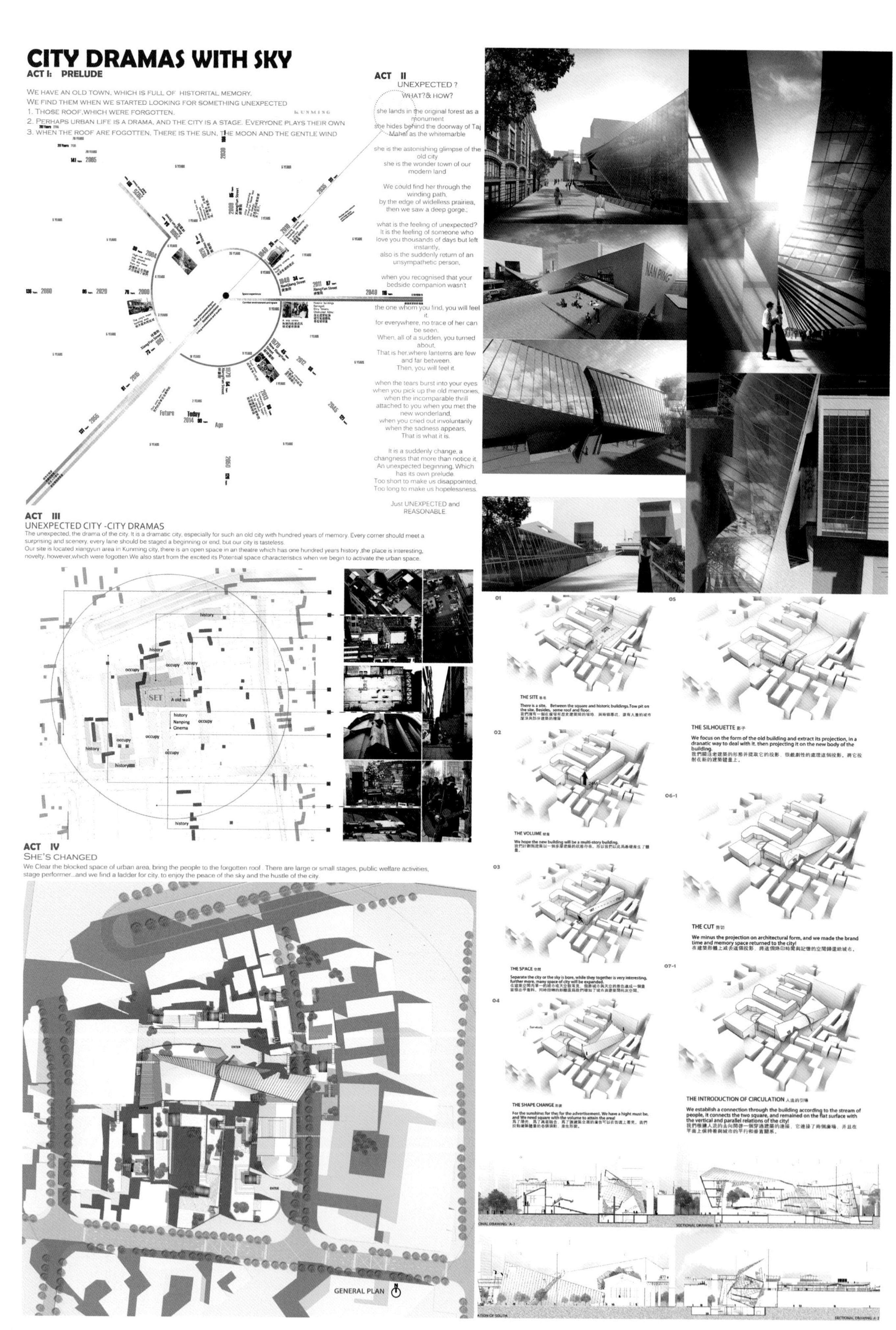

作品名称：出乎意料的城市——城市天空之剧

Unexpected City: City Dramas with Sky

院校名：昆明理工大学

设计人：郭骏超，叶雨辰

指导教师：翟辉，华峰

课程名称：昆明理工大学—研究型设计霍普杯国际竞赛

作业完成日期：2014 年 07 月

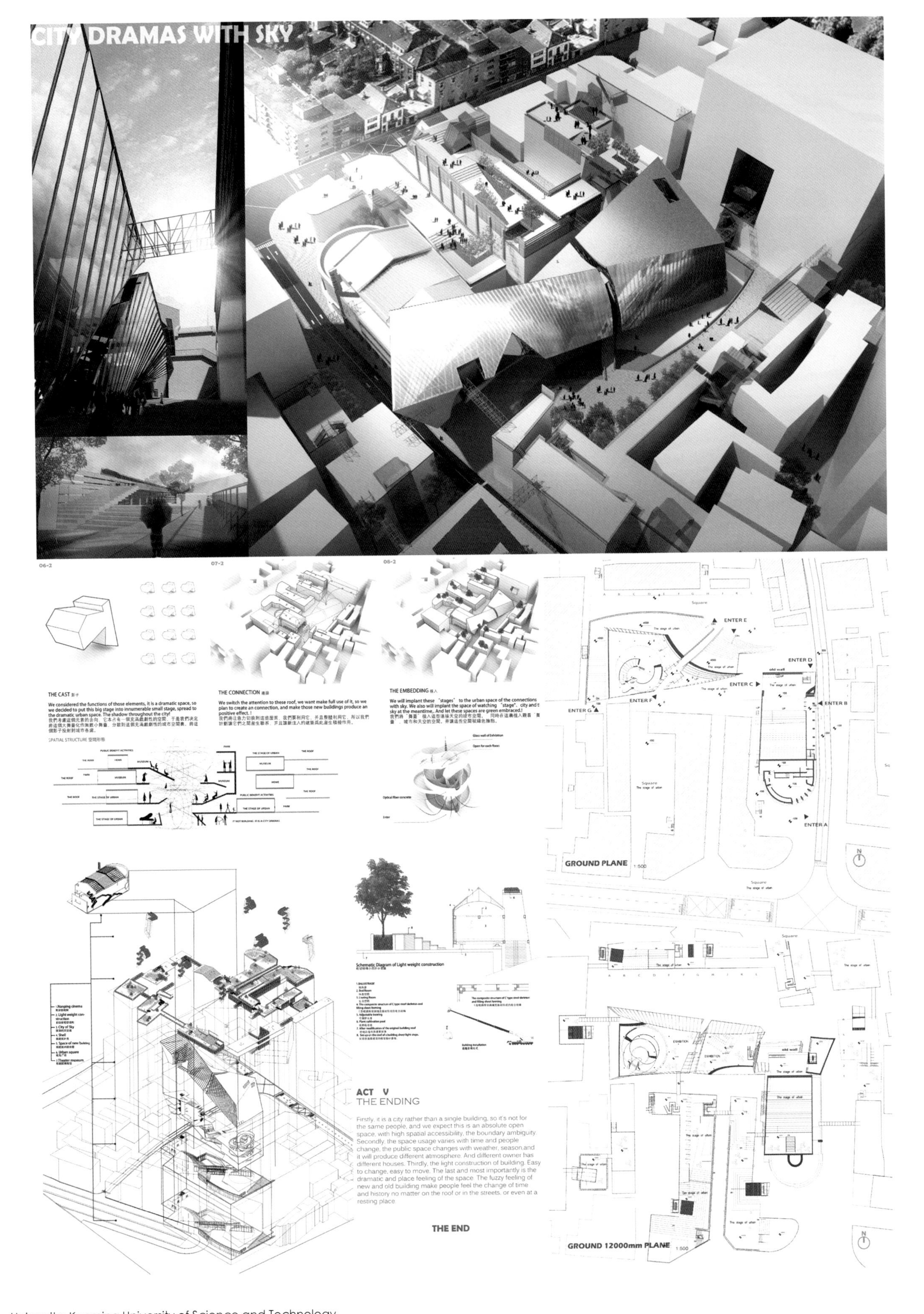

University: Kunming University of Science and Technology
Designer: Guo Junchao, Ye Yucheng
Tutors: Zhai Hui, Hua Feng
Course Name: UIA-HYP cup 2014 International student competition in architecture design
Finished Time: Jul. 2014

BELIEVE-IN CITY

Religion, as an eternal theme of human civilization, reaches peace, love and many other beautiful characters of humanity. In India, the most religion-diverse country in the world, almost every social issue such as politics, ethics, economy and art has a stake in it.This proposal focuses on our habitat of soul, the belief, and depicts a harmonious scene of a city whose spaital typology is rooted in the religion.By providing access to religious activities and mundane ones in different scales of space, people can spend their time with others that share and not share the same religion. Together with the physical livability can we create such as convenient transportation, flood security, completed infra-structural facilities, fresh air and pleasing natural environment. Ultimately, the Believe-in City is the one where poetic life will be live in.

RELEGION OF INDIA

India is the most religion-diverse country in the world. The religion has always been the main theme of the big events.
Different relegions have their own celebrations and traditions, as a result, the patterns of these religious activites are totally different with each other. So the spatial demand are different too. And our goal is to provide them with suitable spaces for BELIEF.

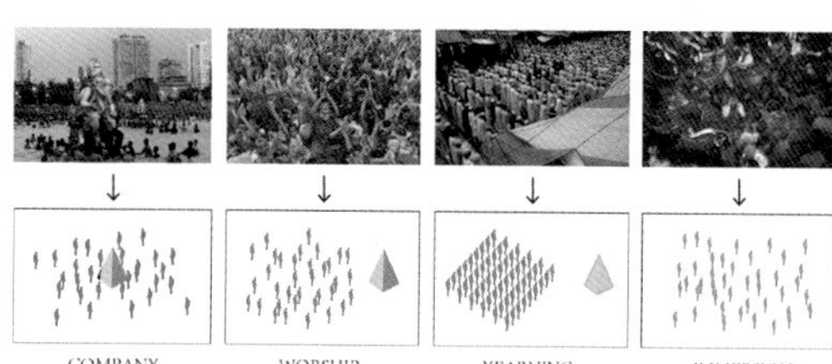

EVOLUTION OF RELIGION & PROPOSAL

Seeing from the history, the religious pattern has gone through a series of changes. At the very beginning, there was only one religion, while people honored the only gods of it. Together with the immigration came the ideas and thoughts of other religions. The city was becoming mixed and the religious space was gradually encroached. Then came the segregation and confrontation which made people of different religions separated from each other.
In our proposal, we redefine the pattern of living with religion.

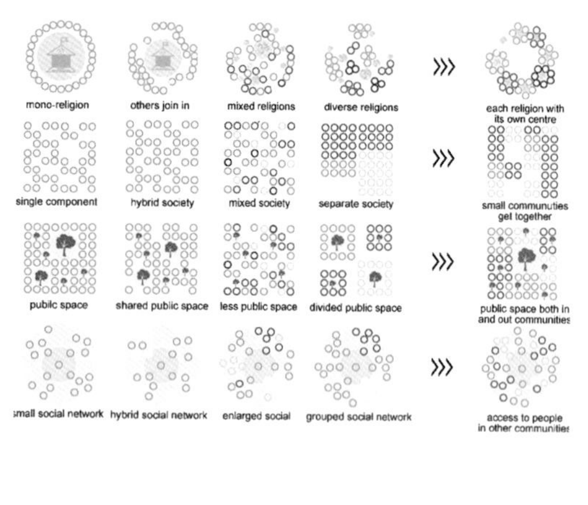

RESULT OF GENERATION

The combination of the demand of religion and the quality of the material environment finally results in a new typology for residence, also a new pattern of life for the future. It reflects the of emotion of spirits and offers a poetic habitat for people' s soul.The Believe-In City maintains the traditional street block pattern in its lower layer, however it stands out for the religious units above.

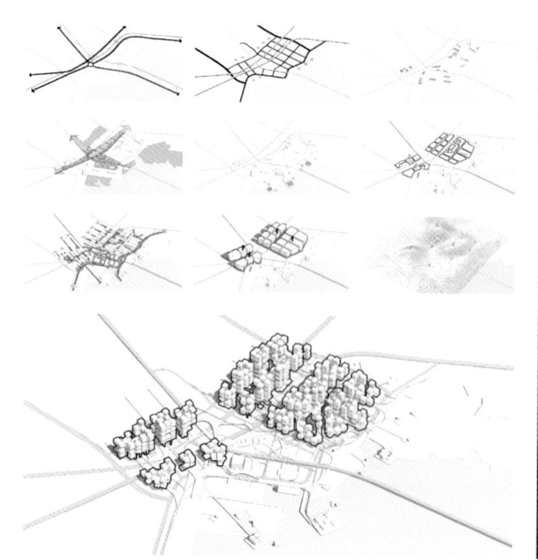

作品名称：相信城市
Believe-in City

院校名：清华大学
设计人：郝田，陈赛宇，孙逸琳，廖思宇
指导教师：朱文一，程晓喜，张弘
课程名称：亚洲垂直城市竞赛
作业完成日期：2014 年 07 月
对外交流对象：新加坡国立大学，香港中文大学，日本东京大学，荷兰代尔夫特理工大学，瑞士苏黎世联邦理工学院，美国加利福尼亚大学伯克利分校，美国宾夕法尼亚大学，美国密歇根大学

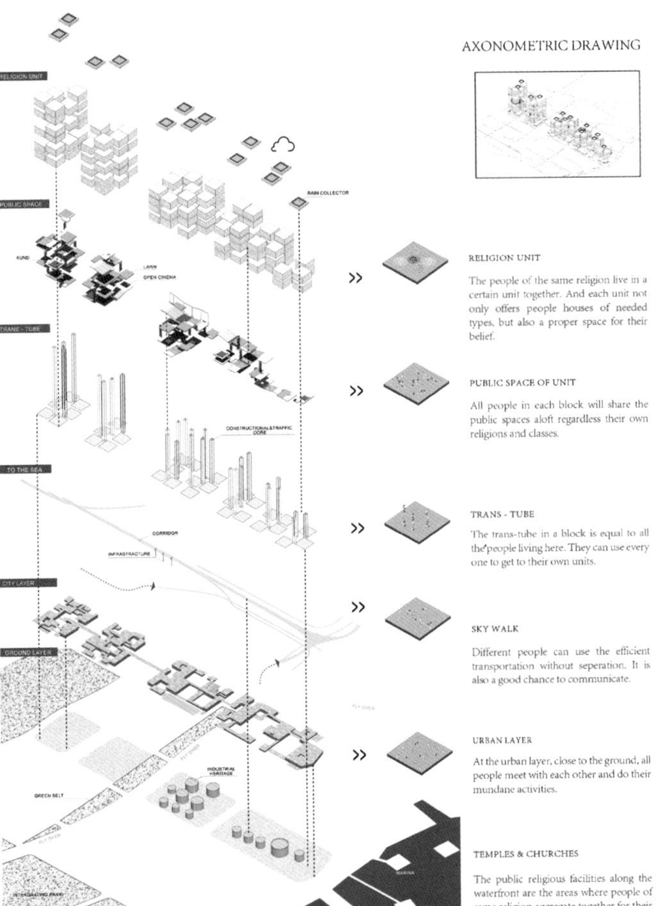

RELIGION UNIT
PUBLIC SPACE
TRANS - TUBE
TO THE SEA
CITY LAYER
GROUND LAYER
RAIN COLLECTOR
CORRIDOR
INFRASTRUCTURE
GREEN BELT
AXONOMETRIC DRAWING
RELIGION UNIT
The people of the same religion live in a certain unit together. And each unit not only offers people houses of needed types, but also a proper space for their belief.
PUBLIC SPACE OF UNIT
All people in each block will share the public spaces aloft regardless their own religions and classes.
TRANS - TUBE
The trans-tube in a block is equal to all the people living here. They can use every one to get to their own units.
SKY WALK
Different people can use the efficient transportation without seperation. It is also a good chance to communicate.
URBAN LAYER
At the urban layer, close to the ground, all people meet with each other and do their mundane activities.
TEMPLES & CHURCHES
The public religious facilities along the waterfront are the areas where people of same religion aggregate together for their own ceremonies.

FORM GENERATION

Each block has two parts, the residential layer and the urban layer, which are organized vertically. Diverse kinds of activities could take place. And people will live a happy and efficient life without disturbing each other.

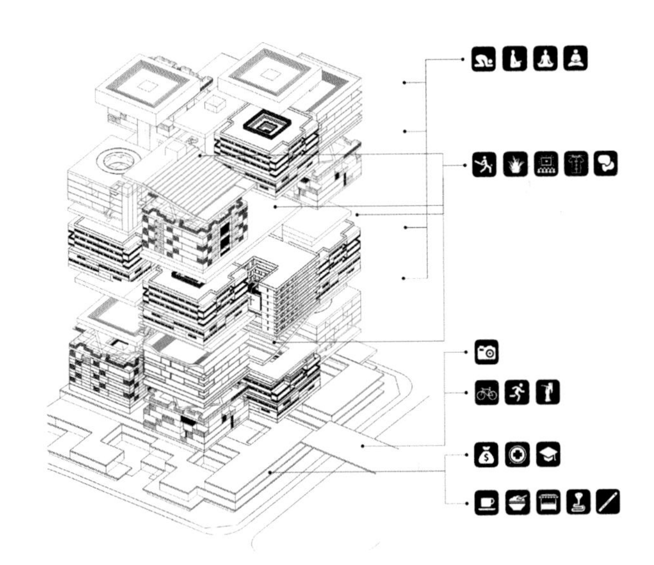

CIRCULATION

In each community, people go through the grey space to the lobby where the public vertical transportation is.

Taking the vertical transportation, people can reach the transfer floor of each unit, which provides public space for social activities and connects to aloft kunds, open cinemas and lawn between units

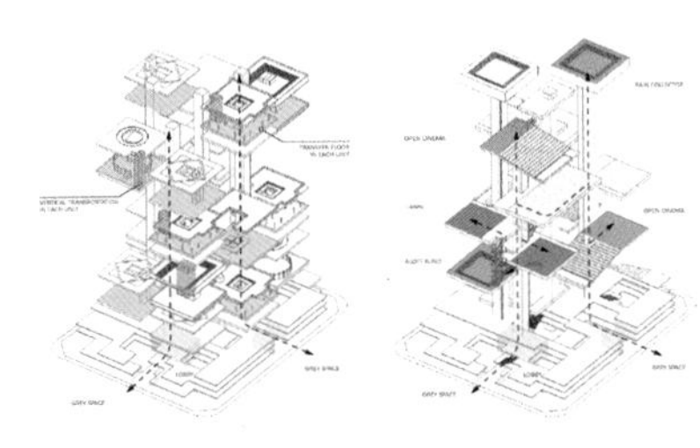

MUNDANE ACTIVITIES

01. City streets being open to mundane entertainment bring the city an inclusive life. 02. The publice platform between units offers people a opportutnity to enjoy films by Bollywood. 03. Mangrove park open to the public, sets a connection to the nature. 04. The linear streets direct to the sea from the west to the east, enable the city to celebrate the Ganesh Festival joyfully. 05. The skywalk speading over the city, provides a connection to transport efficiency as well as interaction opportunities. 06. Section of a street.

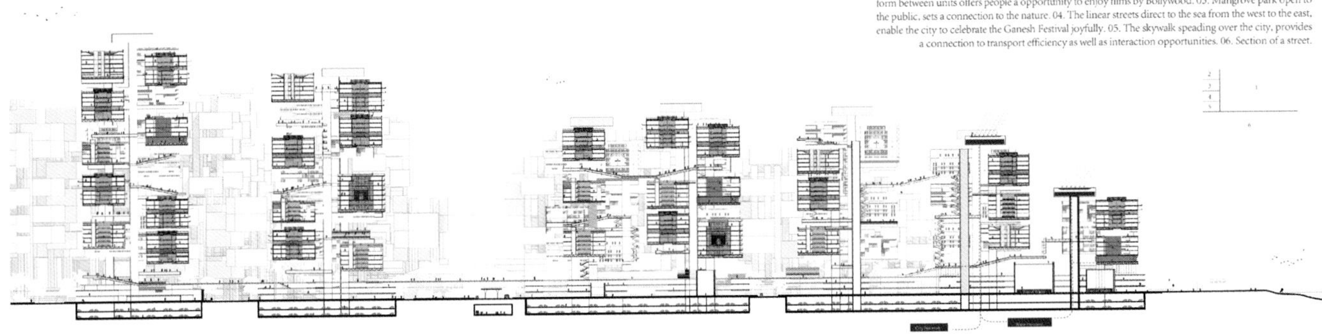

University: Tsinghua University
Designer: Hao Tian, Chen Huanyu, Sun Yilin, Liao Siyu
Tutors: Zhu Wenyi, Cheng Xiaoxi, Zhang Hong
Course Name: Vertical Citles Asia International Competition
Finished Time: Jul. 2014
Exchange Institute: National University of Singapore, The Chinese University of Hong Kong, Tokyo University, Delft University of Technology Eidgenössische Technische Hochschule Zürich, University of California-Berkeley, University of Pennsyivania, University of Michigan

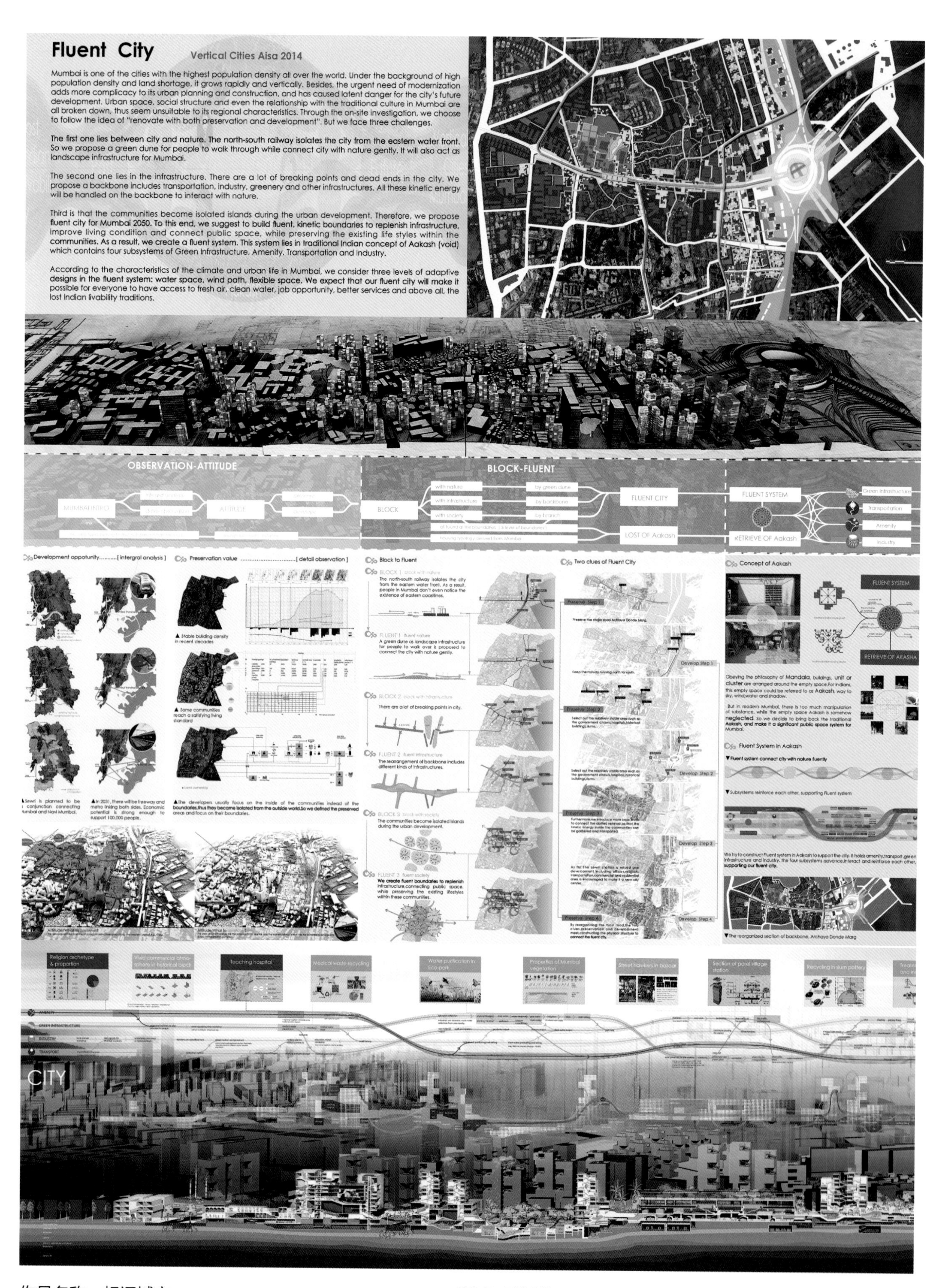

作品名称：畅通城市
Fluent City

院校名：同济大学
设计人：何啸东，李骜，肖璐珩，郑攀，程思，谭杨
指导教师：王桢栋，董屹，黄一如
课程名称：亚洲垂直城市竞赛
作业完成日期：2014年07月
对外交流对象：新加坡国立大学，香港中文大学，日本东京大学，荷兰代尔夫特理工大学，瑞士苏黎世联邦理工学院，美国加利福尼亚大学伯克利分校，美国宾夕法尼亚大学，美国密歇根大学

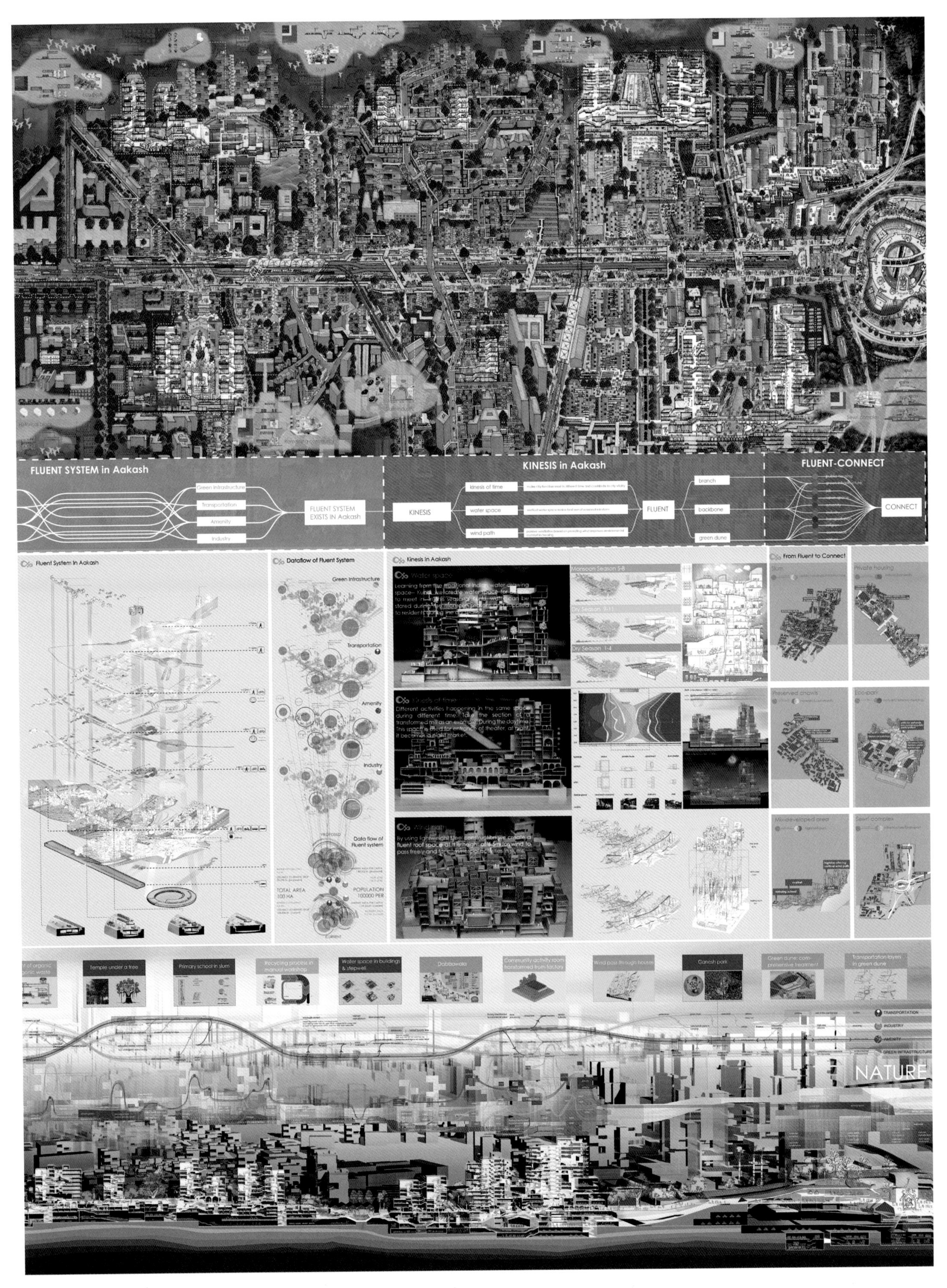

University: Tongji University
Designer: He Xiaodong, Li Ao, Xiao Luheng, Zhen Pan, Cheng Si, Tan Yang
Tutor: Wang Zhendong, Dong Yi, Huang Yiru
Course Name: Vertical Citles Asia Internatinal Competition
Finished Time: Jul. 2014
Exchange Institute: National University of Singapore, The Chinese University of Hong Kong, Tokyo University, Delft University of Technology, Eidgenössische Technische Hochschule Zürich, University of California-Berkeley, University of Pennsyivania, University of Michigan

ON AND UNDER 1

A DIFFERENT APPROACH TO UTILIZE THE REMAINING URBAN SPACE

LOCATION OF DURBAN

1.Durban is a city where east meets west, characterized by its diversification. (xxv UIA, 2014)
2 .Durban has a developed sea and land transport.

PROBLEMS

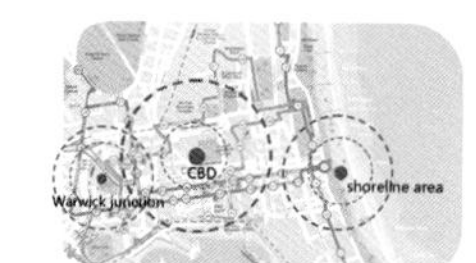

1 Railway Transportation

By the highly-developed railway system,goods are transported from the Durban Port to the hinterland of South Africa.

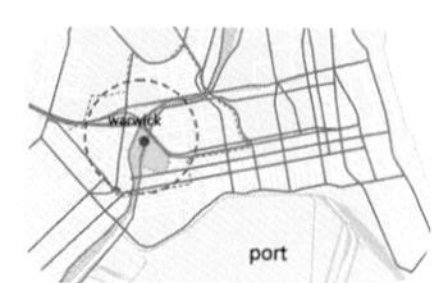

2 Multi-centered Structure

Durban is a city of multiple centers, requiring enhancements of different centers to stimulate the local economy.

3 Urban Arterial Street

In city areas, one-way arterial streets lead to inconvenience in planning public transportation. Connecting railway and highway systems, greater Warwick area functions as a critical link.

4 Commercial spaces

By the northeastern side of cemetery, there are plenty of autonomous but disorganized commercial spots. Integrating of the current commercial space would transform this area into an optimal place for further development of Cultural & Creative industries in greater Warwick area.

SOLUTION—— SPACE RECONSTRUCTION & EXTENSION OF GREATER WARWICK AREA

Long Term Intervention

Spatial Reconstruction in Warwick

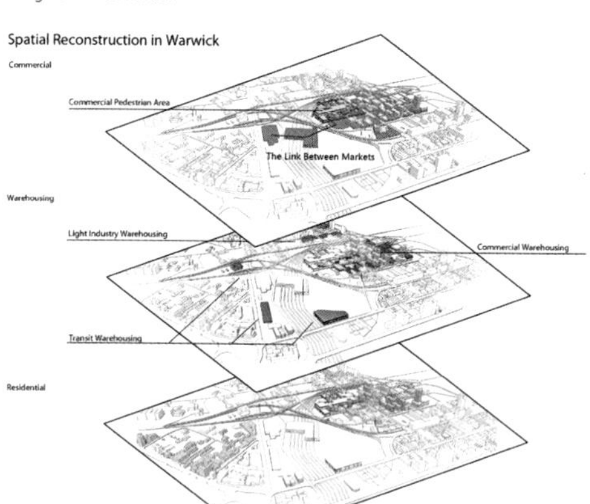

Goal: Industrial upgrading

Space reconstruction of Warwick area accommodates development of the local economy. A commercial pedestrian street is constructed between the north of Victoria Market to the west of Dr Yusuf DaDoo Street. It is an extension of Warwick Market and Victoria Market to facilitate the booming creative industry like traditional handicrafts. Job opportunities, at the same time, are created. Such design is based on the high-loaded passenger flow in this area, providing more possibilities for commercial exchanges.

Space extension comes along with the improvement of urban function. We expect this extension of space to be light and economical efficiently,contributing to the prevention of excessive space explosion.We propose a plan of "Community on the Roof ",creating a new layer above the two or three-story buildings. An integrate community is thus formed on the basis of those low-cost and convenient functional modules,which are believed to be two of the most attractive factors for those novice informal traders and creators.

Roof Community

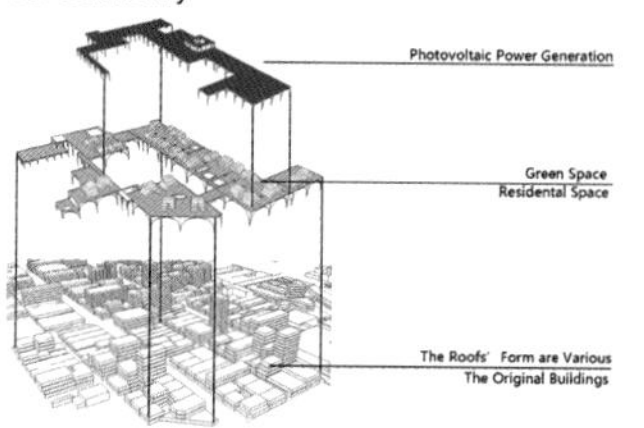

Form an integrate community on basis of low-cost and convenient functional modules,which are believed to be two of the most attractive factors for those novice informal traders and creators.

Public Traffic System

1.The one-way street brings a lot of inconvenience and lowers accessibilities of some area in greater Warwick area.
2.Inadequate bus routes cannot satisfy the traffic needs.
3.The extensive use of the minibus, taxi, and private cars lead to traffic congestion in most area.

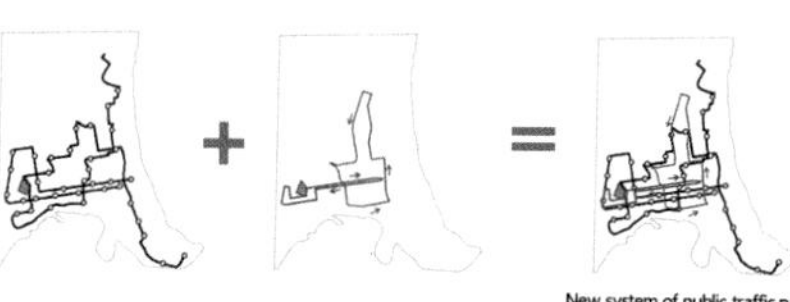

Existing public traffic system. + Add new lines to improve the accessibility. = New system of public traffic,particularly emphasizing on needs of trades and residential communities .

Medium Term Intervention

Problems

1. N3 freeway caused spacial isolation.

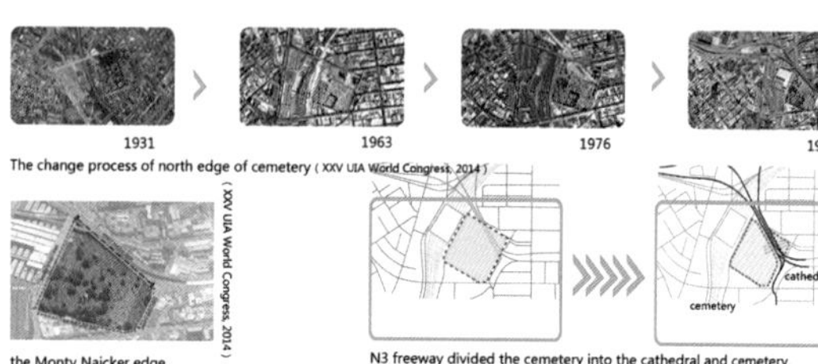

The change process of north edge of cemetery (XXV UIA World Congress, 2014)

the Monty Naicker edge (XXV UIA World Congress, 2014)

N3 freeway divided the cemetery into the cathedral and cemetery.

3. Due to the ineffective utilization of isolated space, different urban areas cannot be connected successfully.

Vacant space　Informal traders　Vehicles　(Google Maps , 2014)

2. In disorder, this boundary spreads across several business streets.

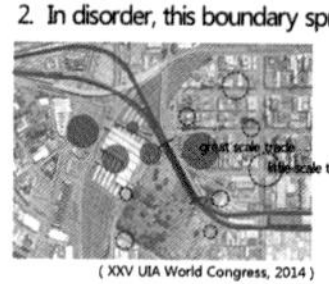

(XXV UIA World Congress, 2014)

N3 monorway runs across the the major conjunction among Victoria Market, Brook Street Market and Herb Market. Insertion of themonorway creates an isolated space alongside the cemetery, leaving no transition between the cemetery and Victoria Market.

Concept of Otherness——Learning From Tokyo

In Tokyo, Japan, land shortage is a challenging problem, which stimulates the emergence of more effective techniques in land use. Durban is now challenged by various problems, such as traffic jam and environmental deterioration caused by space expansion. At the same time, government puts forward a strategy to build a "Compacted City". Accordingly, we could refer to successful practice in Tokyo, utilizing vacant urban spaces by accurate design.

What can we do under the viaduct?

(KAIJIMA Momoyo et al. , 2007)

1 shopping mall

2 housing

3 nursing institution

The Top Layer: N3 motorway
N3 motorway provides a shelter for spaces underneath protecting people from sunshine and raindrops.
The Mid Layer: New Bridge Space
Containers in short term intervention are placed here to open up a new space, connecting streets with the markets. It also develops new spaces for informal trades, entertainment, short term storage, and residence.
Second-hand Containers
Recycled second-hand containers are easy to assemble, cheap and environmental friendly. Their various colors facilitates the roads with lively visual enjoyments.
The Ground Layer: Parking and Green Space
To specificate the original chaotic parking and plant greens in the unused space on this edge.

作品名称：上与下
On and Under

院校名：长安大学
设计人：侯乃菲，万伊涵，杨定宇，王凯旋，杨鑫
指导教师：张磊，王农，张琳，杨宇峤

课程名称：第 25 届 UIA（国际建筑师协会）世界大学生建筑设计竞赛
作业完成日期：2014 年 06 月

ON AND UNDER 2

A DIFFERENT APPROACH TO UTILIZE THE REMAINING URBAN SPACE

Short Term Intervention

In mid term intervention, a narrow space (235 m in length, 12.2 m in width) is created by northern side of the cemetery. We do have realized problems in high cost, risky investment, and business development status (informal and small-amount trades in majority). Considering those potential challenges, we expect to adopt a new approach with greater, faster, and more economical outcomes, exploring more functional possibilities of the newly developed urban space.

(Yang,2005)

In Durban, one of the largest port in Southern Hemisphere, its container throughput has reached a prosperous amount ——2, 568,124TEU in 2012 (Wikipedia, 2012). It is a convincible reason that container becomes the most economical and accessible choice for us.

streetscapes

1 THE ADVANTAGES OF CONTAINER BUILDING

can be built rapidly

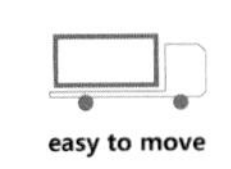

easy to move

recyclable

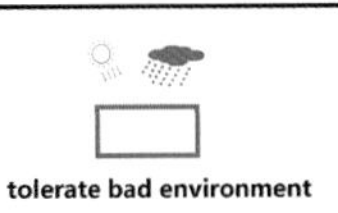

tolerate bad environment

2 VARIOUS ASSEMBLING OF CONTAINERS

Short term intervention would economically satisfy various needs of business, temporary residence and storage. Containers, for its spacial flexibility and structural adaptability, perfectly fit into such requirements. This design is expected as a procedure like "LEGO' s selection & assembling", which invites informal traders into application of containers. Hereby are some of the tentative reconstruction plans.

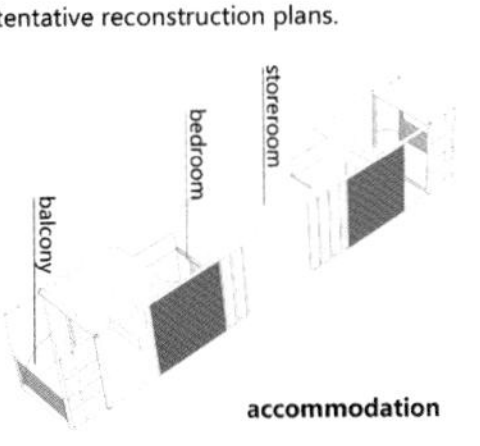

accommodation

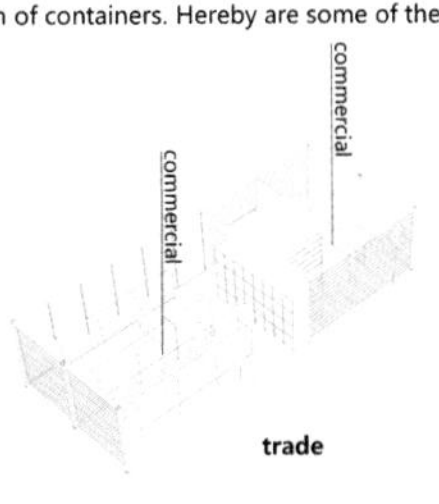

accommodation

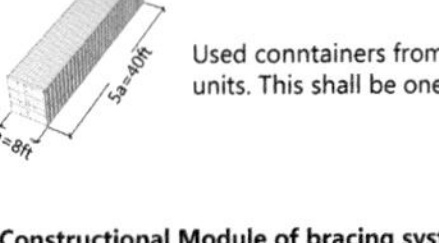

trade

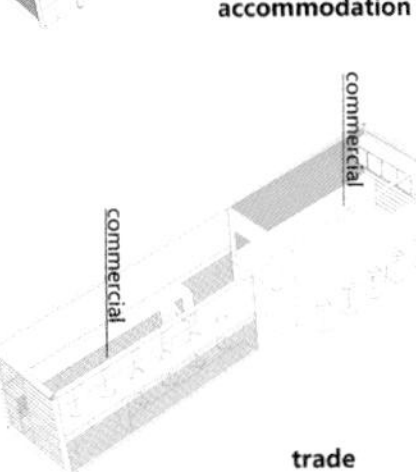

trade

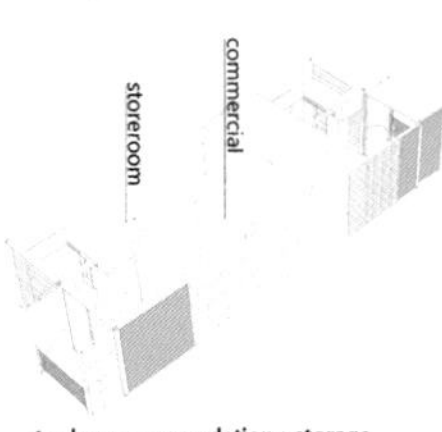

trade

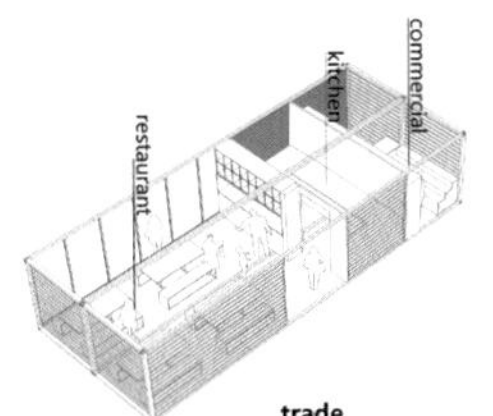

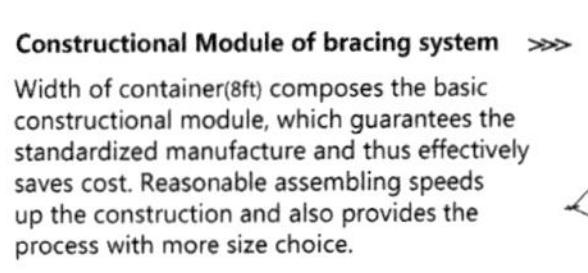

trade + accommodation + storage

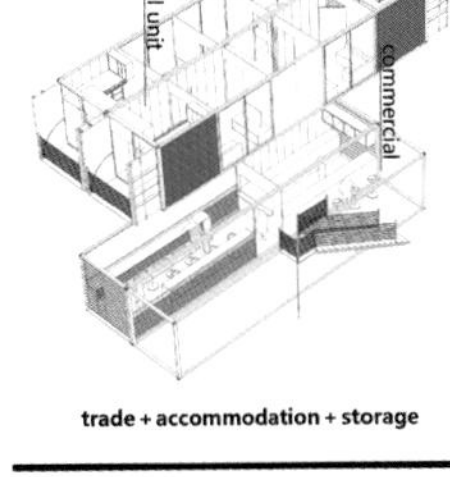

trade + accommodation + storage

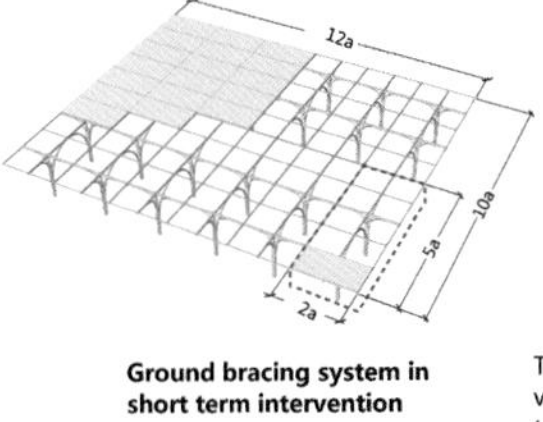

trade + accommodation + storage

trade + accommodation + storage

3 EFFECTIVE CONSTRUCTION & LOW-COST PLAN

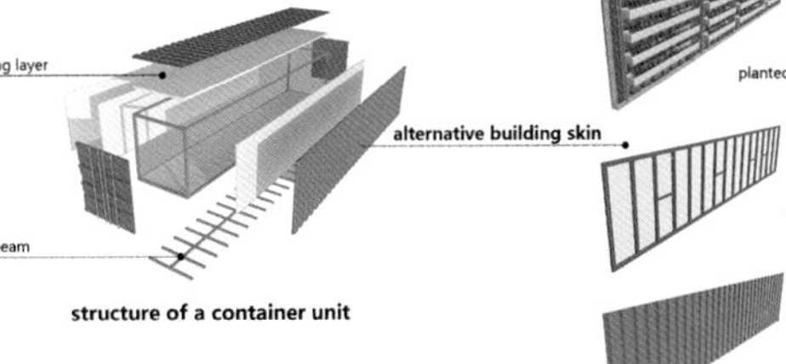

structure of a container unit

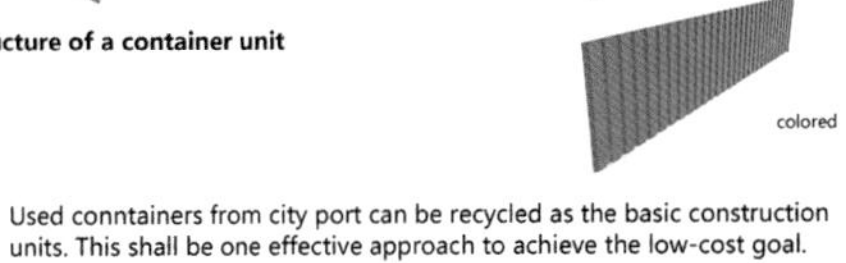

Used conntainers from city port can be recycled as the basic construction units. This shall be one effective approach to achieve the low-cost goal.

Constructional Module of bracing system >>>

Width of container(8ft) composes the basic constructional module, which guarantees the standardized manufacture and thus effectively saves cost. Reasonable assembling speeds up the construction and also provides the process with more size choice.

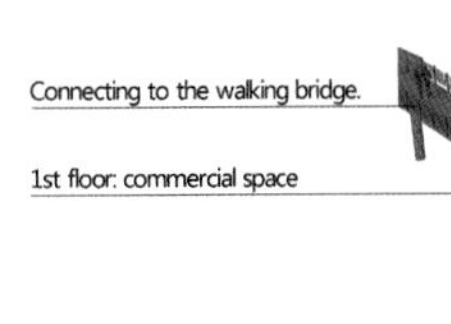

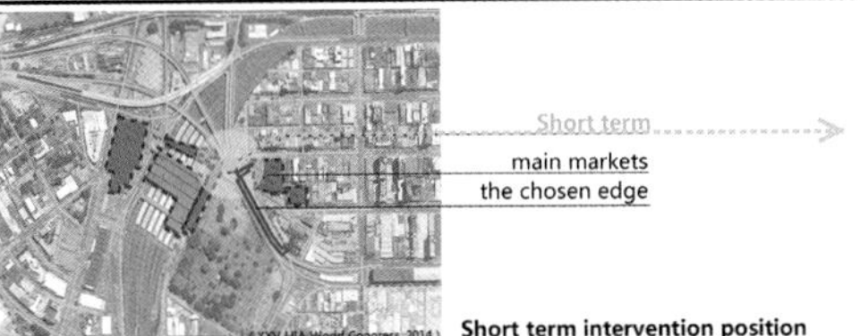

Ground bracing system in short term intervention

The bracing system is highly standardized with various assembling choices and flexibly adapted to different urban spaces.

4 SHORT TERM INTERVENTION PLAN

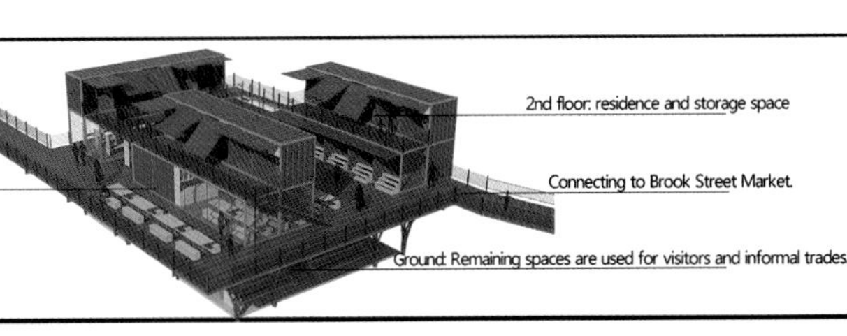

Short term intervention position

5 GENERATION PROCESS

1 It is the only conjunction to Warwick from the north side of the cemetery.

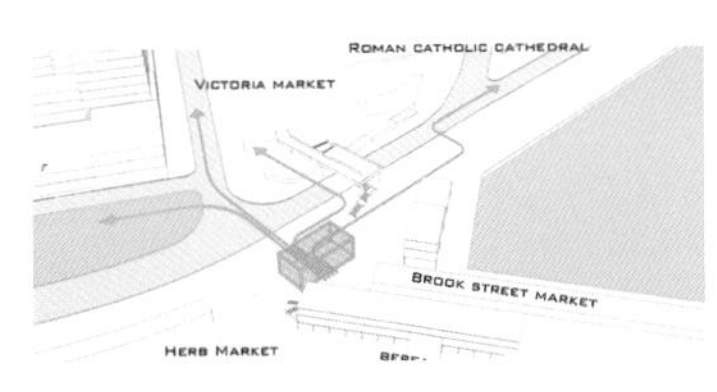

2 It creates a new commercial layer with connection to the walking bridge, which detours the high flow of passengers and reorganizes conjunction spaces.

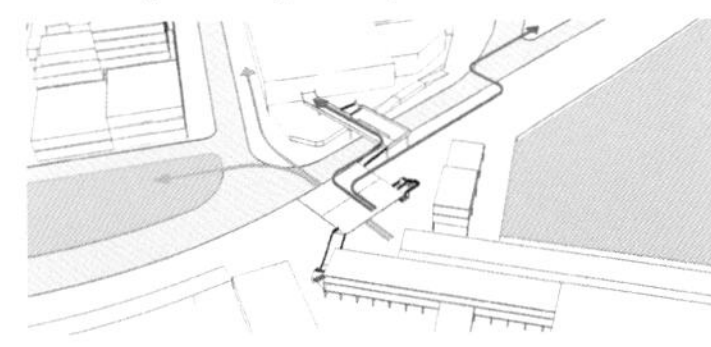

3 The new layer is filled by containers units with two floors: the lower floor for commercial activities while the upper floor for storage and accommodation.

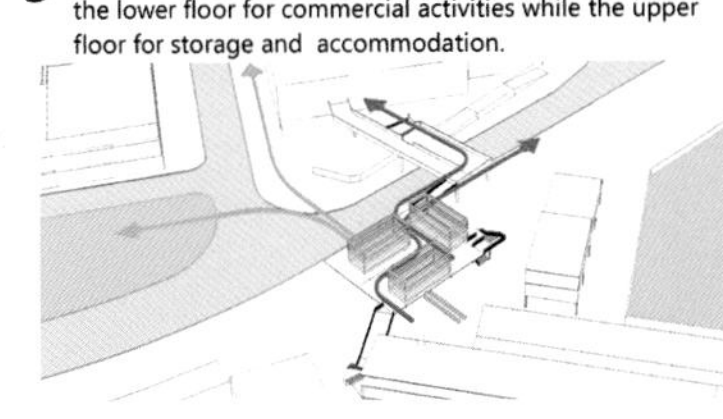

University: Chang'an University
Designer: Hou Naifei, Wan Yihan, Yang Dingyu, Wang Kaixuan, Yang Xin
Tutor: Zhang Lei, Wang Nong, Zhang Lin, Yang Yuqiao

Course Name: The International Union of Architects and 'Architecture Otherwhere' UIA 2014 Durban
Finished Time: Apr. 2014

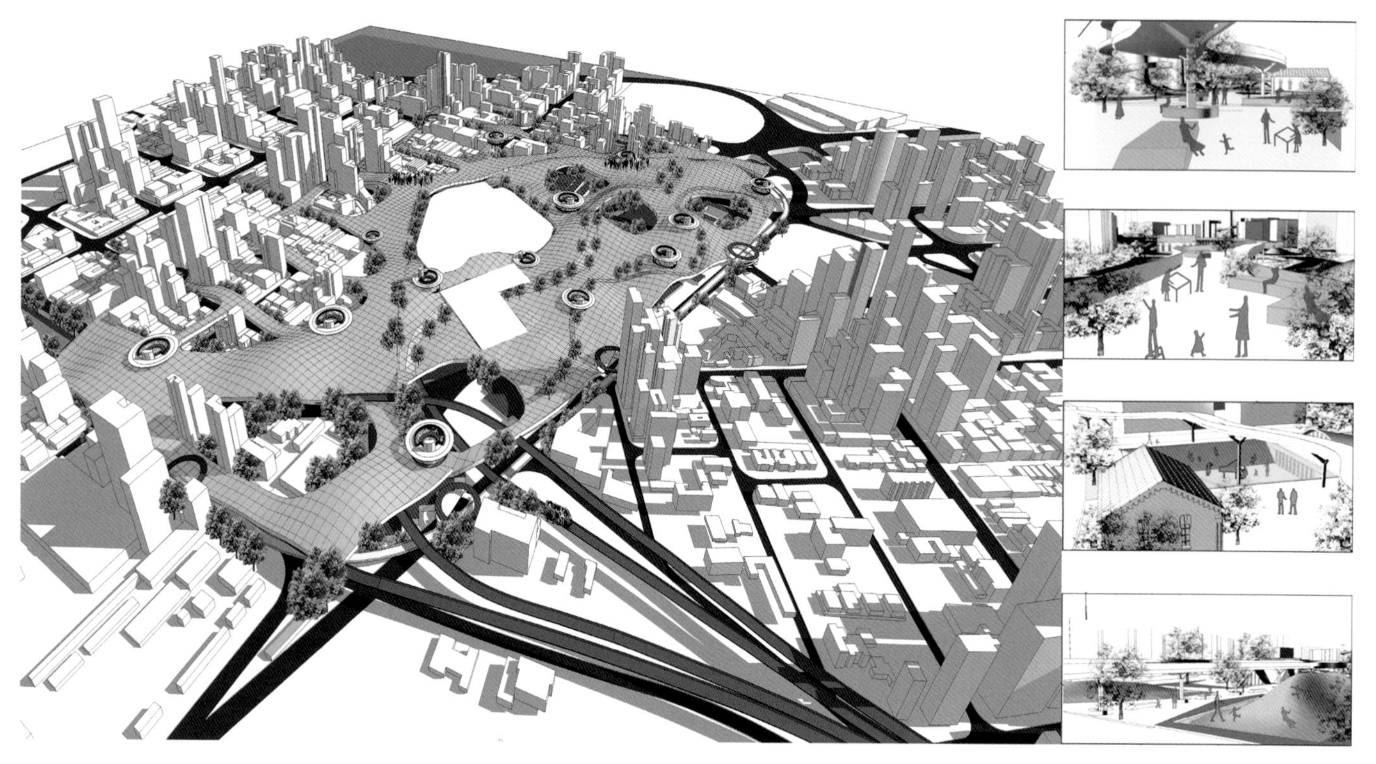

COMBINE

REGENERATION AND METABOLISM

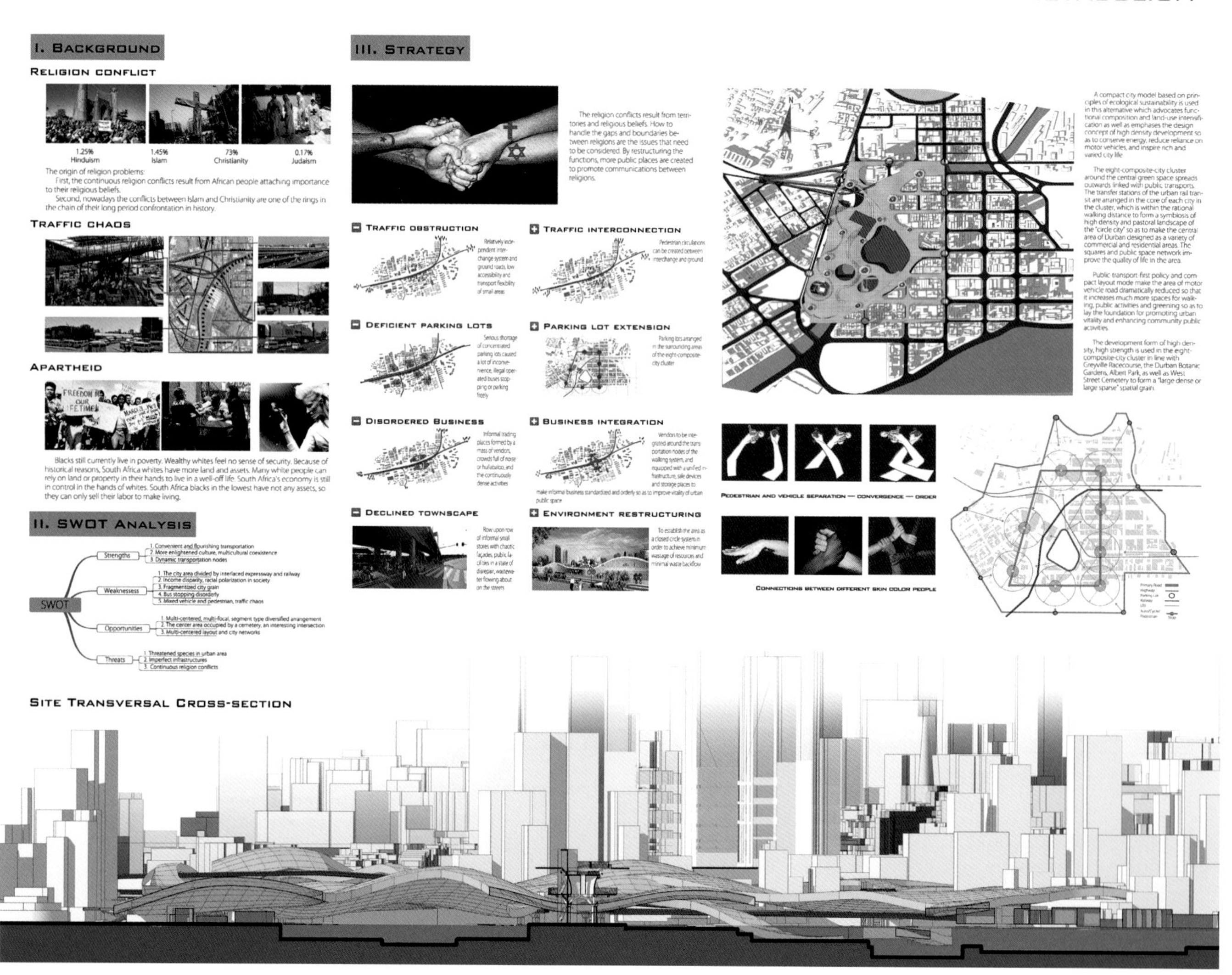

作品名称：新陈代谢与重生
Combine Regeneration and Metabolism

院校名：西安建筑科技大学
设计人：兰青，刘伟，李小同，刘俊，张佳茜，钱雅坤
指导教师：李岳岩，陈静
课程名称：2014UIA 国际大学生建筑设计竞赛（研究生二年级）
作业完成日期：2014 年 03 月

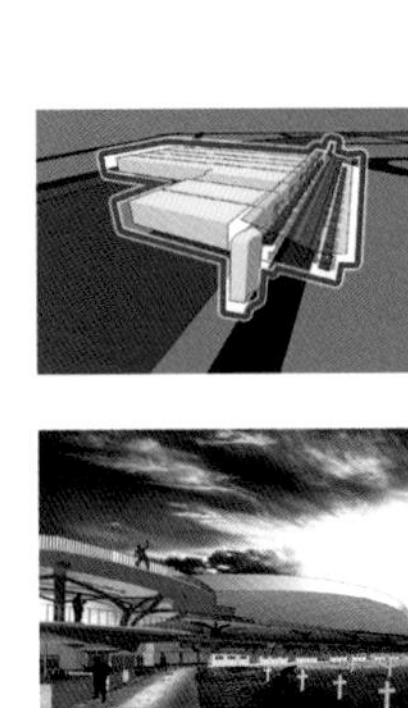
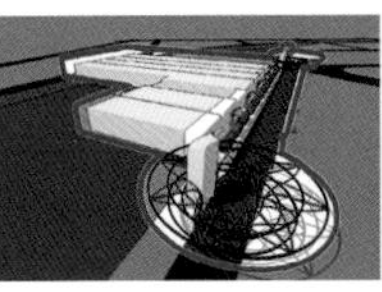

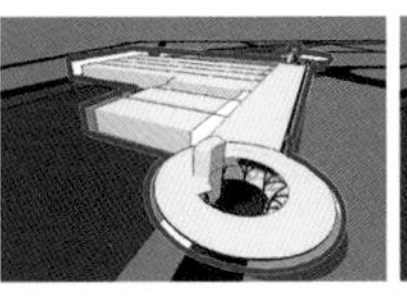
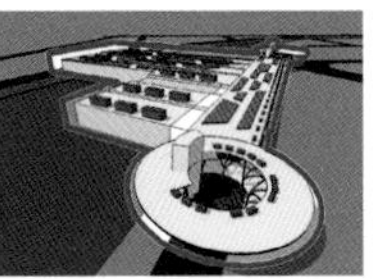

Betterment
Newly-built

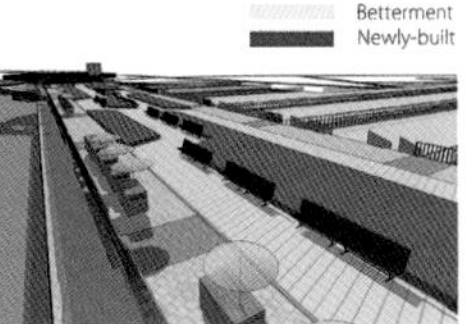

View Brook Street from the cemetery — Brook Street being reconstructed — Passageway above Brook Street, conected to the Junction — Reconstructed rof above Brook Street

BEFORE BETTERMENT — **AFTER BETTERMENT**

colored corrugated steel roof
storehouse
obsolete building
accessible flat-roof
parking lot

Jammed streets, vehicle parking freely, mixed pedestrians and vehicles, terrible townscape, bad vending environment.

Pedestrian and vehicle separation, improved infrastructures, environment reconstruction, standardized marketplace and transport.

structure — 1st floor — 2nd floor — complete

pedestrian flow
view

We designed Warwick District as a variety of commercial and residential area in which there is a green space and public space network to be improved. The access of the area mainly depends on the public transport system so that the road network composed of pedestrian streets, cycle ways, markets, and avenues is greatly expanded. The road network is precisely interlaced with the public transport system, which consists of a unique and interconnected public space and transport network.

COMBINE

REGENERATION AND METABOLISM

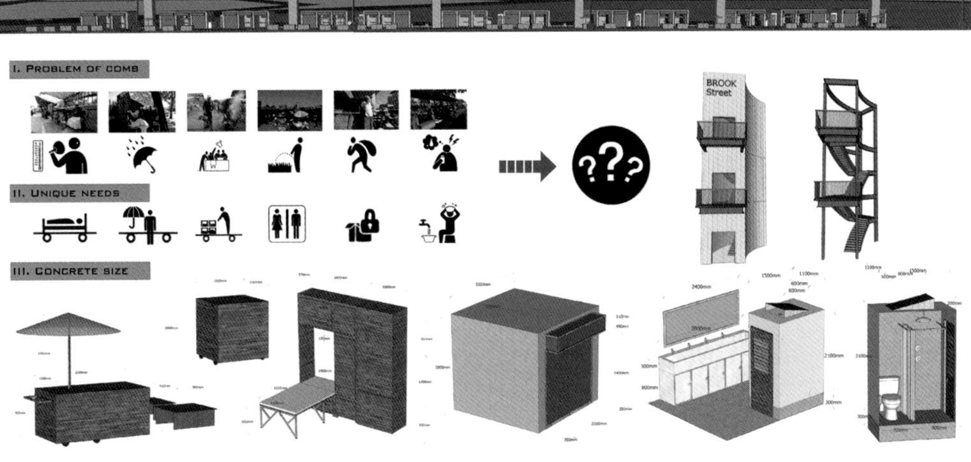

IV. Service condition

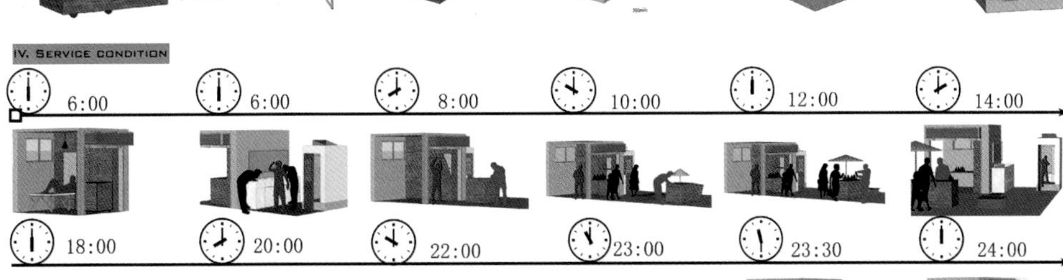

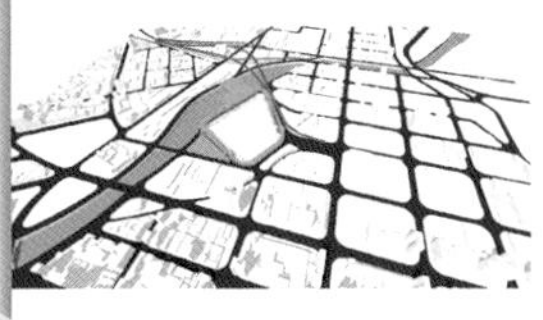

THE FIRST WEEK

In space planning, three annular roads and eight radial avenues form the primary transport framework. The outer ring is planned for buses, taxis, cycles, and pedestrians to use. The middle ring is the urban rail transit to bunch the eight composite cities into a cluster.

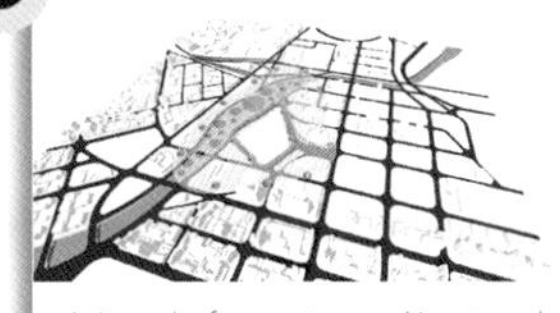

THE FIRST YEAR

A city mode of compactness, multi-center and sustainable development based on integrated and comprehensive public space and transport system.

An interesting public space is around the cemetery to create a diversified local condition.

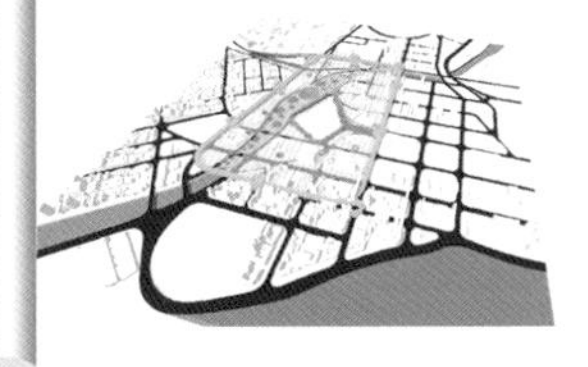

THE FIFTH YEAR

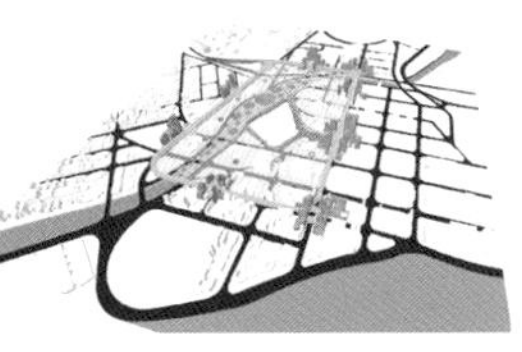

Create a compact city model based on principles of ecological sustainability. Advocate functional composition and land-use intensification. Emphasize the design concept of high density development so as to conserve energy and inspire rich and varied city life.

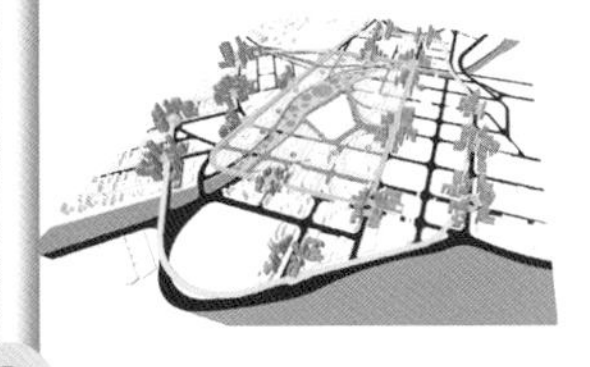

THE TENTH YEAR

All constructions are planned within 350 m to the station of LRT, which facilitates people walking to the station and reduces reliance on cars. Sunk arrangement is used for the inner ring which connects eight car parking lots as well as N3 and N4 expressways. The eight radial avenues connect the three annular roads together as well as the inside cemetery.

University: Xi'an University of Architecture and Technology
Designer: Lan Qing, Liu Wei, Li Xiaotong, Liu Jun, Zhang Jiaqian, Qian Yakun
Tutor: Li Yueyan, Chen Jing
Course Name: UIA 2014 International Student Competition
Finished Time: Mar. 2014

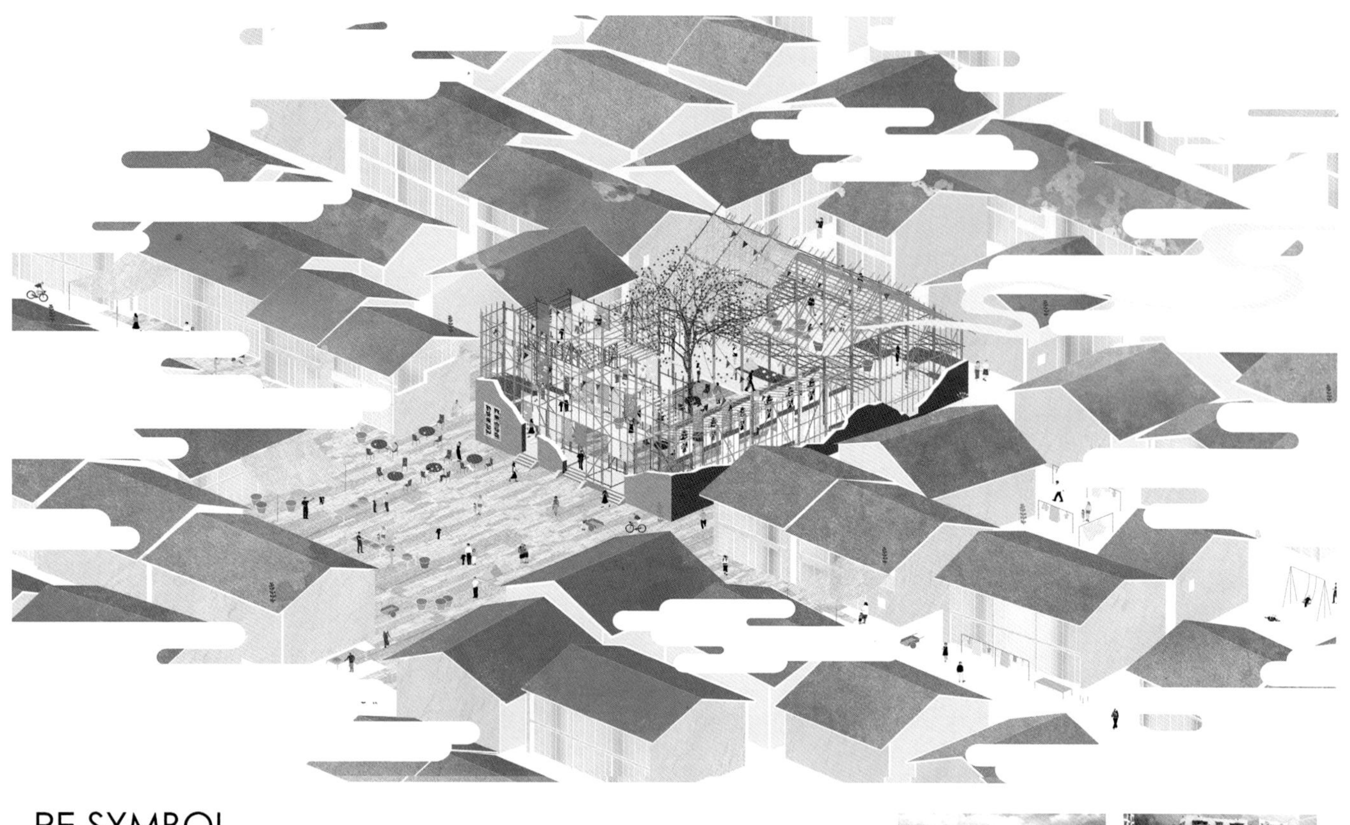

RE-SYMBOL.

This is a monument without being monumental, an established form without being crown yet still loved by its people. In Chinese tradition, an ancestry temple is the heart of a village, both spatially and sociologically. It allows villagers to worship their ancestors, and houses various public gatherings. In this case, a bamboo shelter is fast constructed in memory of an ancestry temple destroyed by earthquake in Qixin Village, China, 5th May, 2014. It re-symbols the village's spiritual meaning, reminds people of the memory, connects to the past, and grows with its village and villagers.

symbol
(before earthquake)

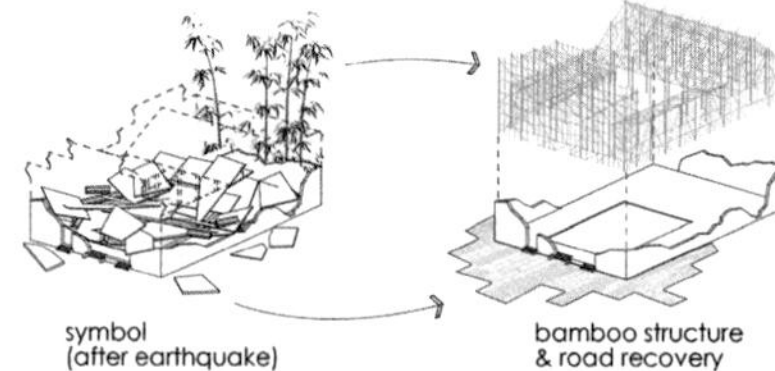

symbol
(after earthquake)

bamboo structure
& road recovery

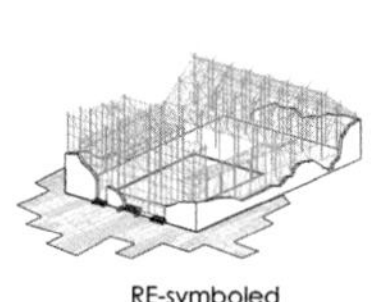

RE-symboled

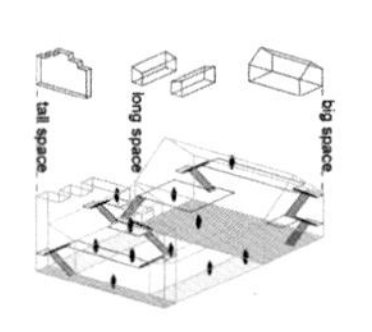

space reorganization

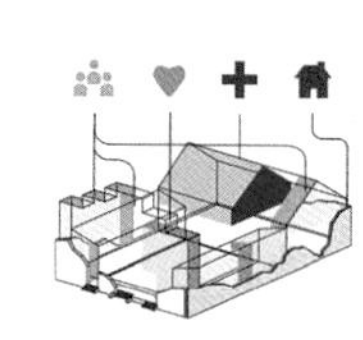

function reorganization

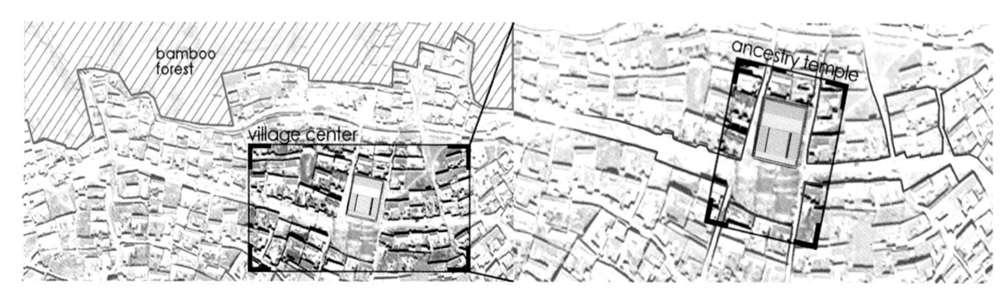

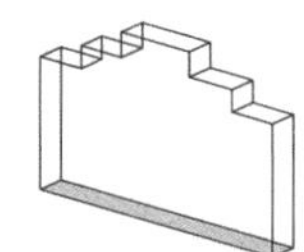

tall volume

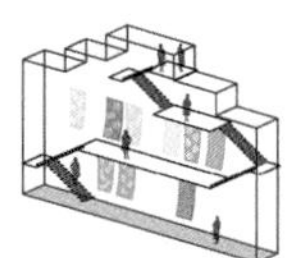

tall volume reorganization

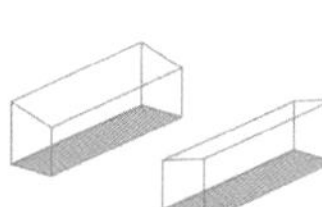

long volume

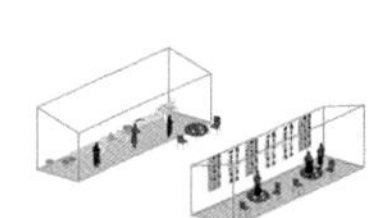

long volume reorganization

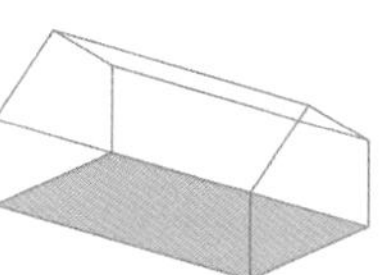

big volume

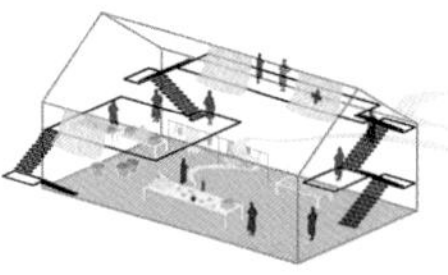

big volume reorganization

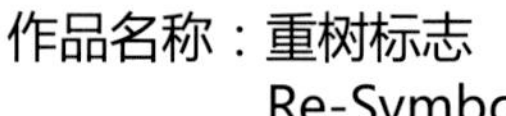

作品名称：重树标志
Re-Symbol

院校名：天津大学
设计人：李斯奇，陈晓婷
指导教师：邹颖
课程名称：第 49 届中央玻璃国际竞赛（硕士二年级）
作业完成日期：2014 年 06 月

RE-SYMBOL ..

The bamboo gradually turns yellow, the tree gradually grows taller, and the people gradually building up their new life. In this process, the building becomes the best witness to the earthquake and the rebirth of the village. Compared to just waiting to be rescued, people can get more power *from devoting themselves into the rebuilding of their most loved symbol. *The completion of this symbol is also the completion of their own redemption. So this is how a symbol gets its second life; the hierarchy and the boundary are dissolved, the material decayed but some spirit remains the same.

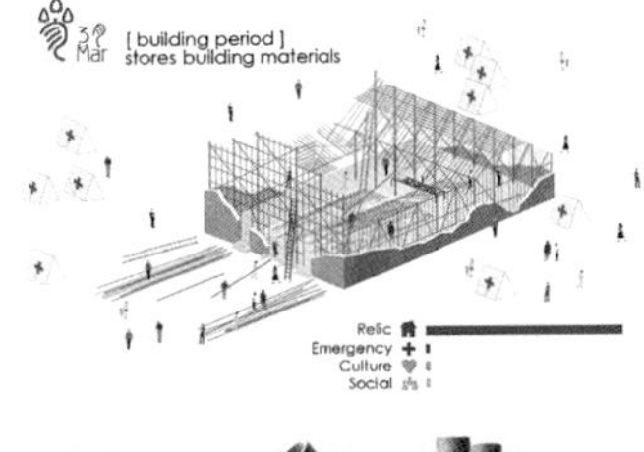

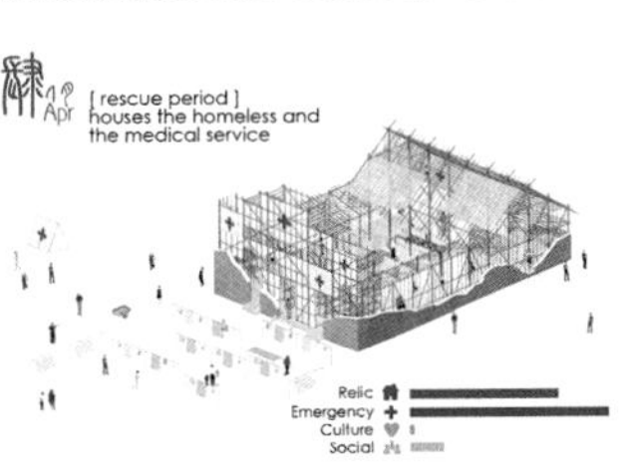

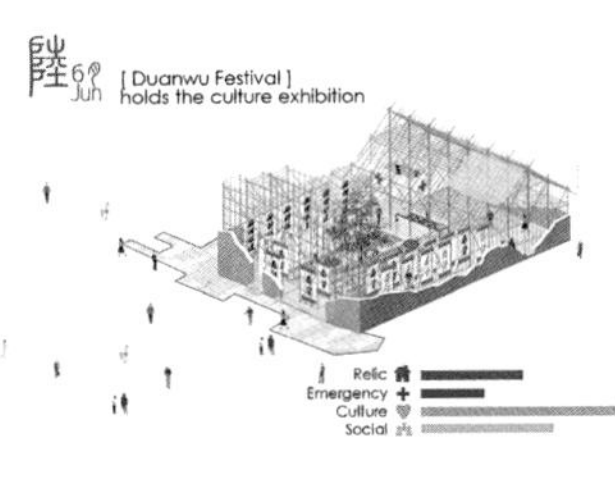

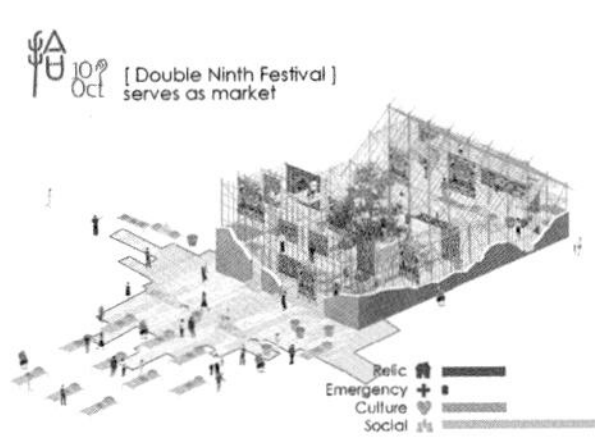

recycled brick & tile

bamboo

rope

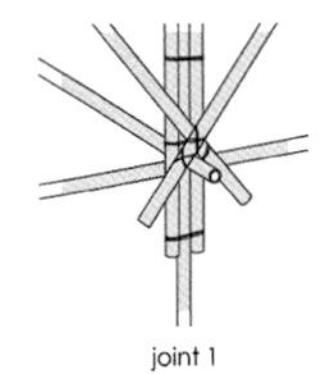

joint 1

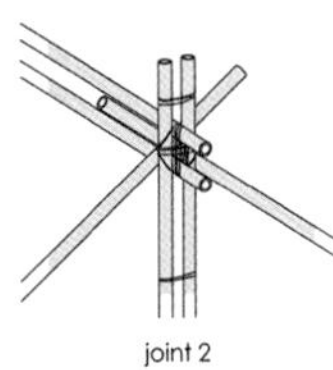

joint 2

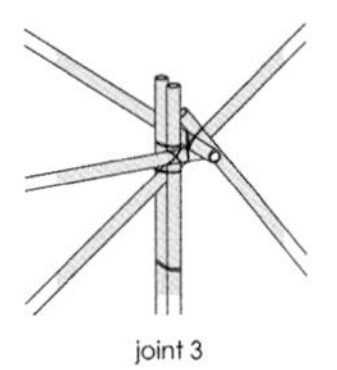

joint 3

University: Tianjin University

Designer: Li Siqi, Chen Xiaoting

Tutor: Zou Ying

Course Name: 49th Central Glass International Architecture Design Competition

Finished Time: Jun. 2014

Movable city Mobile life

01

BACKGROUND

We are facing an age of high technology but still with the problems of increasing number of elders, high density city with few green fields, shortage of resources etc.
1. High tech makes life more convenient. Many works can be solved by AI.
2. The number of people over 65 years gets more and more as years goes by.
3. Intensive city without open space and green field makes people feel boring and sick.
4. Shortage and low efficient utilization of resources makes life harder for new generations.

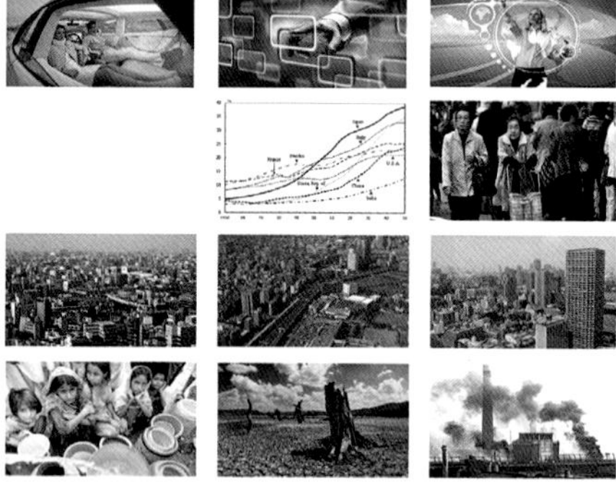

STRATEGY

To deal with the problems, we design an efficient and flexible city. The city shrinks into several big groups, which makes the distance shorter between residence public services. The groups are surrounded by large area of green field and linked by express way. Each group is multi-functional and self-sufficient. The city use wind and solar energy to tackle the problem of resource shortages.

In addition, we try to search for a new kind of lifestyle which gives people more freedom and efficiency. We unite the vehicle and house into a movable house unit. Different from motor homes, it is the basic element that constructs the city.

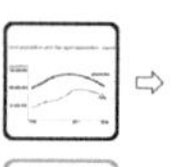
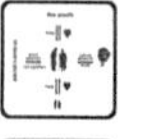

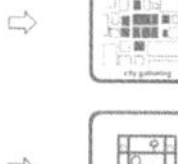
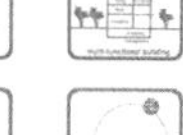
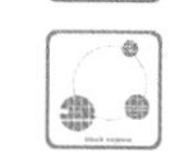
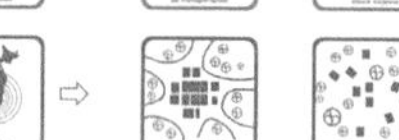

MOBILE LIFE

Actually, the movable city provides a static structure for the movable house units to stay. It is like a transportation hub for vehicles to park. The movable house units are automated driving and multi-functional. They get power from accumulators on their underpans. It gives people more freedom to move to anywhere they want to go with their room. When they travel to another place, they can rest in their sweet bedroom at the same time. When they go to work

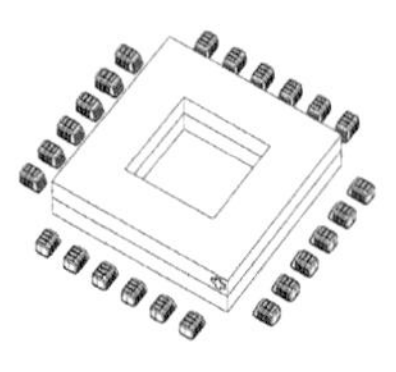
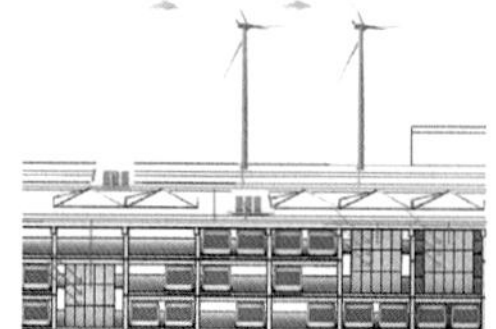
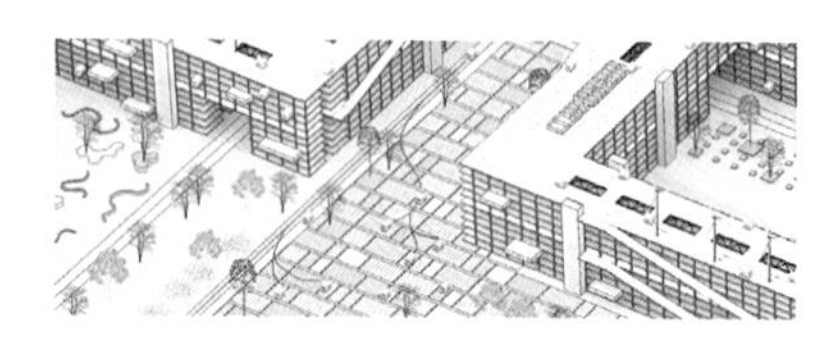

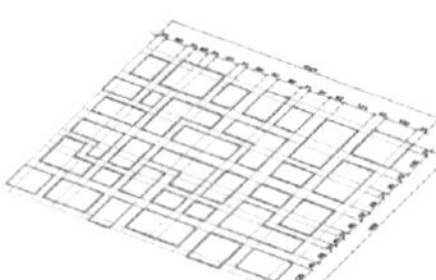

1.Generate the basic grid and arrange the green space and building.

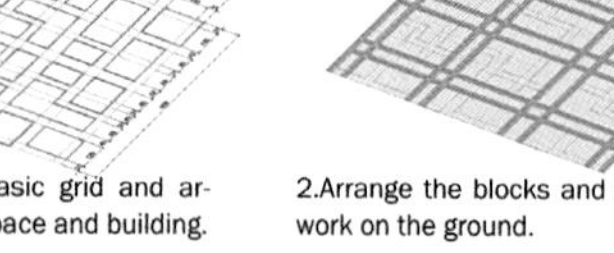

2.Arrange the blocks and road network on the ground.

3.Generate the building blocks at the height of 60m.

4.Change the height of the building blocks to 40m, 60m and 80m according to their location.

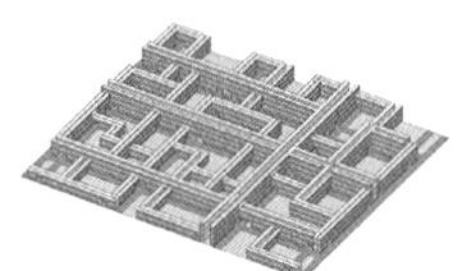

5.Build the main structure of the building.

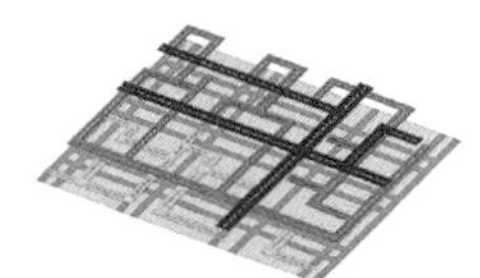

6.Use the roof of each blocks as the main roads.

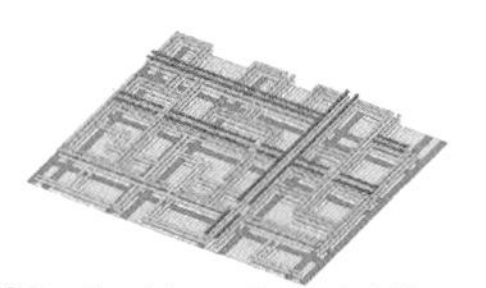

7.Use the slabs on the level of 20m, 40m, 60m as roads in the buildings.

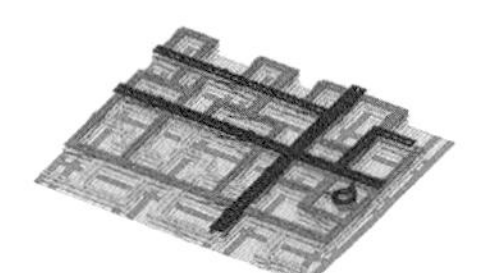

8.Add some straight and spiral ramps to link roads on different levels.

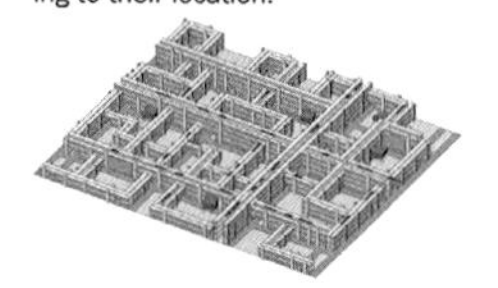

9.Add some public spaces and vertical links for pedestrians.

10.Assemble all the layers to generate the movable city system.

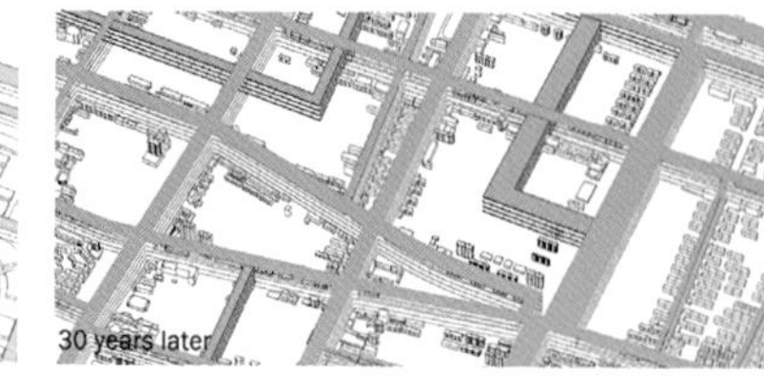

作品名称：移动城市，移动生活
Movable City Mobile Life

院校名：清华大学
设计人：李晓岸，李睿卿
课程名称：UIA 霍普杯 2013 年国际学生竞赛（博士生三年级）
作业完成日期：2014 年 08 月

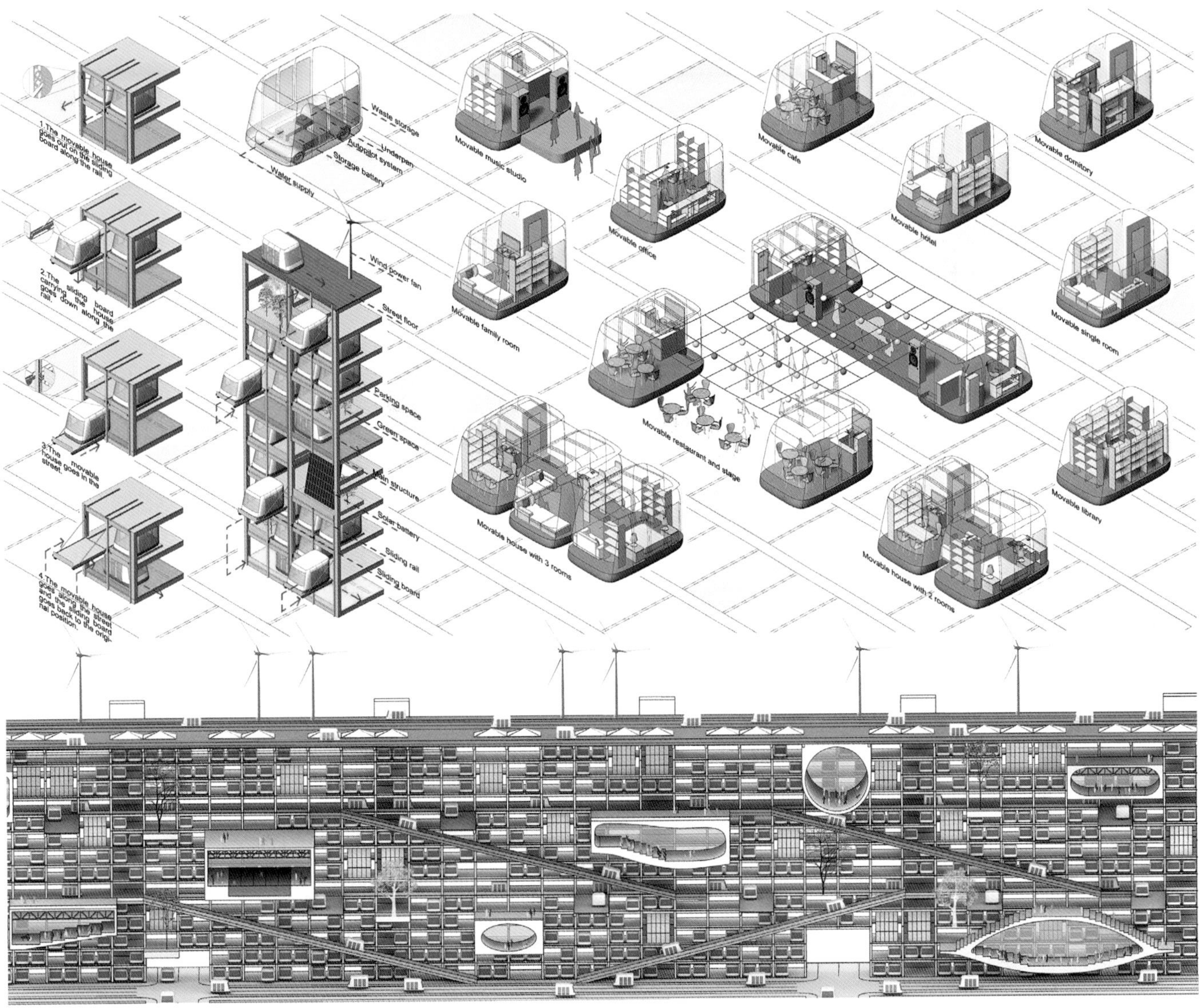

University: Tsinghua University
Designer: Li Xiao'an, Li Ruiqing
Course Name: UIA HUP CUP International Student Competion
Finished Time: Aug. 2014

寺曾相识 西安博物院城市设计及重点建筑设计 1

URBAN DESIGN & ARCHITECTURE DESIGN OF SMALL GOOSE PAGODA AREA

设计说明 design Explanation

现代城市空间与传统寺院空间的碰撞与耦合，不同使用人群间的偶然邂逅与相遇。“出乎意料”的城市空间，便是产生于不同身份属性的场所在特定时间下对话的催化反应。

西安小雁塔始建于唐朝，至今依然延续传统格局，历史遗存丰富。然而周边地段经过数次开发建设已然更加多元与复杂。小学、城市公园、博物馆、市场、城中村等环绕基地。使得这一地段总是发生“出乎意料”的场景。设计选取小雁塔院的“边界”作为设计对象，重塑环形构筑物，来回应复杂的外围环境。

Collision of modern urban space coupled with traditional monastery,encounter of differentgroups .Unexpected urban space, resulting in a catalytic reaction of the dialogue of different identity spaces under a specific time.

We choose Small Wild Goose Pagoda for to be our site.Small Wild Goose Pagoda was built in the Tang Dynasty and remains a continuation of the traditional pattern, rich in historical relics.However,after several development and construction ,the surrounding area is already more diverse and complex. Primary school, city parks, museums, markets, urban villages and so on.Much unexpected scenes always happen around this area.We choose the boundary of pagoda as a design object, reshaping the ring structures , responding to complex external environment.

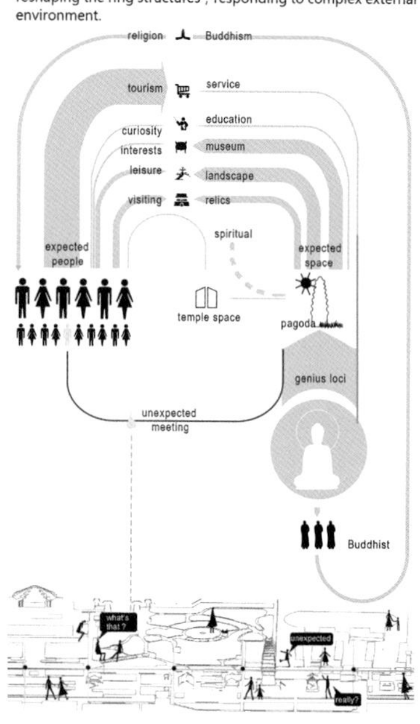

周边要素 analysis of surroundings

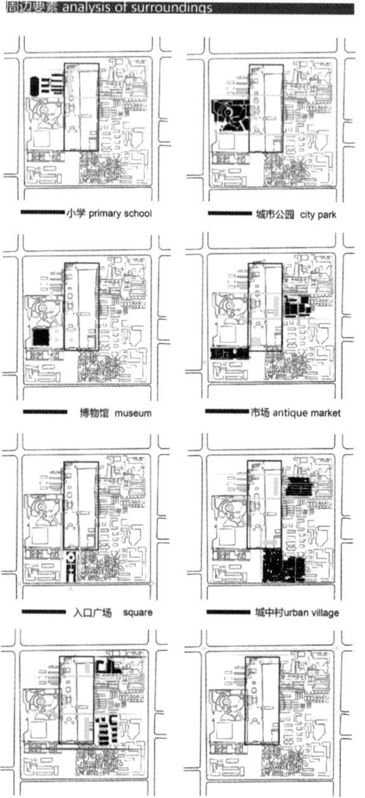

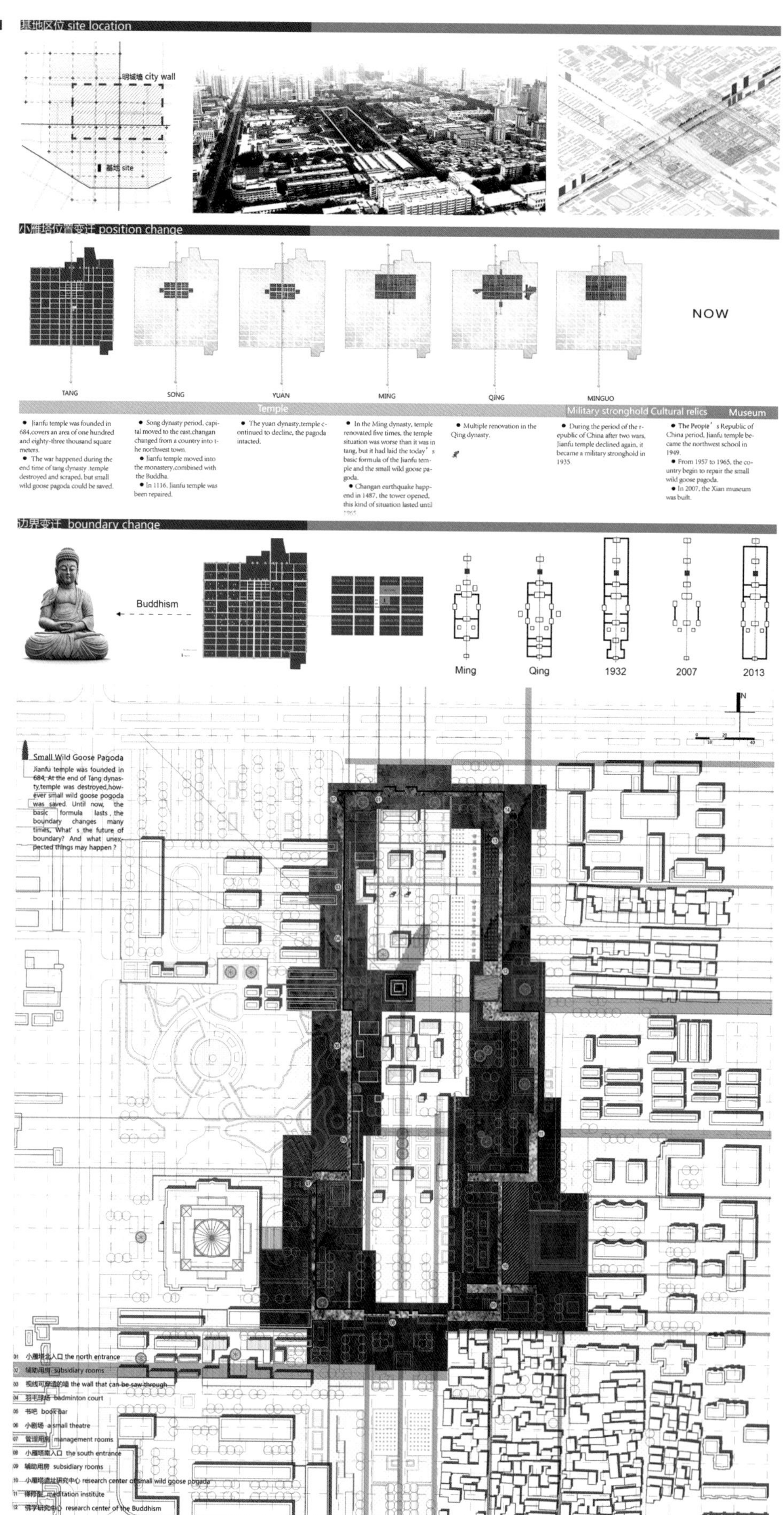

作品名称：寺曾相识

——西安博物院城市设计及重点建筑设计

Urban Design & Architecture Design of Small Goose Pagoda Area

院校名：西安建筑科技大学

设计人：陆星辰，杨骏，岳圆，周曦曦，Michelle Hook，Asher Durham，Matthew Livingston

指导教师：肖莉，常海青，苏静，Albertus Wang，鲁旭，任中琦

课程名称：西安建筑科技大学—佛罗里达大学联合课程设计教学（研究生一年级）

作业完成日期：2014 年 08 月

对外交流对象：美国佛罗里达大学建筑学院

寺曾相识 西安博物院城市设计及重点建筑设计 2

URBAN DESIGN & ARCHITECTURE DESIGN OF SMALL GOOSE PAGODA AREA

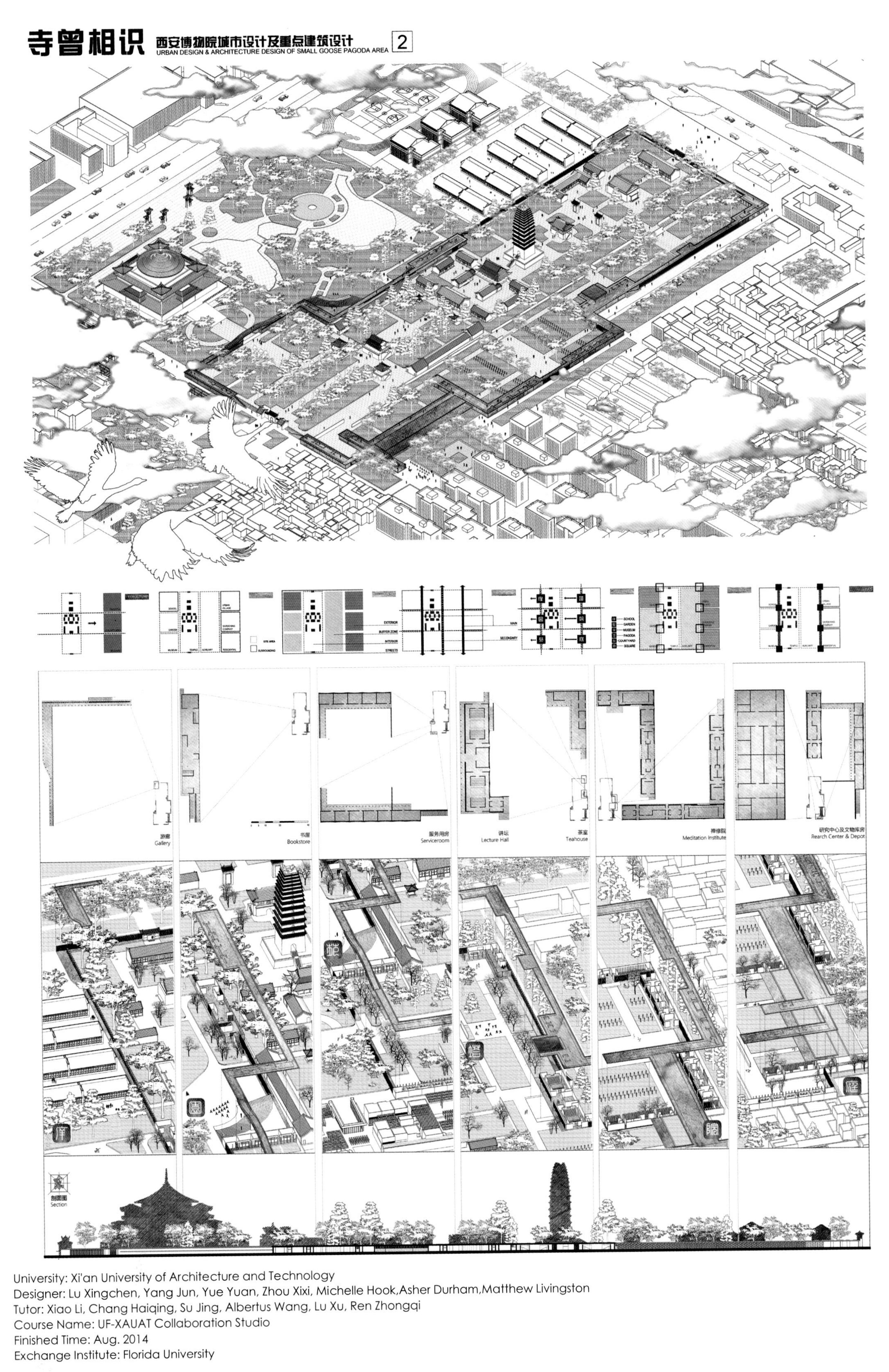

University: Xi'an University of Architecture and Technology
Designer: Lu Xingchen, Yang Jun, Yue Yuan, Zhou Xixi, Michelle Hook,Asher Durham,Matthew Livingston
Tutor: Xiao Li, Chang Haiqing, Su Jing, Albertus Wang, Lu Xu, Ren Zhongqi
Course Name: UF-XAUAT Collaboration Studio
Finished Time: Aug. 2014
Exchange Institute: Florida University

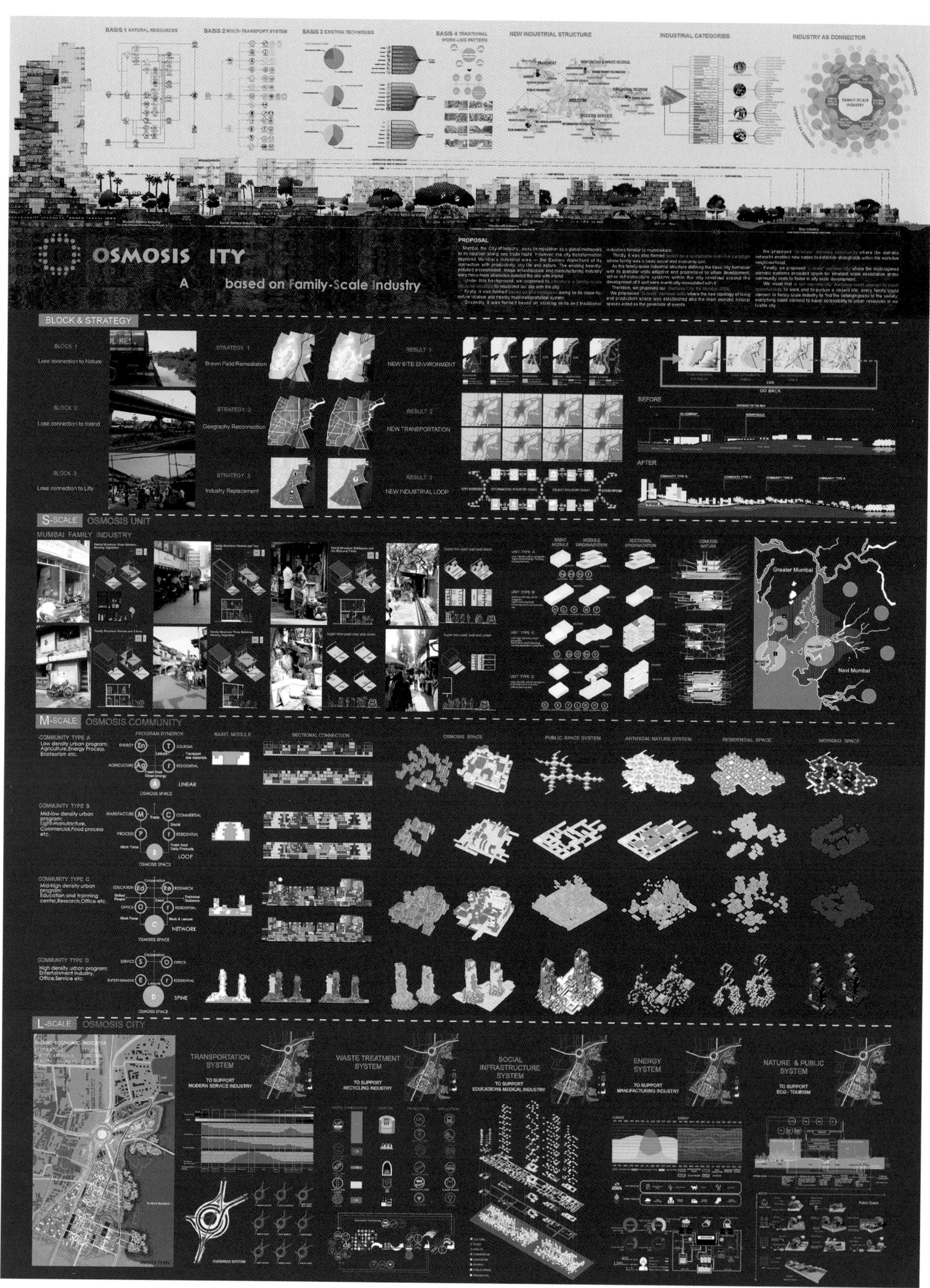

作品名称：渗透城市
Osmosis City

院校名：同济大学
设计人：陆伊昀，陈艺丹，朱恒玉，孙伟，陈伯良，陆垚
指导教师：董屹，王桢栋，黄一如
课程名称：亚洲垂直城市竞赛
作业完成日期：2014 年 07 月
对外交流对象：新加坡国立大学，香港中文大学，日本东京大学，荷兰代尔夫特理工大学，瑞士苏黎世联邦理工学院，美国加利福尼亚大学伯克利分校，美国宾夕法尼亚大学，美国密歇根大学

University: Tongji University
Designer: Lu Yiyun, Chen Yidan, Zhu Hengyu, Sun Wei, Chen Boliang, Lu Yao
Tutor: Dong Yi, Wang Zhendong, Huang Yiru,
Course Name: Vertical Citles Asia Internatinal Competition
Finished Time: Jul. 2014
Exchange Institute: National University of Singapore, The Chinese University of Hong Kong, Tokyo University, Delft University of Technology, Eidgenössische Technische Hochschule Zürich, University of California-Berkeley, University of Pennsyivania, University of Michigan

SEWING

Site Collage

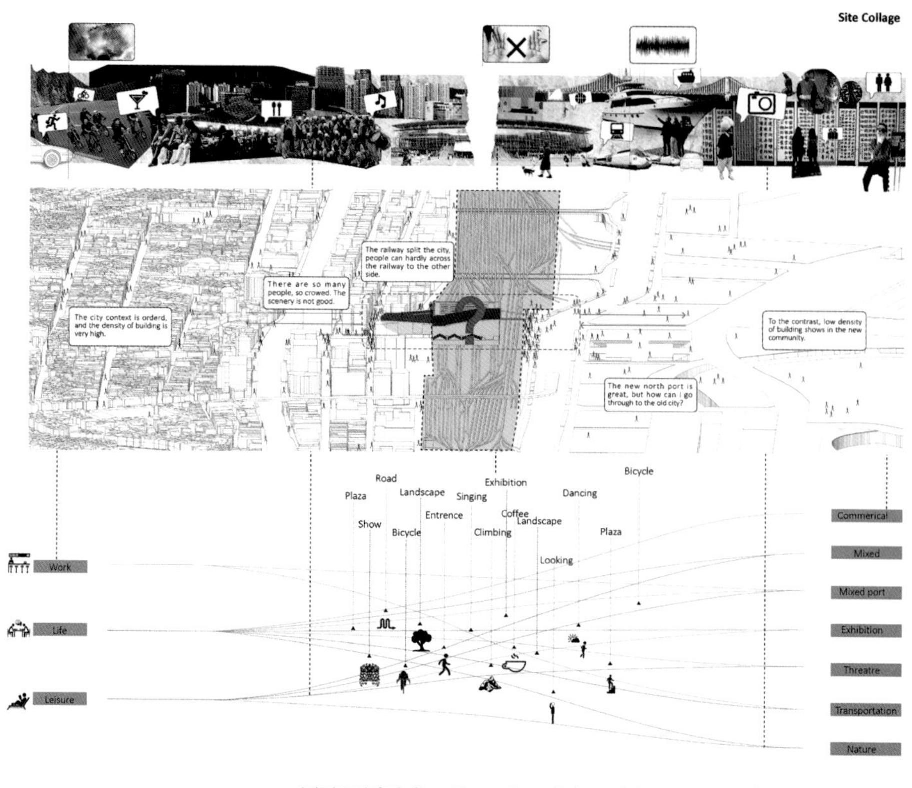

Expectation

1 Consideration of regional value and regeneration of downtown area.

2 Remarking the symbolic in consideration of the history of Busan and this area.

3 The recovery of disconnected urban structures with the surroundings.

4 Environment-friendly idea for co-existence with nature through sustainable technology.

Site Context

Road
Plaza
Landscape
Singing
Exhibition
Dancing
Bicycle
Show
Bicycle
Entrence
Climbing
Coffee
Landscape
Plaza
Looking

Work
Life
Leisure

Commerical
Mixed
Mixed port
Exhibition
Threatre
Transportation
Nature

In this design, the facade of busan station was partly removed,and a mountain shape connectorenwrapping from the square to the digital media center innorth port. Overalpping the busan station and expresswayon the top. By defining a new skyline for hunman instead of controlled by traffic priority,the micro topography enhences the possibility holding activities in public space.

Slope and Human Behavior

Plain
Public spaces like squares and stages attract visitors to gather in broader area.

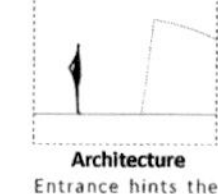

Architecture
Entrance hints the existence of interior spaces beneath the skin.

valley
Continuous rise and fall brings variable spatial experiences for the visitors.

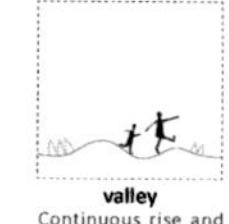

peak
Highest point stands for the climax of spatial consequence.

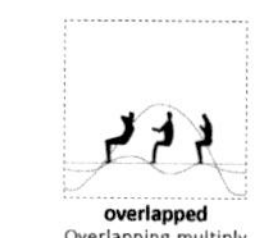

overlapped
Overlapping multiply paths to form interlock relationship.

From Residential Area
From Historic Area
To Train
From CBD
From Subway
Railway Station
Historic City
New Noth Port Area
To Commercial Center
To Office
To Wetland Park
To Port
0 40 80 160

Master Plan

Generation Process

The original train staion

Remove skins

Existing Mountains

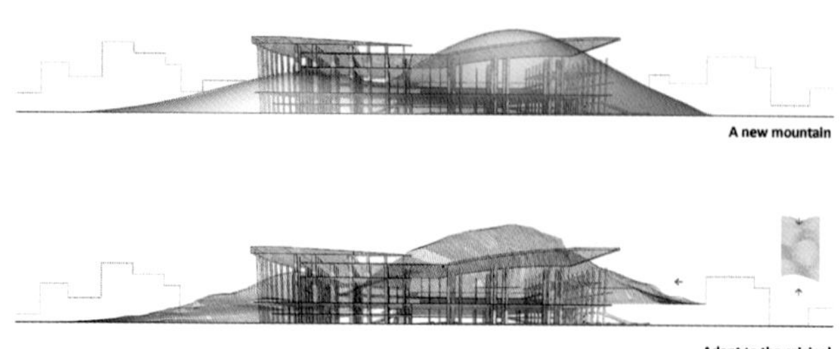

A new mountain

Adapt to the original

作品名称：缝合——釜山车站连接体设计
Sewing—Design of Connecting Body of Busan Station

院校名：北方工业大学
设计人：马赛，夏颖，杨东
指导教师：吴正旺
课程名称：釜山国际建筑设计工作坊（本科四年级）
作业完成日期：2014 年 08 月
对外交流对象：韩国釜庆大学建筑系

SEWING

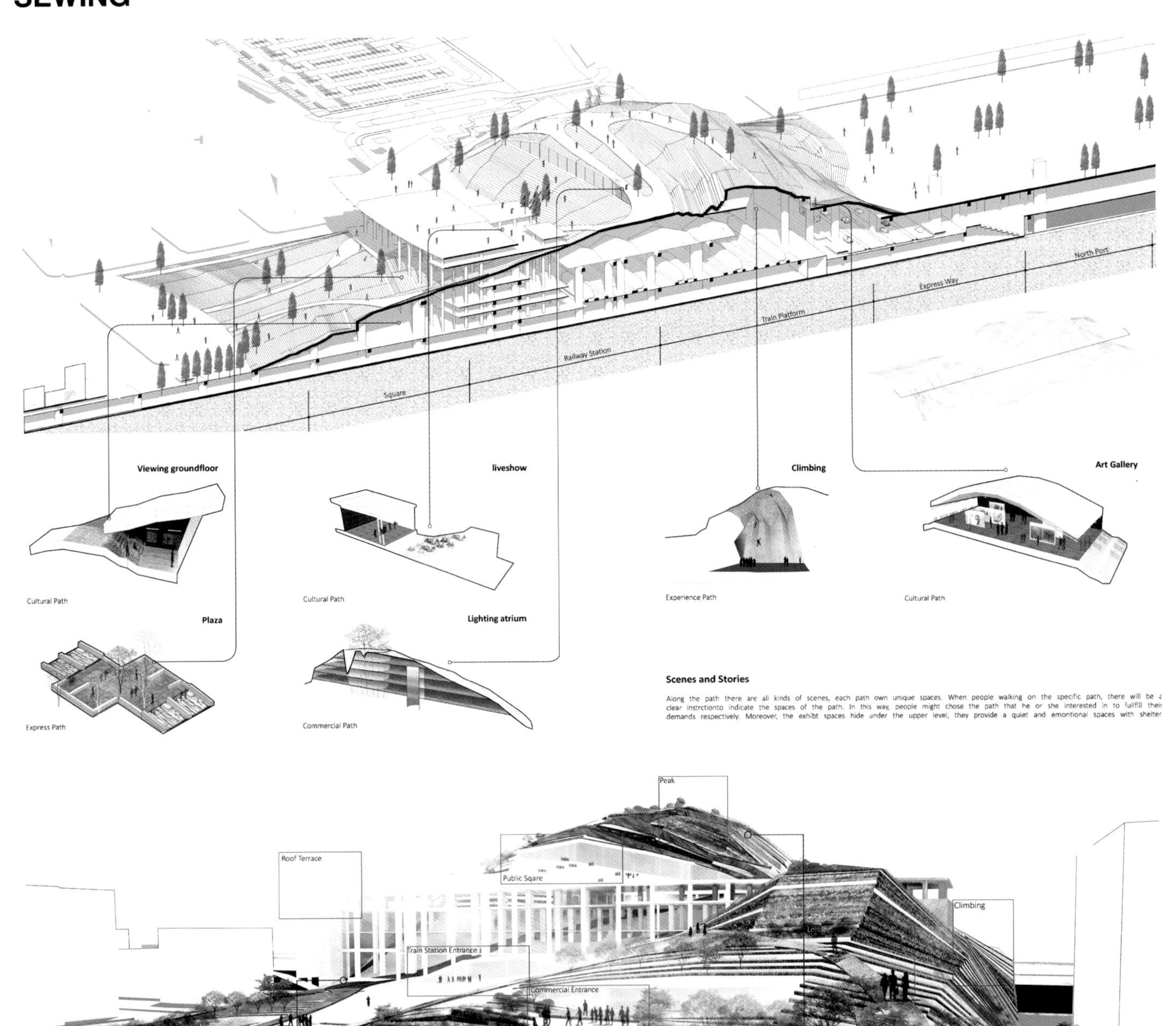

Scenes and Stories

Along the path there are all kinds of scenes, each path own unique spaces. When people walking on the specific path, there will be a clear instrctionto indicate the spaces of the path. In this way, people might chose the path that he or she interested in to fullfill their demands respectively. Moreover, the exhibt spaces hide under the upper level, they provide a quiet and emontional spaces with shelter

University: North China University of Technology
Designer: Ma Sai, Xia Ying, Yang Dong
Tutor: Wu Zhengwang
Course Name: Busan International Architecture Design Workshop
Finished Time: Aug. 2014
Exchange Institute: Pukyong National University

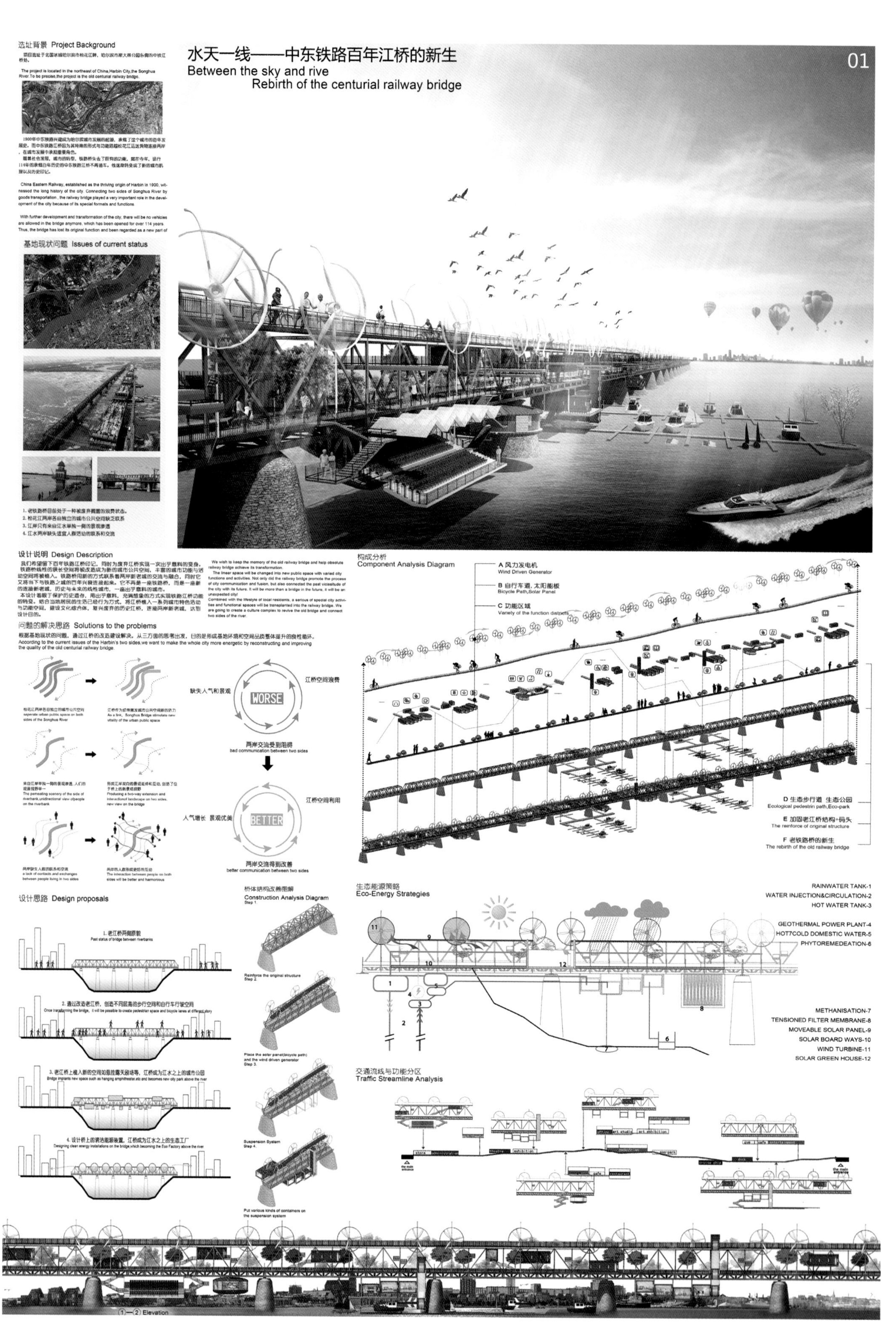

作品名称：水天一线——中东铁路百年江桥的新生

Between the Sky and Rive: Rebirth of the Centurial Railway Bridge

院校名：哈尔滨工业大学

设计人：明磊，何璇，武玥，张之洋

指导教师：孙澄，梁静

课程名称：UIA 霍普杯 2014 国际大学生建筑设计竞赛（本科三年级）

作业完成日期：2014 年 08 月

University: Harbin Institute of Technology
Designer: Ming Lei, He Xuan, Wu Yue, Zhang Zhiyang
Tutor: Sun Cheng, Liang Jing
Course Name: UIA HYP CUP 2014 International Student Competition IN Architecture Design
Finished Time: Aug. 2014

Macau Coloane old shipyard update 1

coastal fishing village on climate adaptation Industrial Revitalization Planrebirth plan

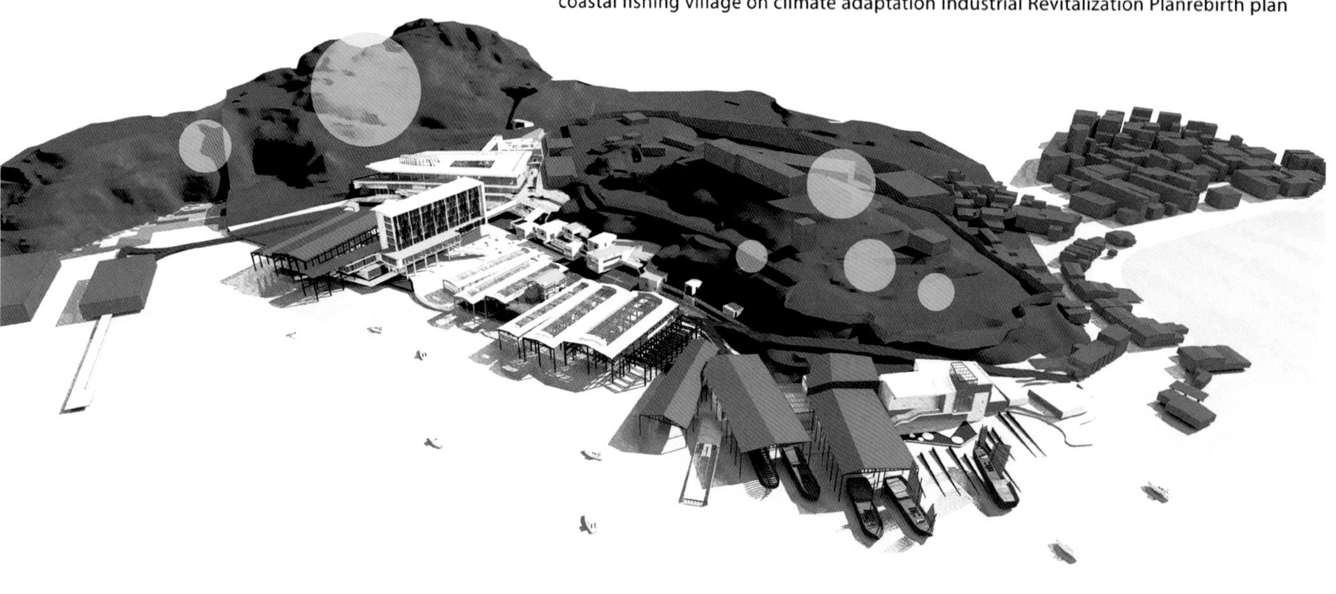

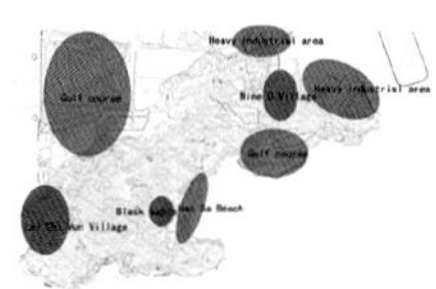

Coloane island was originally an independent island, becoming a whole island with Taipa and Cotai city.

Coloane is a wide area of about 7.6 square kilometers, a population of about 4,300 people. Maintain the natural beauty of the island, such as trees, hills, beaches and natural beach, enchanting and beautiful; which Hac Sa Beach and Bamboo Bay the most famous.

Coloane was fishery-based, so there are some shipyards, cement plants and power plants.

Coloane land, both northwest and east coast reclamation Black-sand beache along the east coast of the golf course and such leisure entertainment area, while others, such as Li Chi Vun village, nine bay village, village of black-sand this kind of original ecological village, includes the heavy industry in the northeast area. Leisure, so to speak, village, pollution and an island.

In addition, Macao prison, nursing homes and psychiatric hospitals, soho team training base, police academy, range, and it fell on the out, this kind of edge feature is the confinement and negative factors. The island's serious lack of cultural content.

Topic Description:

Macau fishing and Shipbuilding industry industries are one of the oldest in Macau, after the War, Macau fisheries decline, Shipbuilding industry also will be cold, Macau Coloane Li Chi Vun fishing village, due to climate change and industrial change, the bustling fishing village has not longer exist, the subject you want to update the old shipyard area through lychee bowl research, conservation and fisheries and Shipbuilding industry technology inherited Macau, and the introduction of a new cultural and creative industries industry updates, while optimizing the area pedestrian environment and the business environment, to enhance the overall community environmental quality and quality of life along the fishing village of Li Chi Vun realize coastal fishing village on climate adaptation industrial Revitalization plan.

Base area of about 21,400 square meters (shipyard area), you can use a subsidiary area of approximately 10,000 square meters (lychee Bowl Road east). There are 17 pieces in the region have been abandoned factory. Shipyard maximum height 11.9m, minimum height 8.5m. The largest single-span distance 15.4m, the smallest single span distance 5.9m.

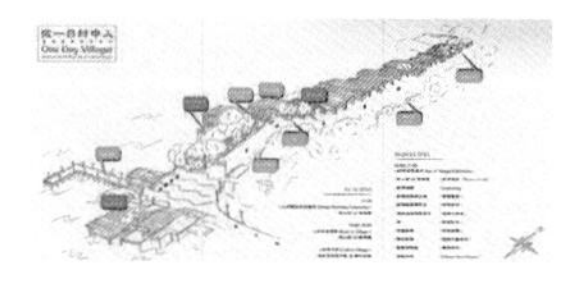

Macao art and design industry were encouraged by the government in recent years, and host all kinds of activities to promote culture and activities. For example: the art market, and "One Day Villager" and Macao arts festival, village, and so on a series of activities, to bring the local art, music, films and fashion and other kinds of development, but due to the time of development is relatively short, still cannot carry out its results, and lack of experience, through which must communicate with adjacent countries or western region and learning.

Red for bus line.
Green for tour bus line.
Yellow for the coastal lane.
These three line make up the base of the roadway system. However, many vehicles are illegally parked because of lack of parking spaces.

a. Coloane fornt square
b. Ma Ji Shi Church
c. Matsu Temple
d. Thean Hou Temple
e. Coloane Port
f. Han's coffee

There are large areas of greenery near the base, but the weed is more, the terrain is steep. South coastal area was planted with neat fake Linden.

Surrounding the base are many closed boundary, in addition to the north of police college, nursing homes, soho team training base, south of prison and private land.
And the south range and cemetery is relatively closed area, lack of openness.

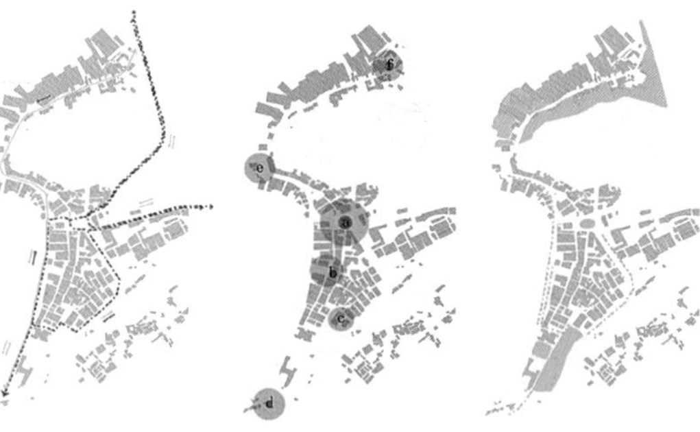

City basic

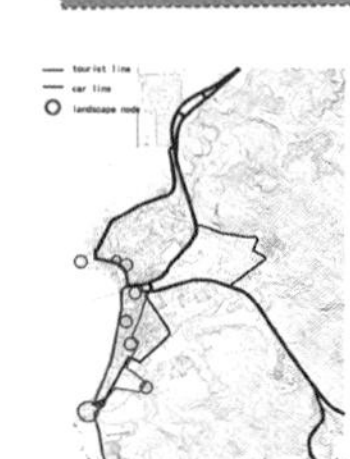

The original planning tourist routes

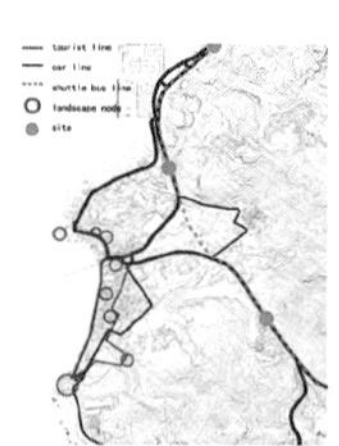

To join the green traffic

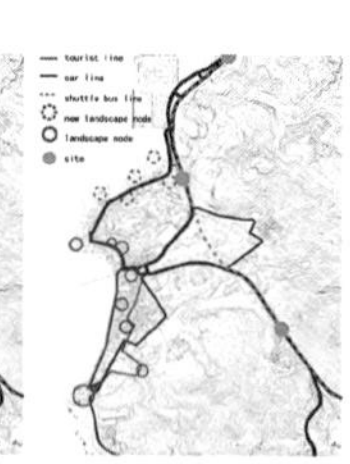

Line into tourist routes

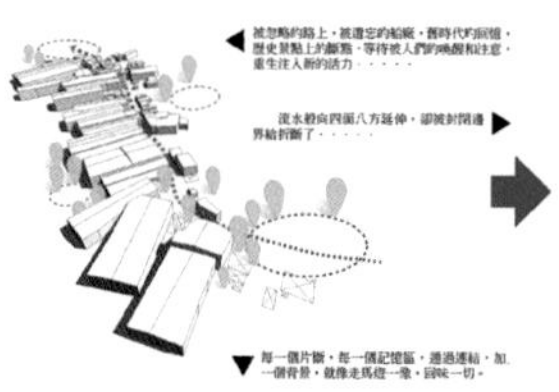

作品名称：澳门路环岛老船厂更新设计
Macau Coloane Old Shipyard Update

院校名：华侨大学
设计人：倪博恒，宋思亮
指导教师：吴少峰
课程名称：华侨大学建筑学院毕业设计
作业完成日期：2014年05月
对外交流对象：中国文化大学建筑及都市设计系

Macau Coloane old shipyard update

coastal fishing village on climate adaptation Industrial Revitalization Planrebirth plan

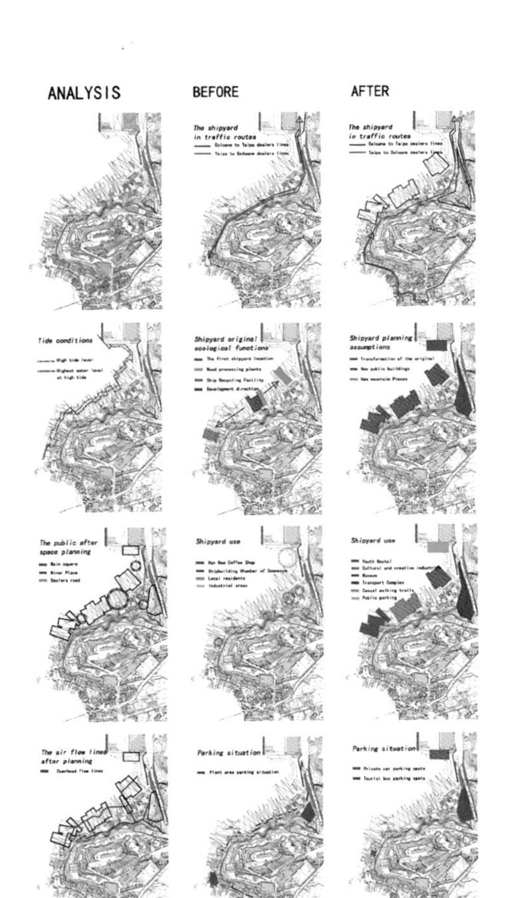

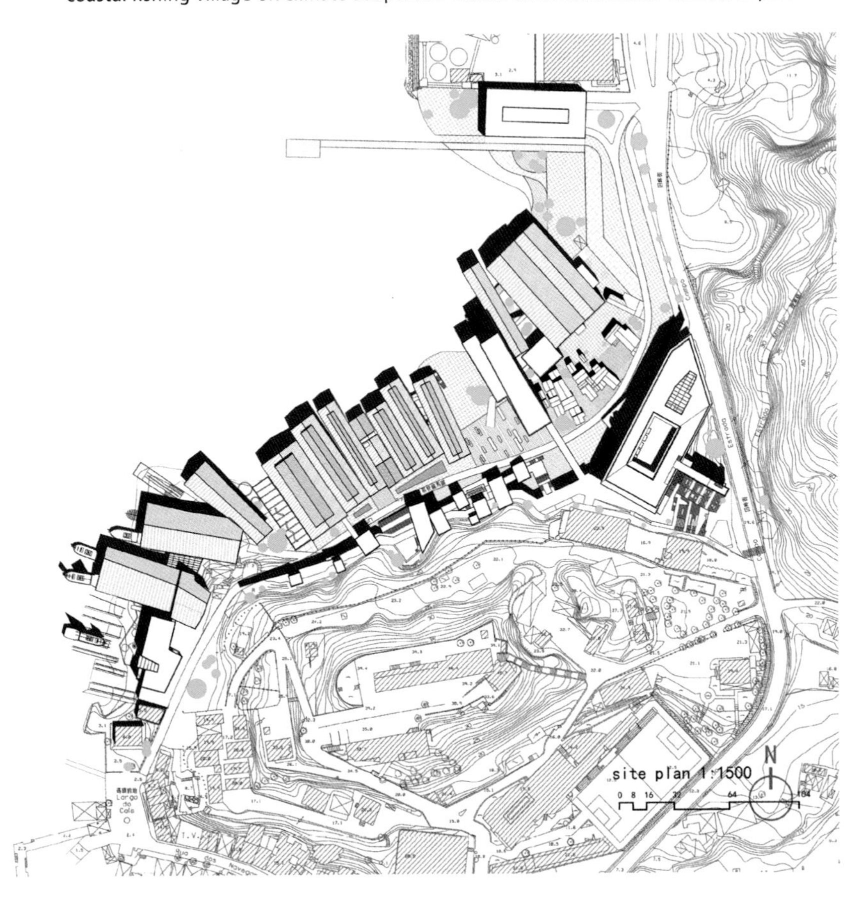

Status

The block now has a station for public transport. But people can not find a shortcut because there are some shelters arround the station. Further more, peopele would not like to get close to the shipyard without an aim. And how to deal with altitude difference of the block will be the a critical question.

keyword

public transport、flowing spece、big pedestrian steps、Visitor Service Center

The way of transform

remove the old illegal building, and confirm the location of the big pedestrian steps on the basis of mountain form. City traffic complex will be build along the mountain line. Our goal is to activate these roads and lead people to the Coloane shipyard.

Status

The block has a wood factory and a abandon shipyard, terrain is low and it has a closed border. Further more, the block has some old ironhide houses, so there are nothing to attract people except a cafe.

Keyword

youth hotel、youth cam、village type、leisure culture、waterside terrace

The way of transform

We keep the structure of the factory, raise the ground to avoid waterflooding, and design a waterside terrace to connect each area which can give it a chance to mix each activities. A youth camp will be build to bring people and offer a plaza which will activate the whole area.

Status

The block has some shipyards. The border is closed and the daylighting is insufficient. The waterside area is affercted by the flux and reflux, so it will be designed as a overhead space. Some area that closed to mountain can be designed as a gallery space to extend the mountain and it has a good view.

Keyword

culture creativity、communication space、workshop

The way of transform

We will find a best way to transfrom the the origin form and connect the interrupted space to create this continuous space

Status

The middle part of the block is higher than the both side, so it has a altitude difference wihch make the shipyard to be a funny space. The block has some shipyard ruins which can be design as a landscape node. The shipyard form is irregular. The block still has some useless factory structure which can be remove.

Keyword

exhibition、shipbuilding craft、landscape node

The way of transform

We plan to keep the form and redesign the structure of the origin shipyard, a exhibiton space will be designed accourding to the origin column. New parts will be built near the shipyard and their correspondingly height will continue the old beautiful skyline.

Urban Transportation Complex

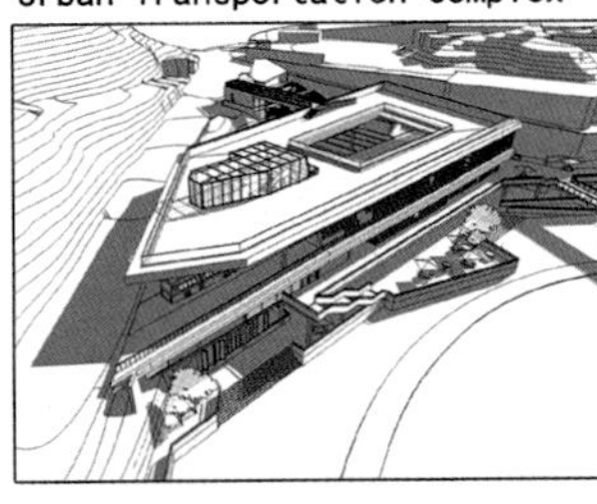

Coloane Youth Hostel

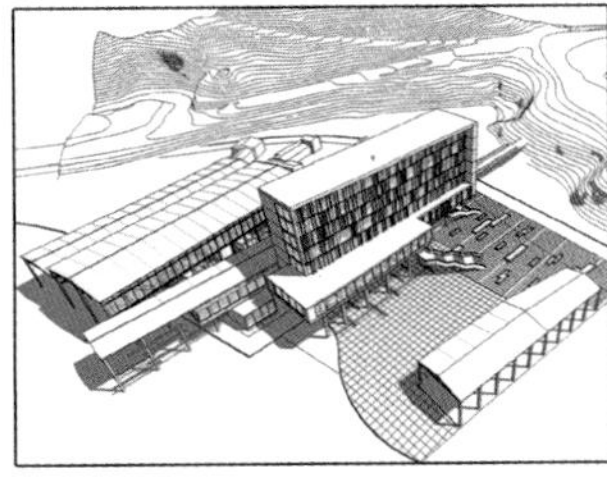

Creative Arts Park

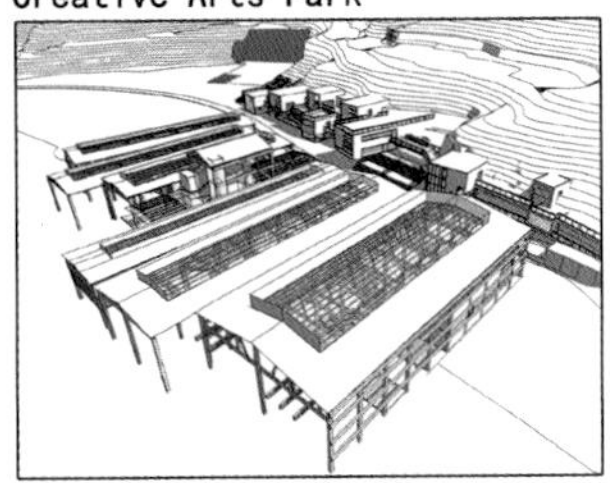

Shipyard and Technology Museum

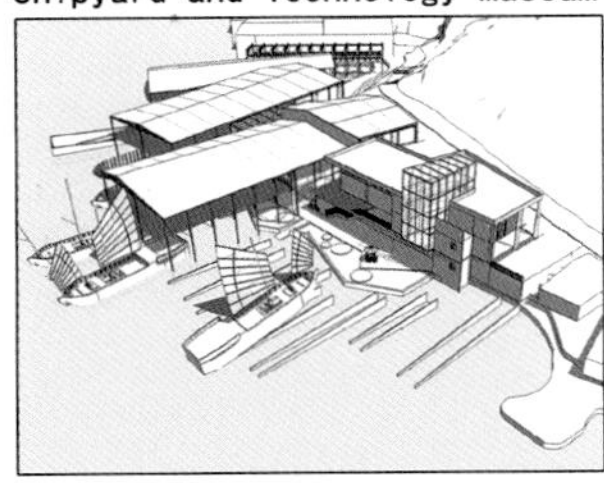

University: Huaqiao University
Designer: Ni Boheng, Song Siliang
Tuto: Wu Shaofeng
Course Name: School of Architecture Huaqiao University Graduation Design
Finished Time: May 2014
Exchange Institute: Department of Architecture and Urban Design Chinese Culture University

THE DISAPPEARED HOUSE Memorial Museum Design

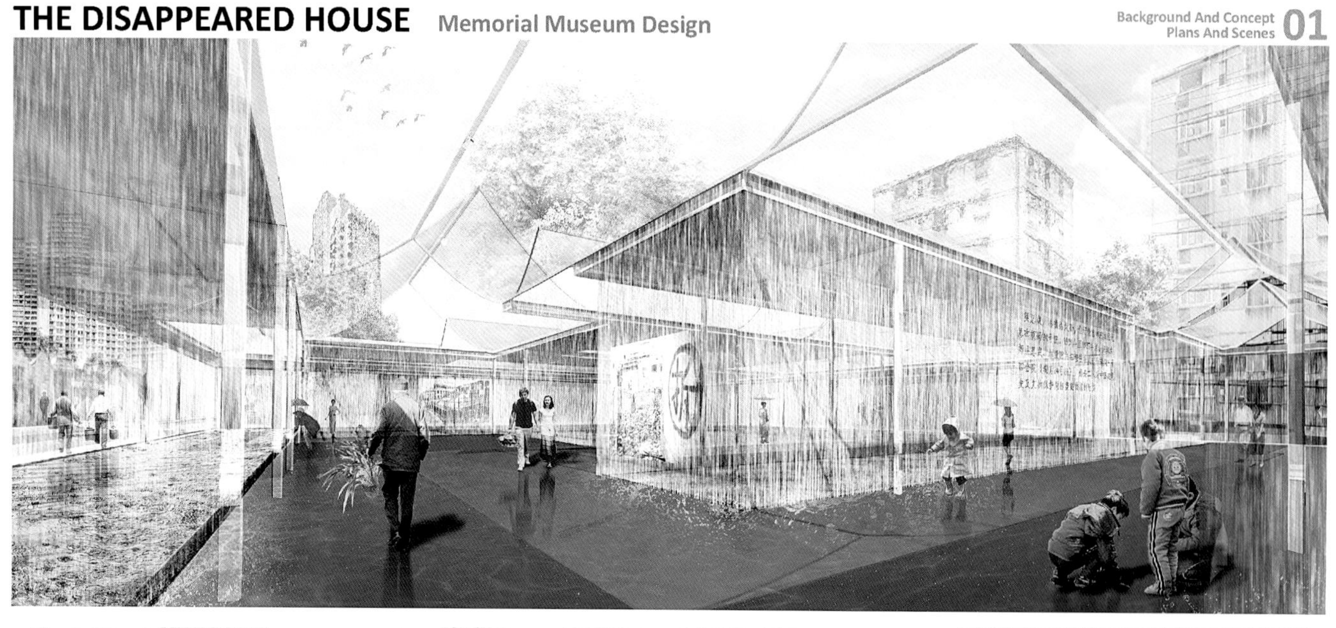

DEMOLITION ERA
DEMOLISHED CELEBRITIES' HOUSE

The expansion of architecture has been a great burden for physical and mental urban environment, that urges us to reduce the effect and release urban spaces by disappearance of architecture in material and understanding. During the development, great demolition, urban renewal and endless occupation of urban space afterwards have become a special kind of disapperance of architecture in this era.This phenomiena **"Demolition Era"** caused by social issue during the renewal of cities in China, especially in **Beijing old city** is worth commemorating. The demolition of a large number of **past celebrities' old houses is a great loss for pubilic space and culture.**

One of the demolished celebrities' houses is the old courtyard house of the chinese architects, liangsicheng and linhuyin, the founders of chinese architecture. The house is located in the southeast of Beijing old city, surrounded by mixed fragments of old courtyards and high-rise buildings.

The **courtyard was demolished** illegally in 2012, and subsequently to be **rebuilt** under government's command, which actually intends to disguise the illegal behaviour with the **fake antique**.

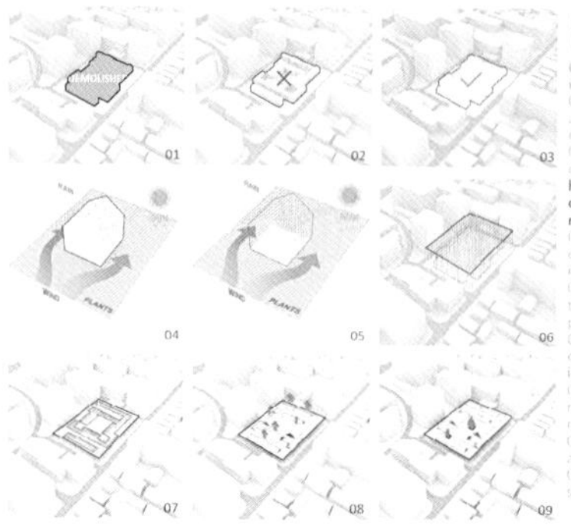

DEMOLISHED HOUSE TO DISAPPEARED HOUSE

01 The old courtyard house is demolished in 2012.

02 If it is to be rebuilt to its original appearance, it will be a fake antique, disguising the reality of demolition.

03 To admit and show the behavior of demolition, we propose **a disappeared house performing as reflection on demolition and memory for the celebrities.**

04 Architecture exists - architecture is distinguished from the surrounding nature.

05 Architecture disappears - architecture becomes a special natural phenomena.

06 To express the disappearance of the courtyard house, we turn the house into a phenomena of rain - rainy space.

07 Turn the original courtyard into rainy space and original house into non rainy space.

08 Shape the flat roof into sloping roof according to the original form.

09 Recover the original under tree space out of rainy space.

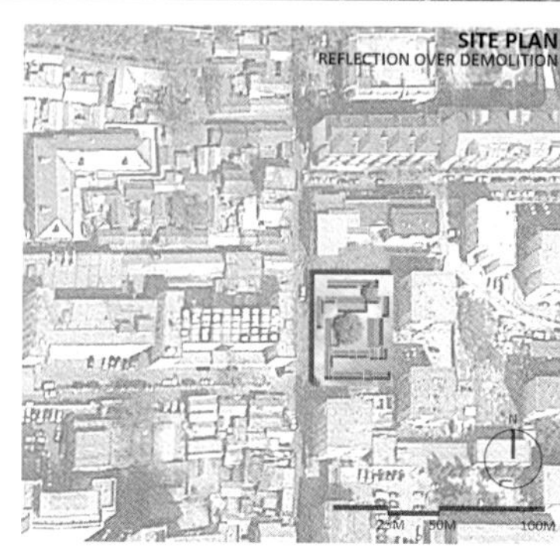

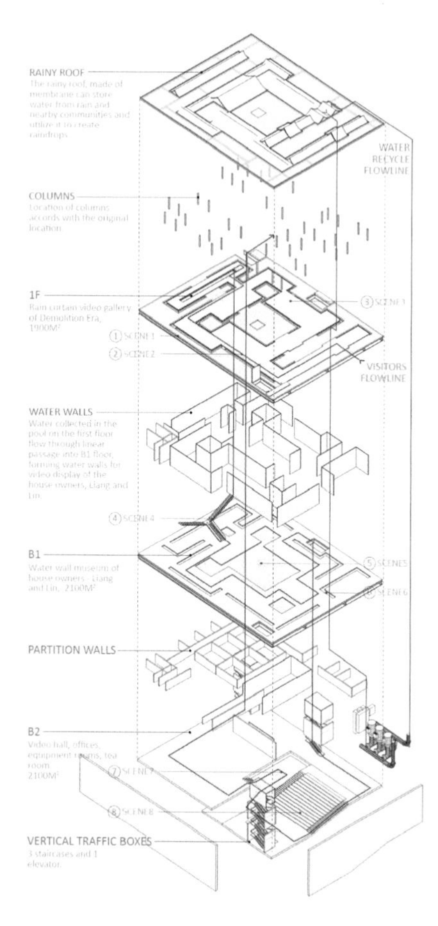

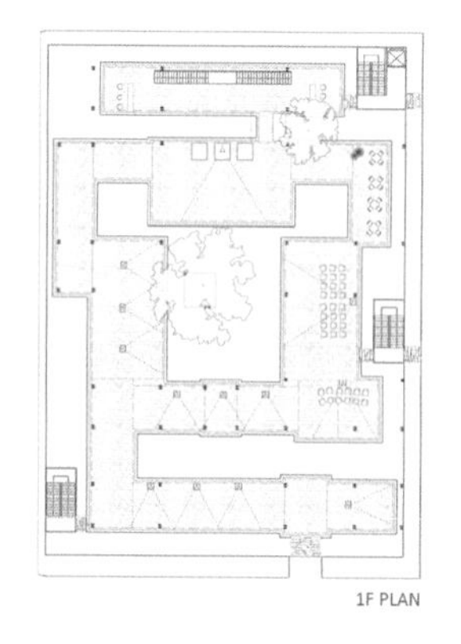

1F PLAN

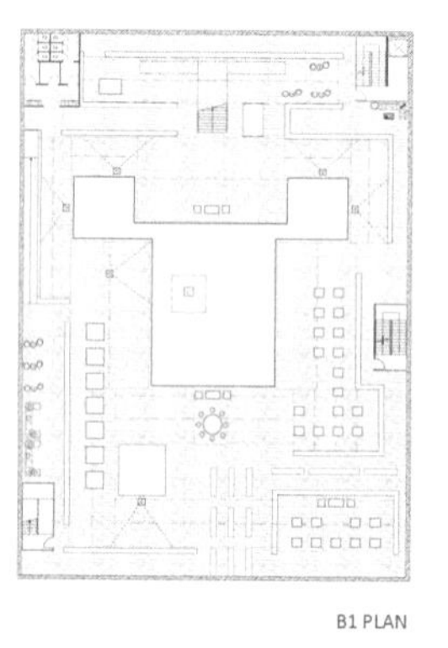

B1 PLAN

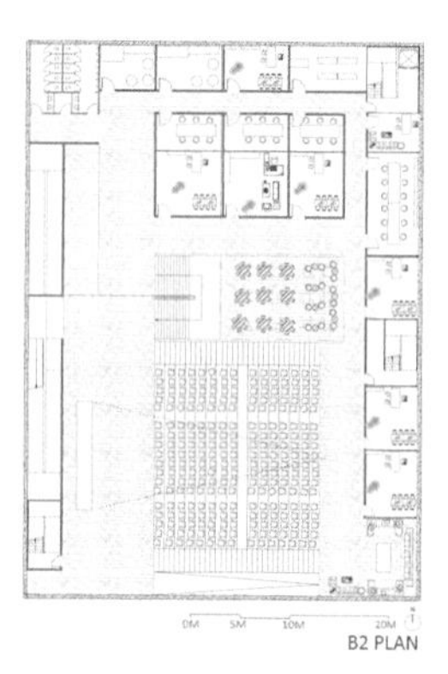

B2 PLAN

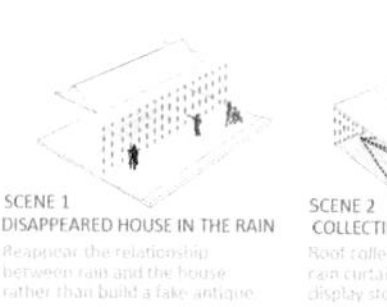

SCENE 1
DISAPPEARED HOUSE IN THE RAIN
Reappear the relationship between rain and the house rather than build a fake antique.

SCENE 2
COLLECTING WATER INTO RAIN
Roof collects water and produce rain curtain as screens to display stories of demolition era.

SCENE 3
COMMUNITY CENTER
In dry season, the building can be convivial plaza for nearby citizens.

SCENE 4
EXHIBITION HALL ENTRANCE
From ground to underground, people come to read stories about Liang and Lin.

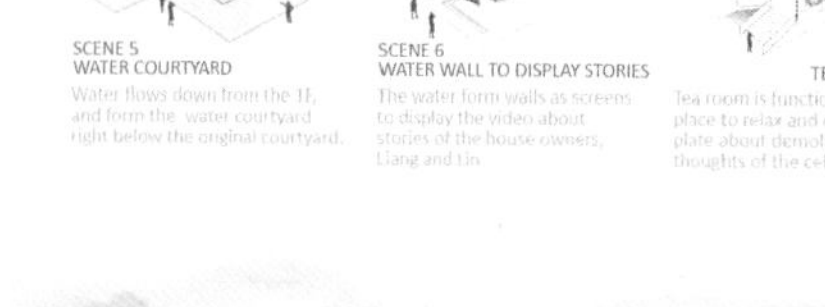

SCENE 5
WATER COURTYARD
Water flows down from the 1F, and form the water courtyard right below the original courtyard.

SCENE 6
WATER WALL TO DISPLAY STORIES
The water form walls as screens to display the video about stories of the house owners, Liang and Lin.

SCENE 7
TEA ROOM
Tea room is functioning as a place to relax and contemplate about demolition and thoughts of the celebrities.

SCENE 8
WATER SCREEN CINEMA
Water flows down from B1, and form water screen to display a documentary of the celebrities.

CIRCULATION AND SCENES

The circulation line for those visitors is designed to start from the ground floor to B1 floor and finally to B2 floor. Each part of the floor is providing visitors with consecutive and storytelling image about the previous celebrities- Liang and Lin. The components of every space are organized together, trying to construct a scene that recount the past story, recall the previous memory of the disappeared house and arouse reflection over the demolition era.

Those different symbolized components includes rain curtain and water roof, rain screen to display visual materials about stories of demolition era, the preserved old tree in the court yard, the previous courtyard in the house enclosed previously by walls and supposedly by rain curtains, the water courtyard in the underground, the water wall to display videos, the old furniture to define previous rooms.

作品名称：被消失的故居
The Disappeared House

院校名：清华大学
设计人：石坚伟，郝田，孙昊德
指导教师：王辉
课程名称：2013 年霍普杯国际建筑设计竞赛（研究生一年级）
作业完成日期：2013 年 08 月

THE DISAPPEARED HOUSE Memorial Museum Design

SECTION OF THE DISAPPEARED HOUSE
The section demonstrates the differently distinctive space and scenes of three floors. The first floor, the water collected in the transparent roof falls down, forming rain curtains and consequently constructing the environment of sorrow and gloominess. The B1 floor displays visual materials on the water walls, forming a more peaceful place for visitors to recall the memory. The B2 floor is a more silent place for visitors to reflection over demolition practice.

DETAILS AND MODEL

The model shows more realistic scenes of how different elements are organized to construct distinctive scenes for the visitors. The first image shows the perspective from the street. Seen from a distant place, what visitors see is the rain curtain and the translucent roof in the air. The second picture shows a more clear image of the rain and the space. The third picture gives an overall impression of three floors. The last image catch a glimpse of the space of the B1 floor.

University: Tsinghua University
Designer: Shi Jianwei, Hao Tian, Sun Haode
Tutor: Wang Hui
Course Name: UIA HUP CUP International Student Competition
Finished Time: Aug. 2013

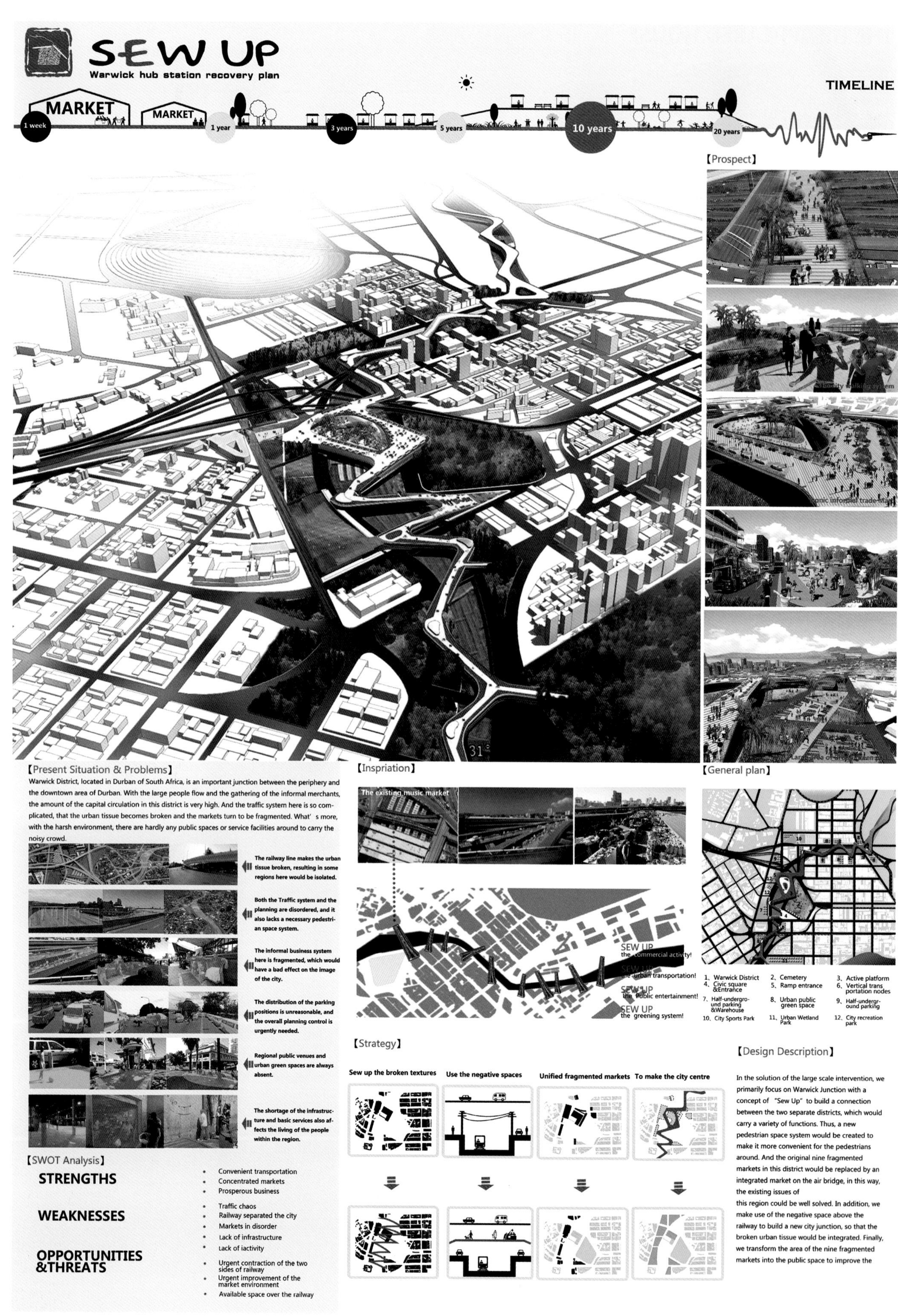

作品名称：缝合

Sew up

院校名：西安建筑科技大学

设计人：宋梓怡，杜怡，李乐，李长春，刘彦京，李乔姗

指导教师：李岳岩，陈静

课程名称：2014UIA 国际大学生建筑设计竞赛（研究生一年级）

作业完成日期：2014 年 03 月

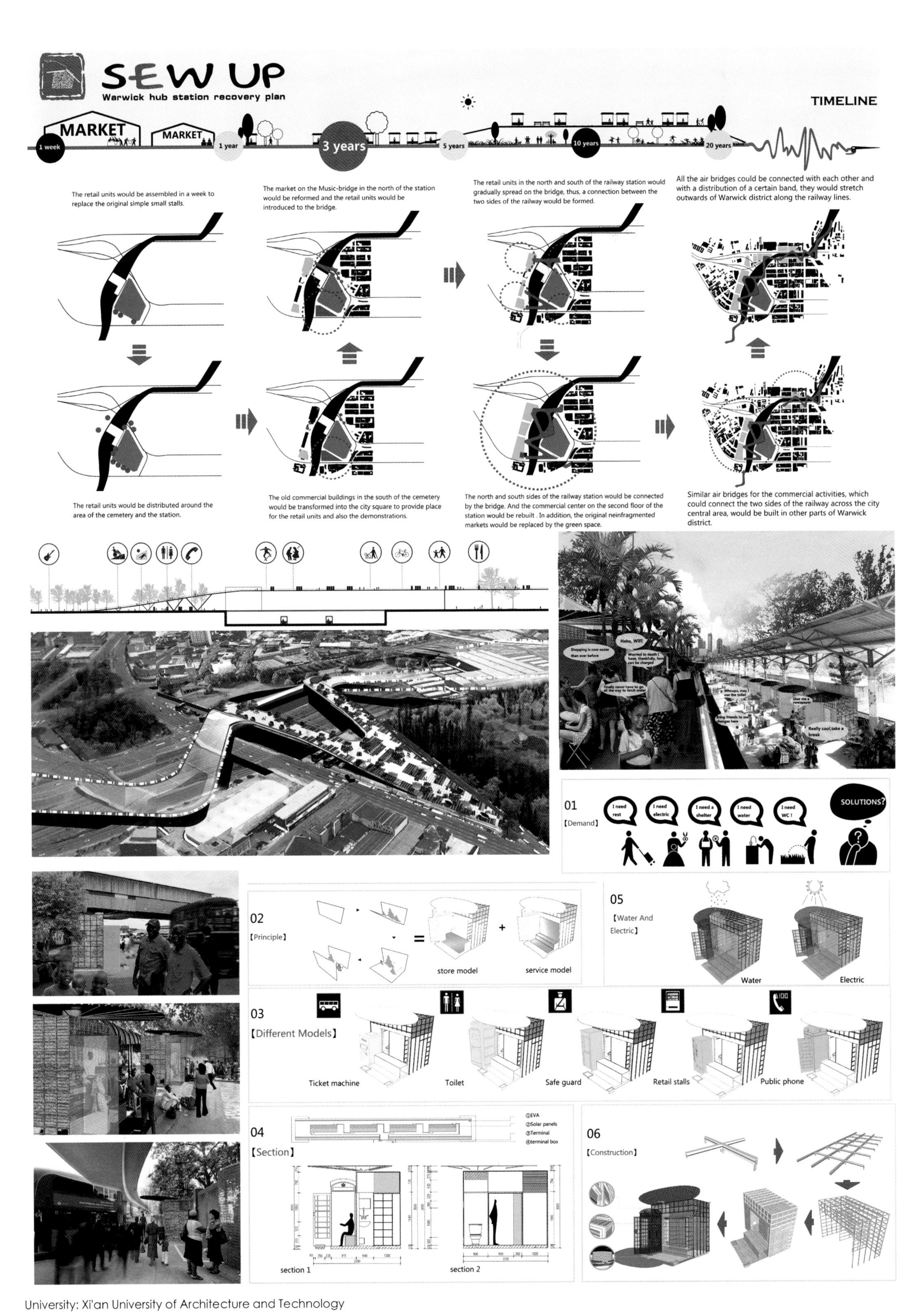

University: Xi'an University of Architecture and Technology

Designer: Song Ziyi, Du Yi, Li Le, Li Changchun, Liu Yanjing, Li Qiaoshan

Tutor: Li Yueyan, Chen Jing

Course Name: UIA 2014 International Student Competition

Finished Time: Mar. 2013

作品名称：浮动网络
Floating Web

院校名：合肥工业大学
设计人：汪宇宸，刘梦茜，林鹏
指导教师：陈丽华
课程名称：Re-thinking the Future Award 2014 国际竞赛（本科五年级）
作业完成日期：2014 年 03 月

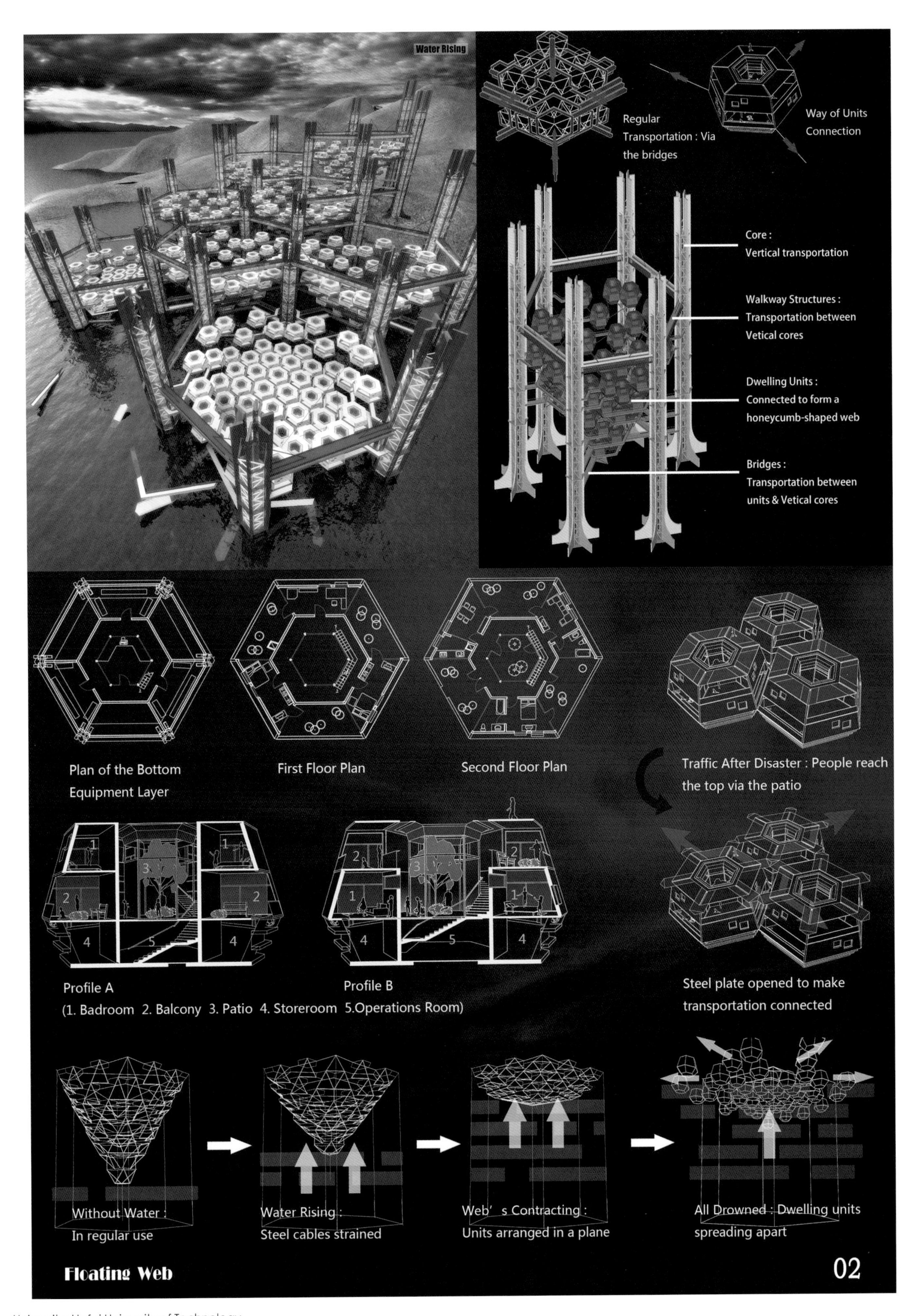

University: Hefei University of Technology
Designer: Wang Yuchen, Liu Mengxi, Lin Peng
Tutor: Chen Lihua
Course Name: Re-thinking the Future Award 2014 International Student Competition
Finished Time: Mar. 2014

A NEW SKY-LINE A retro-active carpet over rooftops

FROM RURAL TO URBAN

Manhattan used to be a spacious place where people could have lots of outdoor spaces. However, with the developement of the city, most of these spaces almost have been replaced by all kinds of skyscrapers.

FUTURE: BACK TO BARE LAND?

Now Manhattan is a place where machinery and artificial experience has replaced the natural things. The project attempts to bring back the natural essence to metropolitan life, and recover what has been lost a long time ago: A spacious area full of unexpectancy and potencial.

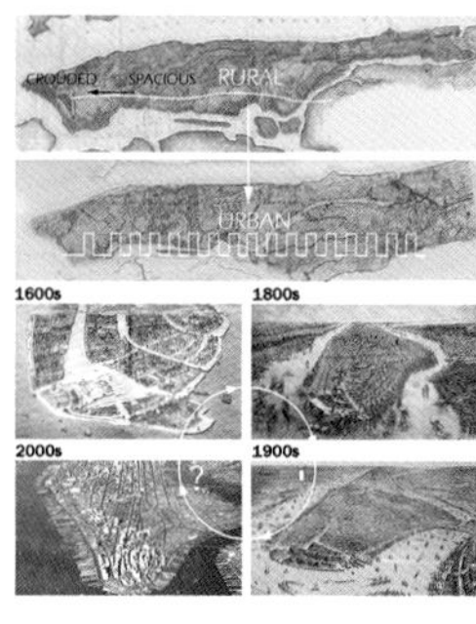

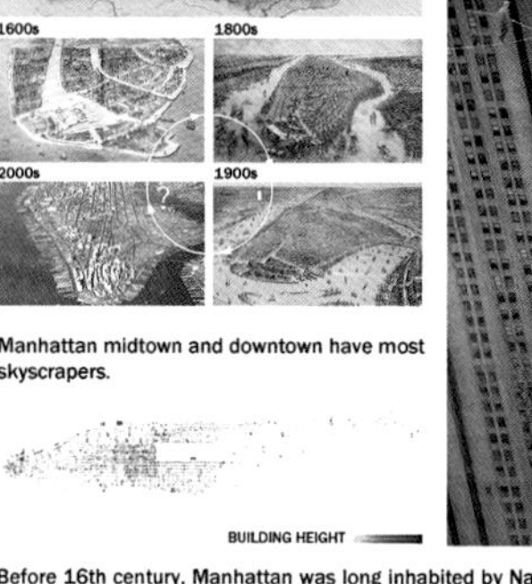

Manhattan midtown and downtown have most skyscrapers.

BUILDING HEIGHT

Before 16th century, Manhattan was long inhabited by Native Americans. The place was flat and totally natural, without human interference but small tribe tents. After 1609, a strict pattern was planned for the city.

From 1823 - 1860 New York turned into a Metropolis, following the previous grid pattern. Architecture went higher seeking for more opportunities.

The project consists in creating a carpet over rooftops of the buildings in the financial district, since there lack space for further expansion. It links buildings trough passways, and above it a flat vast horizon.

BEFORE AFTER TEMPERATURE

PROPOSAL - A COOLING CARPET

The project is also a transparent penetrative system which store and recycle rain water. Not only can it be used to irrigate roof gardens and other urban landscapes, but also the architecture spaces are generated by rain curtains. The project as a whole acts like a cooling system to reduce the UHI effect.

IDEAS FOR THE PROJECT

Roof vegetation, pedestrians, water, sunlight...make up an exuberant interlayer. The top is reflective as the Mono lake. Ventilation for the mayor polluted areas are using huge fans.

Giant bladeless fans

A RAINWATER ESCAPE ROUTE

The frame structure of the "new skyline" have canels to drain water. Openings serve as ventilation and visual connection. People would go from one highrise to another nearby through the "new skyline", without the need of carring an umbrella even in a rainstorm day.

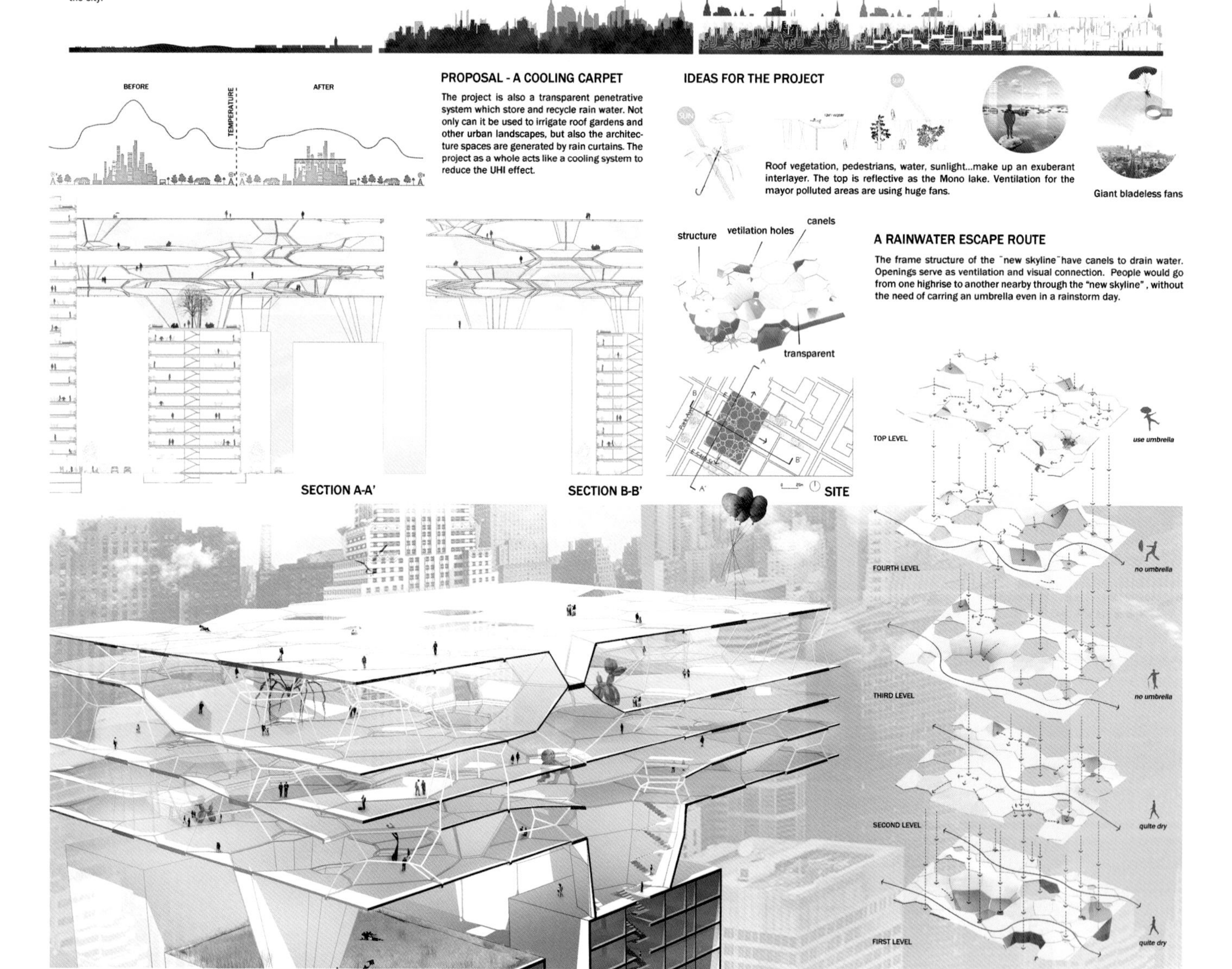

作品名称：新的天际线
A New Sky-line

院校名：天津大学
设计人：王立杨
指导教师：张昕楠，王迪
课程名称：釜山中、日、韩国际学生竞赛（本科三年级）
作业完成时间：2014 年 06 月

A NEW SKY-LINE A retro-active carpet over rooftops

NEW PUBLIC SPACES

The current ground level faces congestion problems. By upgrading the "ground floor to above average building height, we obtain more free space for public activities, that meanwhile unifies the solitary self-sufficient blocks.

BUILDING HEIGHT

DESIGNING PROCESS

Starting from a flat surface, landing like a UFO.

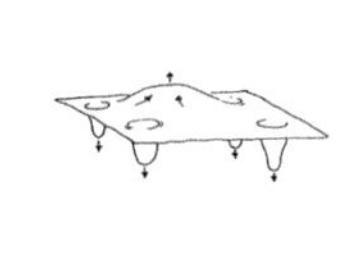

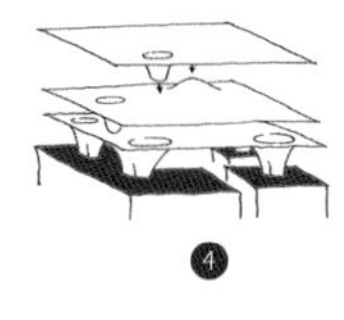

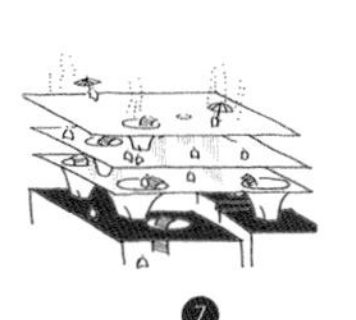

A WATER DRAINING SYSTEM

The new skyline can recollect rainwater, and making use of them to moderate the microclimate of the zone, in addition to irrigating rooftop vegetation. The entry to the new architecture from rooftop is controlled to assure privacy.

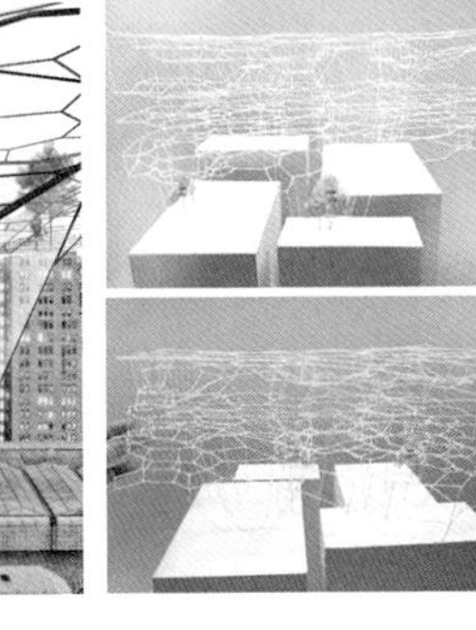

MODEL TESTING

Using 3d pringting pen to lay out the floor structures, and testing how can light architecture deal with different rooftop height of the building´s below it.

AT AN URBAN SCALE

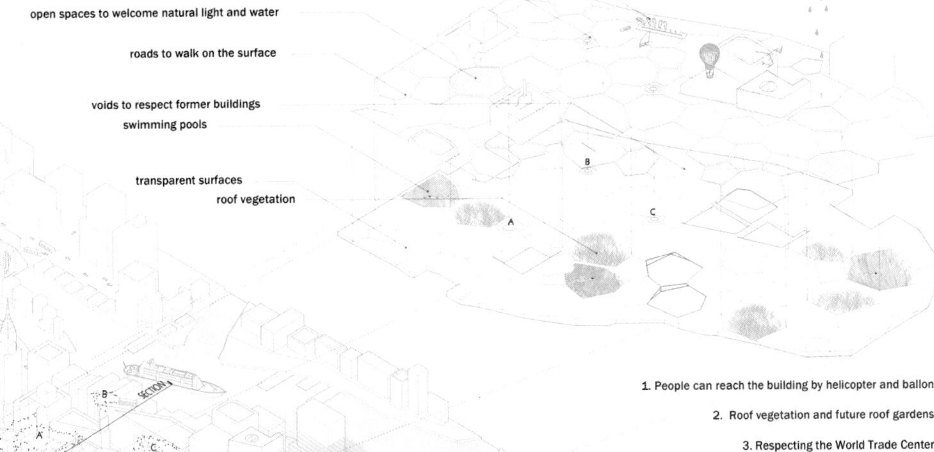

MANHATTAN'S SIMPLIFIED AXONOMETRY

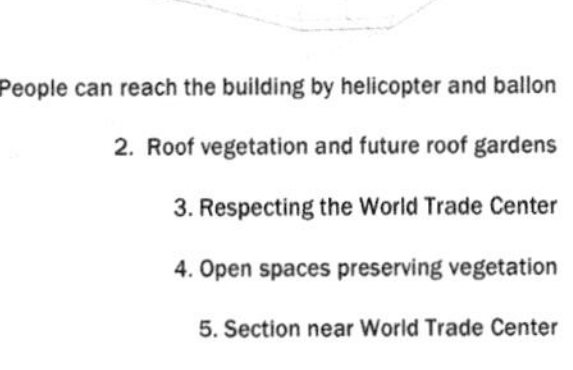

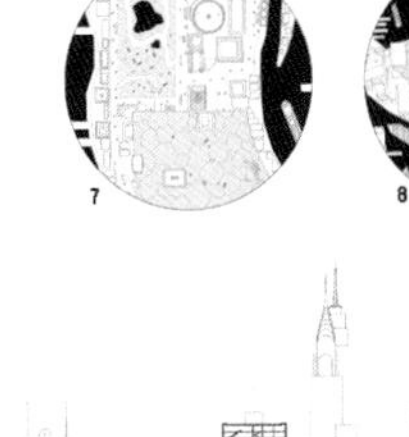

1. People can reach the building by helicopter and ballon
2. Roof vegetation and future roof gardens
3. Respecting the World Trade Center
4. Open spaces preserving vegetation
5. Section near World Trade Center
6. New forms of transportation
7. Top view near Central Park
8. Bird eye view

section A-A'

University: Tianjin University
Designer: Wang Liyang
Tutor: Zhang Xinnan, Wang Di
Course Name: Busan International Architecture Student Competition
Finished Time: Jun. 2014

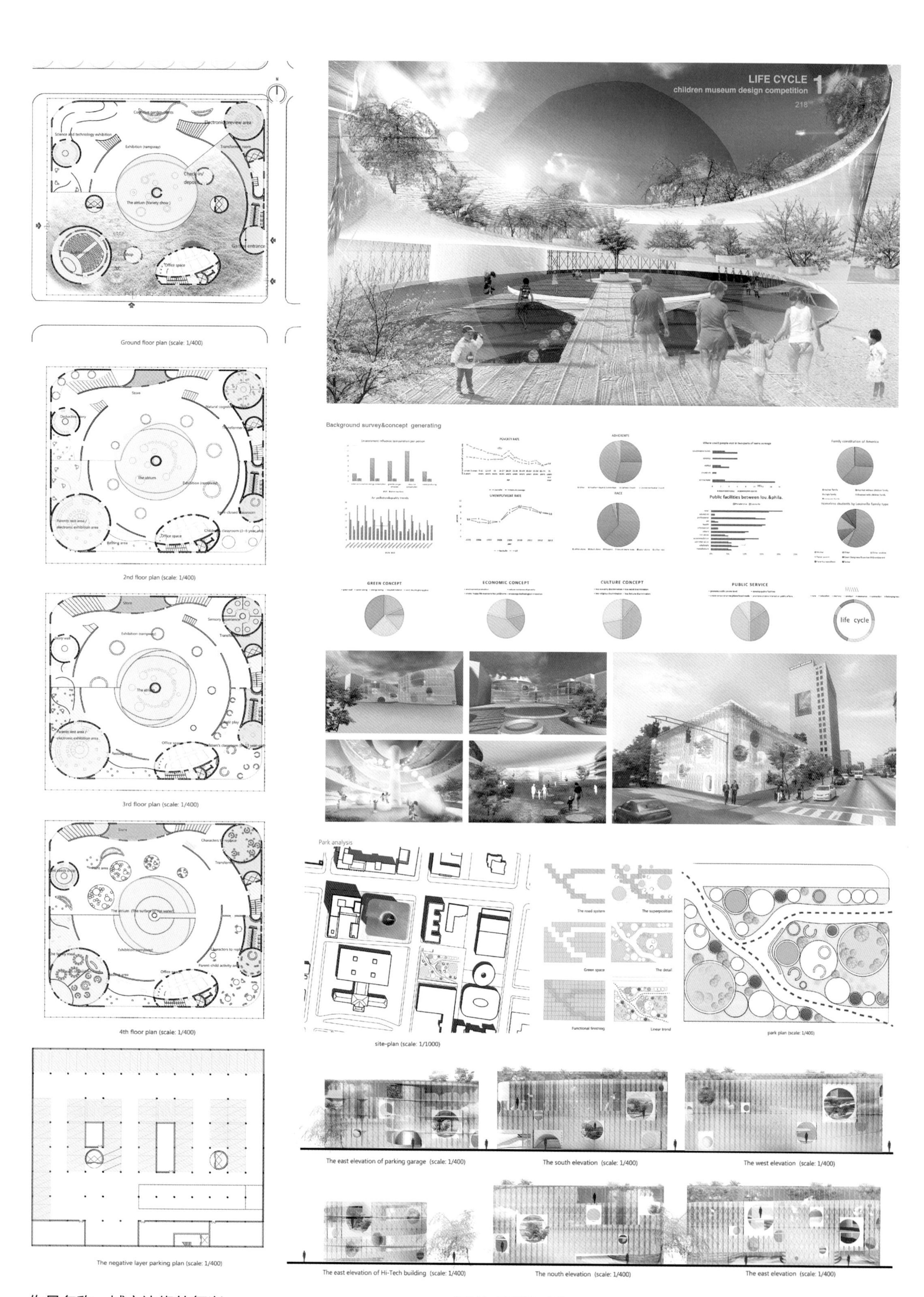

作品名称：城市边缘的复兴
The Regeneration of Edge Louisville

院校名：青岛理工大学
设计人：魏易盟，贺俊，苗天宁
指导教师：郝赤彪，解旭东，程然
课程名称：2014AIA 路易维尔儿童博物馆建筑设计竞赛
作业完成时间：2014 年 02 月

2 LIFE CYCLE
children museum design competition

218

The ultimate concept of CM program is to realize a Life Cycle mechanism on the city level,follow the main line of emotion&Humane Care.We try to use the education mode to combine the children' s growth,adults' emotion and city developing.

About our children' s museum,we are more expecting to create a space atmosphere with freedom,joy,peace,succinct and mystery eastern culture.There would be no discrimination between white&black,rich&poor,everyone will get the same happiness,all the emotion will be pure as well.Children gets happiness,adults would be affected by the CM atmosphere,abandon the useless thinking,recall their initial dreams,come back to daily life with joy&passion.To the space forming,we select more visual transparent wall so that every motion will be noticed by parents.With the people flowing line,the emotion of adults turns to their childhood.The atrium of 3rd floor is the peak of spacial order.We just put pure plants&water,sunshine&wind flow through it.Parents would no longer be adults, all people transform to children,old or yough.This is another mode of cycle and another mode of education.

About the city revitalization,compare to use the function of education by CM to create high quality new force,the pre-school growing of children and their thanksgiving&memory should be more focused. We try to create a virtuous cycle "Children education--Talented people training--Be grateful to city--Build our city--City developing--higher quality children' s education" We consider the thanksgiving as the most important thing.The revitalization of city brings the concept of time,we hope to use the "purest" way--to find out a reaching point between adults&children,combine their emotions together.City,adults and children are the cycle of whole life .

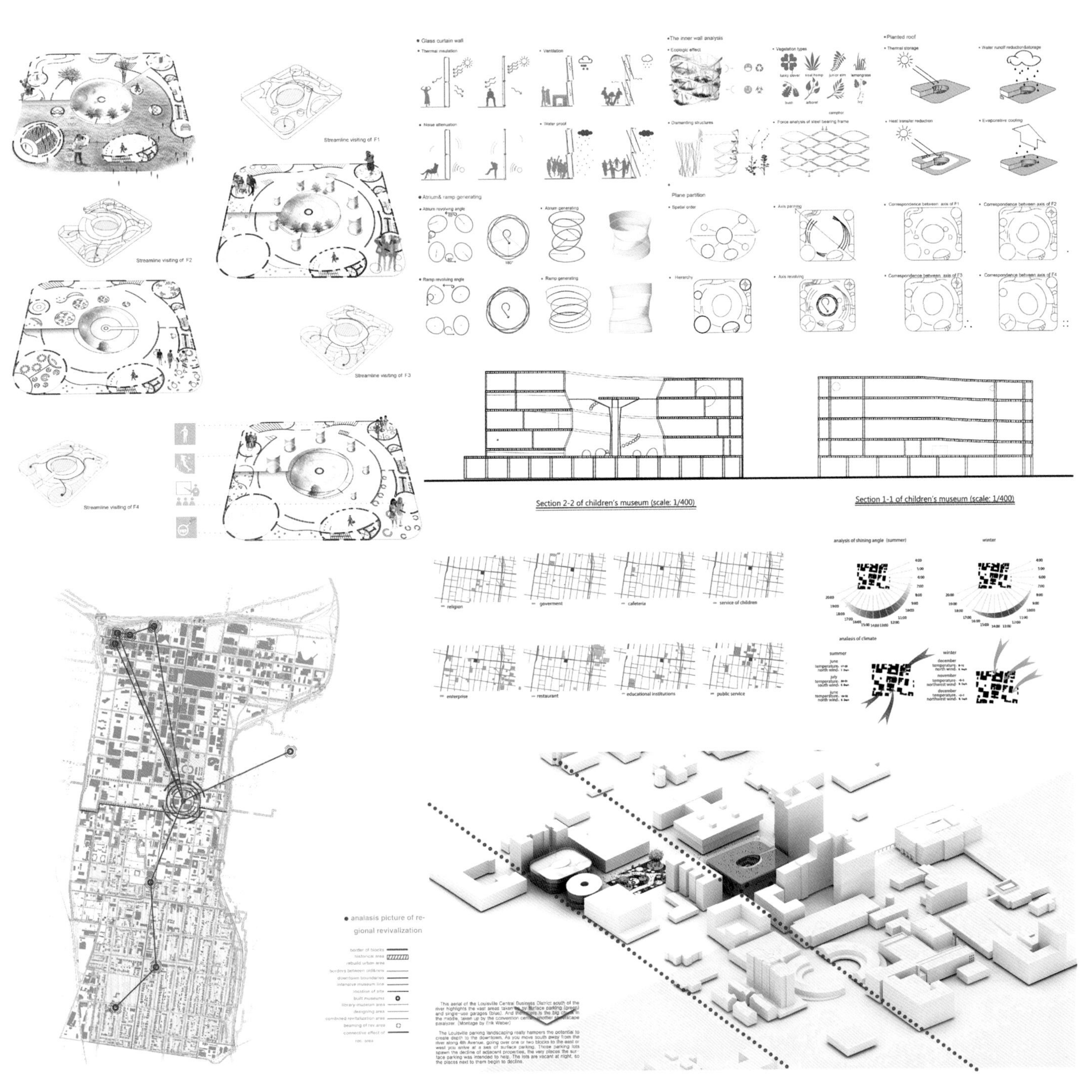

University: Qingdao Technological University
Designer: Wei Yimeng, He Jun, Miao Tianning
Tutor: Hao Chibiao, Xie Xudong, Cheng Ran
Course Name: AIA Children's Museum—The Regeneraiton of Edge Louisville
Finished Time: Feb. 2014

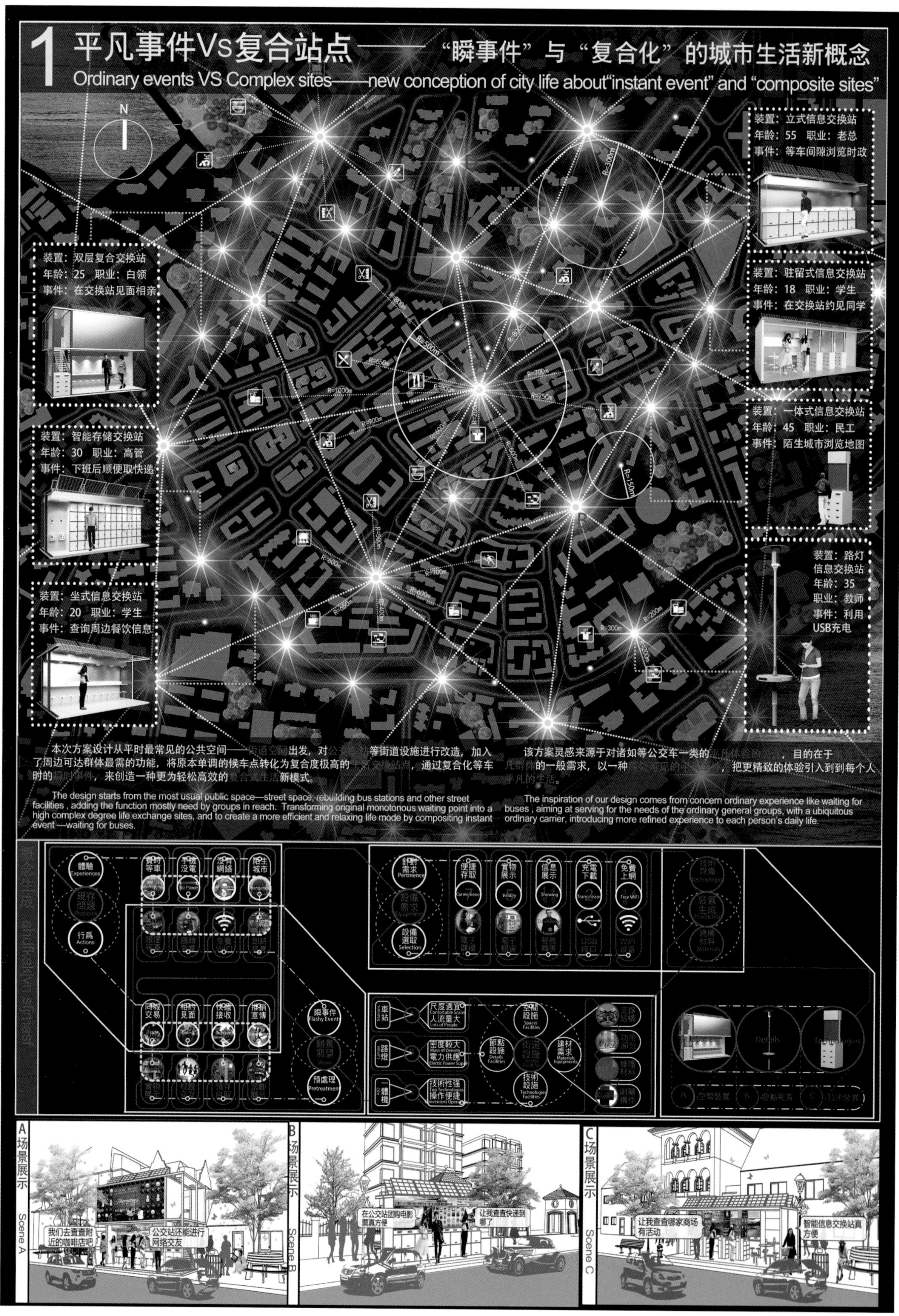

作品名称：平凡事件vs复合站点
Ordinary Events VS Complex Sites

院校名：青岛理工大学
设计人：魏易盟，肖飞，张泽华，李陶，杨明慧
指导教师：郝赤彪，解旭东，程然，耿雪川
课程名称：2014UA 国际大学生建筑设计竞赛
作业完成日期：2014 年 12 月

University: Qingdao Technological University
Designer: Wei Yimeng, Xiao Fei, Zhang Zehua, Li Tao, Yang Minghui
Tutor: Hao Chibiao, Xie Xudong, Cheng Ran, Geng Xuechuan
Course Name: UA Architecture for Everyday Life
Finished Time: Dec. 2014

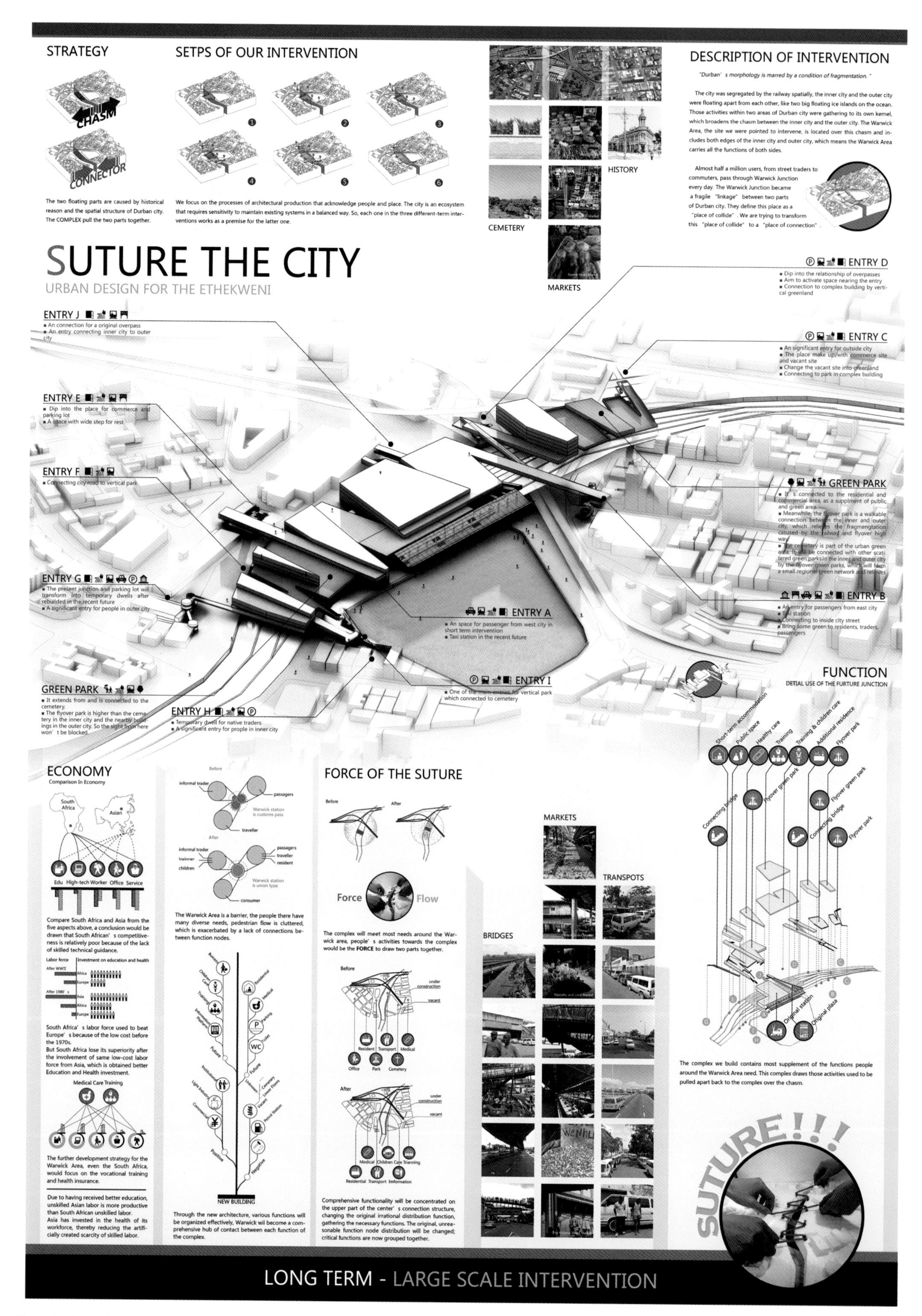

作品名称：缝合城市

——南非德班沃里克枢纽站地区改造城市设计

Suture the City:

Urban Design for the Ethekweni

院校名：西安建筑科技大学

设计人：吴明奇，牛童，冯贞珍，崔哲伦，罗典

指导教师：裴钊

课程名称：2014UIA 国际大学生建筑设计竞赛（本科三、四年级）

作业完成日期：2014 年 03 月

MEDIUM- TERM STRATEGY

The strategy for Medium-term Intervention is an assumption for the Long-term Intervention, as well as the whole project. The edges which are sutured together need to be adjusted first.

MEDIUM TERM INTERVENTION - TRANSITIONAL STAGE FOR THE BROOK STREET MARKET

After the COMPLEX is built, a lot of change will be brought to the Warwick Area. For example, only in the Brook Street Market, the passengers' gateway was moved away from there, and the traders' training and children care agency are added in this Market. The change in the architectural products and functions will lead movement of different target people, which takes time and may cause some conflicts.

Therefore, the COMPLEX cannot come to the site directly, and the change it will bring to the site should be acknowledged by the people who live and work here gradually. In our strategy the exotic COMPLEX lay its little feet around the site in the city first instead of its large body. Those little FEET is the PINs we are about to stick in the city at the short-term intervention. Those PINs are actually some different facilities or agencies whose functions are decided by the needs of its surroundings, which will take place in a comparatively short term. Those street traders generally don' t have fixed stalls. The great mobility of them will easily leads those street traders to move towards and gather around the PINs we stick.

The edge where the medium intervention takes place is one of the edges of the cemetery, as well as the edge of inner city abut to the "chasm", which makes the cemetery a crucial role in this transitional stage. If we want to suture the inner city and the outer city over the chasm, the first thing we need to do is to adjust the EDGE. We need to attach the cemetery to the chasm first, which means the Brook Street Market is a joint between the cemetery and the COMPLEX.

The medium term is a transitional stage for the Brook Street Market, as well as the edge of inner city. The Children Care spot have been built, and the traders who bring children to work have already gathered at the Brook Street Market. In the meantime, other PINs within the Warwick Area have also been stuck in position and exert their influence, which means the new COMPLEX is ready to take place within the Warwick Area over the "chasm" to suture the two parts of Durban city together.

Therefore the short-term intervention is the premise to make sure the medium-term intervention will happen and success to guarantee the complex, which is the main strategy of the Long-term intervention, can work.

From the study we have worked on the economy issue of Durban city, we think the sustainable development approach for Durban' s society and informal traders is to provide education and health investment to those traders on the streets and markets. In the whole project, we are longing to make the education more accessible for informal traders. The PIN in the edges we were pointed to choose at the Short-term intervention is on the Brook Street in the market, the function of which is a place for Children Caring. This PIN is a starter for the EDUCATION CHAIN.

ROOFTOP GARDENS

■ Providing a good view in standpoint, respecting the cemetery.

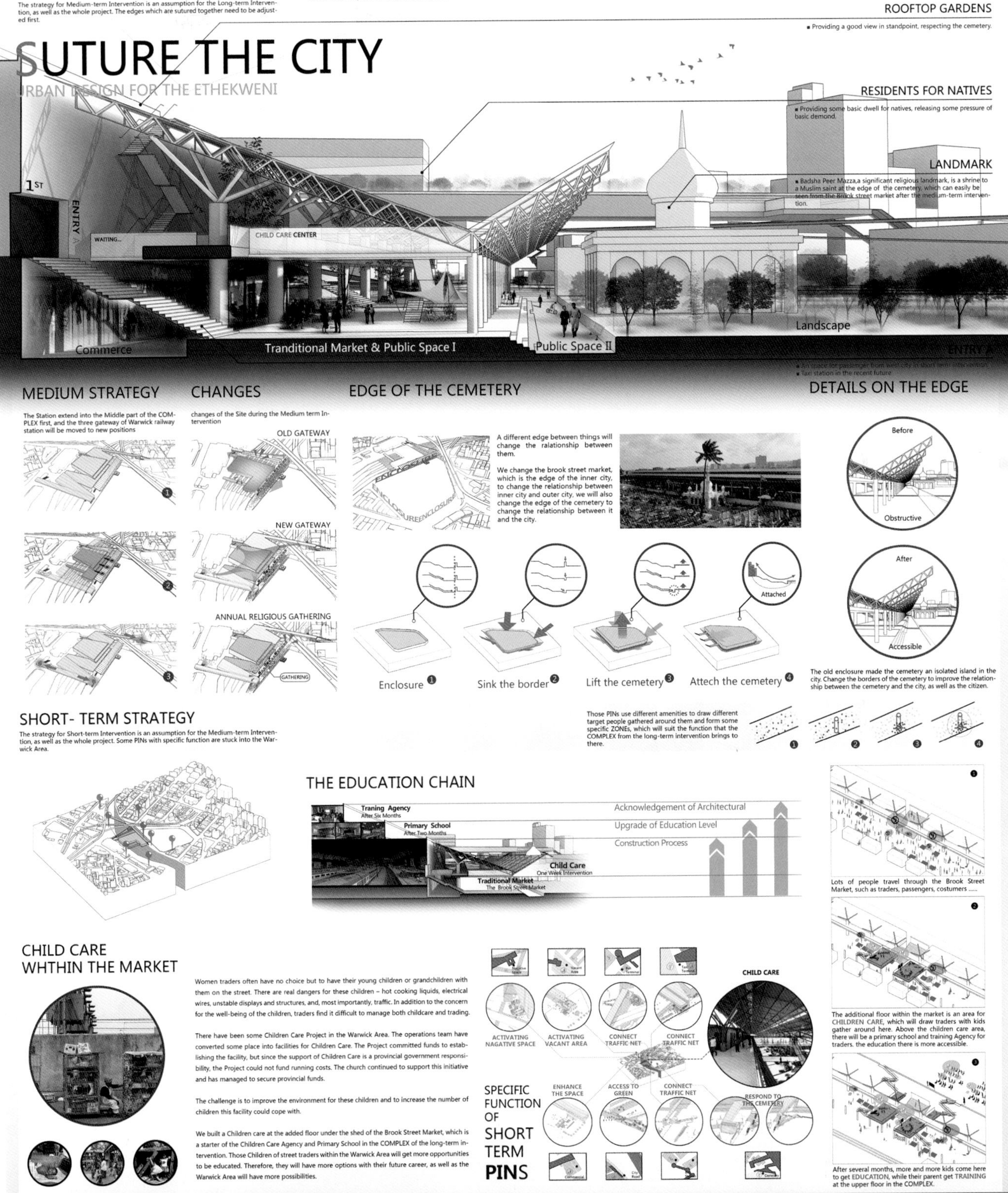

MEDIUM TERM & SHORT TERM- MEDIUM & SMALL SCALE INTERVENTION

University: Xi'an University of Architecture and Technology
Designer: Wu Mingqi, Niu Tong, Feng Zhenzhen, Cui Zhelun, Luo Dian
Tutor: Pei Zhao

Course Name: UIA 2014 Durban Warwick Junction International Student Competition
Finished Time: Mar. 2014

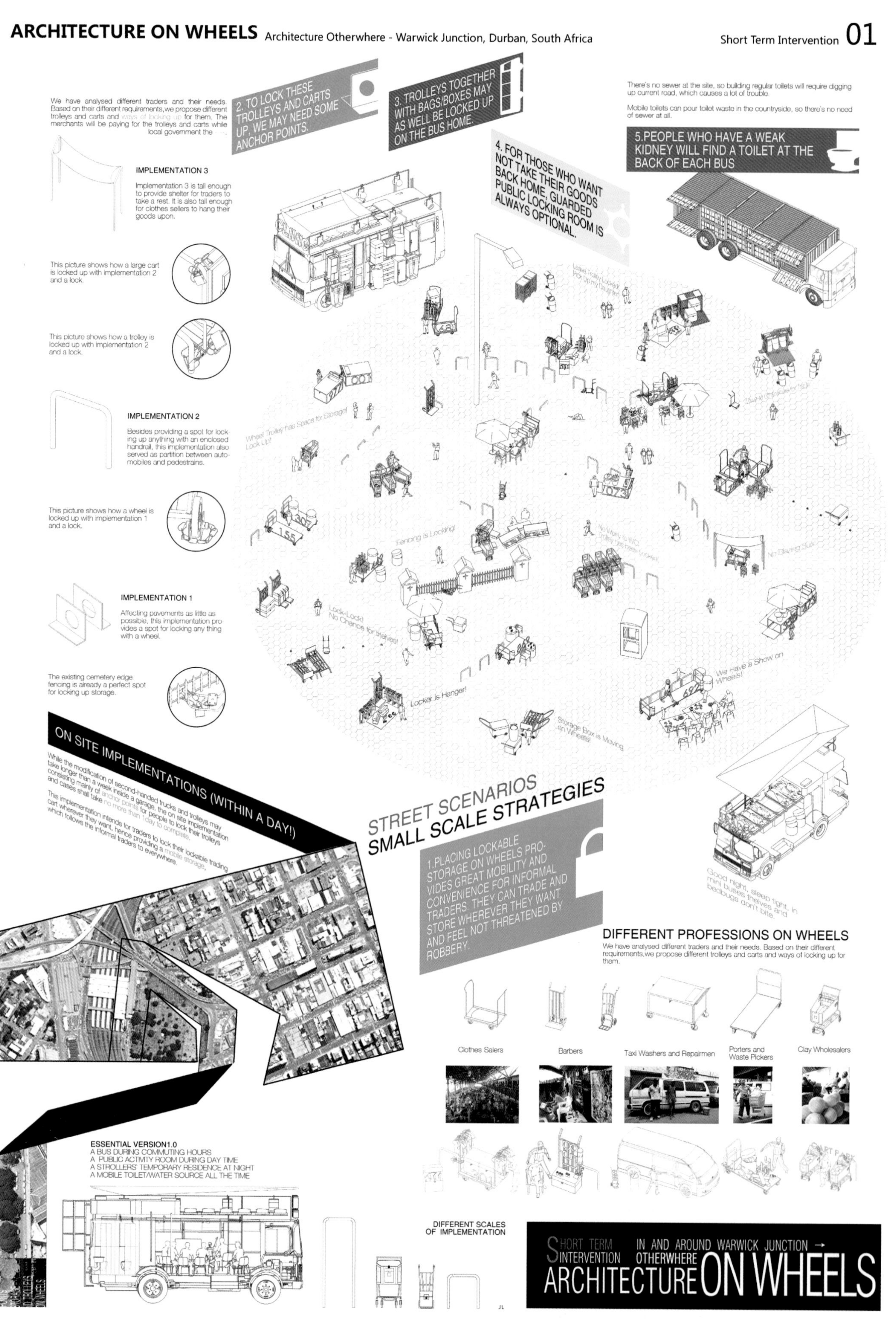

作品名称：辙上建造

Architecture on Wheels

院校名：清华大学

设计人：熊哲昆，刘芳铄，吴旭阳

课程名称：别样的建筑——2014UIA 国际大学生建筑设计竞赛（研究生一年级）

作业完成日期：2014 年 03 月

对外交流对象：2014 南非 UIA 第二十五届国际建筑师大会

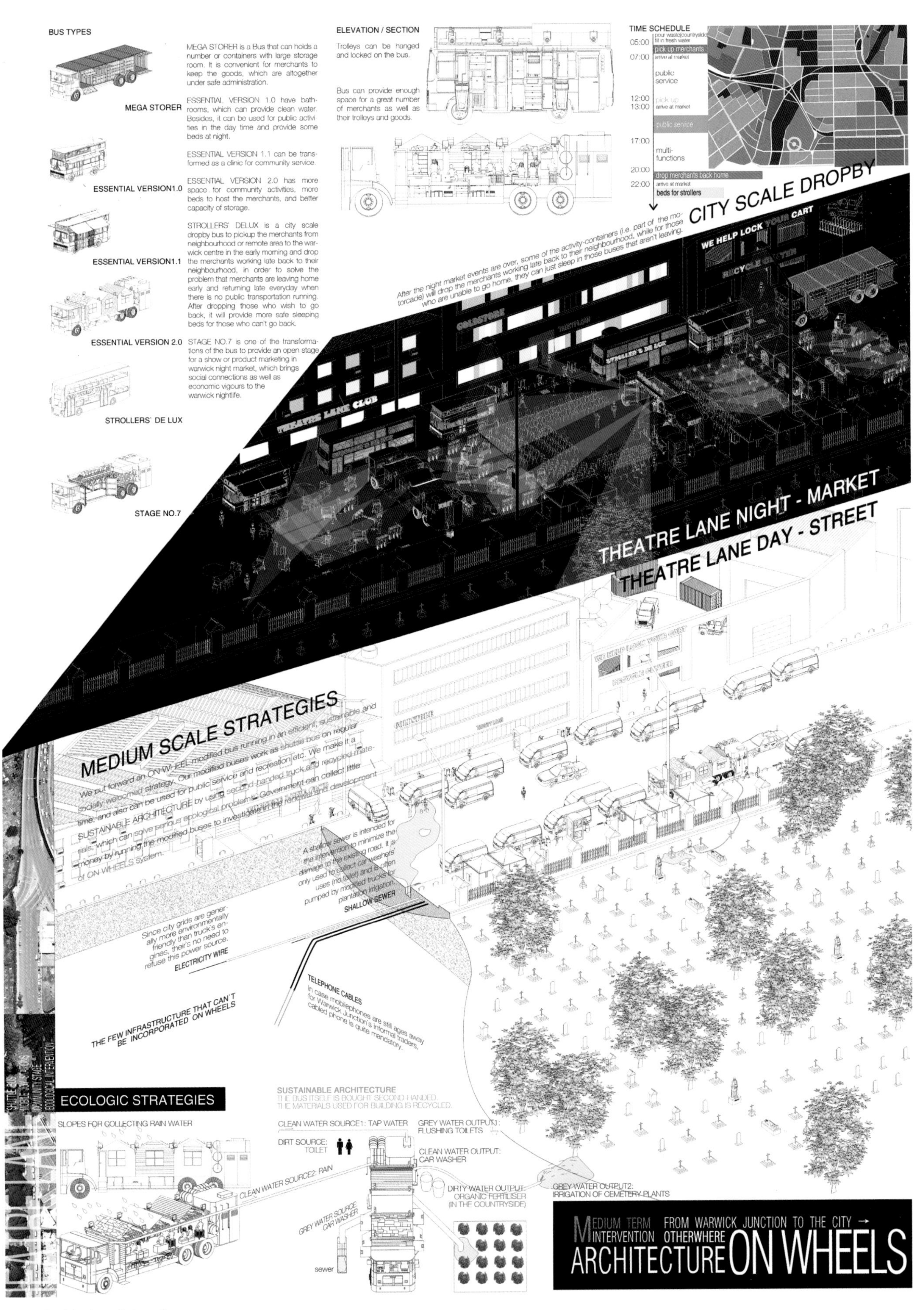

University: Tsinghua University
Designer: Xiong Zhekun, Liu Fangshuo, Wu Xuyang
Course Name: Architecture Otherwhere
Finished Time: Mar. 2014
Exchange Institute: 2014 XXV International Union of Architects (UIA) World Congress

ACTIVATING-BLANKING-FLOATING ANYWHERE

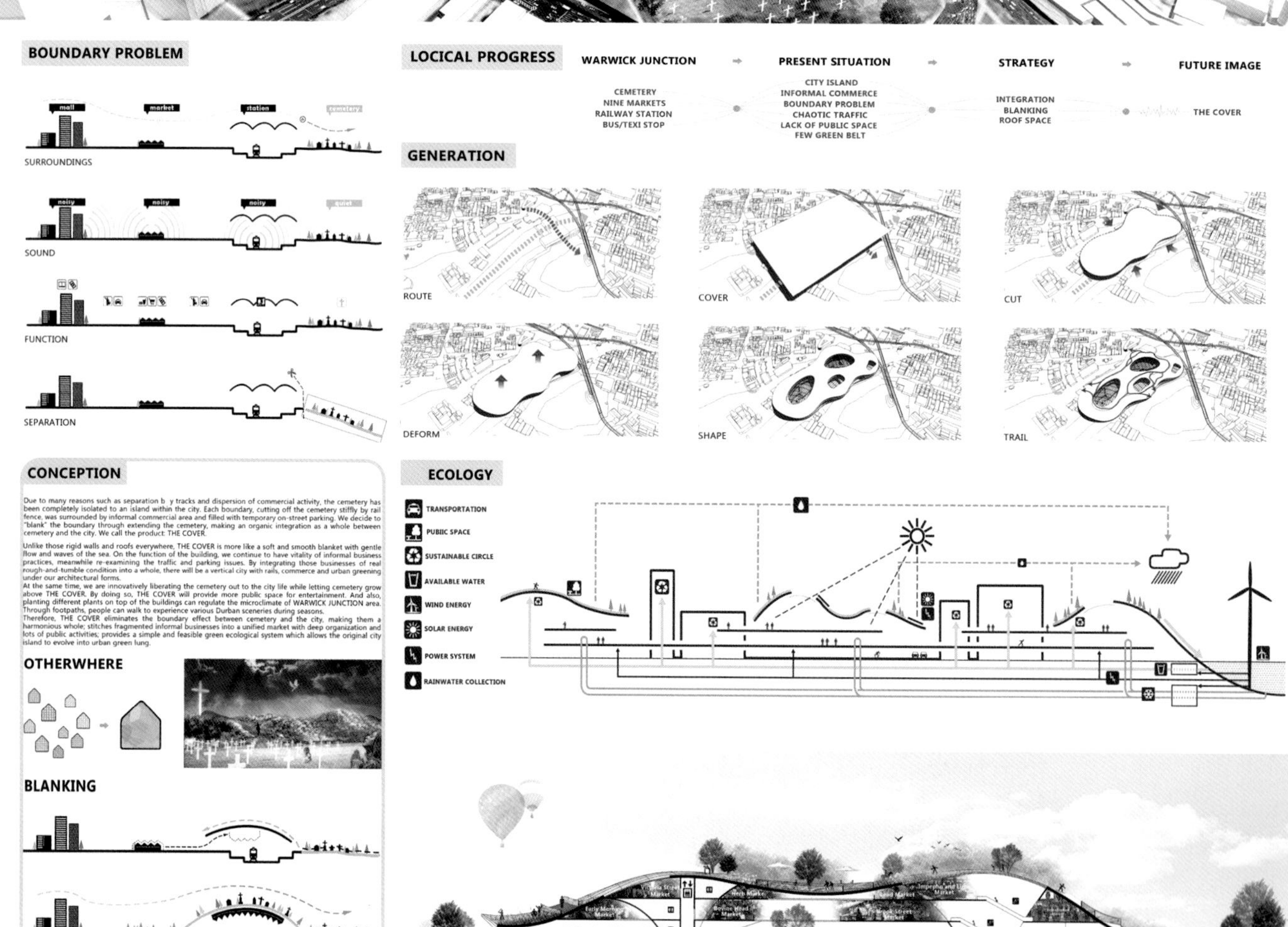

作品名称：激活 · 腾空 · 漂浮各地
Activating-Blanking-Floating Anywhere

院校名：合肥工业大学
设计人：杨梦曦，孙霞，苏朝晶，陈垦，马聃，冯磊
指导教师：李早
课程名称：UIA 国际建筑竞赛（本科五年级）
作业完成日期：2014 年 03 月

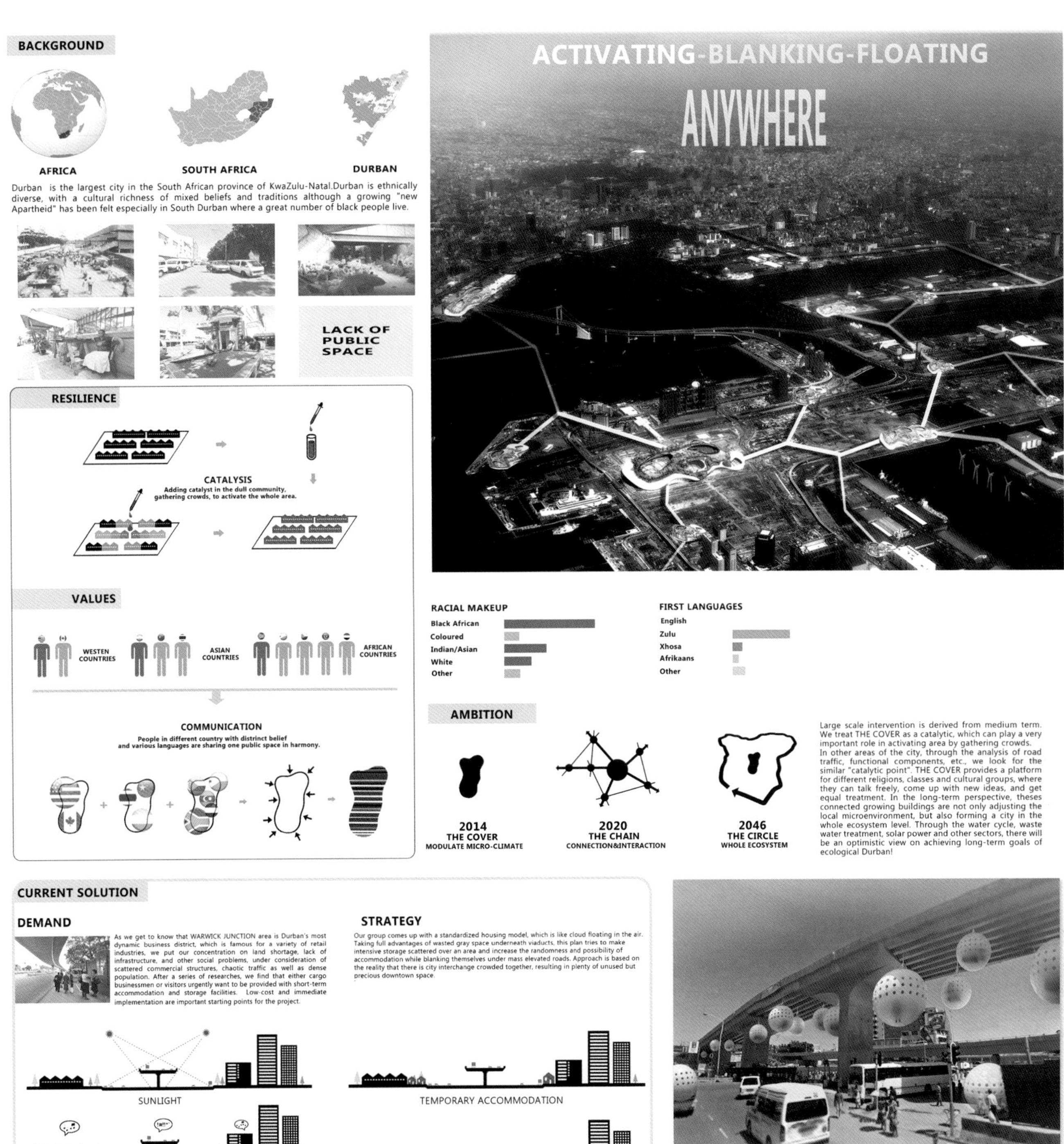

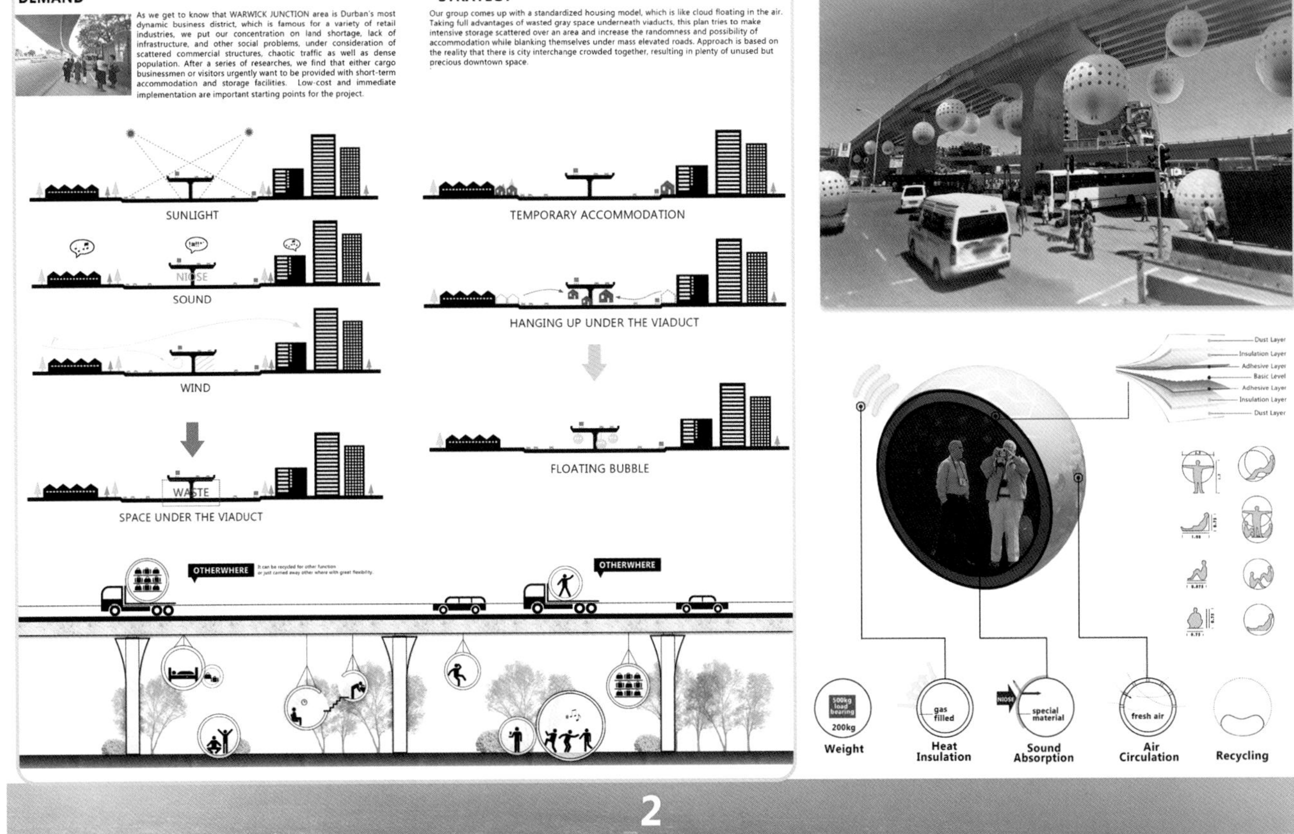

University: Hefei University of Technology
Designer: Yang Mengxi, Sun Xia, Su Chaojing, Chen Ken, Ma Dan, Feng Lei
Tutor: Li Zao
Course Name: UIA International Student Architecture Design Competition
Finished Time: Mar. 2014

延—小客棧設計

INHERITING

THE SMALL INN DESIGN

設計接續街區的歷史文脈，使遊客了解此街區的內涵，獲得新的體驗。新宅與老院、古街相融合，賦予舊宅新的活力，人們穿梭在其中，切身體驗此地的生活與文化。

THE DESIGN INHERITS THE CULTURE OF THE HISTORIC BLOCK , MAKING THE TOURISTS KNOW ABOUT THE HISTORY AND HAVE A BETTER TRIP IN HERE .
THE NEW BUILDING IS CONNECTED WITH THE ARCHAIC YARD AND THE OLD BLOCK ,THEREFORE , THE OLD HOUSE BECOMES VIGOROUS AGAIN .
PEOPLE WILL EXPERIENCE THE LIFE AND CULTURE IN THIS CITY WHILE WANDERING AROUND THE AREA.

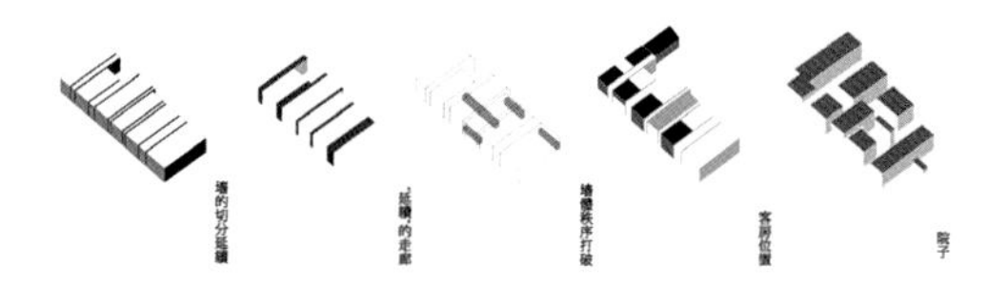

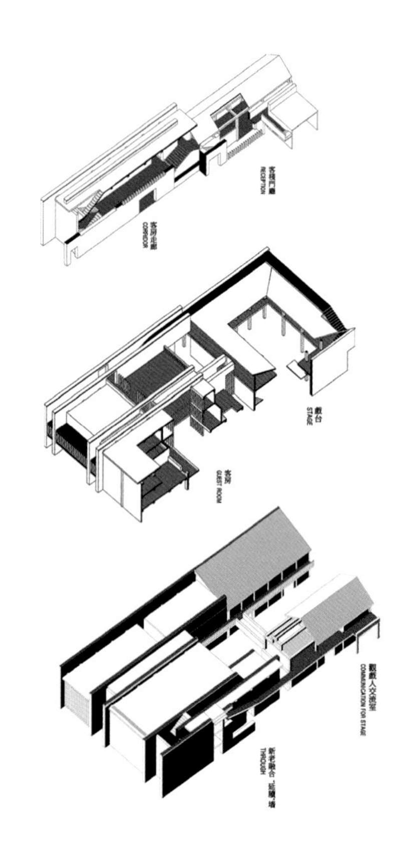

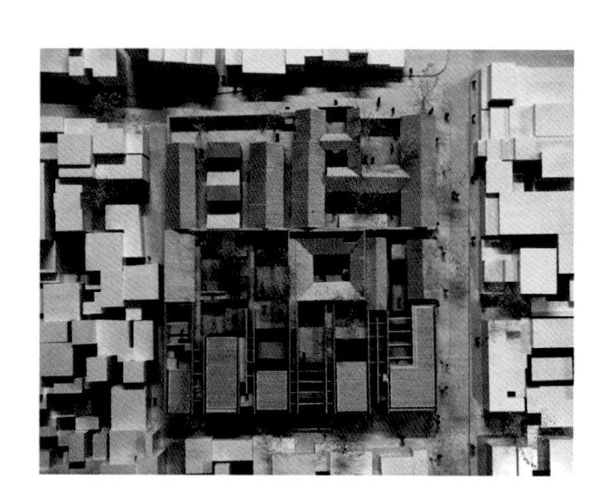

高家大院分析 GAO HOUSE ANALYSIS

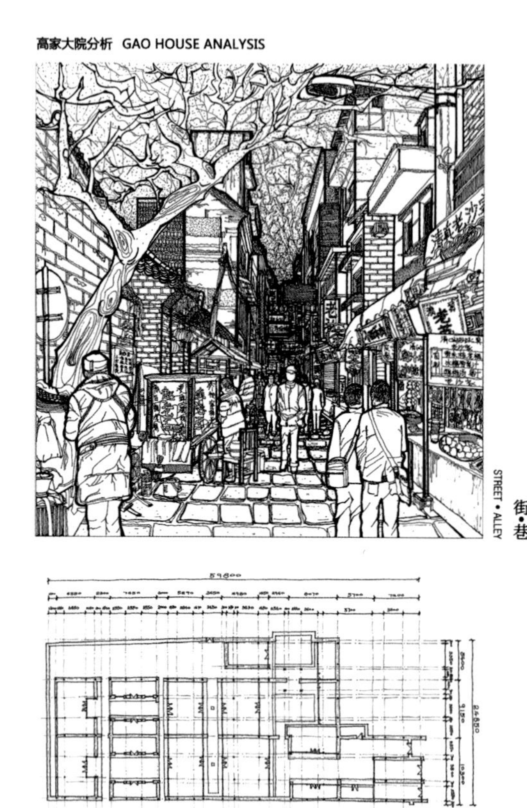

街巷 STREET • ALLEY

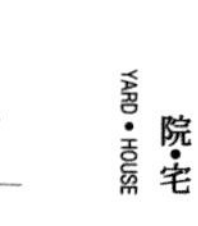

院•宅 YARD • HOUSE

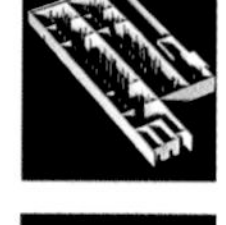
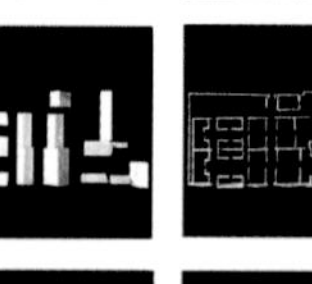
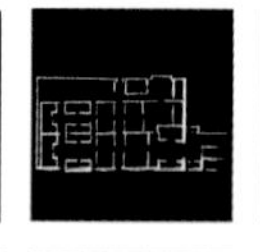
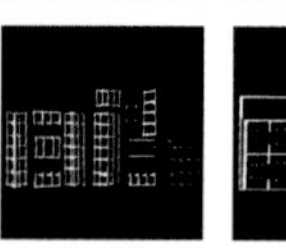
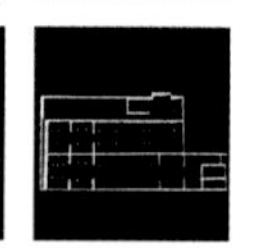

院•宅 YARD • HOUSE

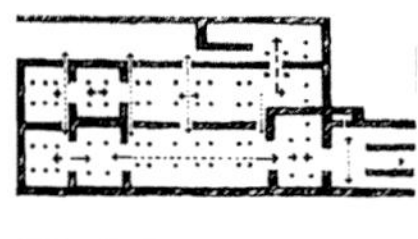
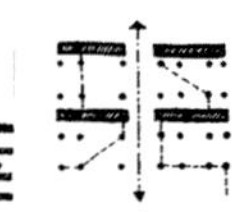

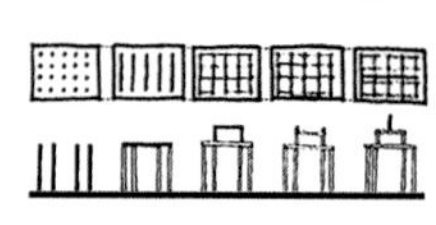
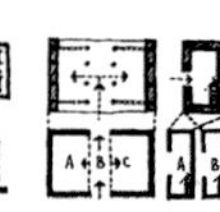

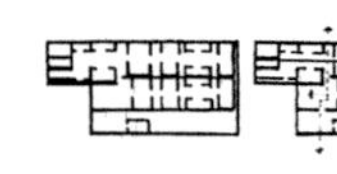
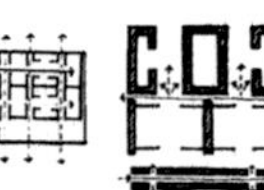
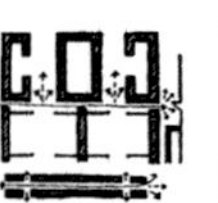

解•讀 INTERPRETATION

作品名称：延——小客栈设计
Inheriting the Small Inn Design

院校名：西安建筑科技大学
设计人：杨子依
指导教师：刘克成，丸山欣也，吴瑞，王毛真，蒋蔚，俞泉
课程名称：西安建筑科技大学—日本早稻田大学联合设计课程（本科二年级）
作业完成日期：2014 年 07 月
对外交流对象：日本早稻田大学

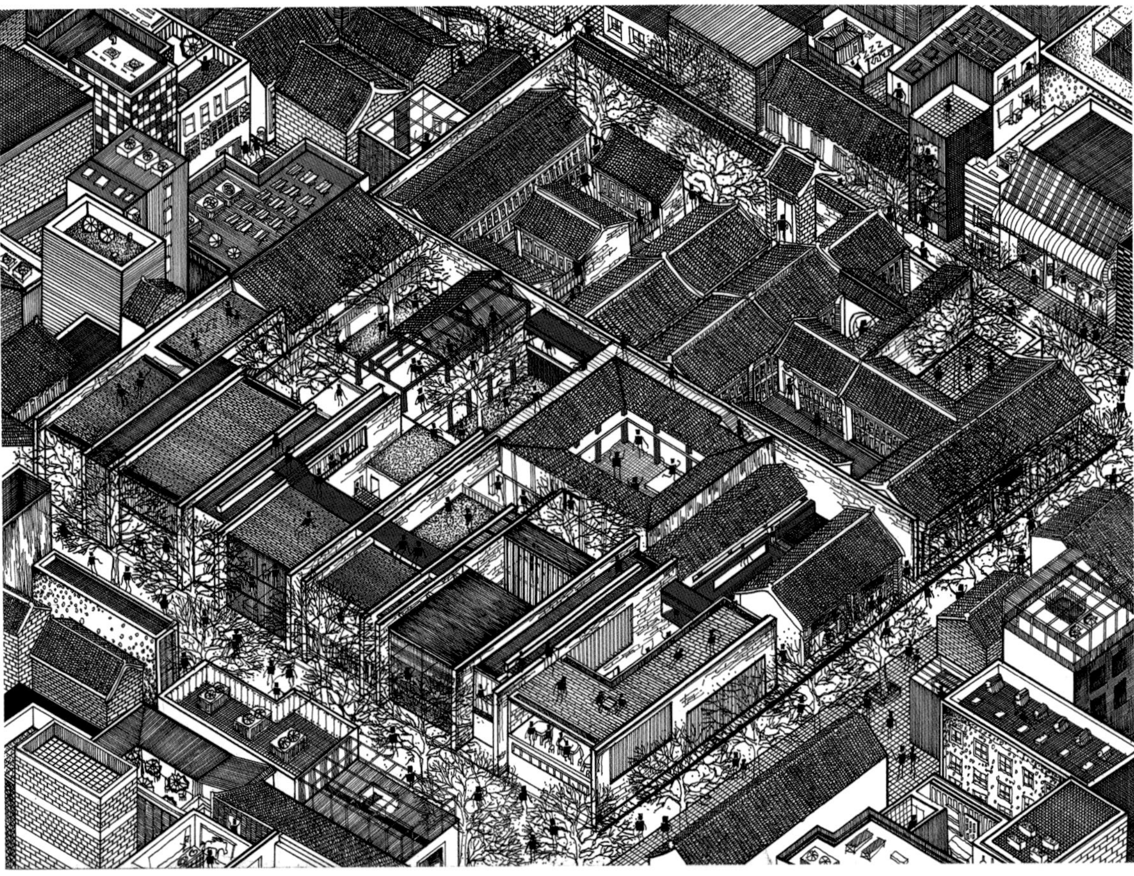

延—小客栈設計

INHERITING

THE SMALL INN DESIGN

THE SITE IS BEHIND THE GAO HOUSE IN THE BEIYUANMEN STREET,XI'AN.AS A PROTECTIVE TRADITIONAL DWELLINGS,THE GAO HOUSE IS NOT ONLY FOR SIGHTSEEING,HOWEVER,THE BEST WAY TO PROTECT IT IS MAKE IT FILL WITH "LIVES".

THIS DESIGN REFORMED THE GAO HOUSE,MAKING THE YARD CONTINUED.THE "LIVE"WILL SHUTTLE THROUGHOUT THE NEW ONE AND THE OLD YARD,WEHEN PEOPLE RAMBLE IN IT.

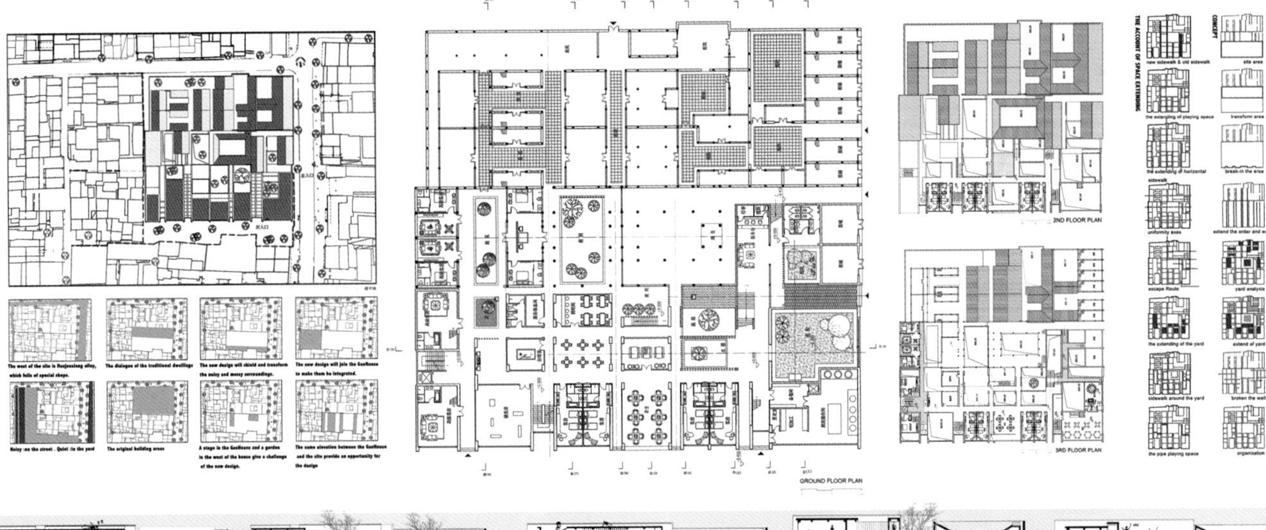

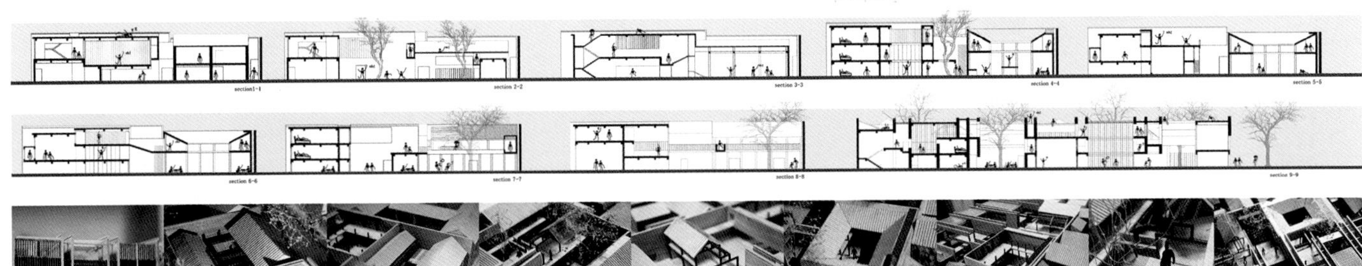

University: Xi'an University of Architecture and Technology
Designer: Yang Ziyi
Tutor: Liu Kecheng, Maruyama Kinya, Wu Rui, Wang Maozhen,Jiang Wei, Yu Quan
Course Name: XI'AN University of Architecture and Technology & Waseda University Joint Design Studio
Finished Time: Jul. 2014
Exchange Institute: Waseda University

公交车站，再平凡不过，我们几乎每天都能看到它，
它像一个个遍布城市中的容器，承载着我们一天的生活，承载着城市中发生的那些事儿 ...
The bus stop, so ordinary, we can almost see it every day.
It is like a container throughout the city, carrying our daily lives, carrying those things occurring in the city ...

小站 那些事儿
Those Things in the Bus Stop

发现问题
Problem Finding

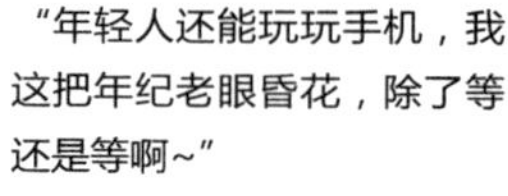

“年轻人还能玩玩手机，我这把年纪老眼昏花，除了等还是等啊~”

——张大爷

"Young people can play with cell phone.The old like me have no other choice but to wait ~"

-- Uncle Zhang

“等公交太无聊了，虽然就几分钟但每天都得等，在无聊中浪费了好多时间。”

—— 小王

"Waiting for bus is too boring。It although needs a few minutes but we have to wait every day, wasting a lot of time in boring process."

-- Mr. wang

我们将青岛市划分为住区、商业、医院、学校、办公、幼儿园、景点7种功能类型，我们分别选取7种功能区域中的一个车站作为基地，并对不同车站等车的人进行了调查，他们普遍抱怨等车太无聊。我们根据不同区域车站服务对象需求的不同选择了几件可以在短时间内完成的活动。

We divide the bases into seven types, and make a survey of the people on different stations . They generally complain that waiting for a bus is too boring. According to the different needs of different crowd in the station of different regions, we choose a few activities that can be finished in a short time.

需求分析
Needs Analysis

概念生成
The Formation of Concept

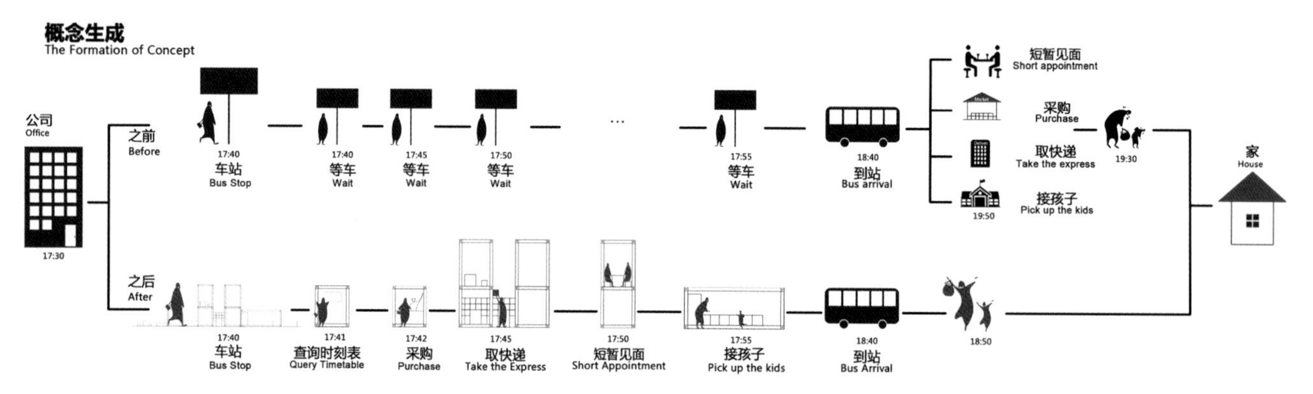

作品名称：小站那些事
Those Things in the Bus Stop

院校名：青岛理工大学
设计人：于立恒，王傲男
指导教师：郝赤彪，王少飞，石新羽
课程名称：2014UA 国际大学生建筑设计竞赛
作业完成日期：2014 年 12 月

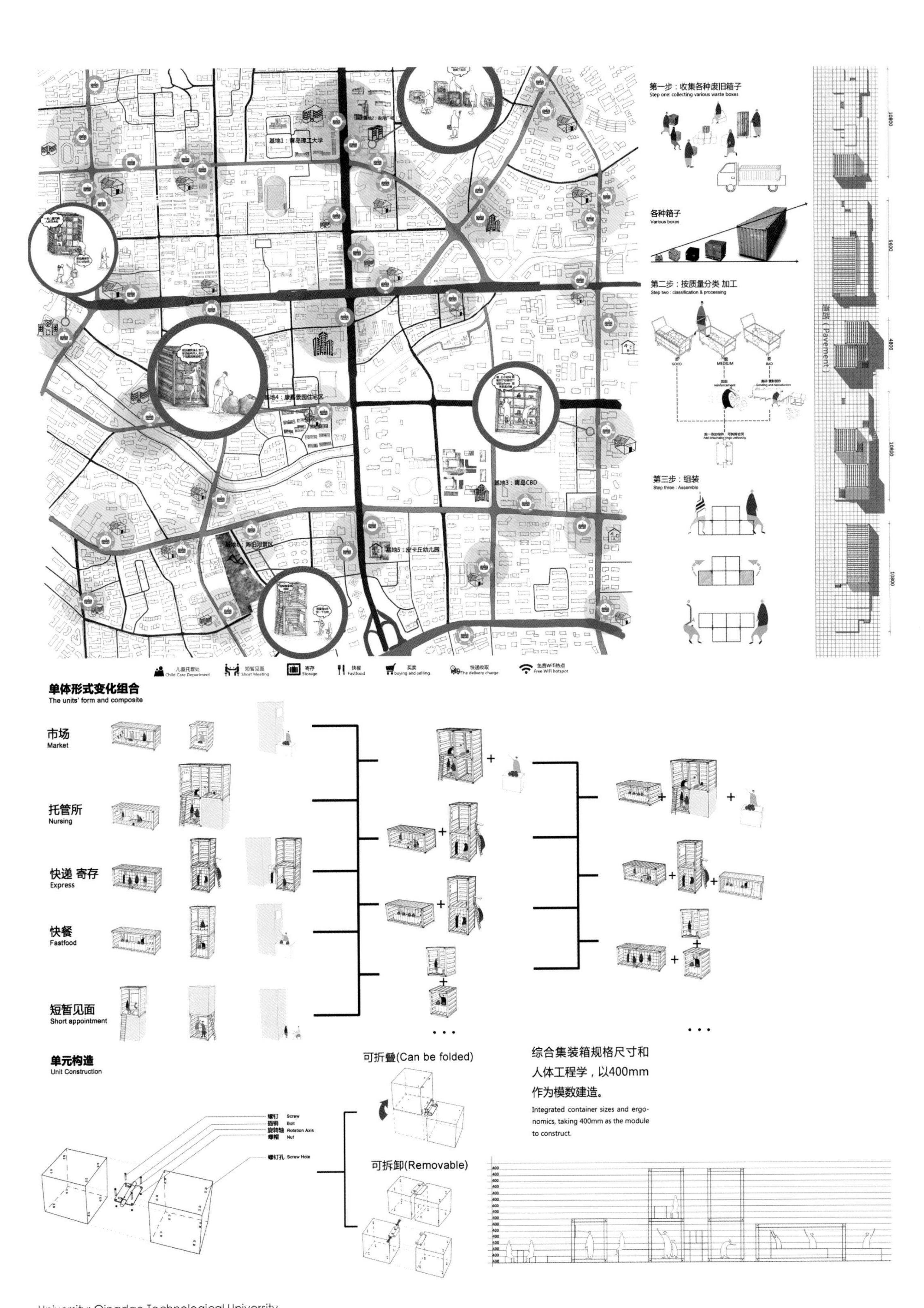

University: Qingdao Technological University
Designer: Yu Liheng, Wang Aonan
Tutor: Hao Chibiao, Wang Shaofei, Shi Xinyu
Course Name: UA Architecture for Everyday Life
Finished Time: Dec. 2014

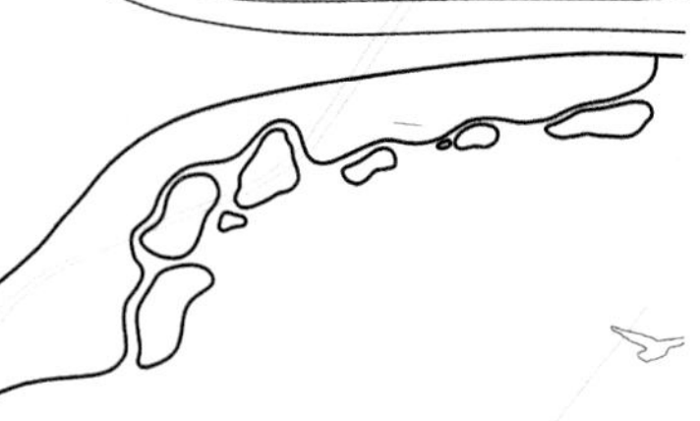

Location & Context

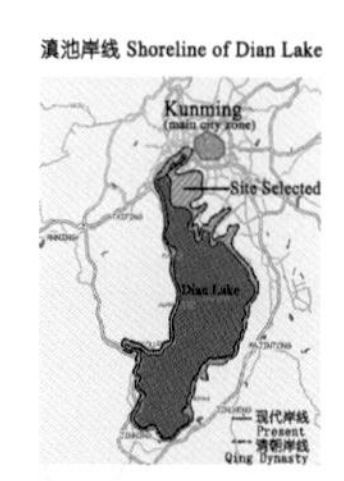

Design Description

本设计选址于昆明滇池围海造田区域，旨在反思历史、弥补过错，立足于现代城市问题，以低能耗、智能化、人性化、可持续作为整体主旨，综合传统聚落、大院经济、田园城市、立体城市、生态城市等优点设计支撑网络和各种功能细胞模块，利用“细胞自动机(cellular automata)”模拟城市的复杂演化过程，构建功能相对完备、密度适中、自给自足、高效综合的城市单位（Unit）——滇池畔的“元胞螺旋城”，营造“情理之中，意料之外”的宜居城市。

Dian Lake is the mother lake of Kunming and has noticeable effect on the warm climate in Spring City. With the rapid expansion of city, reclaiming land from lake lead to climate deterioration. Nowadays, Kunming is not as cool in summer and warm in winter as before. We reflect on history and hope to correct the mistakes. We're seeking a way to resolve modern urban issues, while at the same time reclaiming land to lake and giving back the old beautiful days.

Our project mainly concerns about low energy consumption, intelligent, humanised and sustainable. We assimilated the essence of traditional settlement, Dayuan Unit, Garden city, Vertical City and Eco-city. We design a network including structural system, transportation and other servicing pipes, with functional cells together. Appealing to Cellular Automata, we imitate the complex evolutionary process of the city and build a city unit which is function relatively complete, moderate-density, mixed-use and self-sufficient: The Spiral City with Cellular Automata by Dian Lake, which is a livable city unexpected but reasonable.

1/2 元胞螺旋城
Spiral City with Cellular Automata

自动跟踪太阳的采光装置
Lighting device that can track solar automatically to solve the poor

风能发电
Windmills to take advantage of wind

温室
Greenhouse

立体农场
Vertical farm

立体绿化
Vertical planting

湿地
Wetland

昆明市区 Kunming City

人均用地 Per Capita Land……100㎡/p

人均建筑用地 Per Capita Construction Area……35㎡/p

围海造田 Reclaim Land from Lake

长期的围海造田活动导致滇池湖面积人为的缩小，使得周围地区气候条件恶化。

Reclaiming has shrunk the area of Dian Lake for a long time, leading to climate deterioration.

垃圾污染 Pollution

滇池湖畔人类活动增多也加剧了垃圾对湖水的污染。

The increase of human activity also aggravated the pollution from rubbish.

滇池湿地 Wetland

原有湿地遭到破坏，湖水自净能力减弱。

The damage of original wetland impacted the self-clean ability of the lake.

光辉城市 The Radiant City

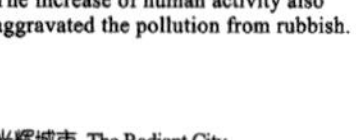

面积 Area……72km²

居住区人均用地 Per Capita residential area……10㎡/p

田园城市 Garden City of Tomorrow

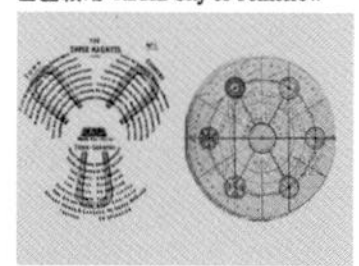

面积 Area……4km²

人口 Population……32,000

人均用地 Per capita land……125㎡/p

概念 Concept

系统分析 Product System

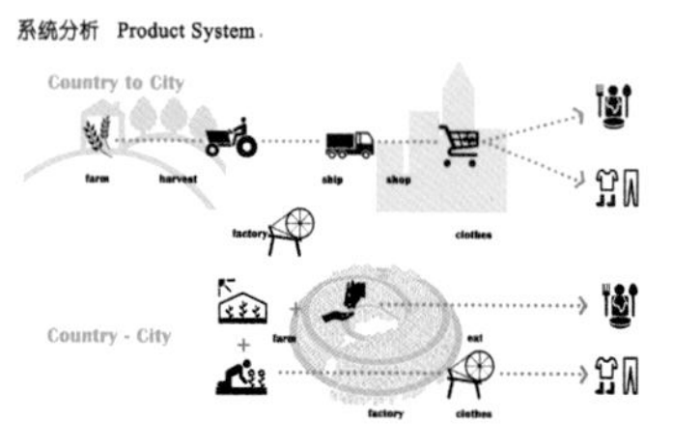

基地改造 Strategies for the site

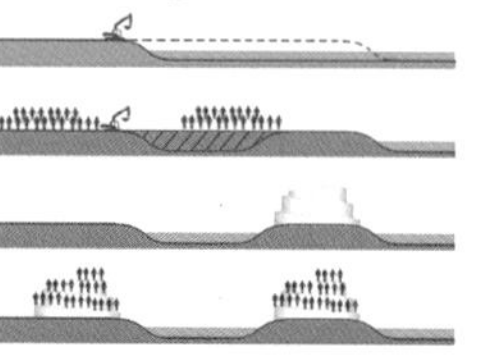

围海造田　清朝开始直到近代的围海造田活动将滇池水面不断缩小。
Reclaim land from lake
Hundreds years of reclaiming land from lake lead to the reduction of water area.

还湖成岛　把填湖的土方挖开堆成岛屿。
Reclaim land to lake
Pile islands and hills with soil that dug in the reclamation.

立体城市　在岛上建立5维的立体城市。
Vertical cities
Build 5D vertical cities on the islands and hills to hold original resident.

品质提升　立体城市自给自足，交通便捷，居民生活品质提升。
Improve the quality of life
self- sufficiency and convenient traffic improve the quality of life.

英格兰威尔士通勤地图
Commuting map of England & Wales

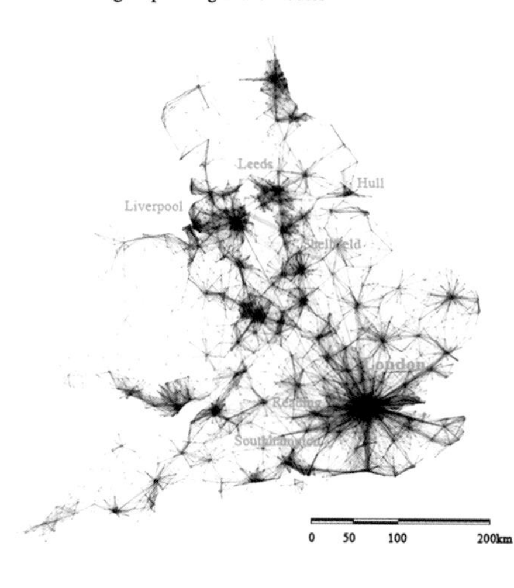

城市尺度
Scale of Kunming City

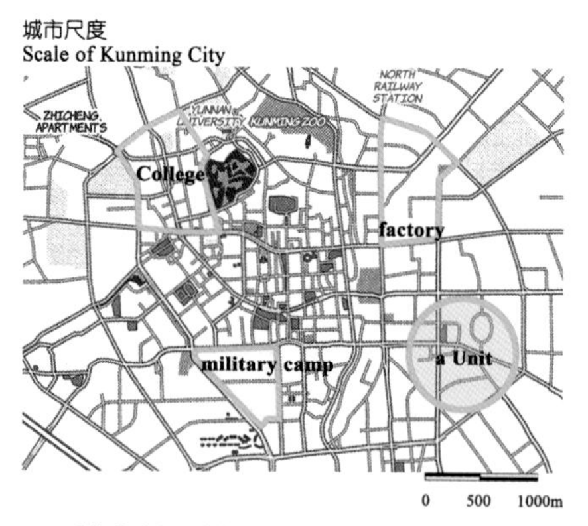

我们的城市尺寸来源于昆明本土“大院”，几个城市单体形成的组团与昆明周边城市的关系又参照了英格兰威尔士通勤地图的尺度。

The scale of our city derives from the economy scale of local community. The compound of residential units and the relationship with towns around Kunming refer to the scale of commuting map of England and Wales.

作品名称：出乎意料的城市——元胞螺旋城
Unexpected City: Spiral City with Cellular Automata

院校名：昆明理工大学
设计人：翟星玥，翟雪竹
指导教师：翟辉，王丽红

课程名称：昆明理工大学—研究型设计霍普杯国际竞赛
作业完成日期：2014 年 07 月

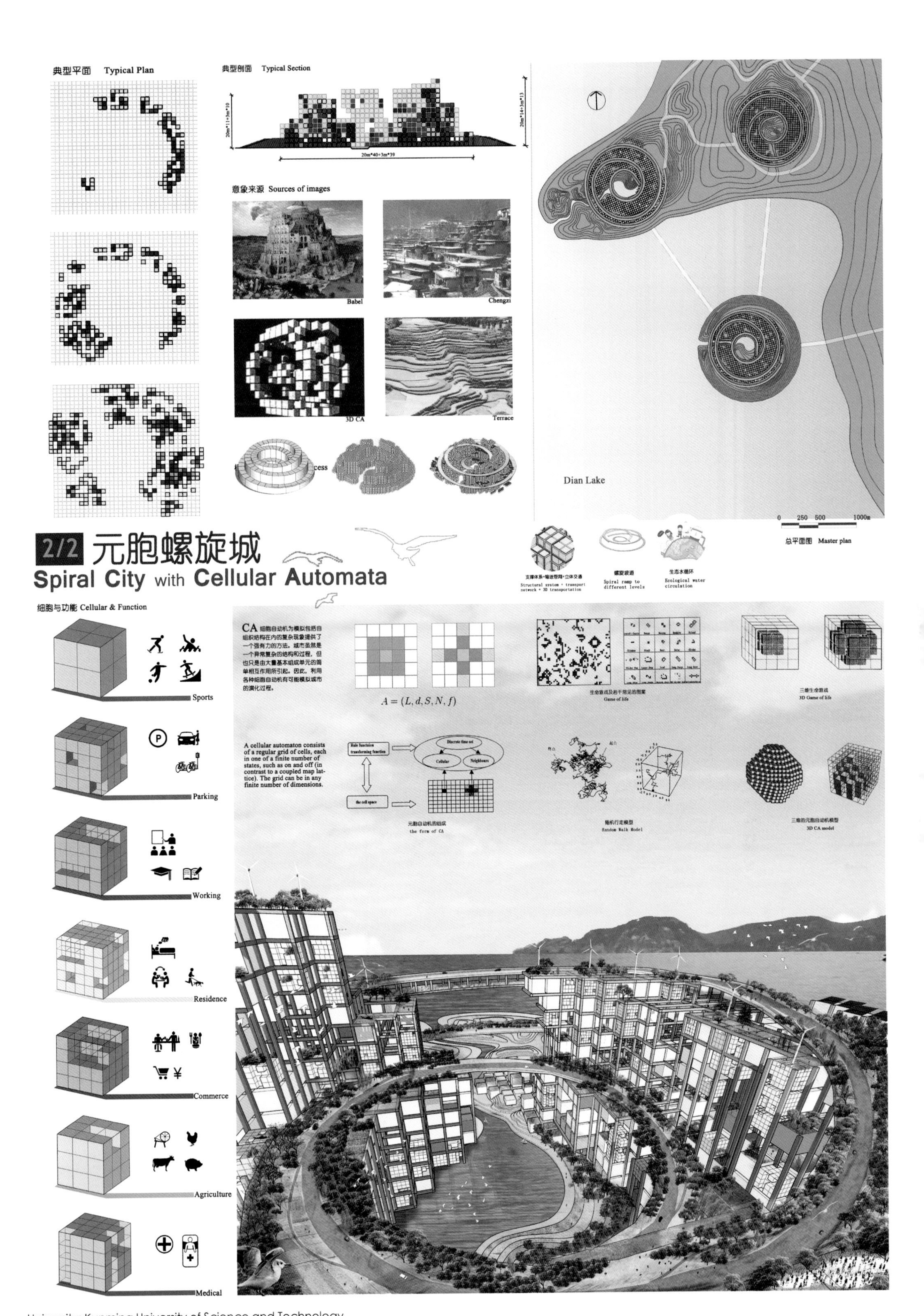

University: Kunming University of Science and Technology
Designer: Zhai Xingyue, Zhai Xuezhu
Tutor: Zhai hui, Wang Lihong
Course Name: UIA-HYP cup 2014 International student competition in architecture design
Finished Time: Jul. 2014

CODE:221

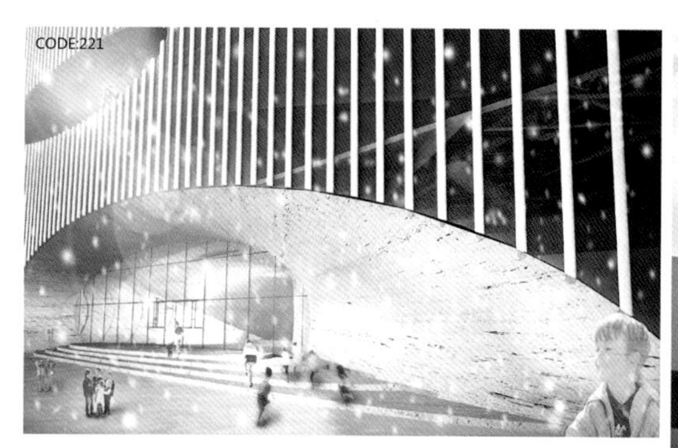

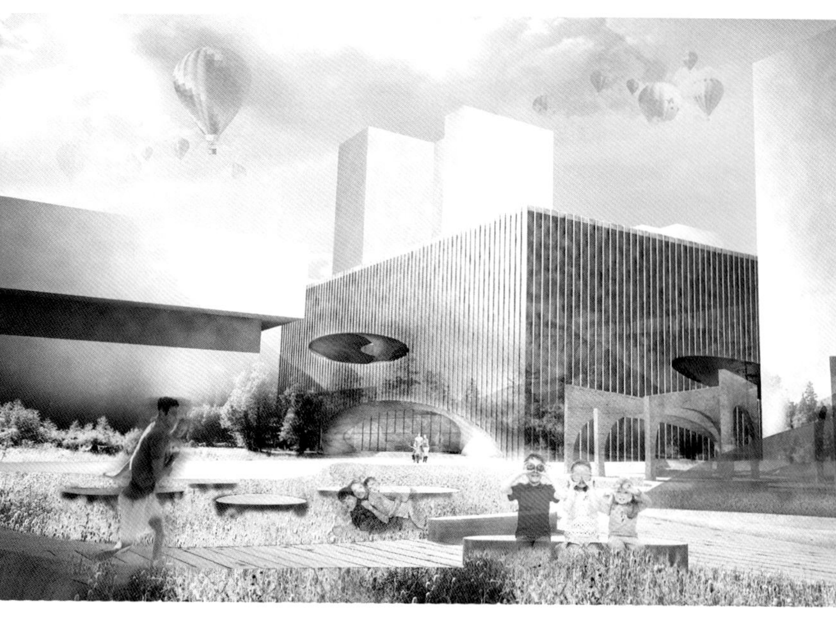

WORMHOLE
Louisville Children's Museum

The idea comes from the wormhole which can connect space-time, It' s like a long tunnel linking the city' s historical development with personal growth. Walking through the tunnel, people are showed three scenes --past ,today and future.Maybe they can be reminded of the pursuit of their childhood dreams, so that people of different ages can find their own childhood in the Children's Museum.So that their memories of louisville will be awaken and ultimately the purpose of revitalizing the downtown edge can be achieved.

Circulation

STAFF ROUTES
TOURIST ROUTES

+18.50
+15.50
+10.00
+2.00
-6.30

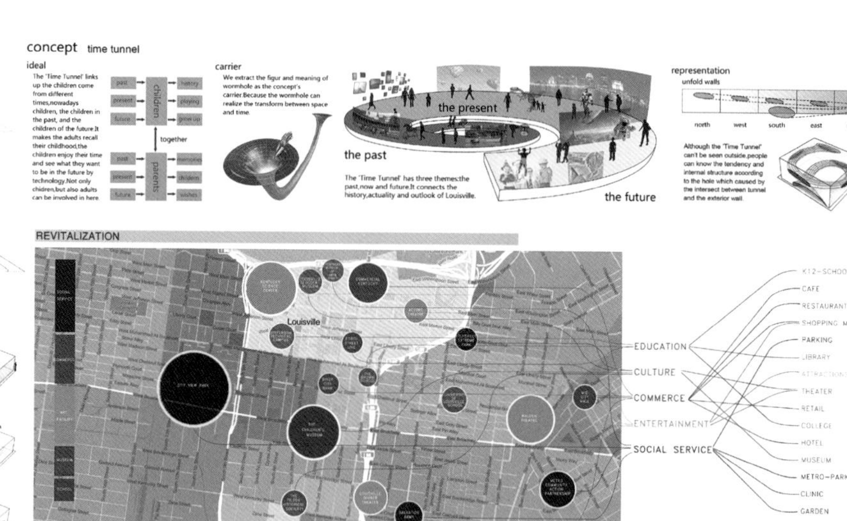

Site plan 1:1000

Broadway
vehicle trance
second entrance
main entrance
4th St
3rd St
2nd St

1 Chidren' s Museum
2 Park
3 High-Tech
4 Parking Garage

Concept & Indoor Scene

A City-A Person
Grow together-People's emotion here

Grow with Louisville
Homeless Children

BACK TO SCHOOL

作品名称：虫洞
Wormhole

院校名：青岛理工大学
设计人：张驰，徐钰茗，王傲男，杨迪雯，冯勃睿
指导教师：郝赤彪，解旭东，程然
课程名称：2014AIA 路易维尔儿童博物馆建筑设计竞赛
作业完成日期：2014 年 01 月

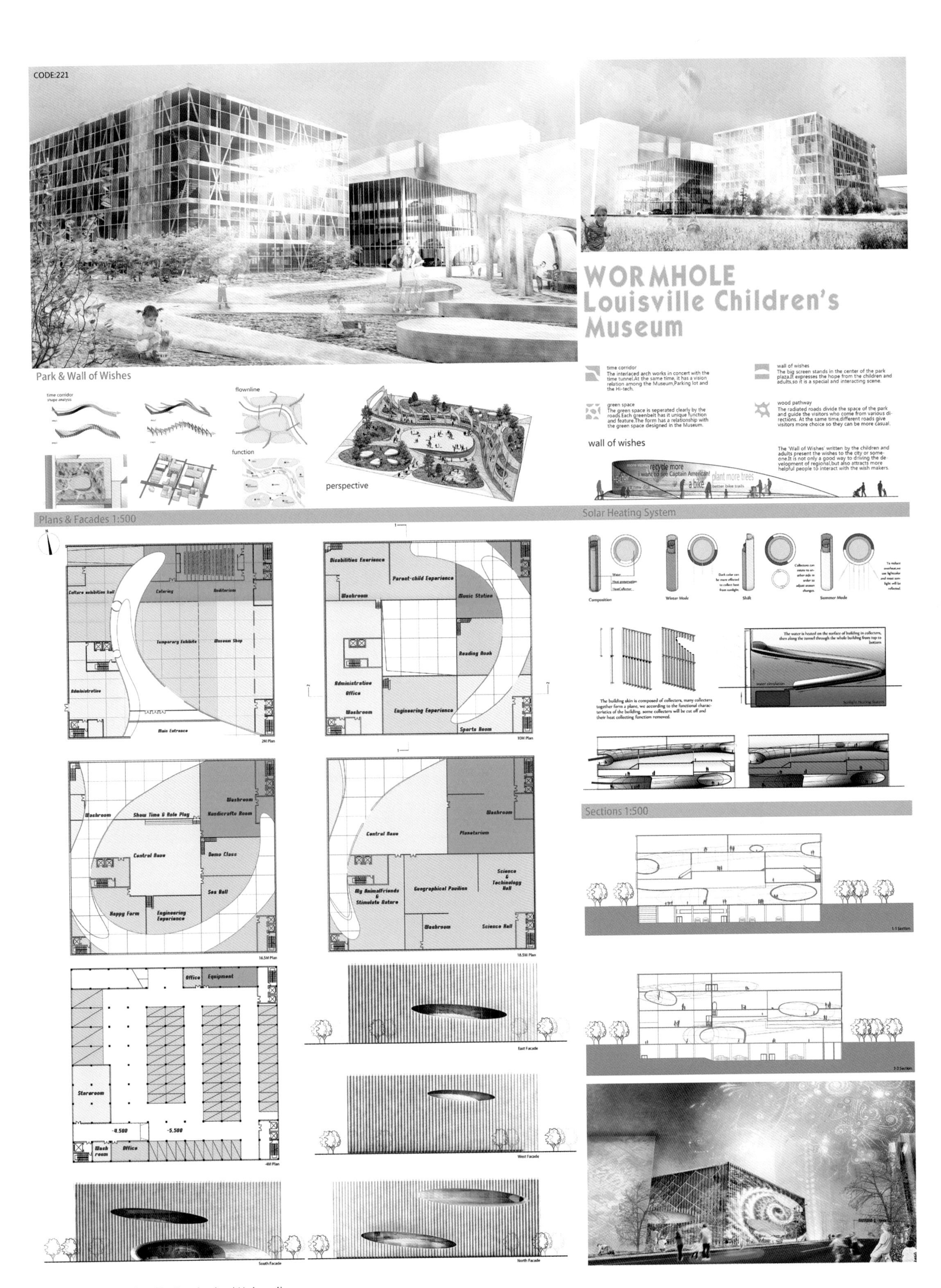

University: Qingdao Technological University
Designer: Zhang Chi, Xu Yuming, Wang Aonan, Yang Diwen, Feng Borui
Tutor: Hao Chibiao, Xie Xudong, Cheng Ran
Course Name: AIA Children's Museum—The Regeneraiton of Edge Louisville
Finished Time: Jan. 2014

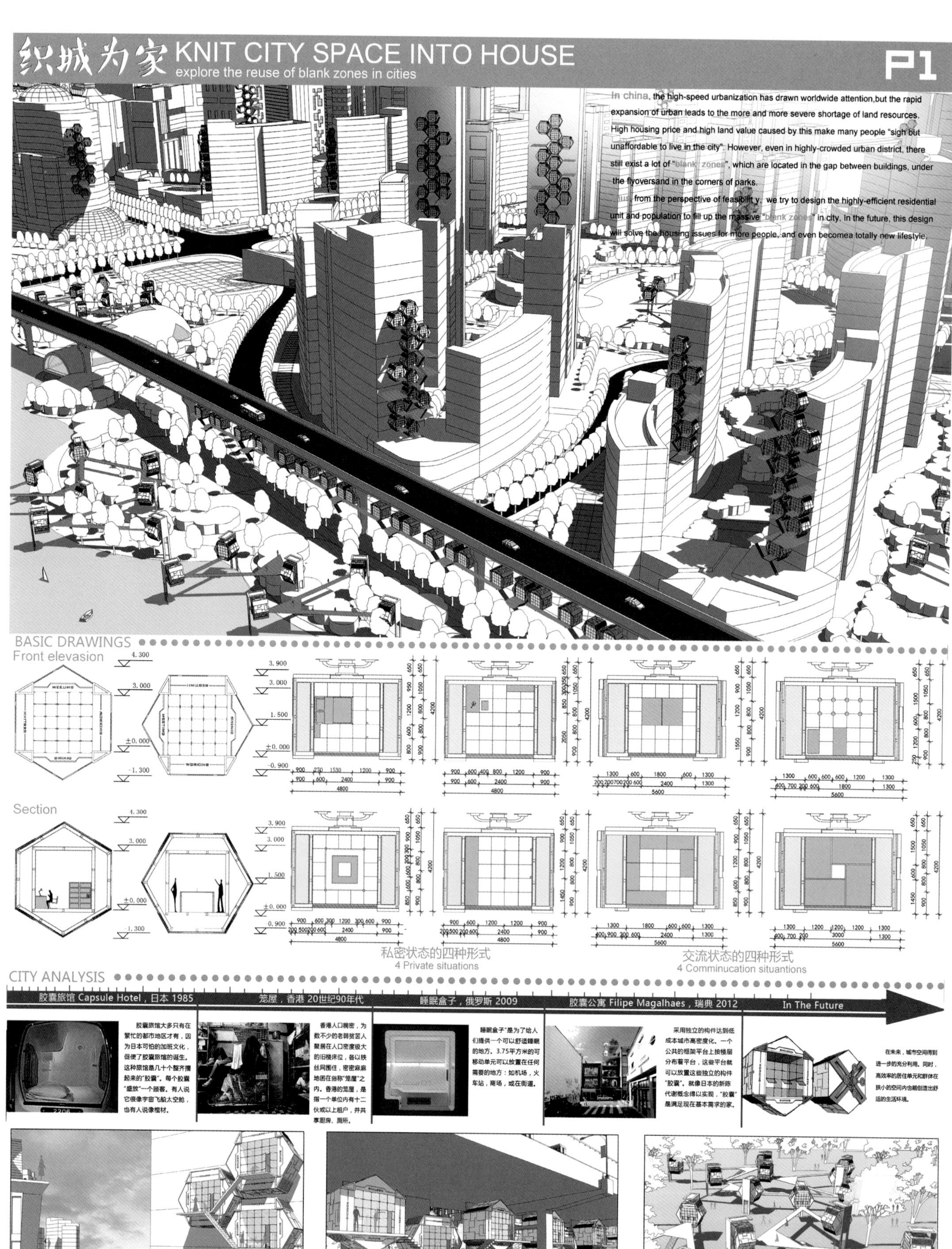

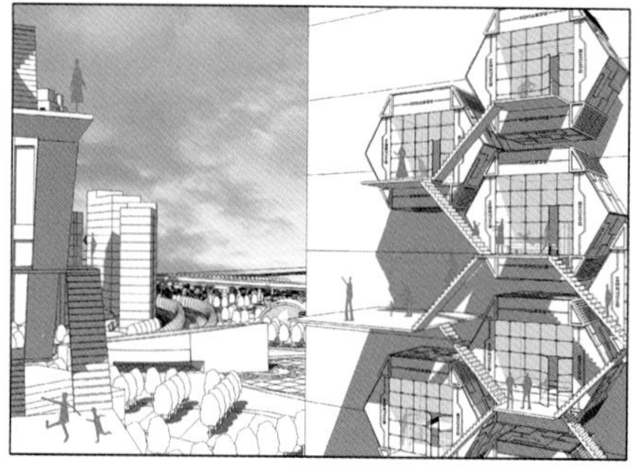

作品名称：织城为家——探索城市空白空间的再利用
Knit City Space into House: Explore the Reuse of Blank Zones in Cities

院校名：中国矿业大学
设计人：张学优，杨瑞东，曹啸，丁纹敏
指导教师：孙良，李明，陈惠芳

课程名称：2014 年霍普杯国际大学生建筑设计竞赛
作业完成日期：2014 年 08 月

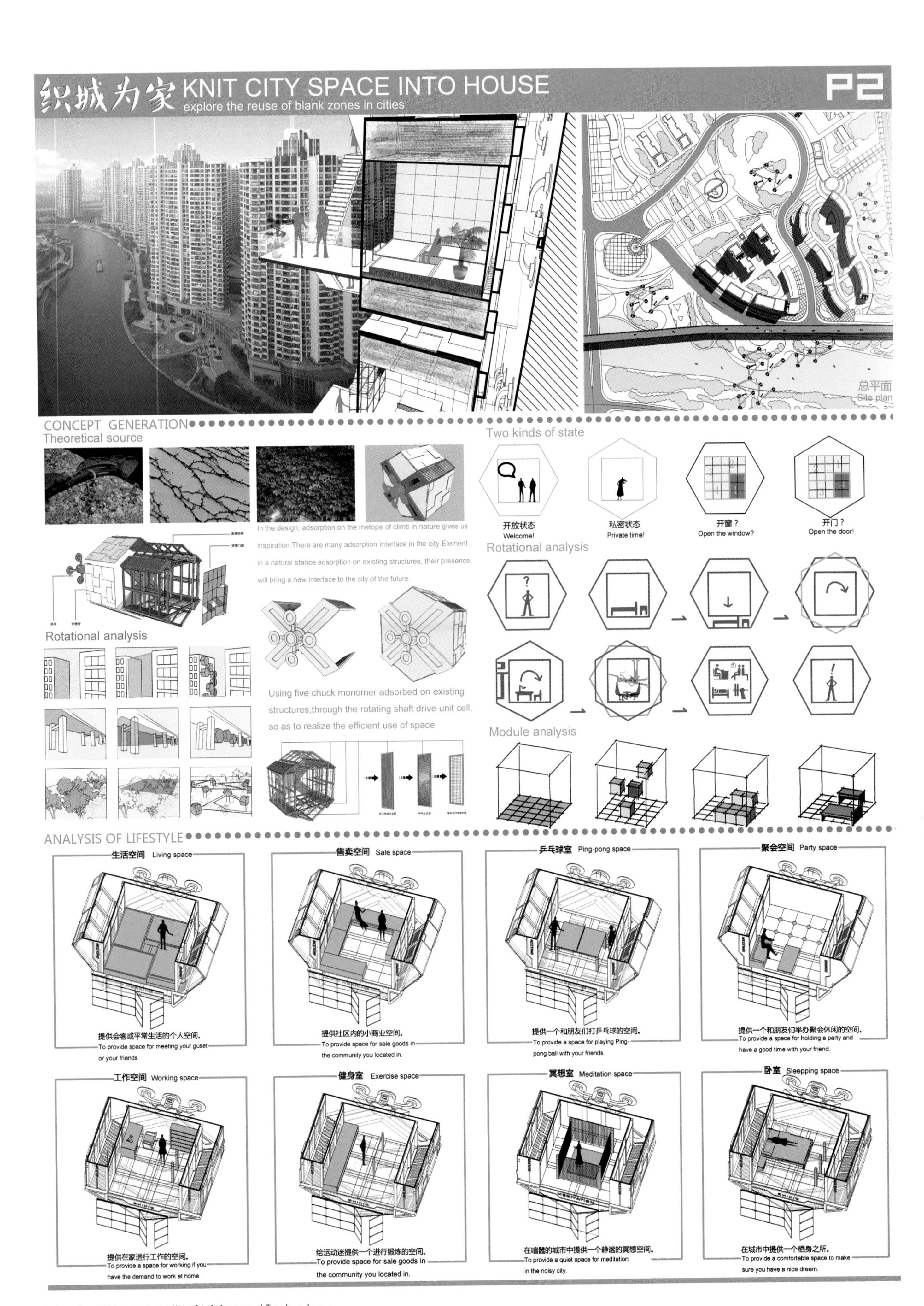

University: China University of Mining and Technology
Designer: Zhang Xueyou, Yang Ruidong, Cao Xiao, Ding Wenmin
Tutor: Sun Liang, Li Ming, Chen Huifang
Course Name: UIA-HYP cup 2014 International student competition in architecture design
Finished Time: Aug. 2014

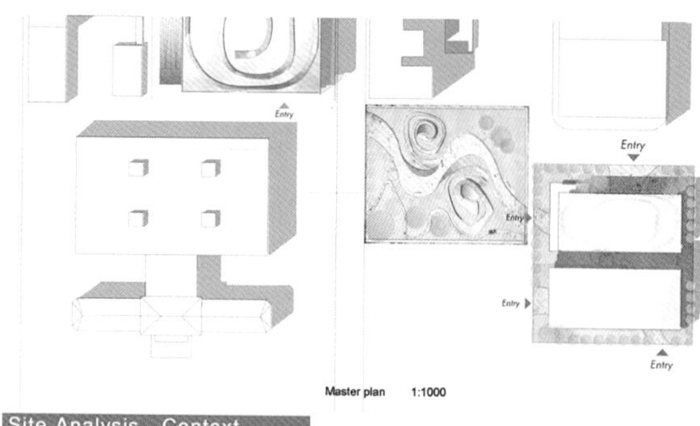

Site Analysis - Context

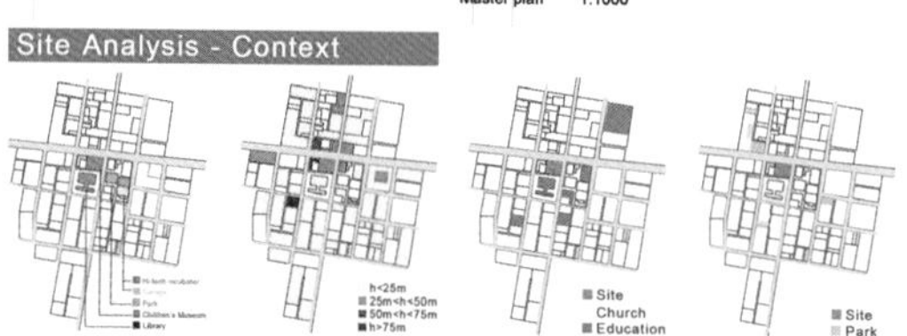

GET INVOLVED!

Louisville Children's Museum Design

Inder to revitalize the downtown edge, we present the slogan of GET INVOLVED to perform our design of the Children's Museum. From several aspects did we implement our idea such as a helical interactive wall rolling the children in, a green layer applying sustainability technology between the children's space inside and the noisy context, a number of room on the layer to get children grow plants devoting themselves to constructing the museum by hand and growing up together with museum. In that case could the museum GET children INVOLVED and as time goes by, all of the people were involving themselves in the museum making it a spirit bond of the Louisville city and the common memory of the generation.

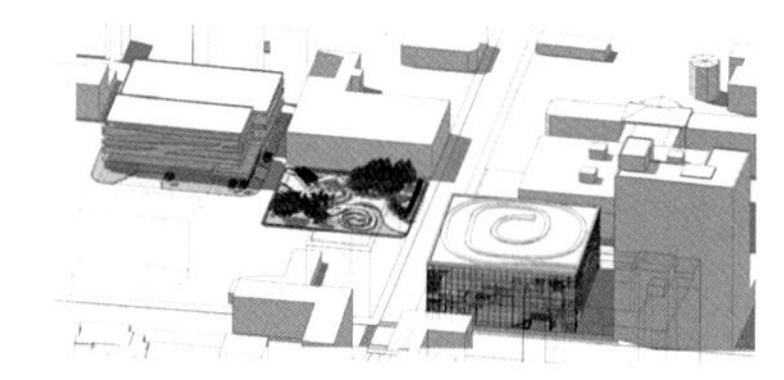

Site Analysis - Traffic

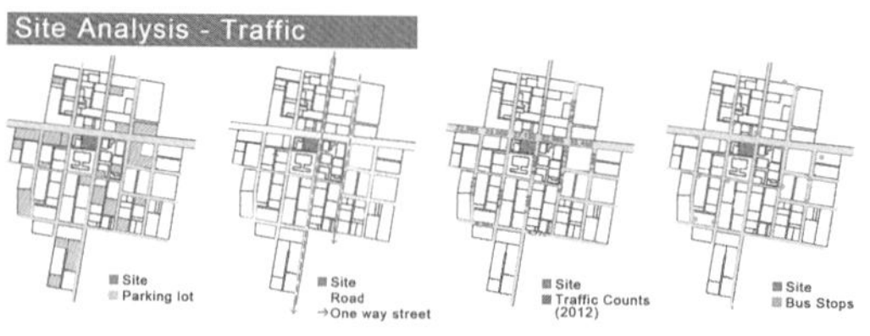

The Interactive Wall

INTERACTIVE WALL

Step 1

Firstly, We designed a INTERACTIVE WALL where kids can play in it and their behavior will form a series of marvelous scenes in the museum.

Step 2

upper floor　bottom floor　top floor

Then we cut it to a certain extent to make room for some functional space. It also make the lower floors more open to the exterior city.

Step 3

traffic　first floor　second floor　third floor　upper layer with light

In accordance with the programm, we give the space into the interactive wall a kind of classification so that it will be more suitable and attractive for kids.

Step 4

The last step is to make the form of the interactive wall more flowing and graceful. We have further studied on the foundamental topic of *What kind of space is more suitable for children*. Then we keep changing our space until it looked quite children-style. Finally we gained the ultimate form.

ROLL UP!

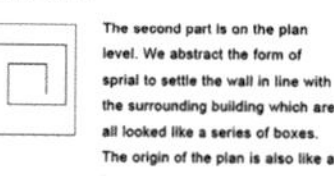

The second part is on the plan level. We abstract the form of sprial to settle the wall in line with the surrounding building which are all looked like a series of boxes. The origin of the plan is also like a box responding to the context.

But the straight line is too stiff and unfriendly to children, so we add some children's elements to it just like the curve and helix.

Then we mixed up the ranctangular outline and the spiral elements and finally formed the structure. It also looked like a drawing made by a kid.

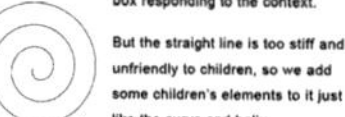

The last step is to form the space on the basis of the abstract original diagram. At the same time we should make the section varied, so we designed several axis and give the curved surface regular wave.

GET INVOLVED!

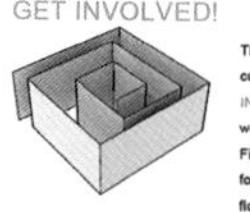

The last part is to ROLL UP the curved surface to GET children INVOLVED. After decided the plan, we are working on a space level. Firstly the straight surface is too stiff for children. We need to make it flowing.

On this step we cut the wall initially and draw the plan like a helix. Then the space looked better, but this kind of structure perform too stereotypical and can not be harmonious with the context.

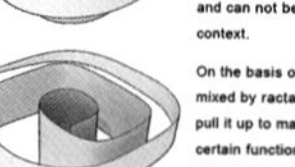

On the basis of a certain curve mixed by ractangular and spiral, we pull it up to make room for the certain functions. But the sections of this form is also a little stiff.

To make the sections of the spatial interactive wall varied, colorful and attractive enough to motivate abundant behaviour, We make the surface flowing and adjust the scale of the space to a children's degree. Eventually achieve the aim to GET children INVOLVED.

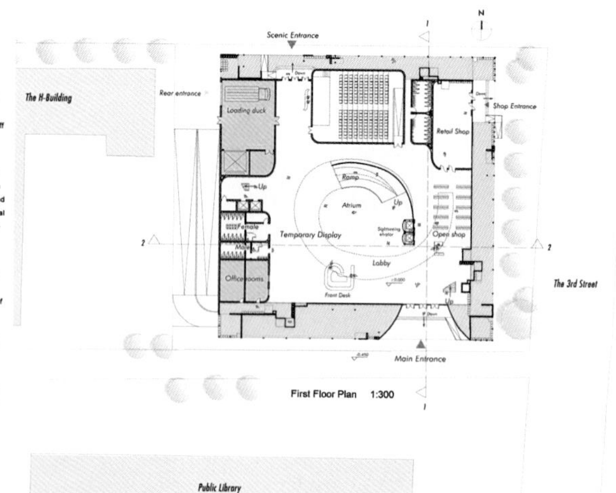

The Green Layer

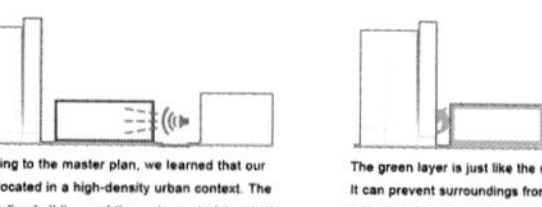

According to the master plan, we learned that our site is located in a high-density urban context. The surrounding building and the main road of the city is just situated closely. Designing a children's space, we choose to design a GREEN LAYER to take step to measure the contradiction.

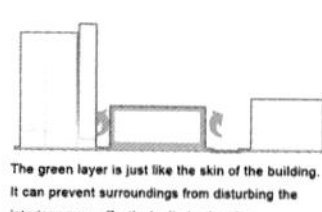

The green layer is just like the skin of the building. It can prevent surroundings from disturbing the interior space effectively. It also has its advantages over sustainability technology and carbon neutral structure.

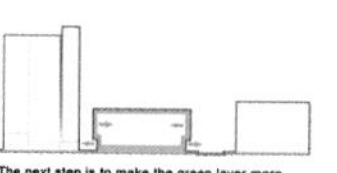

The next step is to make the green layer more harmonious with the context. We choose to let the ground floor open to the city, it also make sense to make cozy room for the homeless children. And the upper floor perform close to interior space considering the purity of the Children's Museum.

We take measures to hammer at developing sustainability technology to make it attain the LEED-certified quality. A number of green features will be used, including:

- A Green Wall: The Museum's south facing facade features a cable and tray system to support a framework for a living, green wall. The Green layer will shade the building from the bright, hot southern sun, helping to cool the building and reduce energy demands.
- A Green Roof: A green roof system will provide for sedums and small scale planting, absorb rainwater (reducing storm run off), improve thermal insulation, and reduce the "heat island" effect.
- Recycled Materials: The new Museum will use recycled structural steel and some recycled brick from demolished buildings.
- A Wind Turbine: The building's wind turbine, a symbol of NCM's commitment to sustainable and renewable energy, will be a source of power.

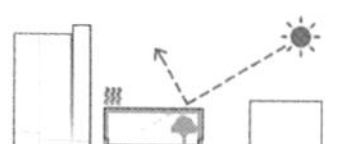

One of the most important virtue of the green layer is that it can prevent sun from shining directly which would injure the eyes of children inside severely. In that case the temperature inward would be quite stabilized which is benefit the carbon neutral structure.

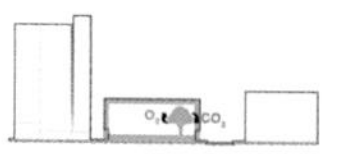

In this layer we encourage children to plant their own plants and we regard this action as a kind of contribution tu construct the museum by children themselves. Accompanying with the photosynthesis of the green plants, the museum would gain a better condition of the air circulation and more oxygen to the children.

The other function of the green layer is that it can make the children's museum a better condition considering the location and context of the building. It can effectively prevent noises and other adverse elements outside from interfering the space inside. Children would enjoy a harmonious and cozy atmosphere on the inside.

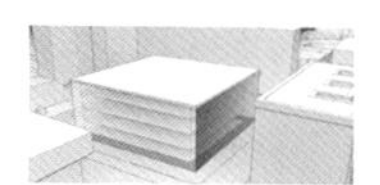

The theme over the green layer to GET INVOLVED children is GROW UP TOGETHER. We assumed that the children is a part of the museum, and they should be involved in contribution and construction to the building. We want to establish the one to one connection between the museum and the children.

We choose the structure of helix to build this layer on account of making it a kind of implication of the interactive wall inside. Children can choose the plants they like to fill in the gap between the curtain wall and the floor. We want get children involved via this running mode.

As time goes by, more and more green plants come out to be fostered in the layer making the facade more vaired and attractive. That's also connected with the city. Picture this scene. All of the People living in the city went to the museum in their childhood. In that case the museum become a spirit bond of the city's memory.

作品名称：被涉及
Get Involved

院校名：青岛理工大学
设计人：张跃，杨舒婷，孙嘉伦，刘俊男
指导教师：郝赤彪，解旭东，程然

课程名称：2014AIA 路易维尔儿童博物馆建筑设计竞赛
作业完成日期：2014 年 02 月

GET INVOLVED!

Louisville Children's Museum Design

Inder to revitalize the downtown edge, we present the slogan of GET INVOLVED to perform our design of the Children's Museum. From several aspects did we implement our idea such as a helical interactive wall rolling the children in, a green layer applying sustainability technology between the children's space inside and the noisy context, a number of room on the layer to get children grow plants devoting themselves to constructing the museum by hand and growing up together with museum. In that case could the museum GET children INVOLVED and as time goes by, all of the people were involving themselves in the museum making it a spirit bond of the Louisville city and the common memory of the generation.

02
CODE 220

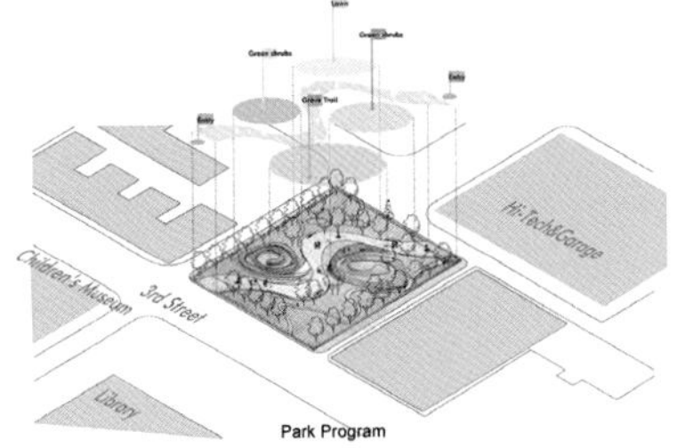

Park Program

Entry
Entry
Entry
Entry

Park Plan

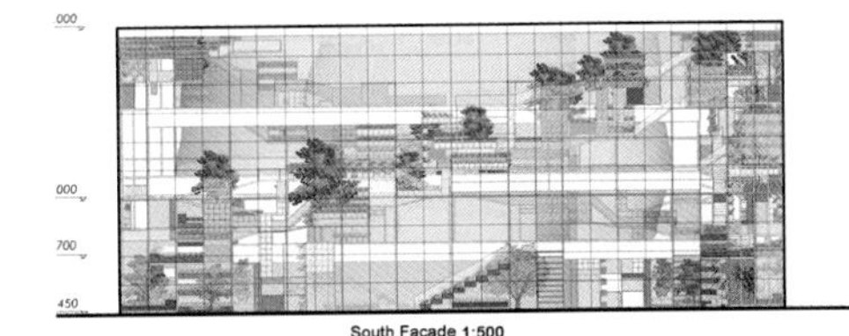
South Facade 1:500

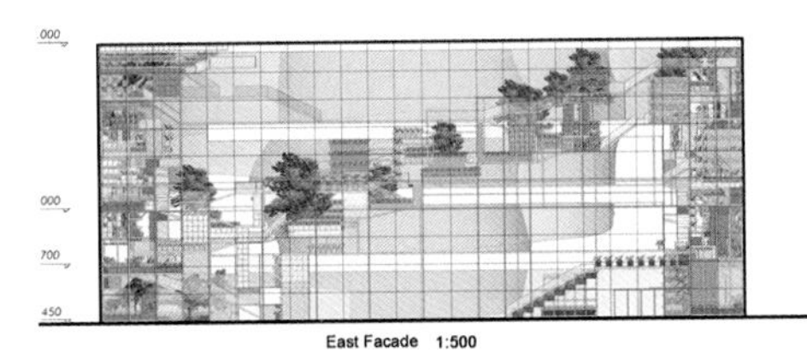
East Facade 1:500

North Facade 1:500

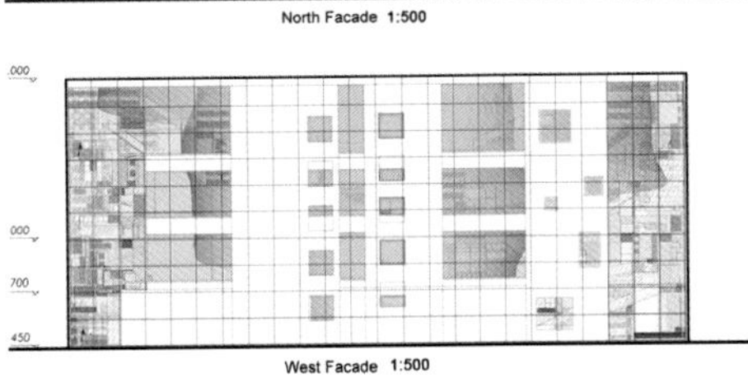
West Facade 1:500

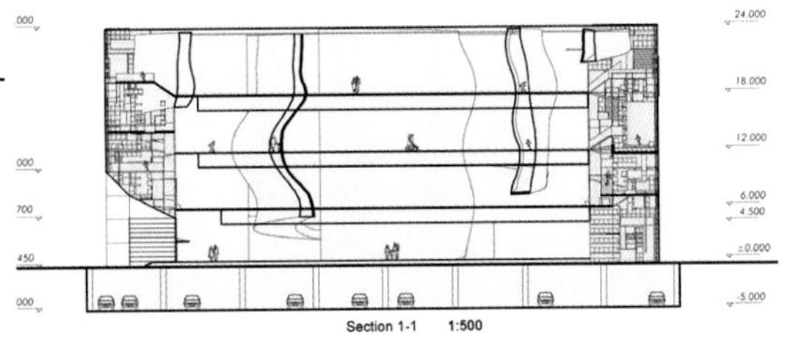

Section 1-1 1:500

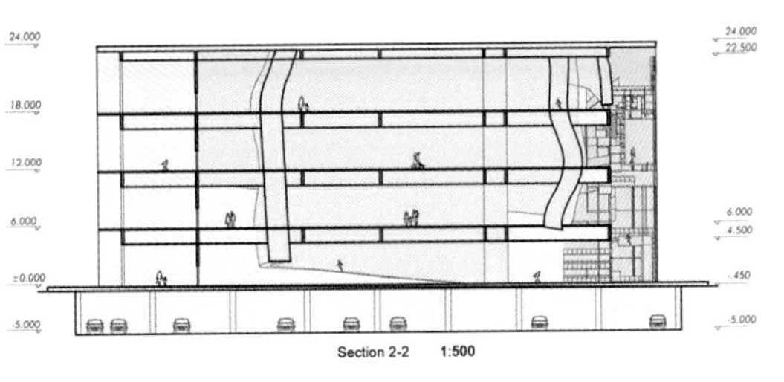

Section 2-2 1:500

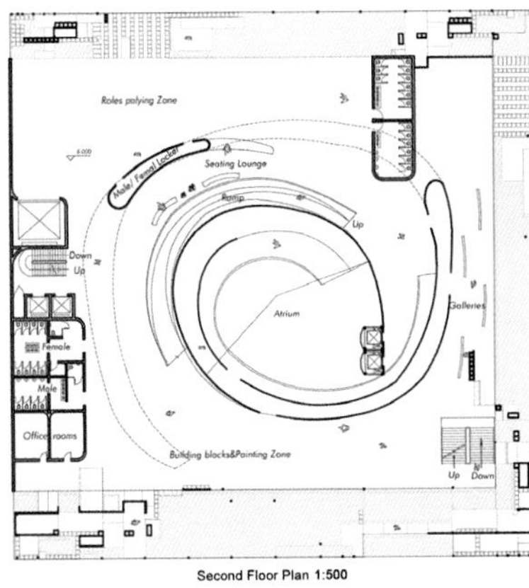

Second Floor Plan 1:500

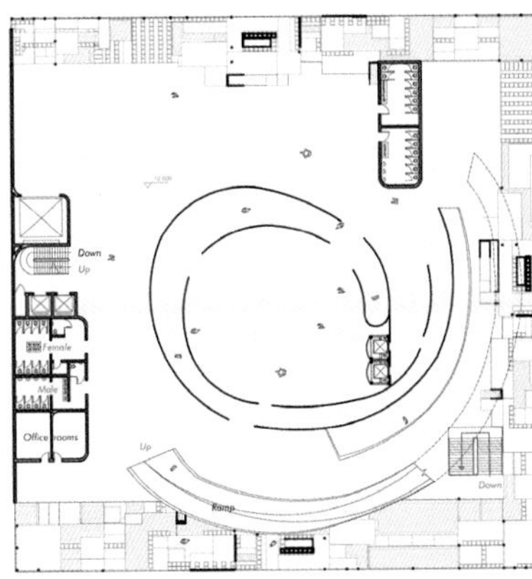

Third Floor Plan 1:500

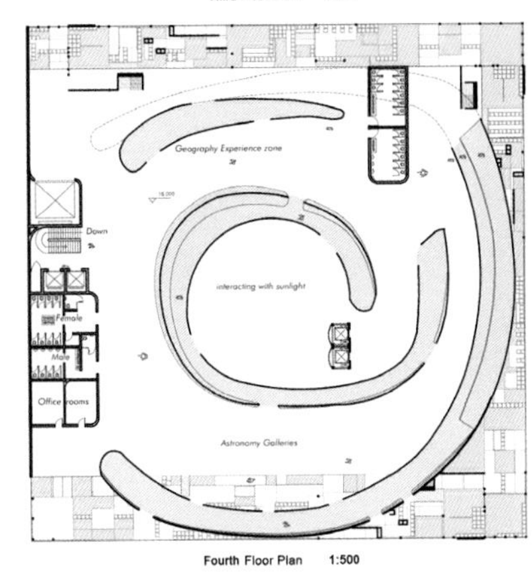

Fourth Floor Plan 1:500

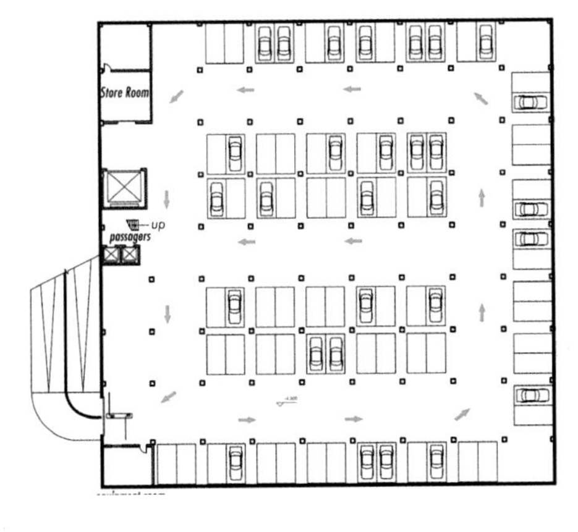

University: Qingdao Technological University
Designer: Zhang Yue, Yang Shuting, Sun Jialun, Liu Junnan
Tutor: Hao Chibiao, Xie Xudong, Cheng Ran
Course Name: AIA Children's Museum—The Regeneraiton of Edge Louisville
Finished Time: Feb. 2014

作品名称：救赎——后工业废弃地的垂直农场转型
Salvation: The Transform from Dersertd Land of the Post Industrial Vertical Farm

院校名：中国矿业大学
设计人：赵倩，丁少华，黄佳喻，李文亮，骆小庆，魏杰，周玉佳，陆家豪
指导教师：孙良，姚刚

课程名称：城市立体农场国际大学生建筑设计竞赛
作业完成日期：2013 年 12 月
对外交流对象：博德西奥（BDCL）国际建筑设计有限公司

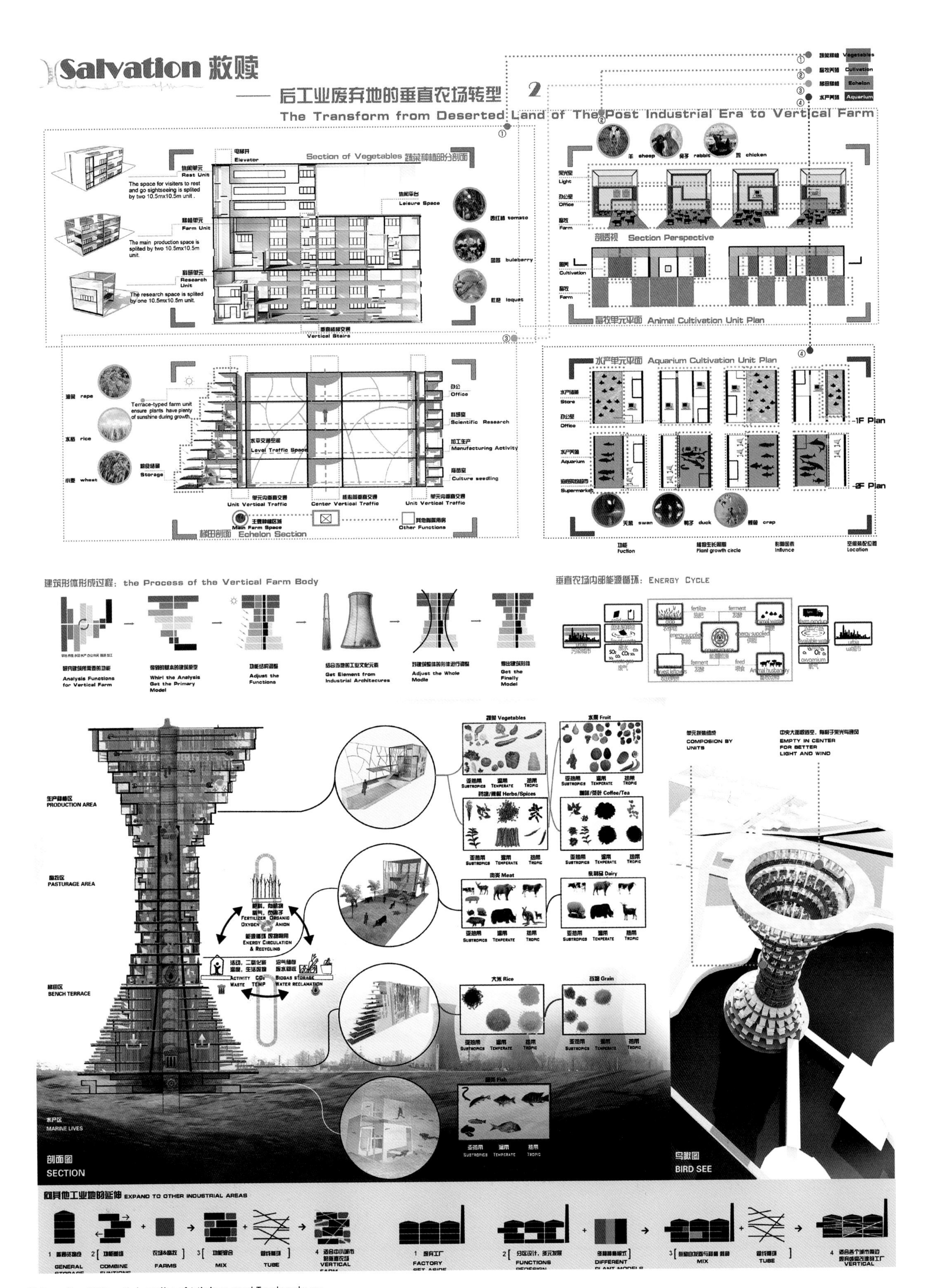

University: China University of Mining and Technology

Designer: Zhao Qian, Ding Shaohua, Huang Jiayu, Li Wenliang, Luo Xiaoqing, Wei Jie, Zhou Yujia, Lu Jiahao

Tutor: Sun Liang, Yao Gang

Course Name: Vertical Farming International College Architecture Design Competition

Finished Time: Dec. 2013

Exchange Institute: BDCL

Unexpected City——Architecture Plan For Citizen Activity Center/城市市民活动中心设计方案 Taiyuan,China

City Coexistent With Breath——植城净息

设计说明：
与城市建筑林立相对立的，城市空气状况实为破败的废墟。“呼吸”作为最本质的生存元素，是智能城市、数字城市等发展的前提。本设计为一个市民活动中心，选址位于太原万柏林区一片拆迁缓慢的废弃厂房区，空气污染问题亟待解决。本设计出乎意料地再生城市呼吸系统：

- 方案源于破土重生的树苗，以基地遗留树木、厂房和道路为历史根系
- 通过引导市民种植，实现全民参与“共同设计”
- 通过道路压电系统，使人群活动成为电源，刺激植物根系以产生大量负离子
- 通过利用日常废弃物和工厂废钢材，实现随手种植和随手建造

Introduction Of Design:
Contrast with cities' buildings,the status of city air can be called ruins.As the most essencial element of live,breathing is the premise of smart city,digital city and so on.We designed an Activity Center,which is located in a slowly disassemble area.The polluted air needs to be purify right now.Unexpectedly,our design can make the whole city breathe.

- Our design comes from revived slaping and we regard trees,factories and roads in the past as root
- By leading people to plant,we make "design together"come true
- By piezoelectric structure on roads,we transform people's activities to electricity and stimulate roots of plants to produce lots of anion
- By using daily wasted things and steel in factories,we make plant and construct anywhere come true

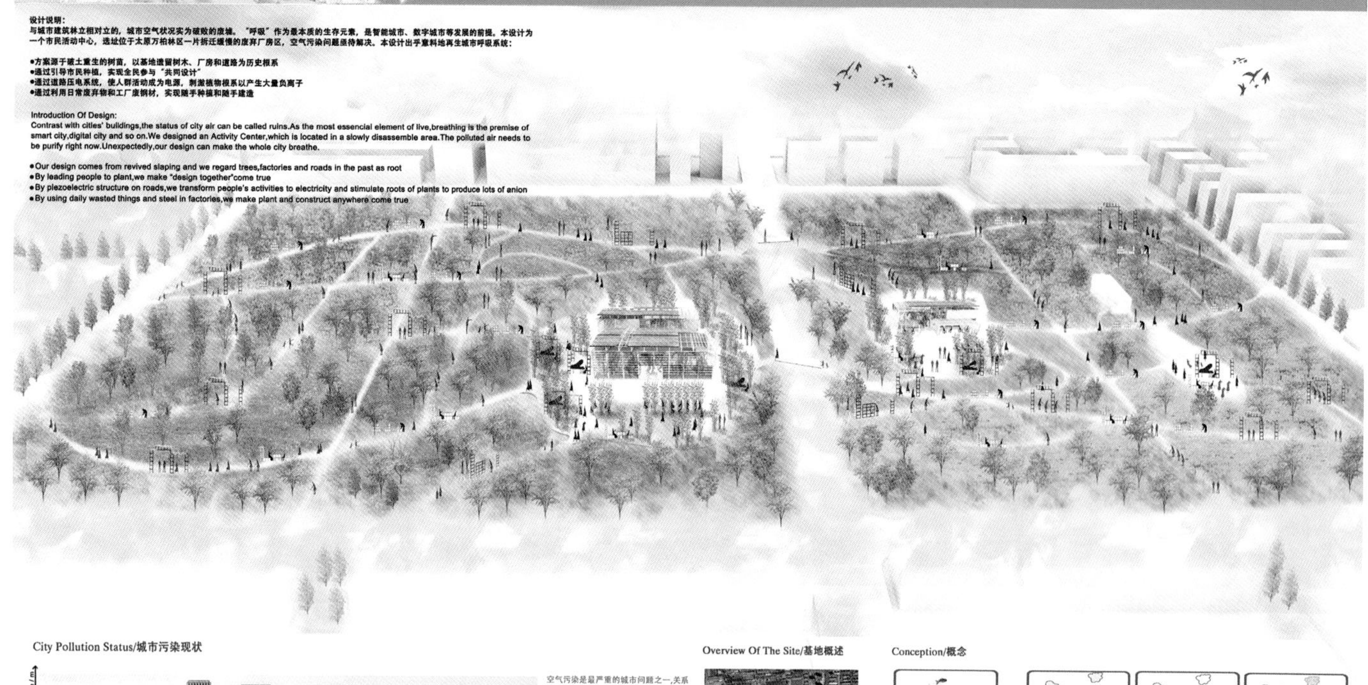

City Pollution Status/城市污染现状

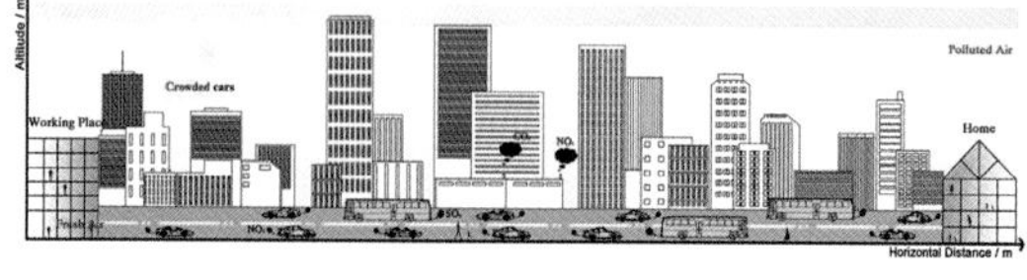

空气污染是最严重的城市问题之一,关系到“呼吸”这一基本生存要素
空气污染关联到一系列城市问题,例如健康问题、社会交流问题、工作效率低下问题等

Air pollution is one of the most ser-ious city problems,which is linked with "breathe",the basic survival el-ement.
Air pollution is linked with a series of city problems such as disease、c-onversation、low efficiency and so on.

Overview Of The Site/基地概述

选址位于太原市万柏林区，基地三面被工厂环绕，污染问题亟待解决

The site is located in Wanb-olin Area,Taiyuan,which is surrounded by factories fro-m three sides.The pollutio-n needs to be handled righ-t now.

Conception/概念

Project's Generation/方案生成

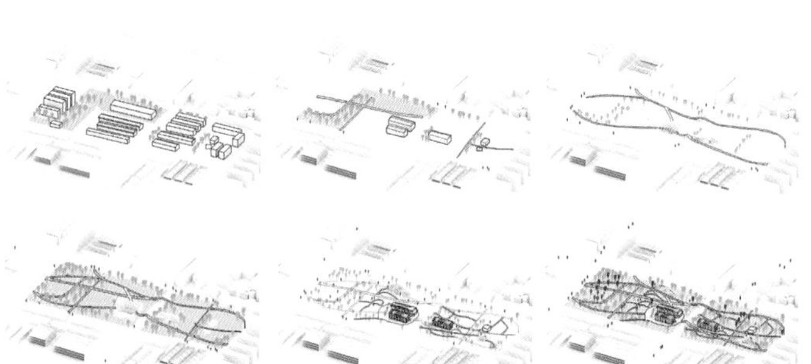

Purification System/净化系统

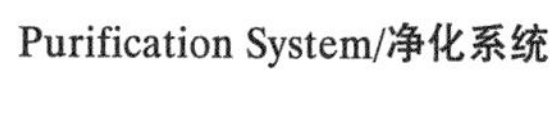
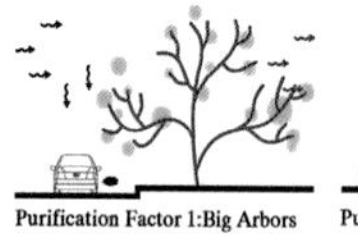
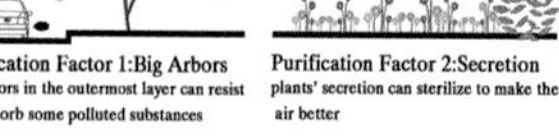

Purification Factor 1:Big Arbors
Big arbors in the outermost layer can resist and adsorb some polluted substances

Purification Factor 2:Secretion
plants' secretion can sterilize to make the air better

Purification Factor 3:Anion
Anion produced by plants can make people keep in good health

Purification Factor 4:Color
Different colors of different plants make different effects to people

Purification Factor 5:Officinal
Some herbs' smell can make people relaxed and happy or even contribute to health

Purification Factor 6:Smell
Different fragrant can make the air more suitable of people

Purification Factor 7:Strongest
The combinations of different plants makes each unit have the strongest purification

Purification Factor 8:Decompress
Plants here make the area a good place for people to decompress

Plan Elements/平面元素

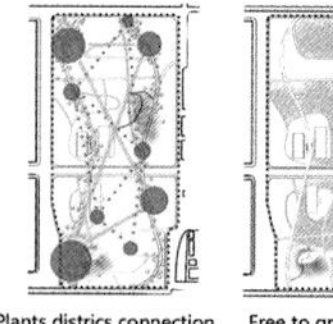
Plants districs connection

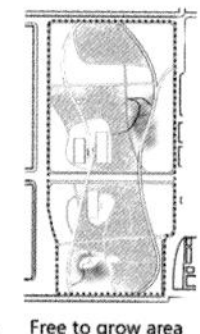
Free to grow area

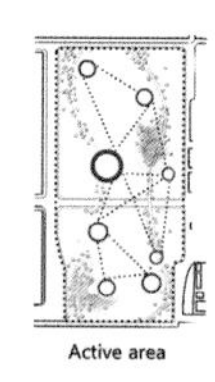
Active area

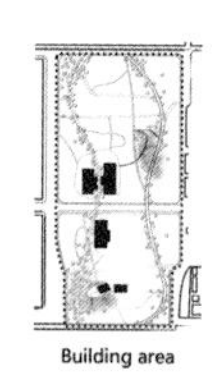
Building area

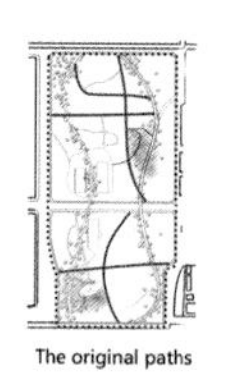
The original paths

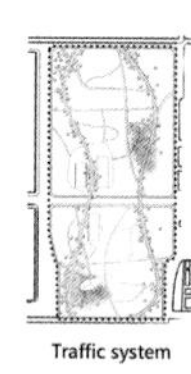
Traffic system

Plant Allocation Principle/植物种植原则

Plant ecological habits & Diversity

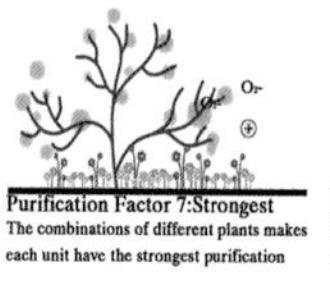
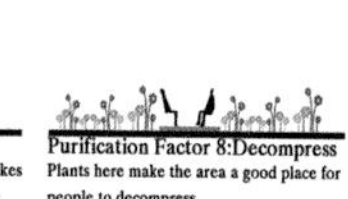
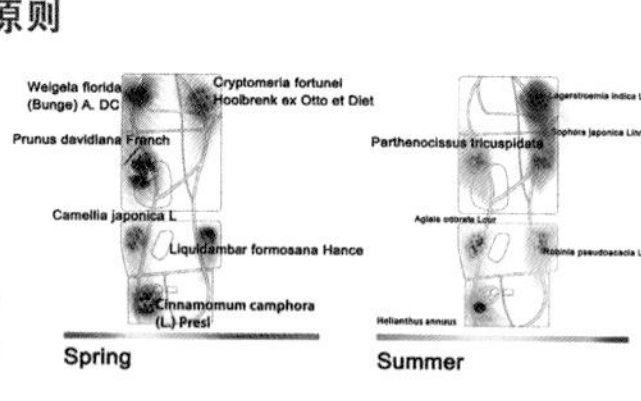

Seasonal aspect & Color

Allocation Modus

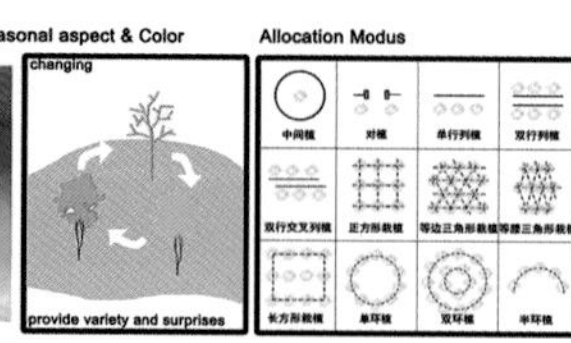

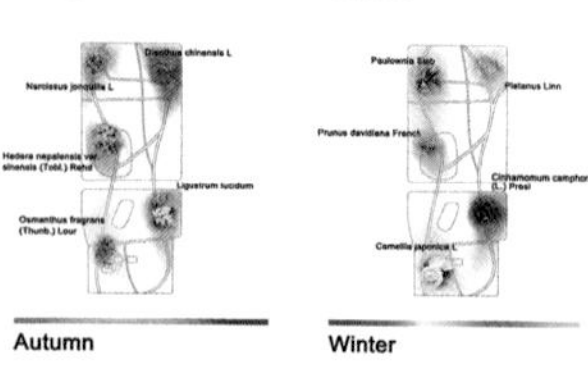

People's Activities/人的活动

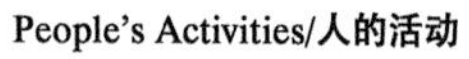

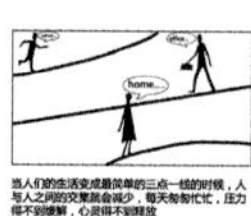

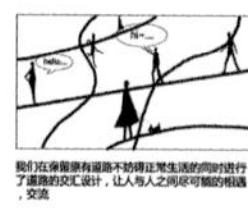

South Evalation/南立面图

作品名称：植城静息
City Coexistent with Breath

院校名：青岛理工大学
设计人：赵云鹤，刘国伟，林汝佳，黄静雯，李陶
指导教师：郝赤彪，王少飞，程然，石新羽

课程名称：2014UIA 霍普杯国际建筑设计竞赛
作业完成日期：2014 年 08 月

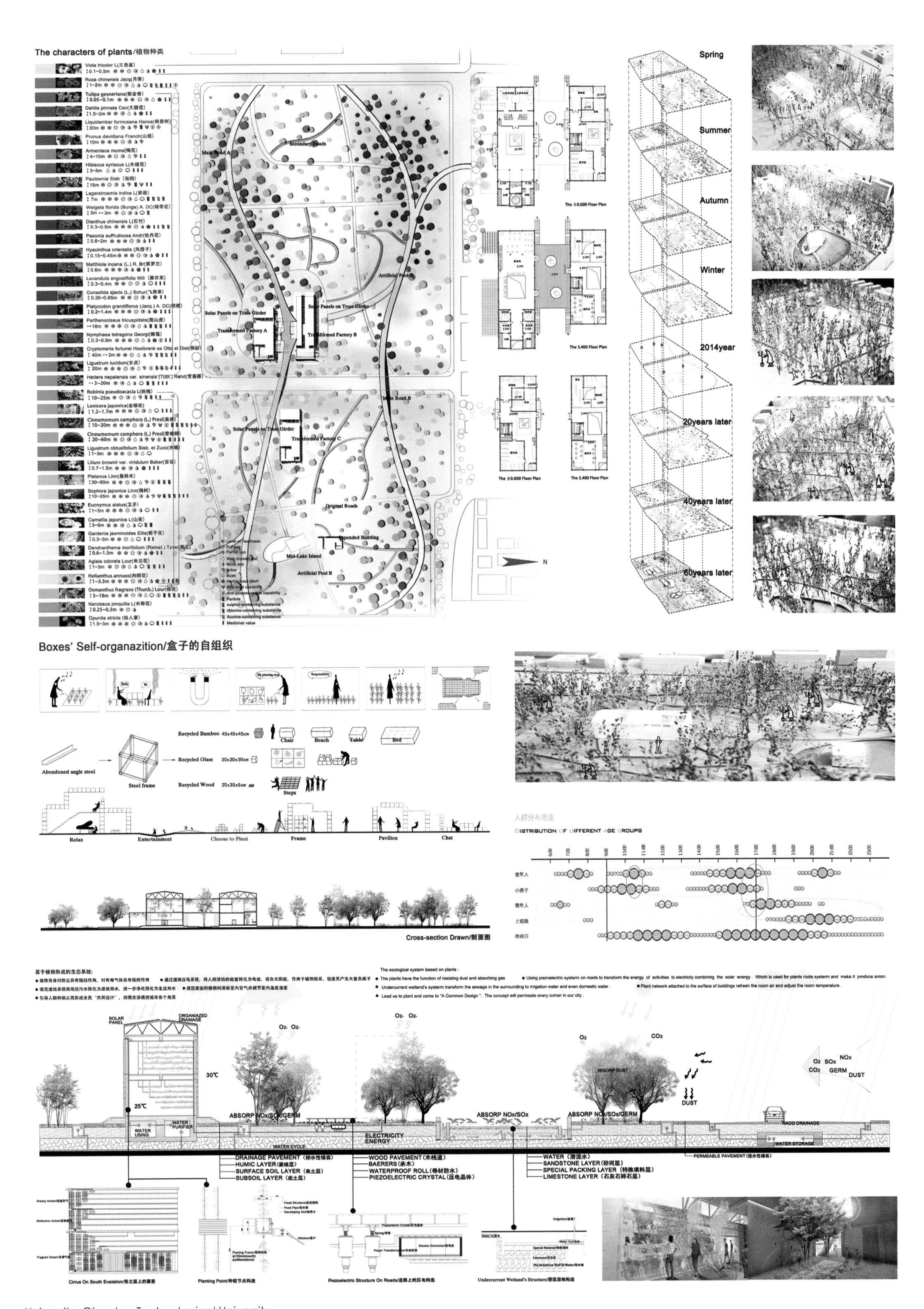

University: Qingdao Technological University

Designer: Zhao Yunhe, Liu Guowei, Lin Rujia, Huang Jingwen, Li Tao

Tutor: Hao Chibiao, Wang Shaofei, Cheng Ran, Shi Xinyu

Course Name: UIA-HYP cup 2014 International student competition in architecture design

Finished Time: Aug. 2014

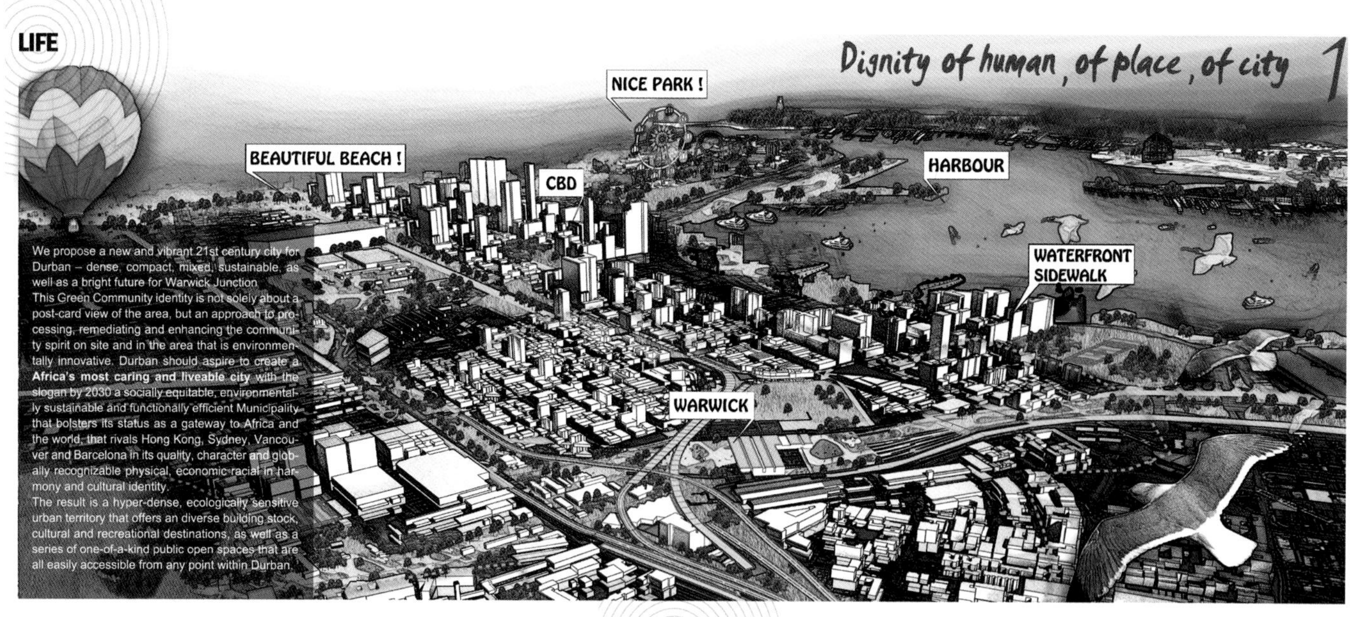

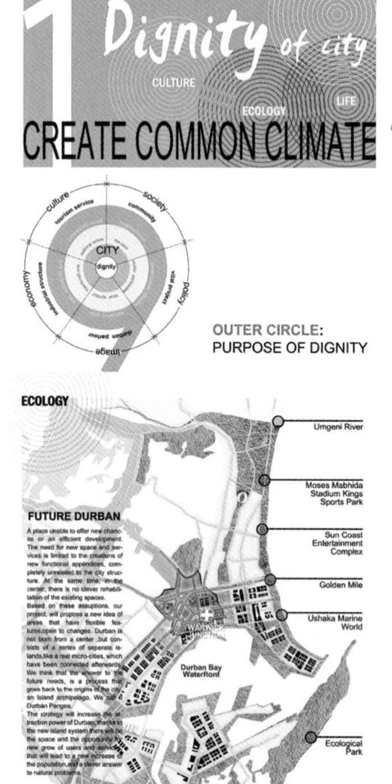

THE REGIONAL SITUATION

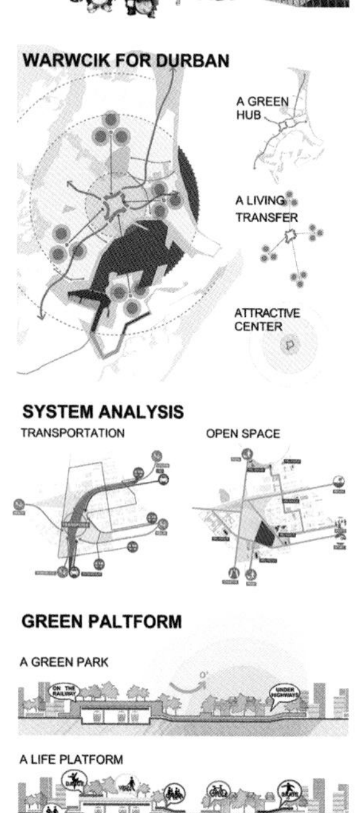

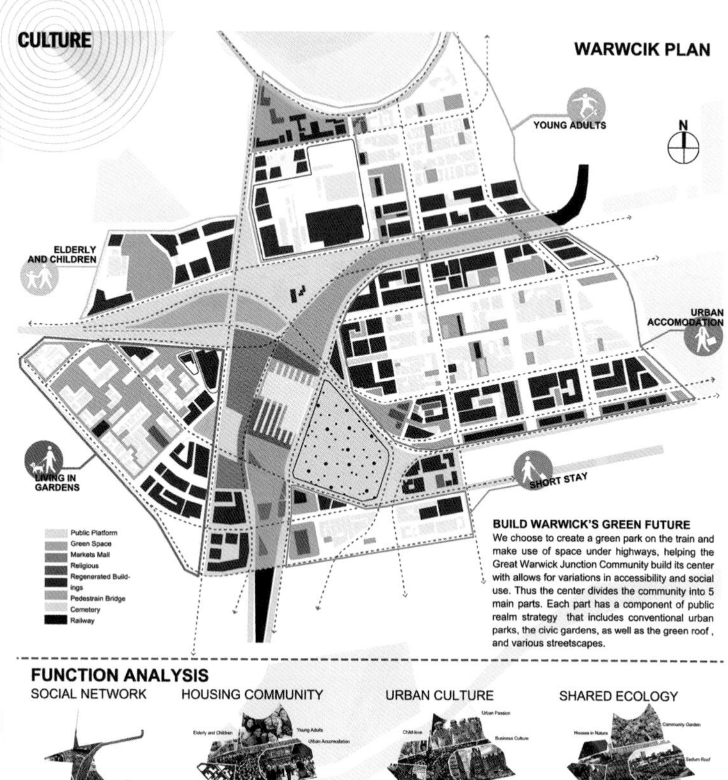

FUNCTION ANALYSIS

SOCIAL NETWORK | HOUSING COMMUNITY | URBAN CULTURE | SHARED ECOLOGY

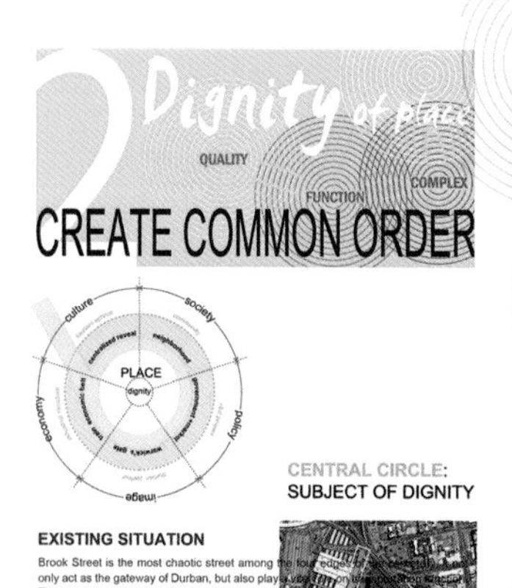

作品名称：城市尊严
Dignity of Human

院校名：西安建筑科技大学
设计人：周正，卢肇松，张士骁，高元，古悦，鞠曦
指导教师：李昊，周志菲
课程名称：第 25 届 UIA（国际建筑师协会）世界大学生建筑设计竞赛
作业完成日期：2014 年 03 月

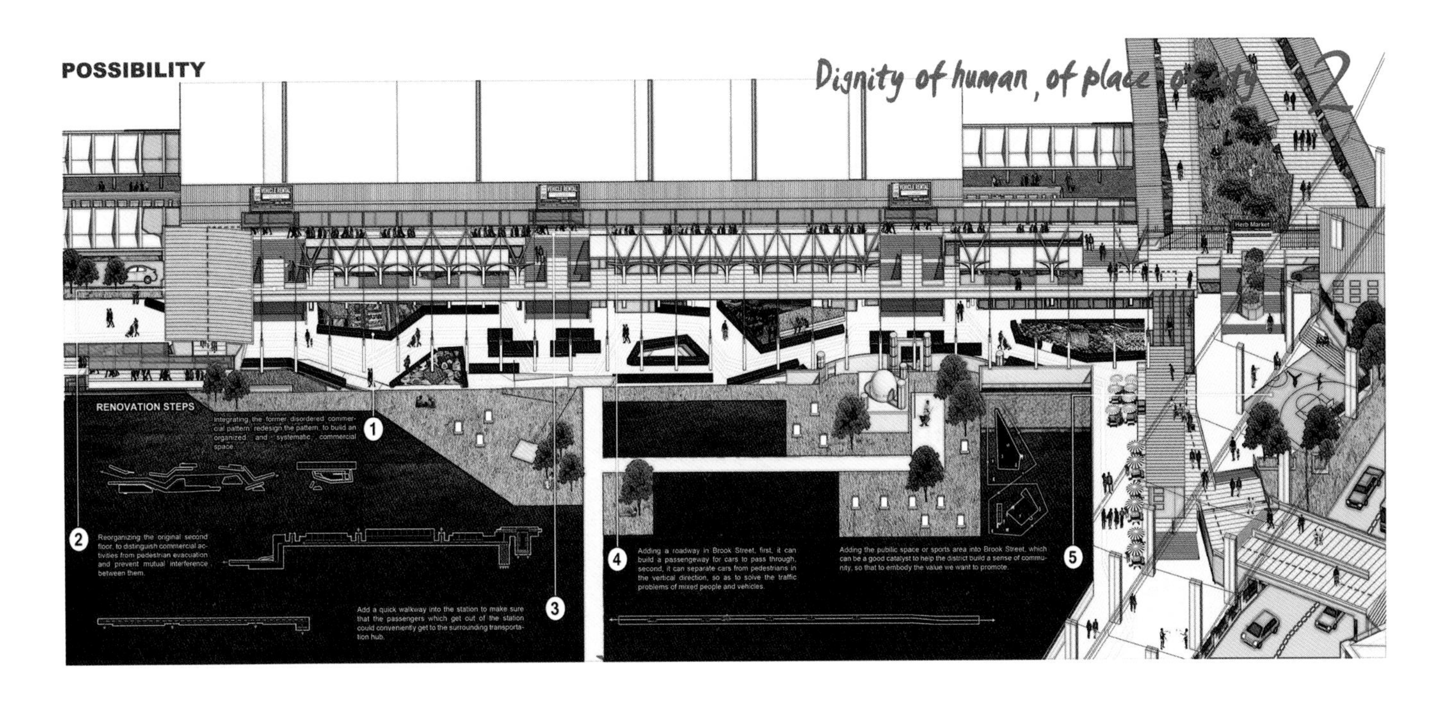

University: Xi'an University of Architecture and Technology
Designer: Zhou Zheng, Lu Zhaosong, Zhang Shixiao, Gao Yuan, Gu Yue, Ju Xi
Tutor: Li Hao, Zhou Zhifei
Course Name: XAUAT - 2014 Durban UIA International Competition
Finished Time: Mar. 2014

2015 年中国建筑院校境外交流优秀作业名单

序号	院校及院系全名	对外交流对象	设计题目	课程名称	学生姓名	第一完成学生所在年级	指导教师姓名	备注
1	北方工业大学建筑与艺术学院	美国南方州立理工大学	延伸——Extending	北方工业大学—美国南方州立理工大学交换生课程设计	朱柳慧	三年级	马欣、 Pegah Zamani	
2	北方工业大学建筑与艺术学院	美国南方州立理工大学	双宅设计	北方工业大学—美国南方州立理工大学交换生课程设计	刘靖雯	三年级	马欣、Saleh Uddin	
3	北方工业大学建筑与艺术学院	美国得克萨斯农工大学	传统小镇的改造——在非奥伦迪诺堡的实践	北方工业大学意大利教学基地春季联合设计课程	李迎、辛奕佳、曹梦溪、金潇	三年级	钱毅	
4	北方工业大学建筑与艺术学院	美国得克萨斯农工大学	传统街区的复兴——在非奥伦迪诺堡的实践	北方工业大学意大利教学基地春季联合设计课程	杜昀瞳、赵翊彤、王瑶、王茜	三年级	钱毅	
5	北方工业大学建筑与艺术学院	美国加利福尼亚州立理工大学波莫纳分校	依存・分离——北京历史文化保护区什刹海地段城市设计	北方工业大学—加利福尼亚州立理工大学波莫纳分校联合设计课程教学	刘靖雯、翟羽峰、许葳葳、迪欧、奥兰德、菲义、周欣蕾、鲁门、Lester Gonzales、May Lee、Sandra Lee、Hnin Nyein、Lynae	三年级	Irma Ramirez、Gwen Urey、安平、高龙、张勃	
6	北方工业大学建筑与艺术学院	美国加利福尼亚州立理工大学波莫纳分校	联结——北京历史文化保护区什刹海地段城市设计	北方工业大学—加利福尼亚州立理工大学波莫纳分校联合设计课程教学	王潇潇、马南晨、张译丹、鲁卡、阿森、萨法利、迪欧、Brian Carr、Frank Chen、Anna Aldrin、Mayrelis Perez、Jose Rojas、Karen Tang	三年级	Irma Ramirez、Gwen Urey、安平、高龙、张勃	
7	北方工业大学建筑与艺术学院	新西兰 Unitec 理工学院	Eco Bubble and community	景观规划设计	成超男、胡凯富、张跃、朱昱泽	三年级	傅凡、杨鑫、Matthew Bradbury	
8	北方工业大学建筑与艺术学院	韩国釜庆大学建筑系	缝合—釜山车站连接体设计	釜山国际建筑设计工作坊	马赛、夏颖、杨东	四年级	吴正旺	
9	北京工业大学建筑与城市规划学院	美国辛辛那提大学 DAAP 学院	Burnet woods 的再思考——提高公共健康	Urban Research	Ashley Combs、Bahareh Rezaee、Stacey Todd、吴奕楠	本科四年级	Wang.X.O（美）、胡斌、廖含文、惠晓曦	
10	北京工业大学建筑与城市规划学院	韩国全南大学建筑学部、台湾大学建筑与城乡研究所	A CommuCity: Sustainable Neighbourhood and Urban Linear Park	IFoU Summer School 2014 Gwangju: Can Big Culture Save the City？ — How to Re-Shape Urban Center in Response to Colossal Cutural Institute	尚晓迪、王璞、林郁文、吕行、Oh Shang Wun、Kim Sang Won、Choi In Seo、Kim Jung Min、Lim Ha Seon、Sohn Suzy	本科五年级	惠晓曦、Park Jung Eun、Youn Dae Han、Lee Hyo Won、Yoo Uoo Sang	
11	北京工业大学建筑与城市规划学院	法国图卢兹国家建筑高等学院	传统建筑的绿色新生	研究生国际交流实践课程	刘孟涵、彭琰、VIEL Kevin、倪振杰	研究生二年级	胡斌、Alxender・Monique（法）、廖含文、惠晓曦	

序号	院校及院系全名	对外交流对象	设计题目	课程名称	学生姓名	第一完成学生所在年级	指导教师姓名	备注
12	北京工业大学建筑与城市规划学院		北京市大兴区低碳农村住宅设计	研究生国际交流实践课程	赵超、郑璇、王丽培、Paul Perkins	研究生二年级	廖含文、胡斌、惠晓曦	
13	北京工业大学建筑与城市规划学院	韩国全南大学建筑学部、台湾大学建筑与城乡研究所	Cultural Diffusion and Reception, and Transitional Urban Space	IFoU Summer School 2014 Gwangju: Can Big Culture Save the City ? — How to Re-Shape Urban Center in Response to Colossal Cutural Institute	姬煜、伍清如、Hsiao Yu-chi、Chung Chieh Lin、Han Su Bin、Kang Su Hyun、Mun Jeong Yun、Kim Yu Jin、Yoo Sun Hee	研究生二年级	张帆、Lee Min Seok、Hendrik Tieben、Oh Se Gyu、Yoo Uoo Sang	
14	北京建筑大学建筑与城市规划学院	美国密歇根大学建筑学院	律动河流——郑州滨水文化休闲区城市设计	北京建筑大学—密歇根大学—思朴国际设计工作营	谷筝、刘玲	本科 5 年级	丁奇、王佐	本科优秀作业特别奖
15	北京建筑大学建筑与城市规划学院	美国南加利福尼亚大学建筑学院	Dashilan Urban Design	北京建筑大学—南加利福尼亚大学联合设计工作营	于英龙、李嘉文、Henry Liu、Said Taghmaoui	研究生 13 级	马英、Yo-ichiro Hakomori、汤羽扬、齐莹、欧阳文、王佐、俞天琦、晁军	
16	北京建筑大学建筑与城市规划学院	美国南加利福尼亚大学建筑学院	Redifining Dashilan：北京市前门大街大栅栏地区历史街区保护与更新设计	北京建筑大学—南加利福尼亚大学联合设计工作营	赵伟、吴少敏、Kenny Chao、Andrea Mendoza	研究生 14 级	马英、Yo-ichiro Hakomori、汤羽扬、齐莹、欧阳文、王佐、俞天琦、晁军	
17	北京建筑大学建筑与城市规划学院	美国密歇根大学建筑学院	文化经验，在水之洲	北京建筑大学—密歇根大学—思朴国际设计工作营	徐松月、王宏侠、张振伟、李雯	研究生 13 级	丁奇、王佐	
18	大连理工大学建筑与艺术学院	成功大学建筑系	不期而遇	Vertical Campus	王博伦	三年级	陈玉霖、吴亮	
19	大连理工大学建筑与艺术学院	成功大学建筑系	Street & Lane	Hybrid Design	王博伦、王雨杨	三年级	赵建铭、吴亮	本科优秀作业特别奖
20	大连理工大学建筑与艺术学院	德国达姆施塔特工业大学	德意志铁路员工学习中心	A wie Aalto	王研	四年级	Prof.Jessen	
21	大连理工大学建筑与艺术学院	东海大学建筑系	东海校园新周边	东海大学三年级设计课题	吴宜珉、林子群	三年级	郭文亮	
22	大连理工大学建筑与艺术学院	德国达姆施塔特工业大学	达姆斯塔特新火车北站	四年级设计	伏音	四年级	Frau.Anna.Jessen	
23	大连理工大学建筑与艺术学院	台湾科技大学建筑学院	relavation-the dragon house	四年级设计	李震寰	四年级	李冰、于辉	
24	东南大学建筑学院建筑系	2014 Vectorworks 学生设计奖	缝合——黟县际村村落改造与建筑设计		邵星宇	本科五年级	夏兵	本科优秀作业特别奖
25	东南大学建筑学院建筑系	日本京都大学、三菱地所	南京地铁马群站城市设计	东南大学设计课程教学	管睿、商琪然	本科四年级	唐芃、沈旸、宗本顺三、惠良隆二	本科优秀作业特别奖
26	东南大学建筑学院建筑系	THE SCHOOL OF ARCHITECTURE LA SALLE, RAMON LLULL UNIVERSITY	传统街区的生长	毕业设计	崔百合	本科五年级	龚恺、鲍莉、刘捷、Josep Ferrando、Jaime Font	本科优秀作业特别奖

续表

序号	院校及院系全名	对外交流对象	设计题目	课程名称	学生姓名	第一完成学生所在年级	指导教师姓名	备注
27	东南大学建筑学院建筑系	日本东京工业大学建筑系	东南大学文化书院	空间构成——住宅	李昂、王君美	本科四年级	葛明、奥山信一	
28	东南大学建筑学院建筑系	瑞士伯尔尼应用技术大学	0 energy mountain hostel	BIM based Architecture design	温子申	本科四年级	朱渊、Odilo Schoch	
29	东南大学建筑学院建筑系	加拿大不列颠哥伦比亚大学建筑学院、加拿大木业	木“扇”亭——当代“传统木榫卯”设计	东南大学—加拿大不列颠哥伦比亚大学—加拿大木业联合设计	杨梦溪、王佳玲、Sara Maria	本科四年级	韩晓峰、朱雷、Anna Lisa Meyboom、Blair Satterfield	本科优秀作业特别奖
30	东南大学建筑学院建筑系	日本东京工业大学	荔枝湾南——广东文化艺术保育和创新基地设计	建筑设计（中日四校联合工作坊）	金海波、洪梦扬、许铎、南泽智规	研究生一年级	唐芃、葛明、王伯伟、孙一民、王方戟、冯江、徐好好、奥山信一	研究生优秀作业特别奖
31	东南大学建筑学院建筑系	澳大利亚新南威尔士大学建筑环境学院	漕运博物馆及丰碑义仓地区城市设计	新南威尔士大学—东南大学—苏州科技学院—清华大学联合跨文化设计课	刘琦、薄尧、陈三才、Ellen Ward	研究生一年级	龚恺、许亦农、马骏华、张愚、夏健、胡莹、王路	
32	东南大学建筑学院建筑系	奥地利维也纳理工大学建筑学院	行与市	东南大学—同济大学—深圳大学—维也纳理工联合设计	时楠、张倩倩、陈凤娇、克里斯托夫	研究生一年级	韩晓峰、葛明、龚恺、莫拉德·亚德力奇	研究生优秀作业特别奖
33	东南大学建筑学院建筑系	美国明尼苏达大学建筑与景观设计学院	镜 / 风吹两成花	6/6 艺术工作室设计——北京大栅栏胡同空间改造	原雯、毛律、王译瑶、Emilie Kopp、孙艳晨、张玥、刘清越、甄明扬、Sean	研究生一年级	韩晓峰、Andrea Johnson、许蓁、苑思楠、Diane Willow	
34	福州大学建筑学院	台湾打开联合团队	嵩口历史文化名镇保护规划	福州大学—台湾打开联合团队联合设计课程教学	杨元传、黄斯、许汉民	本科五年级	邓海、刘国沧、崔育新、陈建东	
35	福州大学建筑学院	意大利威尼斯建筑大学建筑学院	传统建筑的保护与再生——以桂峰村清代茶楼为例	毕业设计	Claudia Stancanelli、柏苏玲	研究生二年级	张鹰	研究生优秀作业特别奖
36	哈尔滨工业大学建筑学院建筑系	美国哈佛大学设计研究生院	逻辑游戏	开放式研究型建筑设计	梅梦月、康可歌	本科四年级	孙澄、姜宏国、刘莹、Ali Malkawi	
37	哈尔滨工业大学建筑学院建筑系	荷兰代尔夫特理工大学建筑学院建筑系	TransferJet 创业中心——基于空间流动模式的办公空间设计	国际联合设计	李彦儒、徐子博、滕东霖	本科三年级	陈旸、邢凯、唐康硕、Giorgio Ponzo	本科优秀作业特别奖
38	哈尔滨工业大学建筑学院建筑系	中原大学设计学院建筑系	街区进化论——基于中坜市中原校区周边街区的演变研究	开放式研究型建筑设计	张睿南、董奕兰、陈聪	本科四年级	董宇、薛名辉、邵郁、张姗姗、杨秋煜、黄承令、黄俊铭	本科优秀作业特别奖
39	哈尔滨工业大学建筑学院建筑系	荷兰代尔夫特大学建筑学院建筑系	空角实心	国际联合设计	徐淼、王宏宇、付豪	本科三年级	陈旸、邢凯、唐康硕、Giorgio Ponzo	本科优秀作业特别奖
40	哈尔滨工业大学建筑学院建筑系	英国建筑联盟学院	Parametric Urbanism	国际联合设计	陈允元、洪烽桓、张正蔚、李栋梁、国建淳	本科三年级	刘莹、姜宏国、Immanuel Koh	
41	哈尔滨工业大学建筑学院建筑系	中原大学设计学院景观系	新陈代谢——从眷村到新城的渐变与更新计划	开放式研究型建筑设计	陈永祥、骆盛韬、尤昊博、王渡、逄佳音	本科四年级	薛名辉、董宇、张姗姗、邵郁、喻肇青、张华荪	本科优秀作业特别奖

续表

序号	院校及院系全名	对外交流对象	设计题目	课程名称	学生姓名	第一完成学生所在年级	指导教师姓名	备注
42	哈尔滨工业大学建筑学院建筑系	无	水天一线——中东铁路百年江桥的新生	UIA 霍普杯 2014 国际大学生建筑设计竞赛	明磊、何璇、武玥、张之洋	本科三年级	孙澄、梁静	
43	合肥工业大学建筑与艺术学院	逢甲大学建筑学系	折居——建筑师事务所及自宅设计	乌托邦与桃花源——建筑师事务所及自宅设计	郭喆	二年级	张登尧	
44	合肥工业大学建筑与艺术学院	英国伦敦大学学院	Post-Waste Cities	英国伦敦大学学院毕业设计	唐洪亚、任国乾、贾宁	研究生二年级	Claudia Pasquero、Marco Poletto	研究生优秀作业特别奖
45	合肥工业大学建筑与艺术学院		Activating-Blanking-Floating Anywhere	UIA 国际建筑竞赛	杨梦曦、孙霞、苏朝晶、陈昱、马聃、冯磊	五年级	李早	
46	合肥工业大学建筑与艺术学院		City Fiber Of Durban	UIA 国际建筑竞赛	高翔、沙成鑫、刘梦柳、封瑞牧、黄宇超、林夏冰、张亚伟	四年级	徐晓燕	
47	合肥工业大学建筑与艺术学院		节点激活——黟县际村村落改造与建筑设计	全球毕业设计大赛	邓慧丽、杨三瑶、汪宇宸	五年级	苏剑鸣、李早、刘阳、任舒雅	
48	合肥工业大学建筑与艺术学院		Floating Web	Re-thinking the Future Award 2014 国际竞赛	汪宇宸、刘梦茜、林鹏	五年级	陈丽华	
49	华南理工大学建筑学院	美国哈佛大学设计学院	Water Experience City	海心沙岛城市设计——国际联合设计课程	宋修教、王琳、李拓欣、李泳妍	2010 级	孙一民、Ruggero Baldasso、苏平	
50	华南理工大学建筑学院	美国哈佛大学设计学院	City Forum	海心沙岛城市设计——国际联合设计课程	俞快、旋凯、郑宇晨、廖绮琳	2010 级	孙一民、Ruggero Baldasso、苏平	
51	华南理工大学建筑学院	美国哈佛大学设计学院	Scenic City	海心沙岛城市设计——国际联合设计课程	招苏恒、叶植滨、刘勉君	2010 级	孙一民、Ruggero Baldasso、苏平	
52	华南理工大学建筑学院	日本东京工业大学建筑学专攻	荔枝湾 ENVELOPE	荔枝湾南地区城市设计——四校联合城市设计课程	陈桂欣、洪梦扬、叶冬青、余润、罔野爱结美	2010 级	孙一民、奥山信一、苏平、钟冠球、徐好好	
53	华南理工大学建筑学院	日本东京工业大学建筑学专攻	Gradational City	荔枝湾南地区城市设计——四校联合城市设计课程	陆诗蕾、郭晓、马俊雄、黄慧婷	2010 级	孙一民、苏平、钟冠球、徐好好	
54	华南理工大学建筑学院	日本东京工业大学建筑学专攻	城市链	荔枝湾南地区城市设计——四校联合城市设计课程	林紫琪、梁俊韬、廖喆璇、李琪、吕一明、小野岛新	2010 级	孙一民、奥山信一、王伯伟、葛明	
55	华南理工大学建筑学院	美国加利福尼亚大学伯克利分校环境设计学院	Jiangmen School of Indistial Design	华南理工大学——加利福尼亚大学伯克利分校城市设计工作坊	蒋梦虞、曾雪娇、俞伊阳、温宇辰、Arijit Sen、Tanvi Maheshwari、Maria Luisa	2010 级（本科）2013 级（研究生）	Peter Bosslmann、孙一民、苏平、徐好好、王璐	
56	华南理工大学建筑学院	美国加利福尼亚大学伯克利分校环境设计学院	The Transformation of Industrial Area	江门水泥厂区城市更新——国际联合设计工作坊	温馨、刘亿瑶、廖喆璇、仇普钊、谷雨蒙、陆诗蕾、Graig Andrew、Swati Sachdeva	2010 级（本科），2013 级（研究生）	Peter Bosslmann、孙一民、苏平、徐好好、王璐	
57	华南理工大学建筑学院	美国加利福尼亚大学伯克利分校环境设计学院	Wetlands and Educational Center	江门水泥厂区城市更新——国际联合设计工作坊	安浩奇、李雨龙、汪丝雨、吴梦笛、曾子扬、张纨素、郑惠婷、David von Stroh、Vanessa Tani	2010 级（本科），2013 级（研究生）	Peter Bosslmann、孙一民、苏平、徐好好、王璐	

续表

序号	院校及院系全名	对外交流对象	设计题目	课程名称	学生姓名	第一完成学生所在年级	指导教师姓名	备注
58	华南理工大学建筑学院	美国加利福尼亚大学伯克利分校环境设计学院	Remediational Plan	江门水泥厂区城市更新——国际联合设计工作坊	董瑶、蔡宁、易凯平、杨雨璇、李拓欣、汤荣强、卓泓铎、陆诗蕾、Rana Haddad、Fabrizio Prati	2010 级（本科），2013 级（研究生）	Peter Bosslmann、孙一民、苏平、徐好好、王璐	
59	华侨大学建筑学院	中原大学建筑系	interface	建筑设计（六）	王睿智	2011 级	陈宇进	
60	华侨大学建筑学院	中原大学建筑系	台中都市会客厅——城市生活体验场	建筑设计（六）	马诗琪	2011 级	杨学文	
61	华侨大学建筑学院	中原大学建筑系	细胞——园艺主题住家实验宅设计	建筑设计（六）	王樱湍	2011 级	张全智	
62	华侨大学建筑学院	澳门土地工务运输局、澳门文化局	东艺、西衍澳门内港码头片区更新活化计划	澳门内港滨水区码头更新与城市设计	程坦	2009 级	费迎庆、郑剑艺	本科优秀作业特别奖
63	华侨大学建筑学院	澳门土地工务运输局、澳门文化局	澳门望德堂文化创意区近代平民排屋“屋顶触媒”	澳门塔石片区城市更新与建筑设计	霍颜颜	2009 级	郑剑艺	本科优秀作业特别奖
64	华侨大学建筑学院	中国文化大学建筑及都市设计系	街巷	泉州历史城区文化剧场共题设计	徐骏	2011 级	连旭、吴少峰	本科优秀作业特别奖
65	华侨大学建筑学院	中国文化大学建筑及都市设计系	Unde the eave	泉州历史城区文化剧场共题设计	吴一迎	2011 级	连旭、吴少峰	
66	华侨大学建筑学院	中国文化大学建筑及都市设计系	Macau Coloane Old shipyard Update	毕业设计（澳门路环造船厂改造）	倪博恒、宋思亮	2009 级	吴少峰	
67	南京工业大学建筑学院	德国莱比锡应用科学大学建筑学院	浮于城墙之上	南京明城墙前湖段缺口连接体设计	王淼、张静、Benjamin Sens、杨锦春、王杰	研究生一年级	郭华瑜、Marina Stankovic、张蕾、李国华、姚刚、段忠诚	
68	南京工业大学建筑学院	德国莱比锡应用科学大学建筑学院	矛盾的共生	南京明城墙前湖段缺口连接体设计	潘江海、黄豪、Linda Schmidt、蒋碧冰	研究生一年级	郭华瑜、Marina Stankovic、张蕾、李国华、胡振宇、姚刚、段忠诚	研究生优秀作业特别奖
69	南京工业大学建筑学院	德国莱比锡应用科学大学建筑学院	城墙博古架	南京明城墙前湖段缺口连接体设计	杨璐、史悦、Chantal Marschall、丁少华、冯赫	研究生一年级	郭华瑜、Marina Stankovic、张蕾、李国华、姚刚、朱冬冬	研究生优秀作业特别奖
70	南京工业大学建筑学院	德国莱比锡应用科学大学建筑学院	古墙新眼	南京明城墙前湖段缺口连接体设计	丁苏煌、周行方、Christoph Weigel、骆小庆	研究生一年级	郭华瑜、Marina Stankovic、张蕾、李国华、姚刚、段忠诚	
71	南京工业大学建筑学院	德国莱比锡应用科学大学建筑学院	纵横明垣	南京明城墙前湖段缺口连接体设计	范轩、雷春光、Miriam Baumheuer、叶怡君	研究生一年级	郭华瑜、Marina Stankovic、张蕾、李国华、姚刚、朱冬冬	
72	清华大学建筑学院	爱尔兰都柏林大学建筑学院	Seamus Ennis 文化中心设计	建筑设计五	黎雪伦	11 级本科	Marcus Donaghy	
73	清华大学建筑学院	美国加利福尼亚大学伯克利分校	Double Negative	Arch100A Fundamental of Architecture Design	赵健程	11 级本科	Yeung Ho Man	本科优秀作业特别奖

序号	院校及院系全名	对外交流对象	设计题目	课程名称	学生姓名	第一完成学生所在年级	指导教师姓名	备注
74	清华大学建筑学院	University of New South Wales-Built Environment	North Bondi Apartment Design	Arch B01 studio	徐菊杰	本科三年级	Bruce Yaxley	本科优秀作业特别奖
75	清华大学建筑学院	美国耶鲁大学建筑学院	时尚中轴 - 北京南城改造	清华大学—耶鲁大学联合教学设计	夏骥、王古恬、吴旭阳	研究生二年级	朱文一、刘健	
76	清华大学建筑学院	美国普林斯顿大学建筑学院、东京大学建筑学院	梅田城市综合体设计	清华大学—普林斯顿大学—东京大学联合设计课程教学	刘剑颖、夏骥	研究生二年级	徐卫国	
77	清华大学建筑学院	日本早稻田大学建筑学院	山水舞台——岩手县大槌町社区信息交流中心设计	311 东日本震灾复兴计划——岩手县大槌町社区信息交流中心设计	吉亚君、蔡长泽	研究生一年级	许懋彦、罗德胤	研究生优秀作业特别奖
78	清华大学建筑学院	2014 南非 UIA 第二十五届国际建筑师大会	Architecture on Wheels/辙上建造	Architecture Otherwhere/ 别样的建筑 -2014UIA 国际大学生建筑设计竞赛	熊哲昆、刘芳铄、吴旭阳	研究生一年级		
79	清华大学建筑学院	Singapore,National University of Singapore, School of Design and Environment 等	Believe In City 亚洲立体城市国际设计竞赛	建筑与城市设计	郝田、陈寰宇、孙逸琳、廖思宇		朱文一、程晓喜、张弘	
80	清华大学建筑学院	UIA 霍普杯 2013 年国际学生竞赛	移动城市移动生活		李晓岸、李睿卿	直博 3		
81	清华大学建筑学院	2013 年霍普杯国际建筑设计竞赛	被消失的故居		石坚伟、郝田、孙昊德	研究生一年级	王辉	
82	山东建筑大学建筑城规学院	INCLUDED 国际组织	落脚城市——T-center 丁家庄移民社区微中心设计	山东建筑大学—新西兰 UNITEC 理工学院联合设计课程教学 2	姜静茹	2011 级	孔亚暐、房文博、Matt Mueller	
83	山东建筑大学建筑城规学院	新西兰 UNITEC 理工学院建筑学院	拱之森林——佛罗伦萨市场设计	山东建筑大学—新西兰 UNITEC 理工学院联合设计课程教学 1	贾鹏	2012 级	房文博、慕启鹏、王月涛、Daniella da Silva	本科优秀作业特别奖
84	山东建筑大学建筑城规学院	新西兰 UNITEC 理工学院建筑学院	A Growing Utopia	山东建筑大学—新西兰 UNITEC 理工学院联合设计课程教学 2	刘丹笛、林晓宇、包依凡、鞠婧	2011 级	慕启鹏、房文博、Francesco Collotti	本科优秀作业特别奖
85	山东建筑大学建筑城规学院	新西兰 UNITEC 理工学院建筑学院	Living in the Park	山东建筑大学—新西兰 UNITEC 理工学院联合设计课程教学 2	韩昆衡、骆逸、孙宁晗、李闻达	2011 级	周忠凯、陈林、Tony Van Raat	本科优秀作业特别奖
86	山东建筑大学建筑城规学院	新西兰 UNITEC 理工学院建筑学院	Urban Regeneration	山东建筑大学—新西兰 UNITEC 理工学院联合设计课程教学 2	崔雅婧、李珏、王梦真、梁正蕾	2011 级	陈林、周忠凯、Tony Van Raat	本科优秀作业特别奖
87	山东建筑大学建筑城规学院	新西兰 UNITEC 理工学院建筑学院	Coexist with Time	山东建筑大学—新西兰 UNITEC 理工学院联合设计课程教学 2	张平昊、张赛、赵洪来	2011 级	张克强、高晓明、Tony Van Raat	
88	山东建筑大学建筑城规学院	新西兰 UNITEC 理工学院建筑学院	佛罗伦萨双子市场	山东建筑大学—新西兰 UNITEC 理工学院联合设计课程教学 1	王露	2012 级	石涛、夏云、Gihan Karunatne	
89	山东建筑大学建筑城规学院	新西兰 UNITEC 理工学院建筑学院	鱼儿园	山东建筑大学—新西兰 UNITEC 理工学院联合设计课程教学 1	孔德硕	2012 级	夏云、石涛、Gihan Karunatne	本科优秀作业特别奖
90	山东建筑大学建筑城规学院	新西兰 UNITEC 理工学院建筑学院	Space to Activate——济南花鸟鱼虫水族市场设计	山东建筑大学—新西兰 UNITEC 理工学院联合设计课程教学 1	李亚男	2012 级	金文妍、刘文、Tony Burge	本科优秀作业特别奖

续表

序号	院校及院系全名	对外交流对象	设计题目	课程名称	学生姓名	第一完成学生所在年级	指导教师姓名	备注
91	山东建筑大学建筑城规学院	新西兰 UNITEC 理工学院建筑学院	鱼舍——济南柏宁水族市场改造	山东建筑大学—新西兰 UNITEC 理工学院联合设计课程教学 1	胡博	2012 级	Yvonne Wang、夏云	
92	苏州科技学院建筑与城市规划学院	澳大利亚新南威尔士大学城市建筑环境学院	生态——形态的重组——漕运博物馆和丰备义仓地区城市设计	新南威尔士大学—东南大学—苏州科技学院—清华大学联合跨文化设计课	徐钰超、吕思扬、Esmonde Yap、Sarah Fayad	本科五年级	胡莹、夏健、马骏华、许亦农、龚恺、王路、谢鸿权	本科优秀作业特别奖
93	苏州科技学院建筑与城市规划学院	澳大利亚新南威尔士大学城市建筑环境学院	柔性改造——漕运博物馆和丰备义仓地区城市设计	新南威尔士大学—东南大学—苏州科技学院—清华大学联合跨文化设计课	徐佳、David Whiteworth、程欣韵、赵文哲	本科五年级	胡莹、夏健、马骏华、许亦农、龚恺、王路、谢鸿权	本科优秀作业特别奖
94	苏州科技学院建筑与城市规划学院	澳大利亚新南威尔士大学城市建筑环境学院	5+——漕运博物馆和丰备义仓地区城市设计	新南威尔士大学—东南大学—苏州科技学院—清华大学联合跨文化设计课	刘欣仪、罗彬、林子圣、廖亦欣	本科五年级	胡莹、夏健、张愚、许亦农、龚恺、王路、谢鸿权	
95	天津大学建筑学院	日本女子大学、淡江大学	Sharing in the sky	住宅工作营	Martin Rasmus、唐奇靓、李桃	本科三年级	张昕楠、王绚	本科优秀作业特别奖
96	天津大学建筑学院	日本女子大学、淡江大学	Crevice in the city	住宅工作营	祁山、何涛	本科三年级	王绚、张昕楠	本科优秀作业特别奖
97	天津大学建筑学院	日本女子大学、淡江大学	Share with light——东京共享住宅	住宅工作营	林碧虹、于安然	本科三年级	王绚、张昕楠	本科优秀作业特别奖
98	天津大学建筑学院	日本女子大学、淡江大学	Intersection	共享住宅	陈默、崔家瑞	本科三年级	张昕楠、王绚	
99	天津大学建筑学院	UCLA 建筑学院	洛杉矶汽车博物馆设计	天津—加利福尼亚大学洛杉矶分校联合设计课程教学	王立杨、邓鸿浩	本科三年级	许蓁、盛强	
100	天津大学建筑学院	美国加利福尼亚大学洛杉矶分校建筑学院	速度与激情——洛杉矶汽车博物馆设计	天津大学—加利福尼亚大学洛杉矶分校联合设计课程教学	王莹莹、应亚	本科	许蓁、盛强	
101	天津大学建筑学院	美国加利福尼亚大学洛杉矶分校	变体——皮特森汽车博物馆	天津大学—加利福尼亚大学洛杉矶分校联合设计课程教学	刘程明、肖楚琦	本科	许蓁、盛强	
102	天津大学建筑学院	美国加利福尼亚大学洛杉矶分校建筑学院	万花景——分型理论下的汽车博物馆设计	天津大学—加利福尼亚大学洛杉矶分校联合设计课程教学	王雪睿、张知	本科	许蓁，盛强	本科优秀作业特别奖
103	天津大学建筑学院		A New Sky-line	釜山中、日、韩国际学生竞赛	王立杨	本科三年级	张昕楠、王迪	
104	天津大学建筑学院		Re-Symbol	第 49 回セントラル硝子国際建築設計競技	李斯奇、陈晓婷	硕士二年级	邹颖	
105	同济大学建筑与城市规划学院建筑系	法国巴黎美丽城国立高等建筑学校	连续 · 共生——上海徐家汇教堂广场地区城市及建筑设计	同济大学—巴黎美丽城国立高等建筑学校联合设计课程教学	吴人洁	本科 5 年级	张凡	
106	同济大学建筑与城市规划学院建筑系	日本大阪大学工学研究课地球综合工学专业	Inside Embrace Outside	上海外滩吴淞路西侧地块设计研究	谢凡、金鑫、金雯	硕士 1 年级	李斌、董春方、李华	
107	同济大学建筑与城市规划学院建筑系	奥地利维也纳工业大学，法国凡尔赛大学，法国斯特拉斯堡大学，比利时布鲁塞尔自由大学	步行者天堂—城市步行系统节点城市设计	同济大学硕士双学位国际设计课程教学	赵玉玲、夏琴、雷少英	硕士 1 年级	孙彤宇、许凯	研究生优秀作业特别奖

序号	院校及院系全名	对外交流对象	设计题目	课程名称	学生姓名	第一完成学生所在年级	指导教师姓名	备注
108	同济大学建筑与城市规划学院建筑系	美国圣路易斯华盛顿大学建筑学院	Shared Living Rooms-Manhattan 2050	同济大学—圣路易斯华盛顿大学联合城市设计课程教学	张林琦、郗晓阳	硕士 1 年级	杨春侠、黄林琳	研究生优秀作业特别奖
109	同济大学建筑与城市规划学院	新加坡国立大学建筑系，香港中文大学建筑系，日本东京大学建筑系，荷兰代尔夫特理工大学建筑系，瑞士苏黎世联邦理工学院建筑系，美国加利福尼亚大学伯克利分校建筑系，美国宾夕法尼亚大学建筑系，美国密歇根大学建筑系	渗透城市	亚洲垂直城市竞赛	陆伊昀、陈艺丹、朱恒玉、孙伟、陈伯良、陆垚	本科 4 年级	董屹、王桢栋、黄一如	
110	同济大学建筑与城市规划学院	新加坡国立大学建筑系，香港中文大学建筑系，日本东京大学建筑系，荷兰代尔夫特理工大学建筑系，瑞士苏黎世联邦理工学院建筑系，美国加利福尼亚大学伯克利分校建筑系，美国宾夕法尼亚大学建筑系，美国密歇根大学建筑系	畅通城市	亚洲垂直城市竞赛	何啸东、李骜、肖璐珩、郑攀、程思、谭杨	本科 4 年级	王桢栋、董屹、黄一如	
111	西安建筑科技大学建筑学院	日本早稻田大学	回舍——北院门小客栈设计	北院门小客栈设计	蒋一汉	本科二年级	刘克成、丸山欣也、吴瑞、王毛真、蒋蔚、俞泉	本科优秀作业特别奖
112	西安建筑科技大学建筑学院	美国佛罗里达大学建筑学院	墙之礼乐——西安博物院城市设计及重点建筑设计	西安建筑科技大学—佛罗里达大学联合课程设计教学	刘思源、方昇辰、邹宜彤、张娜、Jared Lambright、Quang Nguyen、MingJin Hong	研究生一年级	肖莉、常海青、苏静、Albertus Wang、鲁旭、任中琦	研究生优秀作业特别奖
113	西安建筑科技大学建筑学院	2014 亚洲建筑新人战	延——小客栈设计	北院门小客栈设计	杨子依	本科二年级	刘克成、丸山欣也、吴瑞、王毛真、蒋蔚、俞泉	
114	西安建筑科技大学建筑学院	第 25 届 UIA（国际建筑师协会）世界大学生建筑设计竞赛	Suture the City	2014UIA 国际大学生建筑设计竞赛	吴明奇、牛童、冯贞珍、崔哲伦、罗典	本科三、四年级	裴钊	
115	西安建筑科技大学建筑学院	第 25 届 UIA（国际建筑师协会）世界大学生建筑设计竞赛	城市尊严	毕业设计	周正、卢肇松、张士骁、高元、古悦、鞠曦	本科五年级	李昊、周志菲	
116	西安建筑科技大学建筑学院	2014 ICCC 国际学生设计大赛	织・补	2014 ICCC 国际学生设计大赛	白纪涛、王阳、王静、王思睿	本科五年级	张倩、王芳	
117	西安建筑科技大学建筑学院	第 25 届 UIA（国际建筑师协会）世界大学生建筑设计竞赛	Sew up	2014UIA 国际大学生建筑设计竞赛	宋梓怡、杜怡、李乐、李长春、刘彦京、李乔姗	研究生一年级	李岳岩、陈静	

续表

序号	院校及院系全名	对外交流对象	设计题目	课程名称	学生姓名	第一完成学生所在年级	指导教师姓名	备注
118	西安建筑科技大学建筑学院	第 25 届 UIA（国际建筑师协会）世界大学生建筑设计竞赛	Combine Regeneration Metabolism	2014UIA 国际大学生建筑设计竞赛	兰青、刘伟、李小同、刘俊、张佳茜、钱雅坤	研究生二年级	李岳岩、陈静	
119	西安建筑科技大学建筑学院	美国佛罗里达大学建筑学院	寺曾相识——西安博物院城市设计及重点建筑设计	西安建筑科技大学—佛罗里达大学联合课程设计教学	陆星辰、杨骏、岳圆、周曦曦、Michelle Hook、Asher Durham、Matthew Livingston	研究生一年级	肖莉、常海青、苏静、Albertus Wang、鲁旭、任中琦	
120	浙江大学建筑系	西班牙马德里圣帕保罗大学建筑系	杭州拱墅区运河沿岸多媒体中心设计——基于地景建筑设计理念	联合毕业设计	杜信池	本科生 5 年级	陈林、王晖	
121	浙江大学建筑系	西班牙马德里圣帕保罗大学建筑系	运河岸边的城市更新——拱墅区多媒体中心设计	联合毕业设计	邓奥博	本科生 5 年级	陈林、王晖	本科优秀作业特别奖
122	浙江大学建筑系	意大利罗马大学建筑系	废墟上的重生	暑期 workshop	曾伊凡、金通、苏军、翟健、方舒	研究生一年级	贺勇	
123	浙江大学建筑系	美国康奈尔大学环境系	康奈尔物理楼使用后调查	workshop	王静	研究生一年级	王竹、葛坚、王洁、张涛、曹震宇	
124	中央美术学院建筑学院	瑞士卢塞恩应用技术与艺术大学建筑系	Focus on structure-hotel——结构与空间探究	Master Focus Project	袁智敏	本科 4 年级	Niklaus Graber、Bernhard Maurer	本科优秀作业特别奖
125	中央美术学院建筑学院	丹麦奥胡斯建筑学院	好奇的格子（Cabinet of Curiosity）	中央美术学院 / 奥胡斯建筑学院联合课题 Mapping the Void II-a Study of the “Edge” and the Port of Aarhus	沈璐、王琪、Thi Duy An Tran、May Damagard Sorensen	研究生二年级	何可人、王威、Anne Elisabeth Toft、Claudia Carbone	研究生优秀作业特别奖
126	中央美术学院建筑学院	丹麦奥胡斯建筑学院	分类学—边界重构（Taxonomy）	中央美术学院 / 奥胡斯建筑学院联合课题 Mapping the Void II-a Study of the “Edge” and the Port of Aarhus	段邦禹、王金璐、Olga Sigporsdottir、Rosemary Jeremy	研究生二年级	何可人、王威、Anne Elisabeth Toft、Claudia Carbone	研究生优秀作业特别奖
127	中央美术学院建筑学院	丹麦奥胡斯建筑学院	“迹”（Unfolding Harbor Sequence）	中央美术学院 / 奥胡斯建筑学院联合课题 Mapping the Void II-a Study of the "Edge" and the Port of Aarhus	周格、王羽、Siv Bøttcher	研究生二年级	何可人、王威、Anne Elisabeth Toft、Claudia Carbone	研究生优秀作业特别奖
128	中央美术学院建筑学院	挪威奥斯陆建筑与设计学院建筑系	Claim the Territory by Chaos	Work & Workplace	范劼	研究生二年级	Per Olaf Fjeld、Rolf Gerstlauer、Lisbeth Funck	研究生优秀作业特别奖
129	中央美术学院建筑学院	丹麦奥胡斯建筑学院	GRietveld Chair	丹麦制造	邓媛	研究生二年级	Kätte Bønløkke Andersen、Karen Kjærgaard	
130	重庆大学建筑城规学院建筑系	香港大学建筑学院	衔接的场所——在高密度老社区中进化的商业模式	重庆大学—香港大学联合设计	李益、束逸天	2010 级建筑学本科生	田琦、蒋家龙、Ottevaere Olivier	
131	重庆大学建筑城规学院建筑系	香港大学建筑学院	都市绿谷——香港湾仔区高层设计	重庆大学—香港大学联合设计	宋然、李梦郁	2010 级建筑学本科生	田琦、蒋家龙、Ottevaere Olivier	

序号	院校及院系全名	对外交流对象	设计题目	课程名称	学生姓名	第一完成学生所在年级	指导教师姓名	备注
132	重庆大学建筑城规学院建筑系	日本早稻田大学	美好田居生活——蕨地区灾后重建计划	重庆大学—清华大学—早稻田大学—2014 年中日联合设计	赵畅、刘锦、任洋	2011 级建筑学硕士生	卢峰、邓蜀阳	
133	重庆大学建筑城规学院建筑系	德国杜塞尔多夫应用科学大学 PBSA 建筑学院	工业遗产 & 社区复兴	重庆大学与杜塞尔多夫应用科学大学研究生联合设计	高澍、曾柳银、程文楷	2012 级建筑学硕士生	龙灏、褚冬竹、Jorg Leeser、Juan-Pablo Molestina	研究生优秀作业特别奖
134	重庆大学建筑城规学院建筑系	德国杜塞尔多夫应用科学大学 PBSA 建筑学院	多因素控制下的城市更新设计	重庆大学与杜塞尔多夫应用科学大学研究生联合设计	黄一夫、况毅、杨涵	2012 级建筑学硕士生	龙灏、褚冬竹、Jorg Leeser、Juan-Pablo Molestina	研究生优秀作业特别奖
135	重庆大学建筑城规学院建筑系	日本早稻田大学	3.11 东日本震灾复兴计划——岩手县大槌町社区信息交流中心设计	重庆大学—清华大学—早稻田大学—2014 年中日联合设计	傅悦、胡瑶婷	2012 级建筑学硕士生	邓属阳、翁季、蒋家龙	研究生优秀作业特别奖
136	重庆大学建筑城规学院建筑系	德国杜塞尔多夫应用科学大学 PBSA 建筑学院	基于过程研究的城市设计——重庆特钢更新案	重庆大学与杜塞尔多夫应用科学大学研究生联合设计	张奇、张冠、王旭昊、许琦伟、兰显荣、霍翔、邹亚、邹汝波	2013 级建筑学硕士生	杨宇振、Jorg Leeser、龙灏	
137	重庆大学建筑城规学院建筑系	日本北海道大学	生态链接，产业重塑——国际化绿色生态城市规划设计	重庆大学—北海道大学联合设计	王凌云、曾渝京、尚白冰、李超	2013 级建筑学硕士生	龙灏、杨培峰、谷光灿、森傑（日）	研究生优秀作业特别奖
138	重庆大学建筑城规学院建筑系	日本北海道大学	生态乡村，活力再造——国际化绿色生态城市规划设计	重庆大学—北海道大学联合设计	张子涵、李珊、马可	2013 级建筑学硕士生	龙灏、杨培峰、谷光灿、森傑（日）	研究生优秀作业特别奖
139	广州大学建筑与城市规划学院	新西兰奥克兰大学建筑学院	Greening the Village	城市设计	甄穗豪、梁善斐、陈伟敏、张卢峰、谢湘敏	四年级	Manfredo Manfredin、赵阳	
140	广州大学建筑与城市规划学院	新西兰奥克兰大学建筑学院	Activity Center in Downtown Village	城市设计	王冰、杜生涛、林鸿缔、陈汉飞	四年级	赵阳、Manfredo Manfredin	
141	华中科技大学建筑与城市规划学院	法国巴黎瓦尔德塞纳建筑学院	活力之塔	2014 中法建筑联合设计	张姝、肖锐	研究生一年级	汪原、周卫、赵守谅	
142	昆明理工大学建筑与城市规划学院	东海大学建筑系	血脉新生	石岗区客家银发族健康照护与疗育社区环境营造设计交流营	李天依、徐仲莹、蒋蓁蓁、岳斌	研究生 1 年级	翟辉、关华山、邱国维、张欣雁	
143	昆明理工大学建筑与城市规划学院	东海大学建筑系	共生效应	石岗区客家银发族健康照护与疗育社区环境营造设计交流营	穆童、孙春媛、杜雨、苏志勇	研究生 1 年级	翟辉、关华山、邱国维、张欣雁	
144	昆明理工大学建筑与城市规划学院		出乎意料的城市——CITY DREAM WITH SKY	研究型设计	郭俊超、叶雨辰	本科 4 年级	翟辉、华峰	
145	昆明理工大学建筑与城市规划学院		出乎意料的城市——巴别城	设计课程 4	褚剑飞、马杰茜	本科 2 年级	张欣雁	
146	昆明理工大学建筑与城市规划学院		出乎意料的城市——元胞螺旋城	研究型设计	翟星玥、翟雪竹	本科 3 年级	翟辉、王丽红	
147	南京大学建筑与城市规划学院建筑系	澳大利亚墨尔本大学建筑系	Footscary up to 11	毕业设计	杨天仪	本科四年级	Marcus White、华晓宁	
148	南京大学建筑与城市规划学院建筑系	澳大利亚墨尔本大学建筑系	FUTURE MUSEUM	Design Studio	许伯晗	研究生一年级	Donald Bates、华晓宁	

续表

序号	院校及院系全名	对外交流对象	设计题目	课程名称	学生姓名	第一完成学生所在年级	指导教师姓名	备注
149	青岛理工大学建筑学院	韩国光云大学建筑学院	由消极到积极——公共空间改造设计	青岛理工大学—韩国光云大学联合城市设计课程教学	王京鹍、徐懿龙、张颖、Kim Ji Hyeon、Song Bo Keun	研究生	郝赤彪、解旭东、Oh Seok Kyu、Scheck	
150	青岛理工大学建筑学院	韩国光云大学建筑学院	山东路桥底空间改造——立方体的故事	青岛理工大学—韩国光云大学联合城市设计课程教学	刘思洋、张士超、解晓东、于克来、CHOI SEONGBONG、GO HYUNSUN	研究生	郝赤彪、解旭东、Song In Jo	
151	青岛理工大学建筑学院		城市边缘的复兴	2014AIA 路易维尔儿童博物馆建筑设计竞赛	魏易盟、贺俊、苗天宁	2011 级	郝赤彪、解旭东、程然	
152	青岛理工大学建筑学院		The children's museum	2014AIA 路易维尔儿童博物馆建筑设计竞赛	蔡昆洋、张沛、韩梦、谭君睿	2011 级	许从宝、聂彤、徐翀	
153	青岛理工大学建筑学院		Get involved	2014AIA 路易维尔儿童博物馆建筑设计竞赛	张跃、杨舒婷、孙嘉伦、刘俊男	2011 级	郝赤彪、解旭东、程然	
154	青岛理工大学建筑学院		Wormhole	2014AIA 路易维尔儿童博物馆建筑设计竞赛	张驰、徐钰茗、王傲男、杨迪雯、冯勃睿	2011 级	郝赤彪、解旭东、程然	
155	青岛理工大学建筑学院		植城静息	2014UIA 霍普杯国际建筑设计竞赛	赵云鹤、刘国伟、林汝佳、黄静雯、李陶	2012 级	郝赤彪、王少飞、程然、石新羽	
156	青岛理工大学建筑学院		小站那些事	2014UA 国际大学生建筑设计竞赛	于立恒、王傲男	2011 级	郝赤彪、王少飞、石新羽	
157	青岛理工大学建筑学院		平凡事件 vs 复合站点	2014UA 国际大学生建筑设计竞赛	魏易盟、肖飞、张泽华、李陶、杨明慧	2011 级，2012 级	郝赤彪、解旭东、程然、耿雪川	
158	厦门大学建筑与土木工程学院建筑系建筑学	美国天普大学建筑系	Learning from North Broad Street	Architecture Design V Urban Design Studio	包可人	2011 级	professor Brigitte Knowles	
159	厦门大学建筑与土木工程学院建筑系建筑学	美国纽约州立大学布法罗分校建筑系	健康中心设计	Architecture Studio 5	金洪勋	2011 级	Jordan Carver	
160	深圳大学建筑与城市规划学院	意大利米兰理工大学建筑工程与建筑设计学院（Politecnico di Milano）	Cultural Center of Naviglio Grande in Milano	Laboratorio di Progettazione dell'Architettura 1（建筑设计 Studio 1）	陈栩、姜雁彬	2010 级	Prof.Maurizio Meriggi、Prof. Nicola Mastalli	
161	深圳大学建筑与城市规划学院	奥地利维也纳理工大学建筑系（TU Wien）	Gesause 游客中心设计	Desigh Studio HB2	黄敏堃、吴雪茵	2010 级	San-Hwan Lu	
162	四川大学建筑与环境学院建筑系	法国拉维莱特大学建筑学院	夜公园成都——与城市印记的对话	四川大学—法国拉维莱特大学联合设计交流营	李方才、刘思思、邱兴勇、沙澎、施博文、孙中雅、徐丹、周璐、Andreina、Marialucia、Stephanie、Emmanuelle	本科四年级	张鲲、张帆、周波、王霞、张亮	
163	四川大学建筑与环境学院建筑系	日本名古屋工业大学建筑学科	羌文化体验工场	2014 中日国际交流营	仓田骏、陈科臻、杜海辰、帅庆源、杨卓蕾	本科五年级	李沄漳、北川启介、夏目欣昇、	

序号	院校及院系全名	对外交流对象	设计题目	课程名称	学生姓名	第一完成学生所在年级	指导教师姓名	备注
164	四川大学建筑与环境学院建筑系	美国华盛顿大学城市规划专业	乡村社会生态系统规划调查方法研究——以成都市郫县安龙村为例	2014 四川大学—华盛顿大学联合设计课程教学	Schell、Tiernan、Tiantian、XiangWenhu、LiuSiyong、吴婷、胡晓曦、徐思敏、肖磊、潘鹏程、高禹诗	本科三年级、五年级	李伟、Dan Abramson、成受明	
165	四川大学建筑与环境学院建筑系	意大利费拉拉大学建筑学院	旧城市与新发展	四川大学—意大利费拉拉大学联合设计交流营	Bartolini Rossana、Ravaioli Teresa、简祎、陆姗姗、喻干一、王楠天、王季玉	本科四年级	张鲲、张亮、何昕	
166	天津城建大学建筑学院	丹麦 VIA University College 经济与技术学院	风的再生	建筑设计 V	尤佳、陈暄、方沐玮、Dennis Nørlund Schmidt、Rasmus Juhl Højen、Martin Bech Petersen	本科四年级	万达	
167	武汉大学城市设计学院	美国北卡罗来纳州立大学夏洛特分校	Vertical Urbanism: Density, Complexity, Verticality Wuyuan Bay Metro Station TOD Development	Advanced Urban Design Studio	张喻（武汉大学）、Katie Hamilton（UNCC）、Adi Mokha（UNCC）、Gota Miyazaki（UNCC）	本科四年级	林中杰（UNCC）、胡晓青（武汉大学）、张若曦（厦门大学）	
168	武汉大学城市设计学院	美国北卡罗来纳州立大学夏洛特分校	Vertical Urbanism: Density, Complexity, Verticality Wuyuan Bay Metro Station TOD Development	Advanced Urban Design Studio	林婧（武汉大学）、David Perry（UNCC）、Rachel Safren（UNCC）、Will Penland（UNCC）、尹秋怡（厦门大学）、李伟（厦门大学）	本科四年级	林中杰（UNCC）、胡晓青（武汉大学）、张若曦（厦门大学）	
169	武汉大学城市设计学院	英国利物浦大学；ILAUD（International Laboratory of Architecture and Urban Design）；瑞士洛桑理工学院（Ecole Polytechnique Federale de Lausanne）		CPCC 2013（The Second International Workshop Critical Planning for Chinese Cities）	蒋迪、孙璐	2009 级本科，2010 级本科	Abigail-Laure Kern、张点、刘凌波	
170	武汉大学城市设计学院	英国利物浦大学；ILAUD（International Laboratory of Architecture and Urban Design）；瑞士洛桑理工学院（Ecole Polytechnique Federale de Lausanne）	Re programming Dongshan peninsula	CPCC 2013（The Second International Workshop Critical Planning for Chinese Cities）	寇寰	2012 级研究生	Andrew Johnston、孙磊、刘凌波、张点	
171	西安交通大学人居环境与建筑工程学院建筑学系	淡江大学建筑系	自然形态的人工观念——园林	淡江大学建筑设计课程	雷雨虹	四年级（本科 2010 级）	陈珍诚、张定青	

续表

序号	院校及院系全名	对外交流对象	设计题目	课程名称	学生姓名	第一完成学生所在年级	指导教师姓名	备注
172	西安交通大学人居环境与建筑工程学院建筑学系	东海大学景观建筑系	海绵城市——基于台中市筏子溪的暴雨管理	东海大学景观建筑设计课程	李兆琳、黄楚珺	四年级（本科2010级）	黄祺峰、钟德颂、张定青	
173	西安交通大学人居环境与建筑工程学院建筑学系	淡江大学建筑系	淡水国小扩建与改造设计	淡江大学建筑设计课程	陈子奇	三年级（本科2011级）	陈昭颖、赖怡成、张定青	
174	西南交通大学建筑学院	罗马大学建筑学院	Start from the past	西南交通大学—罗马建筑学院暑期联合设计周	麻燕妮、杨硕、李烨晴、单涤非	本科2011级	栗民、沈中伟、Benedetta G. Morelli	
175	西南交通大学建筑学院	意大利罗马大学建筑学院	融入与进发——罗马研究学院设计	西南交通大学—罗马建筑学院暑期联合设计周	王贝、代阳、付一然、王墨瀚	本科2011级	栗民、沈中伟、Benedetta G. Morelli	
176	西南交通大学建筑学院	意大利罗马大学建筑学院	Hill-Pizza della moretti renewal in Italy	西南交通大学—罗马建筑学院暑期联合设计周	李冰怡、廖培然、王兴禹	本科2012级	栗民、沈中伟、Benedetta G. Morelli	
177	西南交通大学建筑学院	美国，俄克拉荷马州立大学农学院园艺景观系	Exhibition Shaping the Walls	西南交通大学—俄克拉荷马州立大学交换课程设计	于泽文	本科2012级	周斯翔、Michael Holmes、沈中伟	
178	长安大学建筑学院		On and Under	第25届UIA（国际建筑师协会）世界大学生建筑设计竞赛	侯乃菲、万伊涵、杨定宇、王凯旋、杨鑫	本科四年级	张磊、王农、张琳、杨宇峤	
179	郑州大学建筑学院	德国安哈尔特应用技术大学建筑学院	汉堡圣尼古拉教堂废墟空间再造	Elective Course （project）选修课	徐维涛	本科2012级	郑东军、Prof. Jasper Cepl	
180	中国矿业大学力学与建筑工程学院建筑系	德国柏林工业大学建筑学院	消隐与显现——徐州市云东二道街地块城市设计（Invisible and Visible-The Urban Design of Yundong second street block,Xuzhou ）	中国矿业大学—德国柏林工业大学联合设计课程教学	唐宽猛、张雅暄、王文卿、李洋、邢艺凡、周保林、孙嘉尉	研13	冯姗姗、刘茜、孙良	
181	中国矿业大学力学与建筑工程学院	德国柏林工业大学建筑学院	阐·示—连墙接栋——徐州云东二道街城市设计	中国矿业大学—德国柏林工业大学联合设计课程教学	胡丽丽、魏杰、王炜航、季旭君、刘姝含、张韬韬、袁梦、吴捷、陈功	研13	刘茜、冯姗姗、孙良	
182	中国矿业大学力学与建筑工程学院	德国柏林工业大学建筑学院	泰山路段城市设计	中国矿业大学—德国柏林工业大学联合设计课程教学	理南南、杨思、马涛、林攀崖、姜韬	研13	冯姗姗、刘茜	
183	中国矿业大学力学与建筑工程学院	2014年霍普杯国际大学生建筑设计竞赛	织城为家——探索城市空白空间的再利用	2014年霍普杯国际大学生建筑设计竞赛	张学优、杨瑞东、曹啸、丁纹敏	建11	孙良、李明、陈惠芳	
184	中国矿业大学力学与建筑工程学院	博德西奥（BDCL）国际建筑设计有限公司	救赎——后工业废弃地的垂直农场转型	城市立体农场国际大学生建筑设计竞赛	赵倩、丁少华、黄佳喻、李文亮、骆小庆、魏杰、周玉佳、陆家豪	研13	孙良、姚刚	
185	中国矿业大学力学与建筑工程学院	博德西奥（BDCL）国际建筑设计有限公司	循环农场——城市立体农场设计	城市立体农场国际大学生建筑设计竞赛	曹琳琳、刘彬峰、胡小华、张琪、于小果、张涛	研13	刘茜、林涛	
186	中南大学建筑与艺术学院	英国考文垂大学艺术与设计学院	Tridimensional Oxygen Bar	建筑创作理论与实践	王颖贺	研究生一年级	石磊、罗明	

注：本表为作业入选名单，与最终确认名单稍有出入。

中国建筑学会建筑教育评估分会第一届第四次理事会代表名单

序号	姓名	单位	职务职称	分会职务	通讯地址	邮编
1	朱文一	清华大学建筑学院	教授	理事长	北京市海淀区清华大学建筑学院	100084
2	王建国	东南大学建筑学院	教授、中国工程院院士	副理事长	江苏省南京市四牌楼 2 号	210096
3	仲德崑	深圳大学建筑与城规学院	院长、教授	副理事长	深圳市南山区南海大道 3688 号	518060
4	孔宇航	天津大学建筑学院	副院长、教授	常务理事	天津市南开区卫津路 92 号天津大学建筑学院	300072
5	刘克成	西安建筑科技大学建筑学院	教授	常务理事	陕西省西安市碑林区雁塔路 13 号	710055
6	刘临安	北京建筑大学建筑与城市规划学院	教授	常务理事	北京西城区展览馆路 1 号	100044
7	孙澄	哈尔滨工业大学建筑学院	副院长、教授	常务理事	哈尔滨市南岗区西大直街 66 号哈工大建筑馆 248 房间	150001
8	孙一民	华南理工大学建筑学院	院长、教授	常务理事	广东省广州市天河区五山华南理工大学建筑学院	510641
9	吴长福	同济大学建筑与城市规划学院	教授	常务理事	上海市四平路 1239 号	200092
10	李保峰	华中科技大学建筑与城市规划学院	院长、教授	常务理事	湖北省武汉市洪山区珞喻路 1037 号华中科技大学南 4 楼	430074
11	沈中伟	西南交通大学建筑与设计学院	院长、教授	常务理事	四川省成都市高新区西部园区西南交通大学建筑学院(老校区: 成都市二环路北一段 111 号)	611756
12	单军	清华大学建筑学院	副院长、教授	常务理事	北京市海淀区清华大学	100084
13	卢峰	重庆大学建筑城规学院	副院长、教授	常务理事	重庆市沙坪坝区沙北街 83 号重庆大学 B 区建筑城规学院	400045
14	赵继龙	山东建筑大学建筑城规学院	院长、教授	常务理事	山东省济南市临港开发区凤鸣路	250101
15	吴越	浙江大学建筑工程学院建筑系	系主任、教授	常务理事	浙江大学紫金港校区建筑系月牙楼 112	310058
16	魏春雨	湖南大学建筑学院	院长、教授	常务理事	湖南省长沙市岳麓区牌楼路湖南大学建筑学院	410082
17	张伶伶	沈阳建筑大学建筑学院	院长、教授	常务理事	辽宁省沈阳市浑南新区浑南东路 9 号	110168
18	丁沃沃	南京大学建筑与城市规划学院	院长、教授	常务理事	江苏省南京市汉口路 22 号南京大学蒙民伟楼 1211	210093
19	李早	合肥工业大学建筑与艺术学院	院长、教授	常务理事	安徽省合肥市合肥工业大学翡翠湖校区 19 号信箱建筑与艺术学院	230601
20	范悦	大连理工大学建筑与艺术学院	院长、教授	常务理事	辽宁省大连市高新区凌工路 2 号	518060
21	韩冬青	东南大学建筑学院	院长、教授	常务理事	江苏省南京市四牌楼 2 号东南大学建筑学院	361021
22	饶小军	深圳大学建筑与城规学院	副院长、教授	理事	广东省深圳市南山区南海大道 3688 号深圳大学建筑与城规学院 A305	100124
23	费迎庆	华侨大学建筑学院	副院长、副教授	理事	福建省厦门市集美区集美大道 668 号	450001
24	戴俭	北京工业大学建筑与城市规划学院	院长、教授	理事	北京市朝阳区平乐园 100 号	116024
25	张建涛	郑州大学建筑学院	院长、教授	理事	河南省郑州市科学大道 100 号郑州大学建筑学院	650500
26	翟辉	昆明理工大学建筑与城市规划学院	院长、教授	理事	云南省昆明市呈贡区景明南路 727 号昆明理工大学建筑与城市规划学院	210009
27	胡振宇	南京工业大学建筑学院	院长、教授	理事	江苏省南京市中山北路 200 号	130118
28	李之吉	吉林建筑大学建筑与规划学院	院长、教授	理事	吉林省长春市净月区新城大街 5088 号	430070
29	王晓	武汉理工大学土木工程与建筑学院	系主任、教授	理事	湖北省武汉市洪山区珞狮路 122 号	361005

续表

序号	姓名	单位	职务职称	分会职务	通讯地址	邮编
30	王绍森	厦门大学建筑与土木工程学院	院长、教授	理事	福建省厦门市思明南路 422 号厦门大学建筑与土木工程学院	510006
31	龚兆先	广州大学建筑与城市规划学院	院长、教授	理事	广东省广州市番禺区大学城外环西路 230 号建筑学院	056038
32	陈晓卫	河北工程大学建筑学院	副院长、副教授	理事	河北省邯郸市光明南大街 199 号	200240
33	张健	上海交通大学船舶海洋与建筑工程学院建筑学系	所长、教授	理事	上海市闵行区石屏路 399 弄 3 号 1102	266033
34	郝赤彪	青岛理工大学建筑学院	院长、教授	理事	山东省青岛市市北区抚顺路 11 号	230022
35	刘仁义	安徽建筑大学建筑与规划学院	副院长、教授	理事	安徽省合肥市金寨路 856 号	710049
36	陈洋	西安交通大学人居环境与建筑工程学院建筑学系	系主任、教授	理事	陕西省西安市咸宁西路 28 号	410000
37	石磊	中南大学建筑与艺术学院	副院长、教授	理事	湖南长沙市韶山南路中南大学铁道校区建筑与艺术学院建筑系	430072
38	程世丹	武汉大学城市设计学院	副院长、教授	理事	武汉大学工学部城市设计学院	100144
39	贾东	北方工业大学建筑与艺术学院	书记、教授	理事	北京市石景山区晋元庄路 5 号北方工业大学建筑工程学院	221116
40	孙良	中国矿业大学力学与建筑工程学院建筑系	系主任、副教授	理事	江苏省徐州市大学路 1 号	215011
41	夏健	苏州科技学院建筑与城市规划学院	教授	理事	江苏省苏州市滨河路 1701 号苏州科技学院建筑城规学院	010051
42	贾晓浒	内蒙古工业大学建筑学院	院长、副教授	理事	内蒙古自治区呼和浩特市新城区爱民路 49 号	300401
43	舒平	河北工业大学建筑与艺术设计学院	院长、教授	理事	天津市北辰区双口镇西平道 5340 号	100102
44	吕品晶	中央美术学院建筑学院	院长、教授	理事	北京市朝阳区花家地南街 8 号	350108
45	关瑞明	福州大学建筑学院	院长、教授	理事	福建省福州市大学新区学园路 2 号	100044
46	夏海山	北京交通大学建筑与艺术学院	院长、教授	理事	北京海淀区西直门外上园村 3 号	030024
47	孟聪龄	太原理工大学建筑与土木工程学院	系主任、教授	理事	山西省太原市迎泽西大街 79 号	310014
48	于文波	浙江工业大学建筑工程学院建筑系	教授	理事	浙江省杭州市潮王路 18 号	264005
49	隋杰礼	烟台大学建筑学院	院长、副教授	理事	山东省烟台市莱山区清泉路 32 号	300384
50	林耕	天津城建大学建筑学院	院长、教授	理事	天津市西青区津静路 26 号	710072
51	杨卫丽	西北工业大学力学与土木建筑学院建筑系	系支部书记兼系副主任、副教授	理事	陕西省西安市友谊西路 127 号西北工业大学西院 17 号楼 030 信箱（曹健 710072 陕西省西安市长安区东祥路 1# 西工大长安校区 883 信箱）	330031
52	姚糖	南昌大学建筑工程学院建筑系	系主任、教授	理事	江西省南昌市红谷滩学府大道 999 号	510090
53	朱雪梅	广东工业大学建筑与城市规划学院	院长、教授	理事	广州市东风东路 729 号广东工业大学建筑与城市规划学院	610065
54	周波	四川大学建筑与环境学院	副院长、教授	理事	四川省成都市一环路南一段 24 号四川大学建筑与环境学院（610041 四川省成都市武侯区盛丰路 30 号海珀香庭 5-801 何昕）	014010
55	马明	内蒙古科技大学建筑与土木工程学院	副院长、教授	理事	内蒙古包头市昆都仑区阿尔丁大街 7 号	710061
56	武联	长安大学建筑学院	院长、教授	理事	陕西省西安市长安中路 161 号长安大学小寨校区	830000
57	王万江	新疆大学建筑工程学院	副院长、教授	理事	乌鲁木齐市友好路 21 号新疆大学北校区	350118
58	严龙华	福建工程学院建筑与城乡规划学院	副院长、教授	理事	福建省福州市大学新区学院路 3 号福建工程学院逸夫楼	350118
59	王薇	河南工业大学土木建筑学院建筑系	系名誉主任、教授	理事	河南省郑州市高新区莲花街河南工业大学土木建筑学院建筑系	450001

后　记

自2012年中国建筑学会建筑教育评估分会成立以来，在住房和城乡建设部人事司赵琦副巡视员、何志方处长、高延伟副处长指导下，各项工作有序开展。作为中国建筑学会建筑教育评估分会的中心工作之一，2014年3月和2015年3月，首辑《2013中国建筑院校学生境外交流优秀作业集》和第二辑《2014中国建筑院校境外交流优秀作业集》顺利出版。作业集的出版取得了良好的交流效果，从一个侧面展示了通过评估建筑院校设计课程及其境外交流合作的丰硕成果，也推进了中国建筑教育评估工作的深化。

中国建筑学会建筑教育评估分会2015年年会暨第一届四次全体理事会于2015年3月21-22日在天津大学举行，并同期举办了“2015年中国建筑院校境外交流学生作业展”，共收到37个院校提交的227份境外交流作业。本作业集收录了其中本科生优秀作业30份，研究生优秀作业19份，国际竞赛作品36份，共计85份境外交流作业。我相信，经过优选的本作业集的出版，将进一步促进中国建筑院校境外交流活动的开展。

尤其需要指出的是，会议承办单位天津大学建筑学院的张颀院长、许蓁副院长等在本书的编辑工作中付出了艰苦的努力；中国建筑学会建筑教育评估分会秘书处专职秘书陈玲以及兼职秘书清华大学建筑学院周政旭博士和商谦博士对此也付出了辛勤的汗水。在此表示诚挚的谢意。中国建筑学会建筑教育评估分会朱文一理事长、张百平秘书长、王柏峰副秘书长及王晓京副秘书长负责组织了本书的总体策划工作。

特别要感谢中国建筑工业出版社徐晓飞主任与张明编辑对中国建筑教育评估事业的鼎力支持。没有他们的通力合作，这本书也不可能如期出版发行。

中国建筑学会建筑教育评估分会理事长　朱文一

2015年11月15日

图书在版编目（CIP）数据

2015中国建筑院校境外交流优秀作业集/全国高等学校建筑学专业教育评估委员会，中国建筑学会建筑教育评估分会编．—北京：中国建筑工业出版社，2016.6

ISBN 978-7-112-19313-4

Ⅰ．①2… Ⅱ．①全… ②中… Ⅲ．①建筑设计－作品集－中国－现代 Ⅳ．①TU206

中国版本图书馆CIP数据核字（2016）第064215号

责任编辑：徐晓飞　张　明
责任校对：陈晶晶　姜小莲

2015 中国建筑院校境外交流优秀作业集
全国高等学校建筑学专业教育评估委员会
中国建筑学会建筑教育评估分会
*
中国建筑工业出版社出版、发行（北京西郊百万庄）
各地新华书店、建筑书店经销
北京锋尚制版有限公司制版
北京顺诚彩色印刷有限公司印刷
*
开本：965×1270毫米　1/16　印张：12¼　字数：300千字
2016年7月第一版　　2016年7月第一次印刷
定价：118.00元
ISBN 978-7-112-19313-4
（28571）